BIOCHEMISTRY OF CUTANEOUS EPIDERMAL DIFFERENTIATION

BIOCHEMISTRY OF CUTANEOUS EPIDERMAL DIFFERENTIATION

Proceedings of the Japan-U.S. Seminar on Biochemistry of Cutaneous Epidermal Differentiation

Edited by
Makoto Seiji and I. A. Bernstein

UNIVERSITY PARK PRESS
Baltimore • London • Tokyo

This publication is supported partly by Japan Society for the Promotion of Science.

UNIVERSITY PARK PRESS
Baltimore · London · Tokyo

Library of Congress Cataloging in Publication Data

Main entry under title:

Biochemistry of cutaneous epidermal differentiation.

Papers from a symposium held in Sendai, Japan, June 2-4, 1976, and sponsored by the Japan Society for the Promotion of Science and the National Science Foundation of the United States.
1. Epidermis—Congresses. 2. Biological chemistry—Congresses. 3. Cell differentiation—Congresses. I. Seiji, Makoto, 1926– II. Bernstein, I.A., 1919– III. Nippon Gakujutsu Shinkōkai. IV. United States. National Science Foundation.
QP88.5.B553 1977 599′.01′858 77-2380
ISBN 0-8391-1115-0

Printed in Japan

Originally published by
UNIVERSITY OF TOKYO PRESS

CONTENTS

IV. BIOCHEMICAL CHARACTERIZATION OF EPIDERMAL CELL CONSTITUENTS

PREFACE

This symposium was motivated by the growing need for investigators active in the areas of epidermal biochemistry to familiarize themselves with the rapidly accumulating knowledge from many and varied disciplines. The idea was first conceived in 1973 by the editors while they were discussing various research problems of mutual interest during a visit by Dr. Seiji to Dr. Bernstein's laboratory at the University of Michigan. The editors felt that a symposium on the biochemical aspects of epidermal differentiation should be held in Japan in a manner similar to the "Symposium on the Biology of Normal and Abnormal Melanocytes" which was held in 1969, sponsored by the Japan Society for the Promotion of Science and the U.S. National Science Foundation. Major investigative contributions to the developing body of biochemical information on epidermal differentiation were being made by investigators in the United States and Japan, and national scientific conferences had been held on this subject in both countries. During the preceding five years, four conferences on "Biochemical Aspects of Epidermal Differentiation" had been organized at the University of Michigan in Ann Arbor, Michigan, by Dr. Bernstein. In Japan, four conferences on "Morphological and Biochemical Aspects of Keratinization" had been held under the auspices of Dr. Seiji. Ten to fifteen investigators doing research on this field were brought together for each of these conferences. Participants in the respective conferences included representatives of nearly all the participating laboratories in each country.

Three years ago, one of the Japanese investigators was invited to the Michigan conference and presented a paper. Similarly, two American investigators were invited to participate in the conference held in Japan. However, there had been no opportunity for a truly international review of all the cogent data. An international seminar would allow such a review to occur in the presence of experts representing all the laboratories in the two countries which had made major contributions to the world's literature on the subject. A small group would be able to evaluate existing information, resolve differences of opinion, and chart a direction which could result in more rapid progress toward the common goals sought by all of the participating laboratories in Japan and the United States.

The Japan-U.S. Seminar "Biochemistry of Cutaneous Epidermal Differentiation" was held in Sendai at the Gonryo-Kaikan, June 2 through 4, 1976, under the joint sponsorship of the Japan Society for the Promotion of Science and the National Science Foundation of the United States.

The editors hope that the publication of this book will promote further international interaction among scientists interested in the biochemical aspects of keratinization in the epidermis.

Makoto Seiji

Department of Dermatology
Tohoku University School of Medicine
Sendai

I. A. Bernstein

Department of Biological Chemistry and
of Environmental and Industrial Health
The University of Michigan
School of Public Health
Ann Arbor, Michigan

OPENING ADDRESS

Mr. Chairman, ladies and gentlemen:

It is a great pleasure for me to speak here today at the opening of this seminar in which more than ten American researchers in skin biology and as many Japanese specialists will report and discuss what they have been studying. First of all, I want to extend my hearty welcome to you all and to express my heartfelt thanks to the U.S. National Science Foundation and the Japan Society for Promotion of Science, under whose joint sponsorship this seminar is being held.

The skin is a tangible and visible object, so its lesions have been known since ancient times and have been studied by many workers. For example, melanization and keratinization, which are specific functions of the skin, have been studied with up-to-date scientific methods, as was shown in that U.S.-Japan symposium held in Japan some time ago.

Moreover, the remarkable progress in the biological methods, especially electron microscopical, histochemical, and immunological examinations, have brought new findings in the biological fields of dermatology.

However, the investigation of the skin seems to be somewhat behind that of the various internal organs. This may be because the skin has been regarded as a covering tissue on the surface of the body rather than as an organ which has much to do with the mechanism of life. Long ago, Hoffmann proposed an "Eisophylaxie" hypothesis, but it has yet to be supported by ample biological data. With the advocation of the views on "systemic" diseases such as lupus erythematosus as a turning point, great importance has been attached to the systemic significance of the skin diseases.

In conclusion, I want to share with you my joy on this occasion when the leading investigators on skin biology are gathered here together to discuss and exchange views on the great significance of the skin to life. I offer you my best wishes for every success in this memorable seminar, brief as it may be.

Minor Itô

Emeritus Professor
Tohoku University

I. MORPHOLOGICAL ASPECTS OF DIFFERENTIATION

Some Aspects of the Keratinization Process of Epidermal and Pilary Epithelial Cells as Observed by Electron Microscopy

Kazumasa Kurosumi

Department of Morphology, Institute of Endocrinology, Gunma University, Maebashi, Japan

SUMMARY

Epithelial differentiation which leads ultimately to keratinization was studied by electron microscopy on thin sections of the ordinary epidermis, intraepidermal sweat ducts, hair follicles and hair roots.

Epidermal keratinocytes contain both filaments and granules. The keratohyalin granules are often irregular in shape, but spherical granules are also found, especially in some particular regions of the animal body and in the epidermis of late fetal or neonatal animals. They are called keratohyalin droplets and are often surrounded by many free ribosomes, but are not associated with tonofilaments. It is suggested that the keratohyalin may be synthesized at the ribosomes and accumulate in the cytoplasmic matrix as droplets without surrounding membrane. At the place where tonofilaments are well developed, keratohyalin may infiltrate into the tonofilament-bundles and form an irregular granule. At last tonofilaments are completely embedded in a dense interfilamentous substance presumably derived from keratohyalin.

The lamellar granules, which are also called either Odland bodies, membrane-coating granules, keratinosomes or cementsomes, are characterized by the internal structure of parallel lamellae. Acid phosphatase activity is localized at the dark band in this granule. The clear band may be made up of lipid, because the granules were disintegrated after the treatment with organic solvent. As this granule has dual composition, its function may also be dual; that is, the hydrolitic enzyme of this granule released in the intercellular space may dissolve the cementing substance and facilitate the desquamation, while the lipid from this granule may spread over the surface of the plasma membrane to protect the cell from the proteolytic enzyme action until the desquamation.

The cells of the epidermal sweat duct also keratinize. Round keratohyalin droplets are found in the luminal cell, whereas the lamellar granules are found in the peripheral cells. The duct cells also desquamate into the lumen, when the keratinization has been completed or almost finished.

Epithelial cells of the internal root sheath of hair follicles contain tonofilaments and trichohyalin granules. The latter are similar in morphology to the keratohyalin granules, though their chemical compositions may differ. The keratinization process of the internal root sheath may be identical with that of the epidermal keratinocytes. However, the keratinization of hair roots may be performed by other mechanisms than in the epidermis, and each layer in the hair shaft may keratinize by a mode different from that of its adjacent layer: in the hair cuticle and medulla the granules are predominant, while the filaments play the most important role for keratinization in the cortex, where no granules can be observed.

Keratinization is the most remarkable characteristic of the epithelial parts of the skin of most vertebrates. Terrestrial animals yield more highly keratinized epidermides than do aquatic animals in general. The mammalian skin contains keratinized portions that are epidermis, hairs, nails (claws and hoofs) and horns. This paper deals with the keratinization of epidermis and hairs in man and some experimental as well as domestic animals. The morphology of the keratinization process as observed with the transmission electron microscope reveals certain differences between the epidermis and hair. In the latter case, the cells of the hair shaft and the inner layer of the hair follicle keratinize, but the mode of keratinization is not the same in cells of various parts of these pilary components. It may be worthwhile to compare the keratinization process of the epidermis to those occurring in various parts of the hair root and hair follicle.

Most keratinizing epithelial cells contain filaments and granules, both of which may be closely related to the keratinization, but some granules of keratinizing cells are known not to be related to the process of keratinization. In the epidermal keratinocytes appear three different kinds of granules: 1) melanosomes, 2) lamellar granules and 3) keratohyalin granules. The melanosomes are not directly related to the keratinization, since they are absent in the epidermal keratinocytes of the albino animals, though they completely keratinize. It is widely accepted that the keratohyalin granules are directly related to the keratinization, but the role of the lamellar granules, which are variously called Odland bodies, membrane-coating granules, keratinosomes and cementsomes, is still under debate. The filaments occurring in various kinds of epithelium, not only the epidermis, are commonly called tonofilaments, and are conjectured to be the main material of the keratinous (horny) substance of the skin and its appendages.

The epithelial cells of the internal root sheath of the hair follicle contain tonofilaments and granules. The latter are called trichohyalin granules and look similar to the keratohyalin granules of the epidermal keratinocytes. The keratinizing process of internal root sheath cells is very akin to that in the epidermis. However, the keratinization of the hair itself is quite different from that in both the hair follicles and epidermis. The cuticle and medulla of the hair contain many granules but no filaments. The hair cortex contains a great amount of filamentous substance but no granules. Detailed discussion of the difference in keratinization mechanism in many types of epithelial cells in various parts of skin will be given in a later passage of this paper.

MORPHOLOGICAL DIFFERENTIATION OF EPIDERMAL KERATINOCYTES IN GENERAL

Histologically the epidermis is the stratified squamous epithelium, whose characteristics are the stratification of epithelial cells and flattening of cells as they migrate toward the surface. As the epidermis is the keratinizing epithelium, the final step of cell differentiation is the death and drying to form hard scales. The piling up of scales to various thicknesses makes the most superficial layer of the epidermis called the horny layer (stratum corneum). The formation of this layer is the so-called keratinization or cornification. As most epidermal cells ultimately keratinize, they are called keratinocytes. There are some nonkeratinizing cells in the epidermis and they are often called dendritic cells. This paper concerns only keratinizing cells, and the dendritic cells may be referred to in another paper published elsewhere.[1]

The bottom of the thick stratified epithelium adjacent to the underlying connective tissue (dermis in this case) is called the basal layer (stratum basale). There is a thin membrane called the basement membrane or basal lamina at the boundary between the basal layer of the epidermis and the dermis. The keratinocytes in the basal layer are usually cylindrical in shape and their long axes are directed vertically to the surface of the epidermodermal junction. At the middle layer of the epidermis called the prickle cell layer (stratum spinosum), the keratinocytes become polyhedral. As they are getting a little flattened at the margin of the former layer, certain granules appear in the cytoplasm, and this layer is called the granular layer (stratum granulosum). The light microscopy had only demonstrated a larger type of granules that is called keratohyalin granules, but another type of granules with smaller size was later found by electron microscopy. They are characterized by fine lamellar structure which is demonstrable only by use of the electron microscope, and the name lamellar granules is often used. The layer next to the granular layer is the horny layer, which is the uppermost part of the epidermis.

The keratinocytes contain filaments within their cytoplasm. The filaments are called tonofilaments and they occur even in the youngest cells of the deepest layer, that is, the stratum basale. As the keratinocytes migrate upwards the amount of tonofilaments obviously increases. The tonofilaments and keratohyalin granules may become the chief elements of the keratin in the cells of the stratum corneum. It is unknown what organelles in the cytoplasm might be engaged in producing tonofilaments.

The main part of the epidermis is the stratum spinosum, whose name is derived from the fact that the keratinocytes in this layer send out many spinous processes. These processes are attached to those from the adjacent cells, end to end or side by side. At the attachment of cell processes develops the so-called desmosome, which is a strong device for cell adhesion. The interconnection between adjacent keratinocytes was called intercellular bridges by light microscopists, but they are not true bridges as observed by electron microscopy. Bundles of tonofilaments run through the processes and end at the desmosomes, but they never go through the desmosomes passing into the next cell. That indicates that the intercellular bridges are not the true intercellular connections, but only attachments between cells.

The desmosome cut perpendicularly to the surface of attachment is seen as a complex of parallel lines. Five dense lines with intervening clear space and additional dark layers are observed (Fig. 1). The central dense line is called the intercellular contact layer[2] and sometimes is split into two incomplete lines. This layer corresponds to the center of the intercellular space. On both sides of this central plate, distinct double membranes are observed and are known to be the plasma membranes with the trilaminar "unit membrane" structure. The inner leaflet of the plasma membrane is seen as thick, because inside this membrane accumulates a somewhat dark substance, forming a specialized dark layer known as the attachment placque[2] at about 40 A inside the inner leaflet of the plasma membrane. Onto this placque tonofilaments may attach or they may make a loop at this portion or a little away from the placque.[3] Near the desmosomes a simple but very close adhesion of two plasma membranes is often observed. Such a specialized strong contact is called the gap junction or nexus (Fig. 1).

Along the basal surface of the cell at the deepest portion of the epidermis are observed enhanced densities known as hemidesmosomes.[4] At the hemidesmosome, a thin dense line is observed just outside the plasma membrane, which may correspond to the intercellular contact layer (Fig. 2). Between this thin membrane and the basal lamina some vertical filaments are observed. The globular material observed in between the

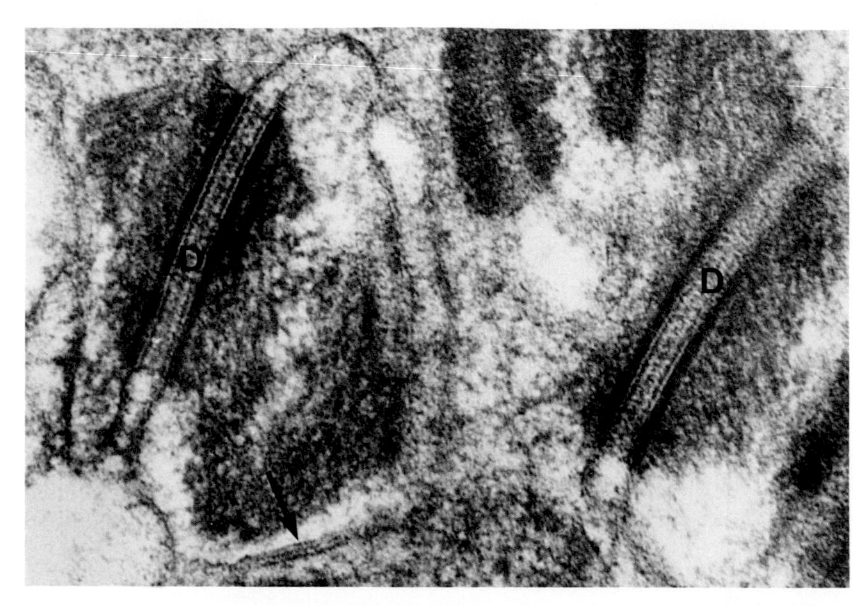

Fig. 1. Stratum spinosum of dog snout skin. Normal sections of desmosomes (D) are observed. They consist of several parallel lines. The arrow indicates a gap junction situated near the desmosomes. ×100,000.

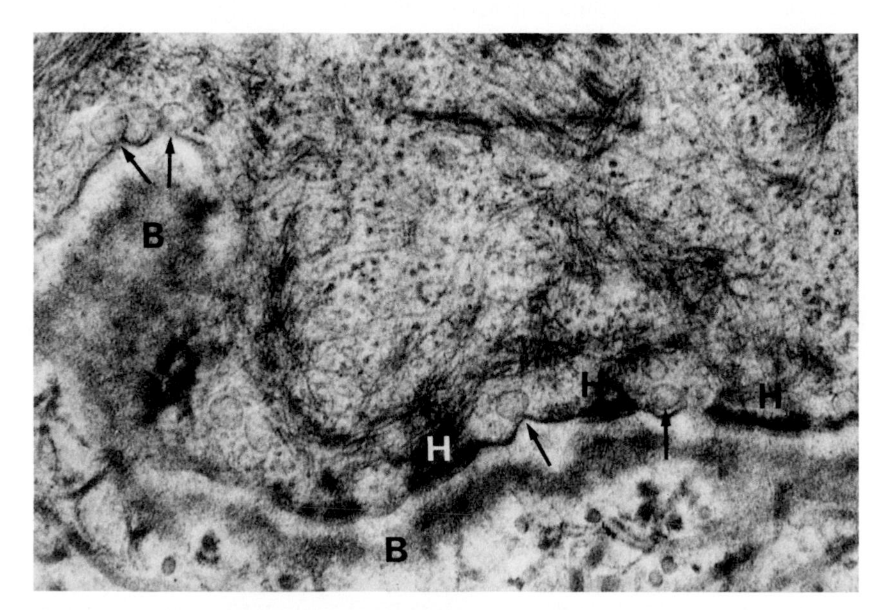

Fig. 2. Stratum basale of the human epidermis. The basal lamina (B) is parallel to the basal surface of the deepest keratinocyte. High density spots along the basal surface are hemidesmosomes (H). Arrows indicate pinocytotic vesicles. ×50,000.

basal surface of the epidermal cell and the basal lamina in the amphibian skin[4] was not found in the mammalian skin. At the basal cell surface not occupied by the hemidesmosomes many vesicles probably formed by the pinocytosis are observed (Fig. 2).

As the keratinocytes mature and differentiate, moving toward the epidermal surface, the plasma membrane becomes thick and desmosomes may change their structure. Thickening of the plasma membrane is one of the salient features accompanied by the advancing keratinization, and may be discussed by other contributors in this seminar in more detail. It is most important to notice that the thickening may be fulfilled by the accumulation of some unknown substance just inside the plasma membrane but not accomplished by the addition of any substance outside the plasma membrane or the so-called coating of the membrane. In the stratum corneum the tonofilaments are so remarkably increased, that the attachment of tonofilaments to the desmosomes become obscure. The intercellular contact layer of the desmosome becomes thick and sometimes spindle-shaped and finally disappears.

LAMELLAR GRANULES

The keratinocytes in the upper part of the stratum spinosum or the stratum granulosum contain many small round or oval granules (Fig. 3). They measure about 100–500 nm in diameter. They are too small to be observed by light microscopy. Therefore, electron microscopy first enabled us to find them. Initially they were erroneously thought to be either small keratohyalin granules,[5] viruses,[6] or deformed mitochondria,[7] but later they were recognized as a distinct entity among the cell organelles of the keratinocytes. Thus they have been called with various names, such as lamellar granules,[8] membrane-coating granules (MCG),[9] Odland bodies,[10] keratinosomes,[11] and cementsomes.[12] Among these various names, "lamellar granule" may be most suitable as far as the present state of knowledge goes, because this term indicates only the morphological characteristic of the granule. All other names for this peculiar granule imply certain presumed functions, though its function is still under debate.

High resolution electron microscopy of a section of this granule cut at a suitable angle reveals its internal structure with a characteristic lamellar pattern (Fig. 4). It consists of alternating light and dark lamellae (plates) each about 20–35 A thick. The granule is covered with a limiting membrane showing the well-known unit membrane structure. Sometimes it is shown that the dark lamellae can be divided into two types of different thicknesses: the thicker is 25–30 A, while the thinner is 10–15 A in thickness. These two different plates may appear alternately.[13–16] The origin and fate of lamellar granules were repeatedly studied by many authors. It has

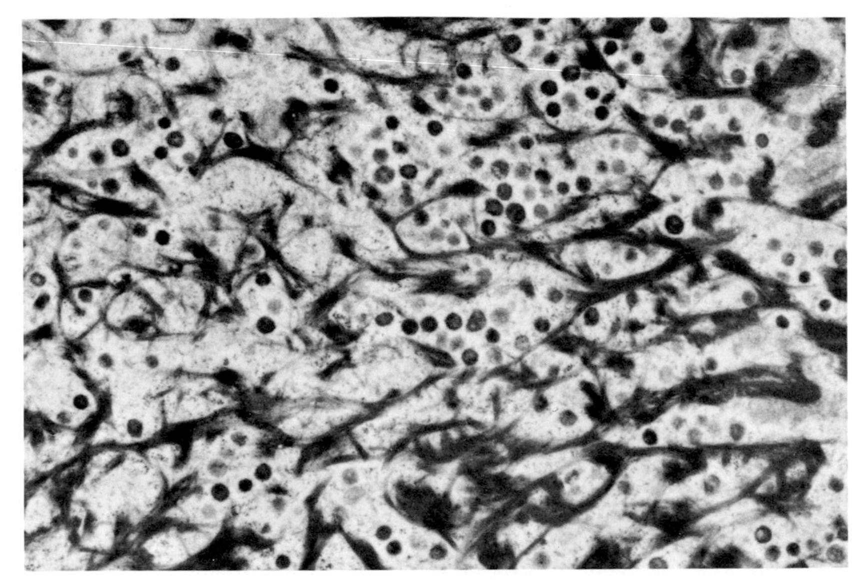

Fig. 3. A part of the keratinocyte at the upper spinous layer of the dog snout skin. Many round lamellar granules are scattered among the network of tonofibrils. ×20,000.

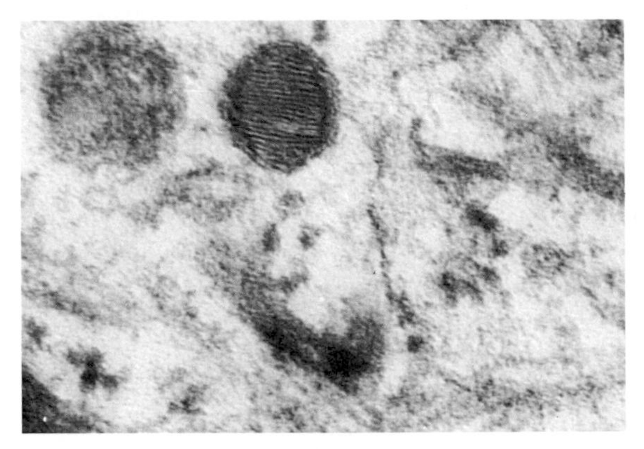

Fig. 4. Lamellar granules of the keratinocyte of the dog snout skin. The granule seen on the right shows lamellar internal structure. ×100,000.

been suggested that the granules may be elaborated by the Golgi apparatus.[9,14,16,17–19] In the stratum granulosum, lamellar granules tend to migrate toward the peripheral part of the keratinocytes, and become attached to the inner side of the surface plasma membrane. In most cases the lamellar granules are gathered beneath the plasma membrane of the distal (outer) surface of the keratinocytes. The interior of the granules may be released out of the cell through a small orifice.[9,13,16,20,21]

The chemical composition of lamellar granules is still under debate, though some investigators argue that the granules contain lysosomal enzymes,[22] while others postulate that the granules are of phospholipid.[12] On the function of these granules several different views were proposed: for example Matoltsy and his coworkers[9] suggested that the substance of lamellar granules extruded into the intercellular space may coat the surface of the plasma membrane and give rise to thickening of the plasma membrane as the keratinocytes migrate toward the skin surface and gradually keratinize. From this view they called the granules the membrane-coating granules (MCG). Wilgram[11] published a view that the lamellar granules might exert a role in desquamation of cornified keratinocytes. The findings of Wolff's study that the granule contains hydrolases may support this view. But the name ''keratinosome'' proposed by Wilgram[11] implies a presumable function of this granule which might be essential for keratinization. However, it is known that the cells constituting the hair root and internal root sheath of the hair follicle contain no lamellar granules, though they can completely keratinize. The name ''keratinosome'' has not been so widely used, because these granules are thought not essential for keratinization.

Hashimoto[12] suggested that the lamellar granules might contain phospholipid and function to combine the keratinocytes with each other; therefore, the granules were considered to play a role in intercellular adhesion, and thus called ''cementsome.'' If the lamellar granules have a function effective for the tight adhesion between keratinized cells, they must be abundant in the tissue of hairs, because desquamation does not occur in the hair. But, in fact, no lamellar granules are observed in cells of the hairs and internal root sheath, as mentioned above. It is most likely that the granules may be related to the desquamation of keratinized cells, because they appear in great number in the keratinocytes of the epidermis, where the desquamation actively occurs; but in the hairs the situation is reversed.

In our cytochemical study at the level of electron microscopy, acid phosphatase activity is shown at the dark plates in the lamellar granules, but the light plates are not stained by the reaction products (Fig. 5). The substance extruded to the intercellular space is also enhanced in electron density after the reaction for acid phosphatase (Fig. 6), showing the presence of acid hydrolases in the substance of lamellar granules, as shown first by Wolff and Holubar.[22]

On the other hand, the light plates may be composed of another substance, because it has no positive reaction for acid phosphatase. The presence of lipid was repeatedly assumed to be another component of the granule. It is likely that the lipid may reside in the light plates. Hashimoto's[12]

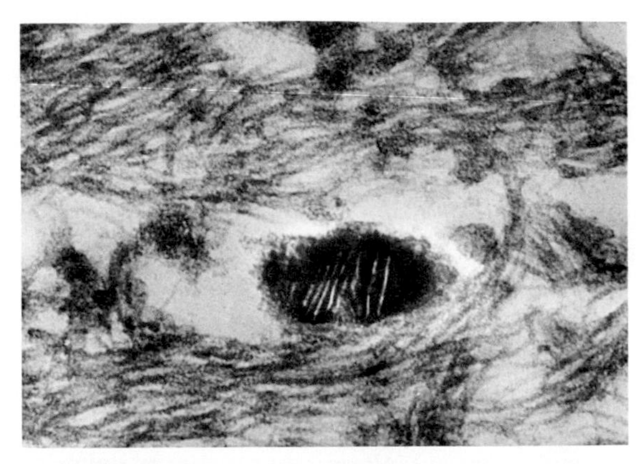

Fig. 5. Acid phosphatase reaction of a lamellar granule of the mouse epidermis. The dense reaction products have precipitated upon the dark bands of the granule. ×100,000.

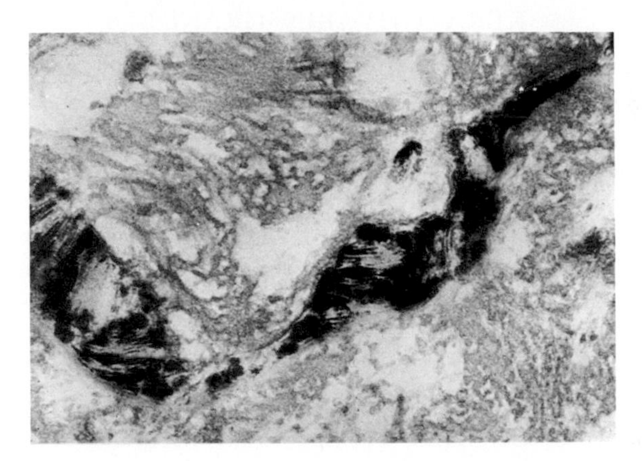

Fig. 6. The histochemical examination on the stratum granulosum of the mouse skin. Dark reaction products for the acid phosphatase activity with lamellar structures are seen in the intercellular space. ×90,000.

description said that the lamellar granules were digested by phospholipase C, and Suzuki[23] in our laboratory demonstrated the disappearance of lamellar granules after treatment with chloroform-methanol. It became clear that the lamellar granules contain both protein and lipid. Most of the protein constituent may be some hydrolytic enzymes which may form the dark plates, whereas the phospholipid which may be the major part of the lipid moiety forms the light plates. As the granules are dual in chemical composition, the function may also be dual. The acid hydrolases may digest and dissolve the intercellular cementing substance and promote the separation of the keratinized cells, that is, the desquamation. However, the second component of this granule, the phospholipid, may be spread out over the surface of the cell membrane soon after its release and form a thin film of oil, which may act as a barrier preventing digestion by hydrolases. The horny cells have never been broken to pieces nor dissolved away until the time of desquamation, because the surface plasma membrane coated by the lipid moiety of lamellar granules is strong enough to protect the cell against the digestion.

The substance of lamellar granules which were called membrane-coating granules may actually coat, at least partially, the surface plasma membrane of keratinocytes. But thickening of the plasma membrane as the cell keratinizes is mostly due to the apposition of dense material to the inside surface of the plasma membranes. Such a membrane-thickening substance may come from cytoplasmic organelles or inclusions other than lamellar granules.

KERATOHYALIN GRANULES

The keratohyalin granules could be observed by light microscopy, and gave the name of epidermal layer containing these granules, that is, the stratum granulosum. By light microscopy, however, the structural detail could not be resolved, so that the granules were all seen as round granules. On the contrary, electron microscopy revealed most keratohyalin granules to be irregular in shape (Fig. 7). They are very high in electron density and not covered with a membrane, unlike other kinds of intracellular granules, such as lamellar granules, melanosomes, secretory granules, and lysosomes.

In the epidermis of late fetal or neonatal rats, regular round keratohyalin granules are frequently observed.[24] They are surrounded by a great number of ribosomes, but tonofilaments are either very few or absent near these round granules (Fig. 8). It may be better for these round granules to be called keratohyalin droplets, and they are synthesized by the free ribosomes and appear as droplets in the relatively liquid matrix of the cytoplasm. If the keratohyalin appears closely adjacent to tonofilament bundles, the shape of accumulated keratohyalin masses may become irregular. Conversely, at the place where tonofilaments are very few, the keratohyalin appears as a droplet. In the skin of dog snout and in some cases of human disease, for example, congenital ichthyosiform erythroderma, the keratohyalin droplets were abundantly observed.[16,25] In the lower granular layer of the mouse plantar skin, a large keratohyalin granule was observed and its one side was regularly round and associated with ribosomes, while the other surface was irregularly elongated and continuous to the bundles of tonofilament.[24] Features suggesting the infiltration of keratohyalin substance into the tonofilament bundles were observed, and thus the keratohyalin was known to become an interfilamentous

Fig. 7. Stratum granulosum of the adult rat epidermis. The dark irregular bodies are keratohyalin granules. They grow larger as they approach the upper part of the epidermis. N: nucleus. ×8,000.

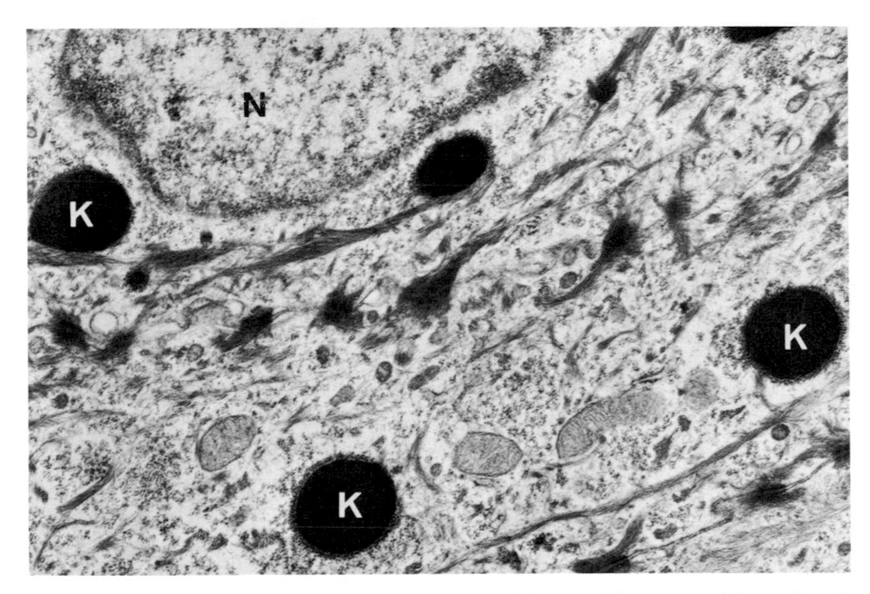

Fig. 8. Stratum granulosum of the fetal rat epidermis. Many round keratohyalin granules (K) are observed. Upper granules are associated to the bundles of tonofilaments and are a little deformed, while lower granules are round and surrounded by many ribosomes. N: nucleus. ×16,000.

substance (Fig. 9). The fully keratinized cells of the horny layer are packed with filaments and the interfilamentous substance, and their appearance under the electron microscope has been called the keratin pattern, in which the filaments look clear while the interfilamentous substance looks dark (Fig. 10). The so-called keratin pattern may be visualized by the intermingling of tonofilaments which are seen as relatively clear and the dark interfilamentous substance derived from the keratohyalin substance.

EPIDERMAL SWEAT DUCTS

Intraepidermal ducts of eccrine sweat glands also keratinize. The epidermal ducts of the human, mouse, and dog sweat glands were observed. The sweat duct consists of two kinds of epithelial cells: the luminal cells and peripheral cells.[26] The luminal cells face the lumen of the sweat duct, and their free surface is provided with conspicuous microvilli. These cells are specially differentiated but the peripheral cells directly surrounding the luminal cells are similar to the ordinary keratinocytes in every aspect for their shape and localization, that is crescent-shaped surrounding the duct when the latter is cut transversely.

In the mouse epidermal sweat duct, the peripheral cells form a single layer below the lower granular layer, but they are one to three layers from

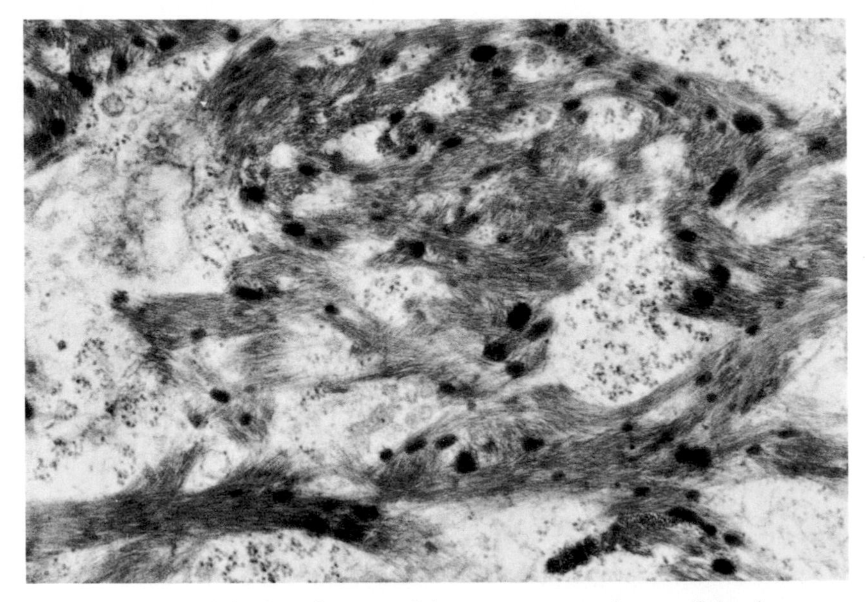

Fig. 9. A part of the keratinocyte of the stratum granulosum of the dog snout skin. Small dark keratohyalin granules are seen on the filament bundles, and they are infiltrating into the bundles of tonofilaments. ×30,000.

Fig. 10. Stratum corneum of the mouse skin. The fully keratinized portion indicates that the filaments are clear and the interfilamentous substance is dark, showing the so-called keratin pattern. ×100,000.

the upper part of the granular layer, going upwards, while the luminal cells are in a single layer throughout the entire course of the intraepidermal sweat duct.[27] The microvilli of the free surface of luminal cells are relatively short and blunt as compared with those of the secretory cells of the sweat gland. The luminal cell contains fairly abundant tonofilaments but they are evenly scattered throughout the cell and do not form bundles (or tonofibrils) which are seen in the ordinary keratinocytes. The keratohyalin granules found in the luminal cells are always spherical in shape (Fig. 11). They are variable in size very widely from 0.1 to 1 μm in diameter. The electron density of these granules is very high and no limiting membrane is provided. Free ribosomes are often gathered around these spherical granules.[26] These round granules are the same as keratohyalin droplets seen in some keratinocytes of specialized regions of the skin or in the fetal or neonatal epidermis.[16,24,25] The keratohyalin droplets appear at the place where the tonofilaments are very few or absent. The frequent occurrence of round keratohyalin droplets in the luminal cell of the epidermal sweat duct may indicate the relatively low density in distribution of tonofilaments in this cell.

The lamellar granules are not found in the luminal cell, but they appear abundantly in the peripheral cells (Fig. 12). They may be extruded into

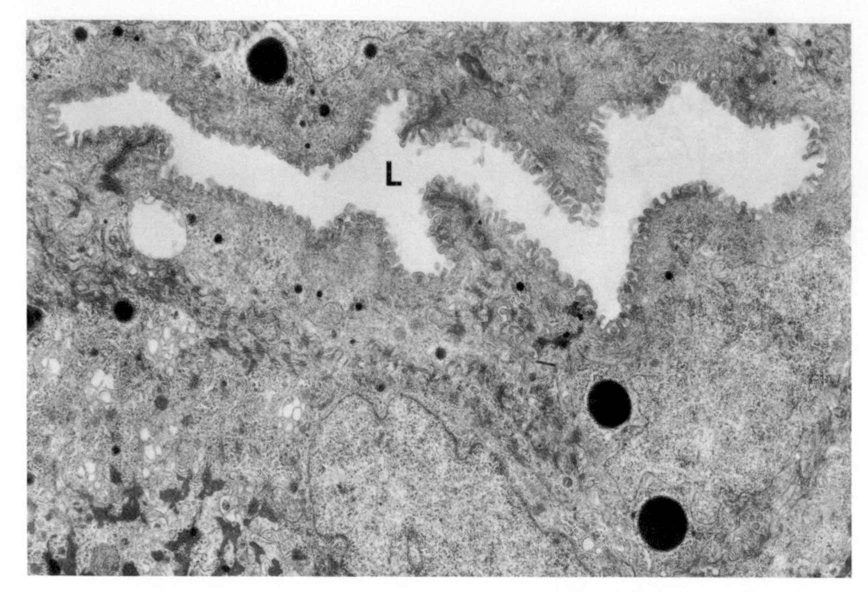

Fig. 11. An epidermal sweat duct of the mouse is shown. A rather irregular lumen (L) is seen in the upper part of the figure. The surface of the luminal cell is provided with short and blunt microvilli. The luminal cell contains many round keratohyalin droplets of various sizes. ×8,000.

the intercellular space between the luminal and peripheral cells or the space between adjacent peripheral cells.[27] The thickening of the plasma membrane may occur at the cell borders between the peripheral cells or between the peripheral and luminal cells. However, the plasma membrane of the luminal cell facing the duct lumen never becomes thickened.[27] Tonofilaments in the peripheral cells are often arranged to form bundles like those in the usual epidermal keratinocytes. There are some round or oval keratohyalin granules, but some irregularly shaped keratohyalin granules are also observed, especially where the granules are in close contact with tonofilament-bundles. The keratinization of peripheral cells is often complete and a typical keratin pattern can be observed, but the luminal cells do not completely keratinize. At the uppermost part of the stratum corneum, however, the luminal cells may keratinize fairly well, and they may be shed off into the lumen of the duct, though they are premature in keratinization.[27,28] Such a desquamation of sweat duct cells is not restricted to the luminal cells, but some peripheral cells also desquamate. The lamellar granules occurring in the peripheral cells may be responsible for such a desquamation of cornified duct cells.

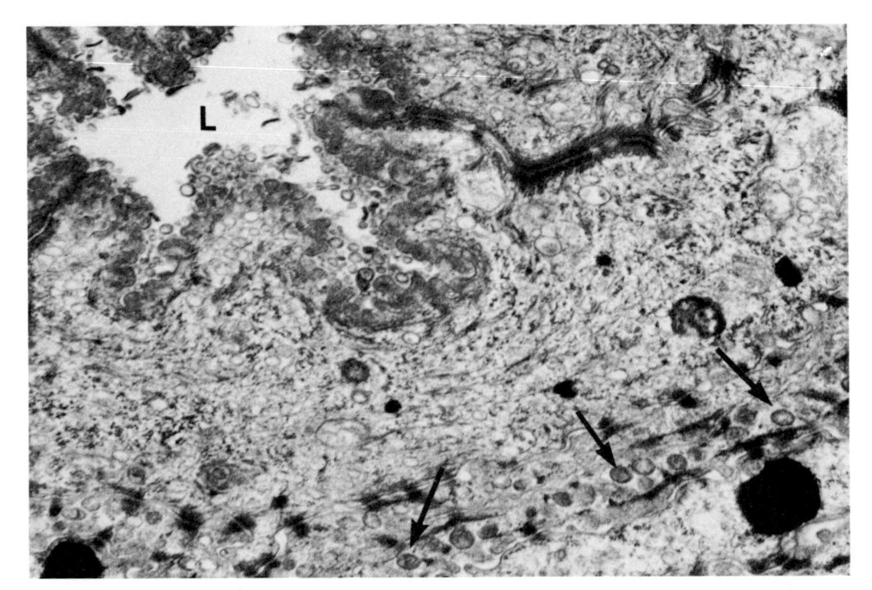

Fig. 12. The luminal part of the human epidermal sweat duct. The intercellular margin is provided with atypical desmosomes. Arrows indicate lamellar granules situated near the cell surface at the border between the luminal and peripheral cells. L: duct lumen. ×16,000.

In the mouse luminal cells found at the level of the lower horny layer, the microvilli is almost empty but the plasma membrane is left intact (Fig. 13). In the luminal cells of the human eccrine duct, there are many clear vesicles situated along the luminal surface of the cell (Fig. 14). Such vesicles have not been observed in animal sweat ducts. They may be related to the reabsorption of some material from the sweat, as they look like the so-called pinocytotic vesicles. Zelickson[29] suggested that the luminal cell of the human sweat duct is specially differentiated and comparable to the cells of the proximal convoluted segment of the renal tubule, and postulated that the luminal cell might play an active part in reabsorption of sweat. Hashimoto *et al.*[28] described the presence of vesicles in the luminal side of the inner (luminal) cell cytoplasm, and postulated that these vesicles might pass from the inner cell into the lumen through disruption of the plasma membrane and also as a result of pinching-off of microvilli. Though the latter mechanism of secretion discharge, which has been called microapocrine secretion, may possibly occur in the duct as well as the secretory portion of the eccrine sweat gland,[30] because the cell debris is often seen in the duct lumen. However, the vesicles found in the luminal cells of the epidermal duct do not correspond to the secretory

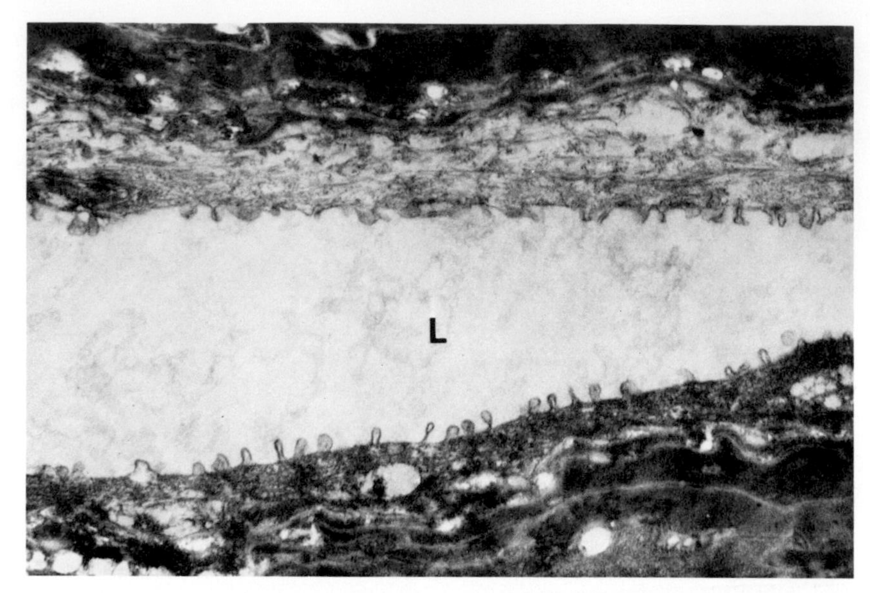

Fig. 13. An epidermal sweat duct of the mouse at the level of the lower horny layer. The luminal cell of the duct is blackened as it keratinizes. The microvilli are seen empty but the membrane is intact. L: duct lumen. ×18,000.

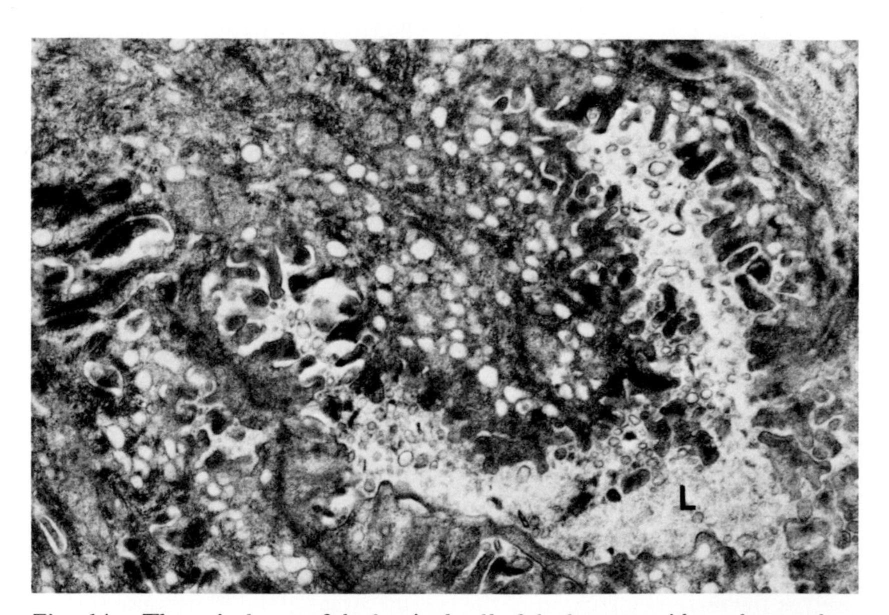

Fig. 14. The apical part of the luminal cell of the human epidermal sweat duct. In the lumen (L) small fragments of cell debris are observed. Many clear vesicles are gathered in the cytoplasm near the lumen. ×15,000.

granules. The true nature of these vesicles of the human sweat duct cells will be studied more in the near future.

KERATINIZATION IN THE HAIR FOLLICLES AND HAIR ROOTS

The hair is the most conspicuous appendage of the skin of mammals, including man. The part of the hair embedded in the skin tissue is called the hair root, and it is surrounded by the tissue named the hair follicle. The hair root consists of three layers, that is, the hair cuticle situated externally, the hair cortex at the middle forming the main substance of the hair, and the hair medulla that is the central core of the hair shaft.

a. The hair follicle

The hair follicle is divided into three main parts: the innermost layer is the internal root sheath, the middle is the external root sheath, and the last layer at the outermost region is the connective tissue follicle (Fig. 15). The total tissue including both the internal and external root sheaths is called the epithelial follicle. The internal root sheath as well as the hair root keratinize, while the external root sheath and the connective tissue follicle do not keratinize.

The internal root sheath is also divided into three layers: the outermost layer is the Henle layer, the middle is the Huxley layer, and the innermost layer is the sheath cuticle. The sheath cuticle and hair cuticle are closely attached and interlocked to each other. The keratinization advances in different steps among these three layers: the Henle layer is the earliest to begin keratinization, the sheath cuticle proceeds next, and the Huxley layer keratinizes most slowly (Fig. 16). Cells of these three layers at slightly advanced stages in keratinization contain conspicuous granules and filaments. The former are round, oval or spindle-shaped granules and called trichohyalin granules. As the tonofilaments are arranged parallel to the long axis of the hair root and follicle, the trichohyalin granules also tend to be elongate in shape, being oriented longitudinally. As the staining behavior of the trichohyalin granules in light microscopic preparations differs from that of the keratohyalin granules of the epidermal keratinocytes, they are suggested to be different in chemical composition from each other, but their ultrastructural morphology is very similar to each other.

At the relatively early stage of keratinization, tonofilaments look as if they extend in both directions parallel to the follicle axis from the upper and lower ends of lenticular or spindle-shaped trichohyalin granules. Birbeck and Mercer[31] argued that the tonofilaments of internal root sheath cells might be transformed from the trichohyalin granules. Our findings on the

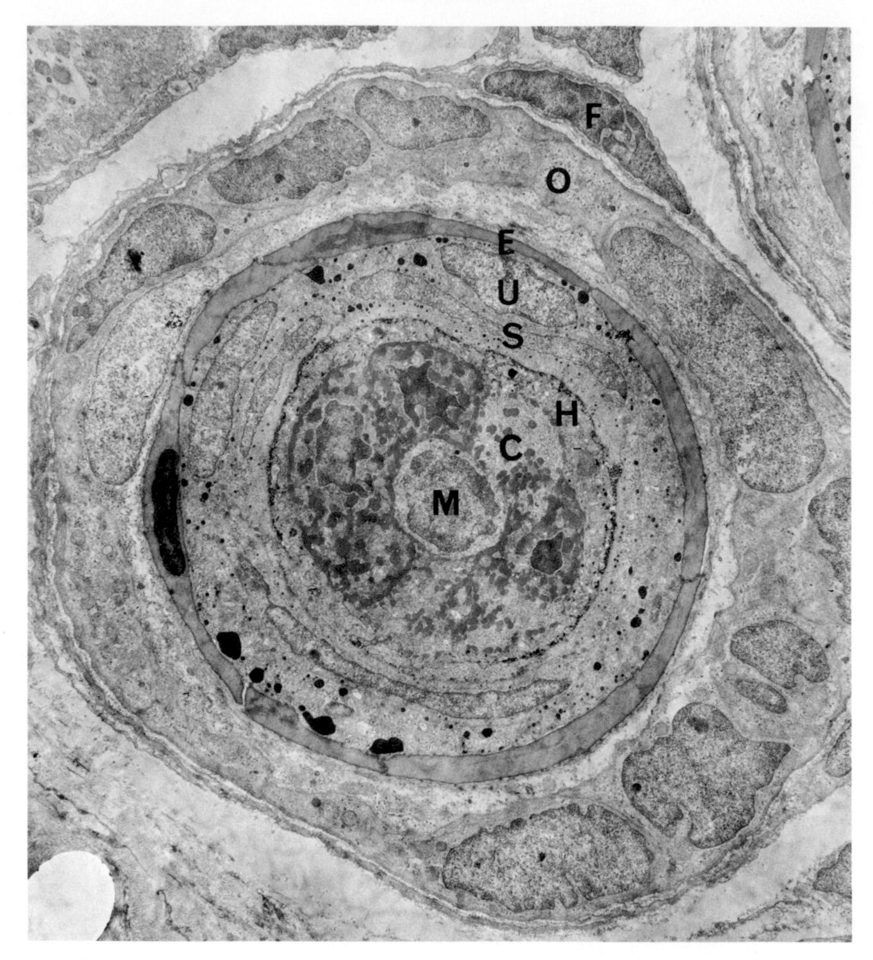

Fig. 15. A survey electron micrograph of the cross-cut profile of the rat hair follicle containing the hair root. M: hair medulla, C: hair cortex, H: hair cuticle, S: sheath cuticle, U: Huxley layer, E: Henle layer, O: external root sheath, F: connective tissue follicle. ×2,000.

rat hair follicles indicate, however, that the tonofilaments appear before the advent of trichohyalin granules at the lower part of the hair follicle (Fig. 17). The relationship between the tonofilaments and trichohyalin granules may presumably be identical with that between the tonofilaments and keratohyalin granules in the epidermal keratinocytes, since the basal layer cells contain only tonofilaments but no keratohyalin granules. Though the chemical composition of the trichohyalin is different from that of keratohyalin, the keratinization process of the internal root sheath of the hair follicle may be the same as that of the epidermis.

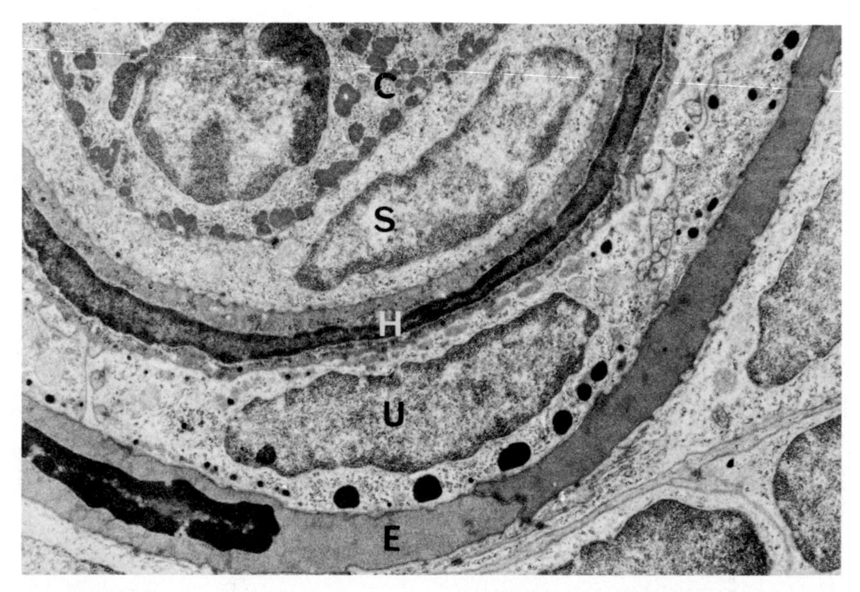

Fig. 16. A part of a cross section of the rat hair follicle. The labelling is the same as Fig. 15. ×8,000.

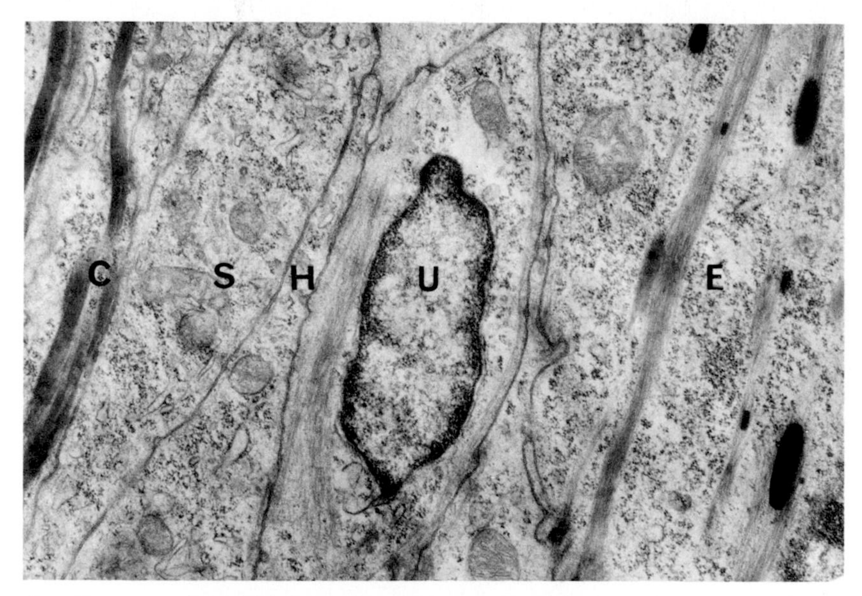

Fig. 17. A longitudinal section of the rat hair follicle. In the Huxley layer only filaments are observed, while the trichohyalin granules appear in the Henle layer. The labelling is the same as Fig. 15. ×14,000.

b. The hair root

The surface of the hair shaft is covered by scale-like flat cells called the hair cuticle. In a rather thin hair the cuticular cells are arranged in a single cell layer (Figs. 15, 16), but in many cases of relatively thick hair, the cells of the hair cuticle are stratified in several layers (Fig. 18). At any rate, cells of the outermost layer of the hair cuticle are interlocked with those of the sheath cuticle.

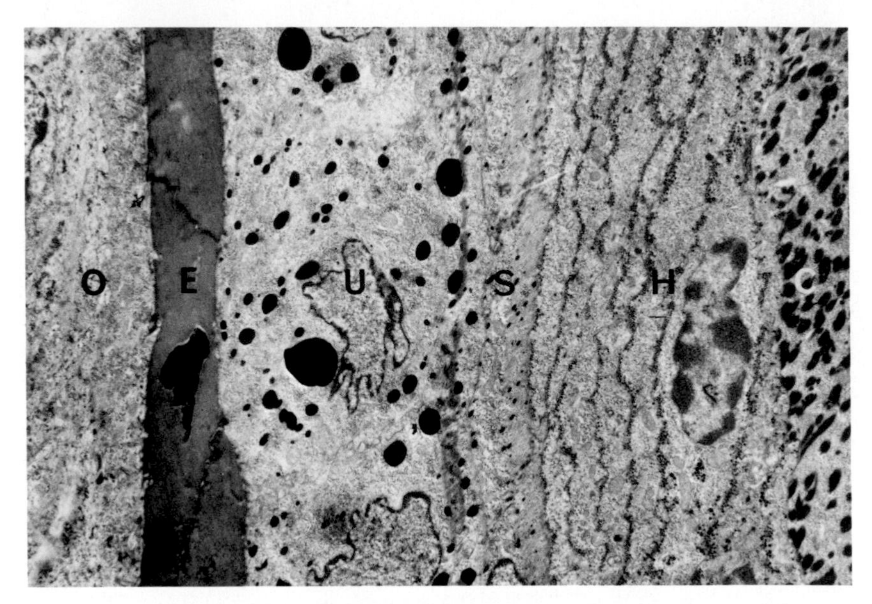

Fig. 18. A longitudinal section of the rat hair follicle and hair root. The hair cuticle consists of several layers of epithelium. The labelling is the same as Fig. 15. ×3,000.

The cuticular cells of the hair itself contain irregularly shaped dark granules, which are arranged along the cell surfaces (Figs. 19, 20). They are more numerous along the lateral cell surface, and on the medial surface they are very few in number (Fig. 19). These granules are roughly round in a rather earlier stage (Fig. 19), but they become irregular in shape as the cell matures (Fig. 20). The origin of such cuticular granules has not yet been clearly shown, though some assumptions can be made as follows: The flattened sacs of the Golgi apparatus of this cell sometimes contain a relatively dark substance, but it has not yet been proven whether it is the precursor of the specific granules of the hair cuticle. On the other hand, the cells of this layer contain a great number of free ribosomes which are scattered throughout the cytoplasm, but they are not gathered around

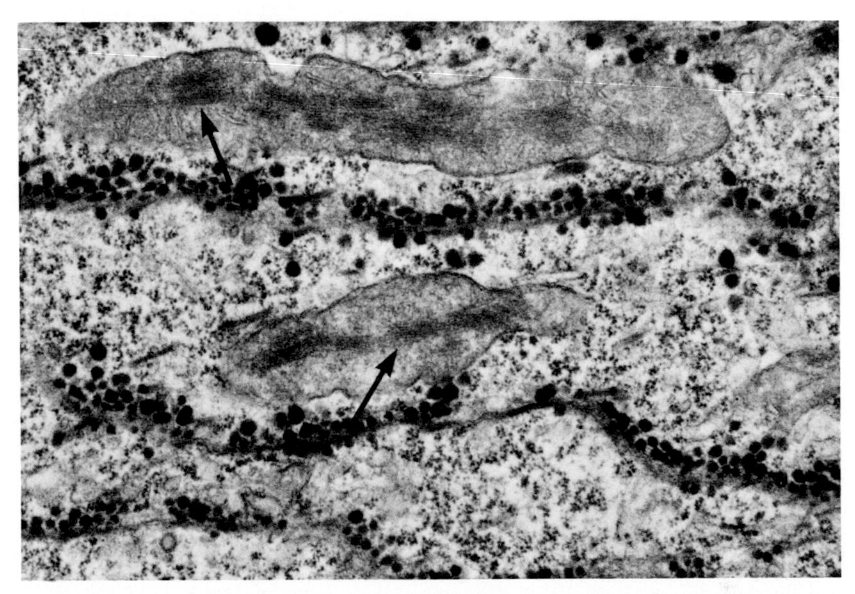

Fig. 19. A part of the rat hair cuticle, whose cells contain many dark granules arranged along the cell surface. The top of this figure orients to the center of the hair. Free ribosomes are abundant in the cytoplasm. Mitochondria contain filamentous inclusions (arrows), but no filaments exist in the cytoplasmic matrix. ×20,000.

the dark granules, which are eventually arranged along the cell surface.

The irregular shape of granules at a rather advanced stage may be brought about by the fusion of small round granules of the initial form. As the cells grow further, the granules may turn to long tortuous strands. No filaments are observed at any stage of development in this layer of the hair, though some mitochondria contain filamentous inclusions (Fig. 19).

Orfanos and Ruska[32] studied the extended portion of the human hairs above the epidermis and showed the fine structure of full grown hair cuticle. It was seen to be almost homogeneous but fine-granular when higher magnification was achieved. Filamentous structures like the keratin of either hair cortex or epidermis were not observed. Just inside the surface plasma membrane, dark zones are recognized. The dark zone along the lateral cell surface is thicker than the corresponding zone along the medial surface. The former was described as about 60–80 nm in thickness, while the latter is said to be less than 30 nm. The middle, less dense layer was reported to be about 250–350 nm thick. These subsurface dark zones are presumably related to the preferred arrangement of granules along the surface plasma membrane. The fact that the granules are more numerous

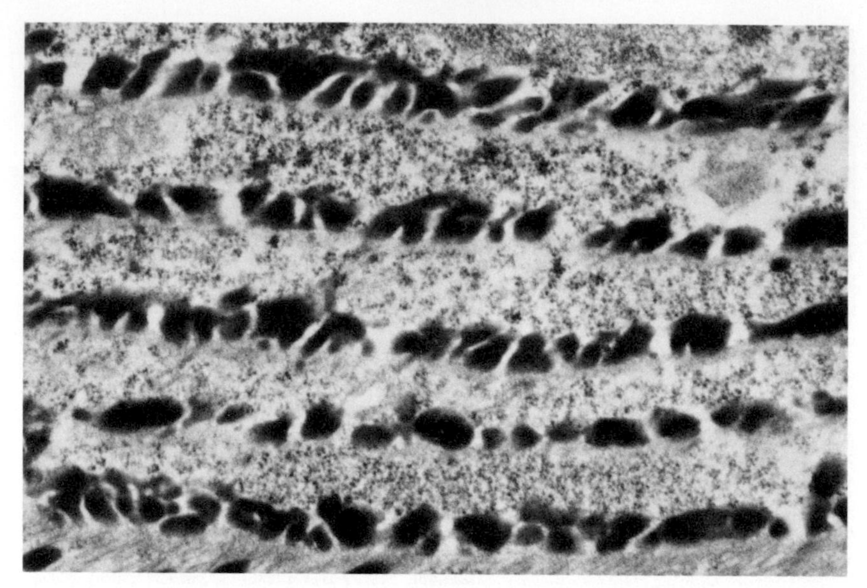

Fig. 20. Higher magnification view of the granules of the rat hair cuticle. The granules are irregular in shape, which suggests the fusion of granules. ×23,000.

along the lateral surface than the medial surface at the premature state of hair cuticle coincides well with the fact that the lateral dark zone is thicker than the medial dark zone in the full grown hair.

The cells of the hair cortex are filled with many filament-bundles running longitudinally (Fig. 21). In the cross section filament-bundles are seen as round, oval, or somewhat elongate profiles (Fig. 22). As the cells grow, the filaments may increase in number and the bundles fuse to one another. The borders between filament-bundles are partially left after the keratinization has completed.[33] No dense granules comparable to trichohyalin nor keratohyalin granules are observed in the hair cortex, but the so-called keratin pattern can be recognized in the fully keratinized hair cortex after using an adequate method of preparation.[33,34]

The hair medulla is a slender column usually of a single cell thickness (Fig. 21). Medullary cells contain many round dense droplets of various sizes, which are called medullary droplets. In both longitudinal and transverse sections, medullary droplets are seen round (Figs. 21, 22). Therefore, the globular shape of these droplets is proved. Tonofilaments are almost absent in the medullary cells and, if any, they are not correlated with the medullary droplets. It is not known whether the differentiation of hair medulla may be called keratinization or not. The medullary cells are vacuolated as they have fully differentiated, but the fate of droplets has not been well understood. According to Roth,[35] the droplets as well as cell organ-

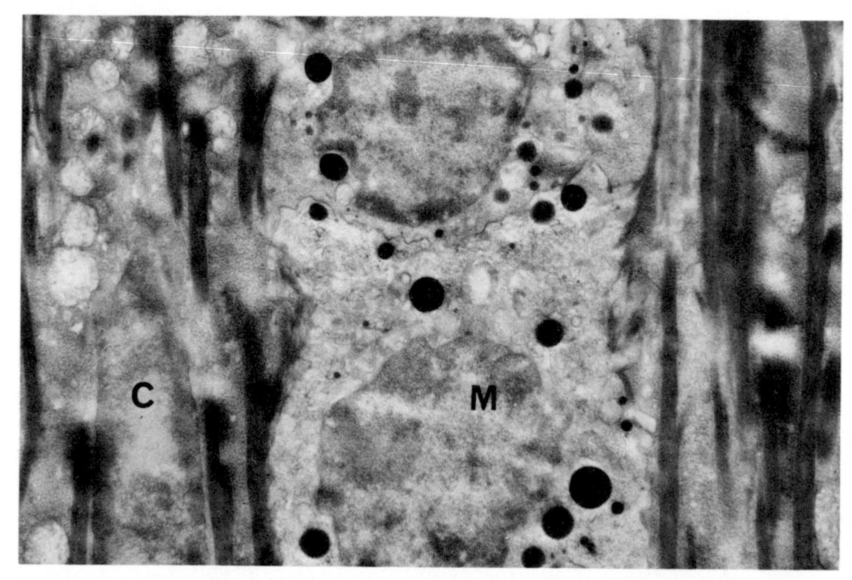

Fig. 21. A longitudinal section of the central part of the rat hair root. The cortex (C) contains longitudinal filament-bundles. The medulla (M) contains round dense droplets. ×8,000.

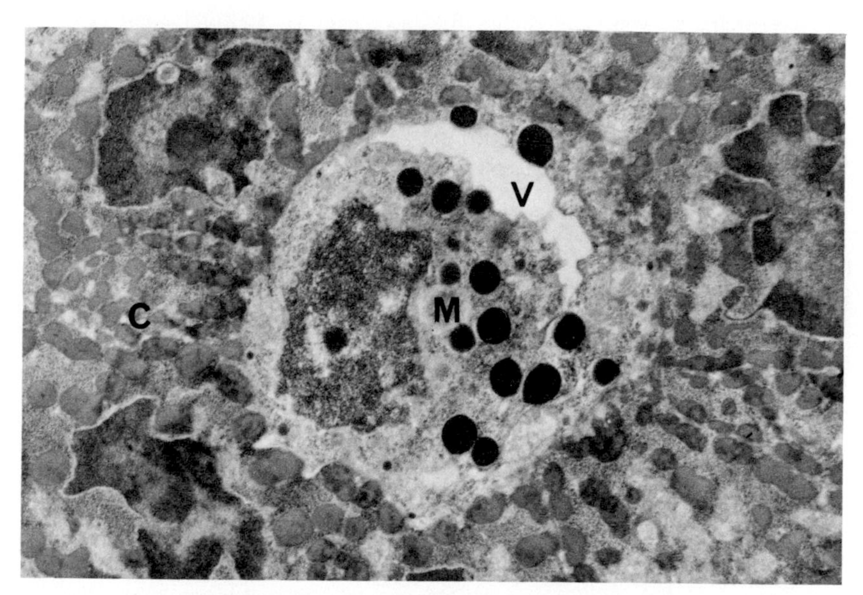

Fig. 22. A cross section of the central part of the rat hair root. The cortex (C) contains round or irregular profiles of cut filament-bundles. The medulla (M) contains round dense droplets and a clear vacuole (V). ×9,000.

elles finally disappear, and the vacuoles become filled with air. As the cytoplasmic matrix of dried medullary cells at the level of hair shaft above the epidermis contains filamentous material, it may be said that the medulla also do keratinize. However, many things are left to be studied concerning such a keratinization (differentiation) process of the hair medulla.

REFERENCES

1. Kurosumi, K. Paraneuronic and related cells in the mammalian skin. In Chromaffin, Enterochromaffin and Related Cells (R. E. Coupland and T. Fujita, eds.), ASP Biological and Medical Press, Amsterdam, 1976, In press.
2. Odland, G. F. The fine structure of the interrelationship of cells in the human epidermis. *J. Biophys. Biochem. Cytol.*, **4**: 529–538, 1958.
3. Kelly, D. E. and Luft, L. H. Fine structure, development, and classification of desmosome and related attachment mechanism. In Electron Microscopy (R. Uyeda, ed.), Maruzen Co., Tokyo, 1966, vol. 2, pp. 401–402.
4. Kelly, D. E. Fine structure of desmosomes, hemidesmosomes, and an adepidermal globular layer in developing newt epidermis. *J. Cell Biol.*, **28**: 51–72, 1966.
5. Selby, C. C. An electron microscope study of thin sections of human skin. II. Superficial cell layers of footpad epidermis. *J. Invest. Derm.*, **29**: 131–149, 1957.
6. Wettstein, D. von, Lagerholm, B., and Zech, H. Cellular changes in the psoriatic epidermis. *Acta Dermato-Venereol.*, **41**: 115–134, 1961.
7. Odland, G. F. A submicroscopic granular component in human epidermis. *J. Invest. Derm.*, **34**: 11–15, 1960.
8. Odland, G. F. and Reed, T. H. Epidermis. In Ultrastructure of Normal and Abnormal Skin (A. S. Zelickson, ed.), Lea & Febiger, Philadelphia, 1967, pp. 54–75.
9. Matoltsy, A. G. and Parakkal, P. F. Membrane-coating granules of keratinizing epithelia. *J. Cell Biol.*, **24**: 297–307, 1965.
10. Wilgram, G. F., Caulfield, J. B., and Madgic, E. B. A possible role of the desmosome in the process of keratinization. In Epidermis (W. Montagna and W. C. Lobitz, eds.), Academic Press, New York, 1964, pp. 275–301.
11. Wilgram, G. F. Das Kerationosom: ein Faktor in Verhornungsprozess der Haut. *Der Hautarzt*, **16**: 377–379, 1965.
12. Hashimoto, K. Cementsome, a new interpretation of the membrane-coating granule. *Arch. Derm. Forsch.*, **240**: 349–364, 1971.
13. Frithiof, L. and Wersäll, J. A highly ordered structure in keratinizing human oral epithelium. *J. Ultrast. Res.*, **12**: 371–379, 1965.
14. Oláh, I. und Röhlich, P. Phospholipidgranula im verhornender Oesophagusepithel. *Z. Zellforsch.*, **73**: 205–219, 1966.
15. Martinez, I. R., Jr. and Peters, A. Membrane-coating granules and membrane modifications in keratinizing epithelia. *Am. J. Anat.*, **130**: 93–120, 1970.

16. Suzuki, H. and Kurosumi, K. Lamellar granules and keratohyalin granules in the epidermal keratinocytes, with special reference to their origin, fate and function. *J. Electron Micr.*, **21**: 285–292, 1972.
17. Frei, J. V. and Sheldon, H. A small granular component of the cytoplasm of keratinizing epithelia. *J. Biophys. Biochem. Cytol.*, **11**: 719–724, 1961.
18. Weinstock, M. and Wilgram, G. H. Fine-structural observations on the formation and enzymatic activity of keratinosomes in mouse tongue filiform papillae. *J. Ultrast. Res.*, **30**: 262–274, 1970.
19. Bonneville, M. A., Weinstock, M., and Wilgram, G. F. An electron microscope study of cell adhesion in psoriatic epidermis. *J. Ultrast. Res.*, **23**: 15–43, 1968.
20. Farbman, A. I. Electron microscope study of a small cytoplasmic structure in rat oral epithelium. *J. Cell Biol.*, **21**: 491–495, 1964.
21. Hashimoto, K., Gross, B. G., and Lever, W. F. The ultrastructure of the skin of human embryos. I. The intraepidermal eccrine sweat duct. *J. Invest. Derm.*, **45**: 139–151, 1965.
22. Wolff, K. und Holubar, K. Odland-körper (Membrane-coating Granules, Keratinosomen) als epidermale Lysosomen. *Arch. klin. exp. Derm.*, **231**: 1–19, 1967.
23. Suzuki, H. and Ishikawa, K. Fine-structural changes in epidermal keratinocytes after chloroform-methanol treatment (Japanese text). *Jap. J. Derm. Ser. A.*, **83**: 616, 1973.
24. Kurosumi, K. Studies on the keratinization process in the epidermis and hair follicles. *J. Electron Micr.*, **16**: 213, 1967.
25. Suzuki, H., Kurosumi, K., and Miyata, C. Electron microscopy of spherical keratohyalin granules. *J. Invest. Derm.*, **60**: 219–223, 1973.
26. Kurosumi, K. and Kurosumi, U. Electron microscopy of the mouse plantar eccrine sweat glands. *Arch. Histol. Jap.*, **31**: 455–475, 1970.
27. Suzuki, H. An electron microscopic study of keratinization in the mouse intraepidermal sweat duct. *Arch. Histol. Jap.*, **35**: 265–282, 1973.
28. Hashimoto, K., Gross, B. G., and Lever, W. F. Electron microscopic study of the human adult eccrine gland. I. The duct. *J. Invest. Derm.*, **46**: 172–185, 1966.
29. Zelickson, A. S. Electron microscopic study of epidermal sweat duct. *Arch. Derm.*, **83**: 106–111, 1961.
30. Matsuzawa, T. and Kurosumi, K. The ultrastructure, morphogenesis and histochemistry of the sweat glands in the rat foot pads as revealed by electron microscopy. *J. Electron Micr.*, **12**: 175–191, 1963.
31. Birbeck, M. S. C. and Mercer, E. H. The electron microscopy of the human hair follicle. Part 3. The inner root sheath and trichohyaline. *J. Biophys. Biochem. Cytol.*, **3**: 223–230, 1957.
32. Orfanos, C. und Ruska, H. Die Feinstruktur des menschlichen Haares. I. Die Haar-Cuticula. *Arch. klin. exp. Derm.*, **231**: 97–110, 1968.
33. Orfanos, C. und Ruska, H. Die Feinstruktur des menschlichen Haares. II. Der Haar-Cortex. *Arch. klin. exp. Derm.*, **231**: 264–278, 1968.
34. Rogers, C. E. Electron microscope studies of hair and wool. *Ann. N. Y. Acad. Sci.*, **83**: 378–399, 1959.

35. Roth, S. I. Hair and nail. In Ultrastructure of Normal and Abnormal Skin (A.S. Zelickson, ed.), Lea & Febiger, Philadelphia, 1967, pp. 105–131.

Discussion

Dr. Goldsmith: Are there examples of structural proteins in other tissues which are initially formed as these round bodies?

Dr. Kurosumi: There are round keratohyalin granules in the oral and esophageal epithelia, but such structures are not found in other tissues.

Dr. Steinert: You have suggested that the inner route sheath filaments are similar to the epidermal tonofilaments whereas a great deal of biochemical evidence suggests that they are quite distinct. The inner route sheath filaments contain the amino acid, citrulline, which makes them distinct from the epidermal tonofilaments. Would you like to comment on this?

Dr. Kurosumi: Oh, you meant that they are different chemically?

Dr. Steinert: Yes. Furthermore, the inner route sheath filaments appear as hollow tubes in sections.

Dr. Baden: The problem in trying to define function for this material is the tremendous variation that occurs. Furthermore, in some diseases you don't even see the granules, yet the stratum corneum looks normal. Would you like to comment on this?

Dr. Kurosumi: I said that in hair we never see any such keratohyalin substance but the hair cortex keratinizes completely. Keratinization is not uniform in all keratinizing epithelia. Some keratinize without keratohyalin, for example, the hair cortex. So keratohyalin substance may be a very important component but it is not completely necessary for keratinization.

Progression of Events of Epidermal Differentiation in Wound Healing

George F. ODLAND

University of Washington, Seattle, Washington, U.S.A.

SUMMARY

By means of light and transmission electron microscopy of healing incised superficial skin wounds, it can be shown that there is an orderly progression of events related to epidermal cell differentiation in the repair of wounds. Stages of cellular synthesis attended by formation of basal lamina, desmosomes, and cytoplasmic filament synthesis are followed by evolution of lamellar granules, thickening of plasma membrane, and then development of keratohyalin granules, and the conventional filament/matrix complex of a mature cornified cell.

Over the past two decades, light and electron microscopic analyses of the epidermis have helped to develop the view that there are several stages identified by morphologic markers attesting to differentiation (see reviews by Odland,[1] Breathnach,[2] and Matoltsy and Parakkal[3]). We now recognize events of early differentiation as manifested by the appearance in basal epidermal cells of desmosomes and hemidesmosomes, cytoplasmic filaments (so-called tonofilaments), and synthesis of basal lamina (Hay and Revel,[4] Briggaman, Dalldorf and Wheeler[5]). In the spinous layer, there appear submicroscopic lamellar granules (0.2 to 0.3 μm in diameter), which appear to migrate to the extracellular compartment in the vicinity of the stratum granulosum. Higher in the epidermis, there evolve small keratohyalin granules which generally increase in size as the cells move outward. A final sequence is marked by disappearance of nucleus and organelles of synthesis, by thickening of the plasma membrane, and by a flattening of the cell with apparent decrease in volume attended by the consolidation of its filamentous component which is embedded in a pervasive matrix material.

Disruption of the time scale of epidermal cell maturation by wounding should test the validity of the sequence of steps in differentiation inferred from analysis of static structure.

It will be the objective of this presentation to review previously published observations (Odland and Ross[6]) on the fine structure of incised superficial wounds in the skin of man, in order to sort out on the time scale of epithelial repair the sequential appearance of structural markers of epidermal differentiation.

The methodology for the wounding and tissue processing has been described in detail elsewhere (Odland and Ross[6]). Wounds were sampled at several time intervals (1, 2, 3, 5, and 7 days) following superficial incision to a depth of 1/2 to 1 mm in the flexor aspect of the forearm.

Differentiation in migrating epidermis

At about 24 hours after incision, sheets of epidermis commence migrating beneath the wound crust over a fibrin scaffold in the wound base. Wound closure is effected by 2–3 days. The appearance of the migrating cells differs from normal by 1) increased volume, 2) by surface pseudopodial processes projecting into the wound milieu, 3) by assuming a flattened or squamous profile, and 4) by demonstrating phagocytic behavior towards elements of the wound base. Figure 1 illustrates a cross section of the advancing epithelial sheet 48 hours after wounding. The cells of the sheet remain attached to one another by a few desmosomes. In this wound series, wound closure in two days heralds the end of migration.

The following cytologic features are interpreted as markers of early differentiation in basal epidermal cells. Distribution of filaments and ribosomes is scant in the cytoplasm of the foremost cells, but a few cells back from the advancing margin, the basal epidermal cell has evolved ribosome rosettes and whorls, cytoplasmic filaments, and, resting on the fibrin matrix, has established half desmosomes against the temporary wound scaffold (Fig. 2).

Differentiation in wounds after epithelial closure of wound defect

As the wound age progresses to between three and five days, epithelial closure is complete, and the new epidermis over the wound bed becomes thickened and hyperplastic. The suprabasal cell layers, increased in number compared to normal, exhibit continued bleb formation at the basal surface and among adjacent lower cell layers, until the fibrin matrix is replaced by collagenous connective tissue (Figs. 4, 5). Nonetheless, there are numerous ribosomes and continued cytologic evidence of filament formation.

In the middle layers of the hyperplastic epithelium near the central, and presumably least old, part of the regenerated epidermis, cytoplasmic

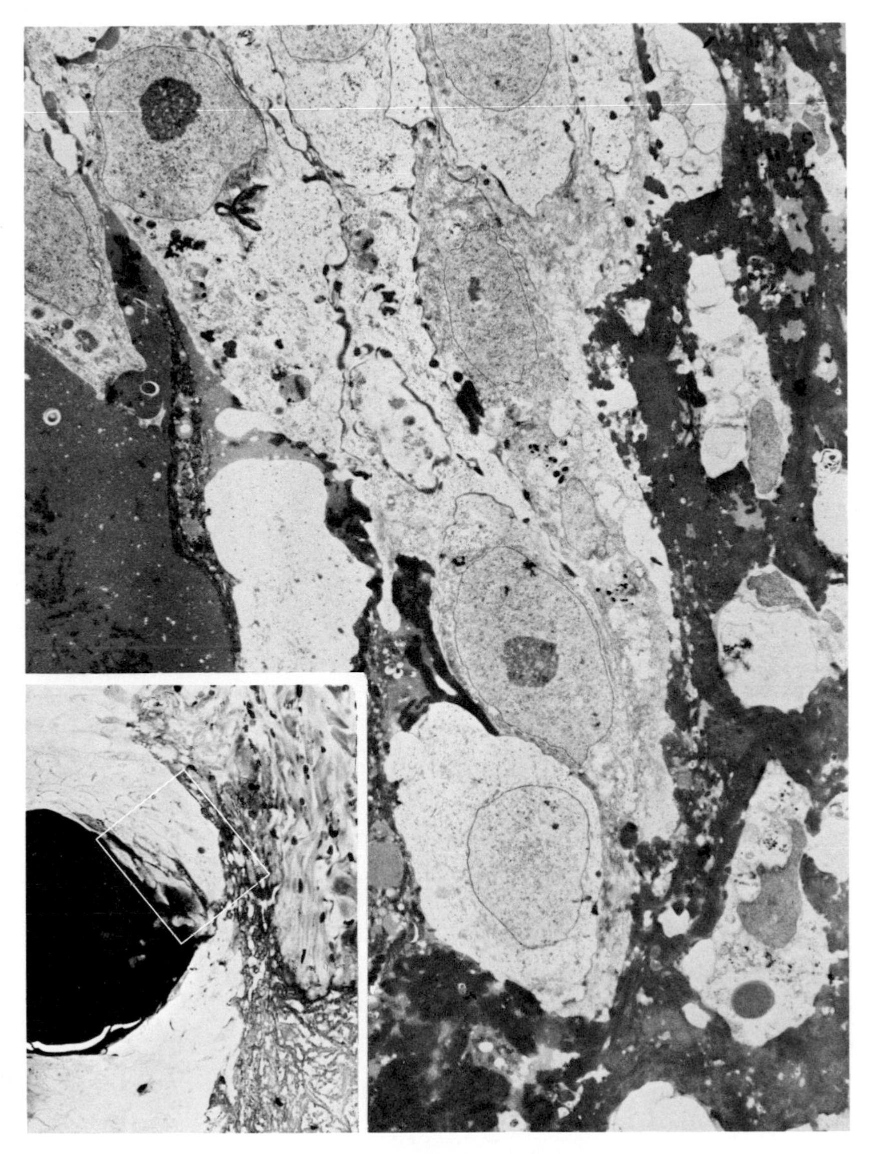

Fig. 1. Electronmicrograph of advancing epithelial sheet in two-day-old wound revealing phagocytosis by epidermis, desmosomes, pseudopodial projections of epidermal cells into the serofibrinous wound matrix. Inset is a photomicrograph of the same specimen. ×3000.

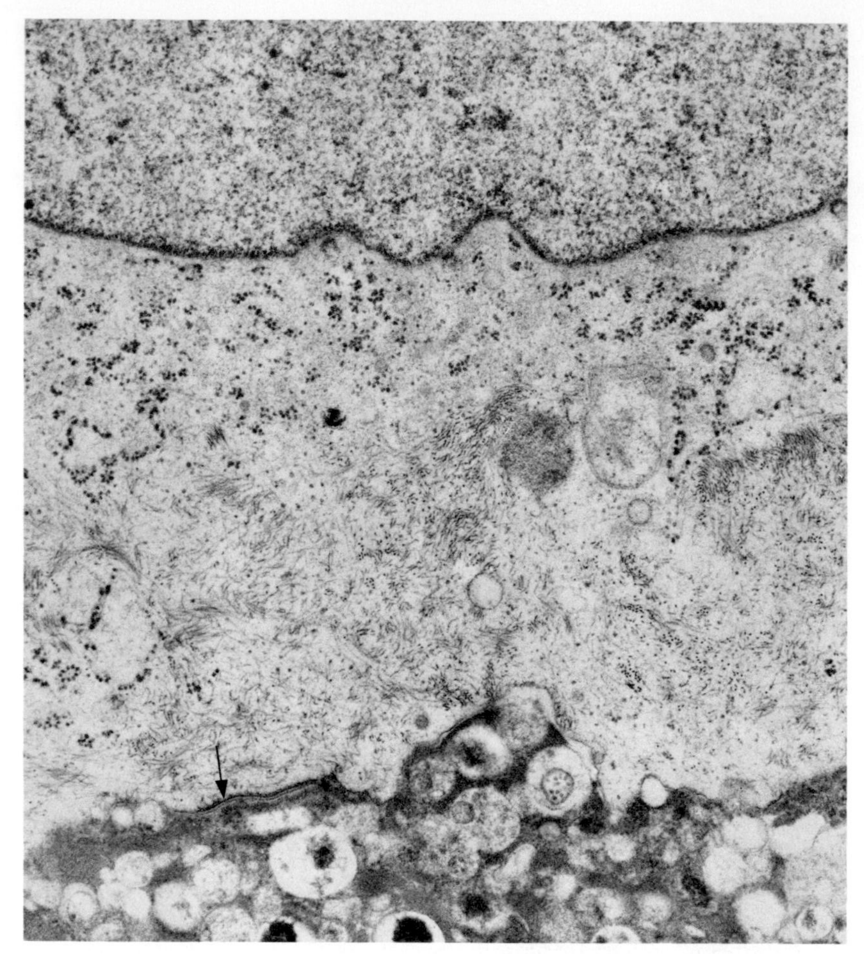

Fig. 2. Electronmicrograph of basal surface of an epidermal cell a few cells back from the advancing margin of the epithelial sheet showing ribosomes, cytoplasmic filaments, and half-desmosomes (→), as well as basal lamina structure against the supporting fibrin net. Note areas of clot lysis around lysosomal neutrophil granules. ×13,000.

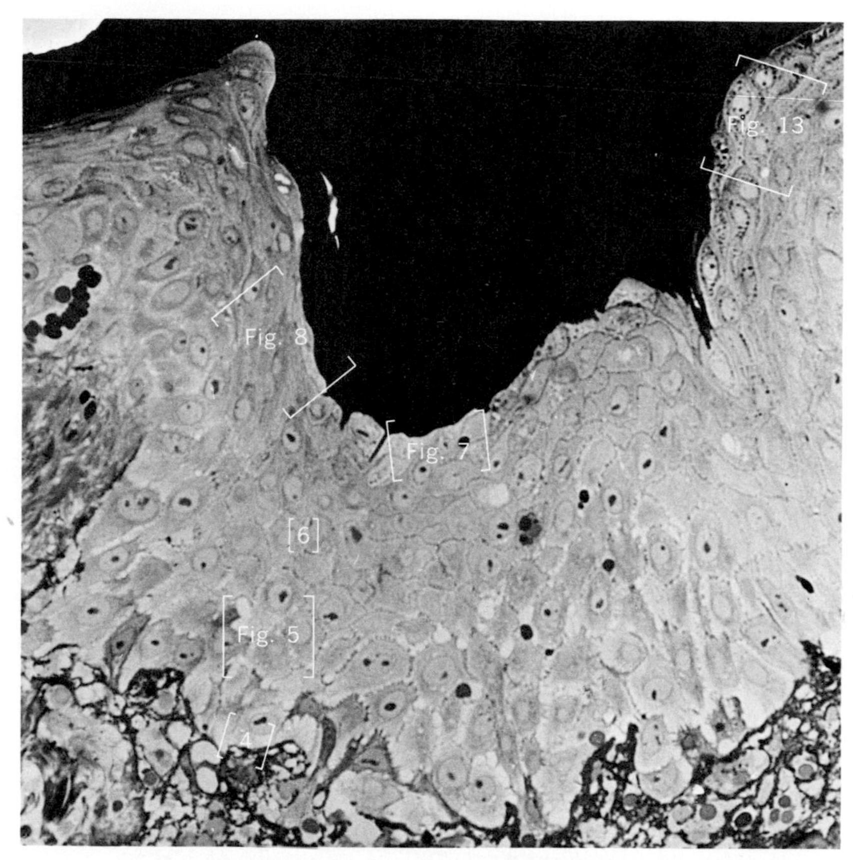

Fig. 3. Photomicrograph of three-day-old wound showing hyperplasia of regenerated epithelium over underlying fibrin net. Dark area in center is wound crust or scab. Bracketed areas indicate figure numbers and locations comparable to those from which ensuing electron-micrographs are taken. $\times 50$.

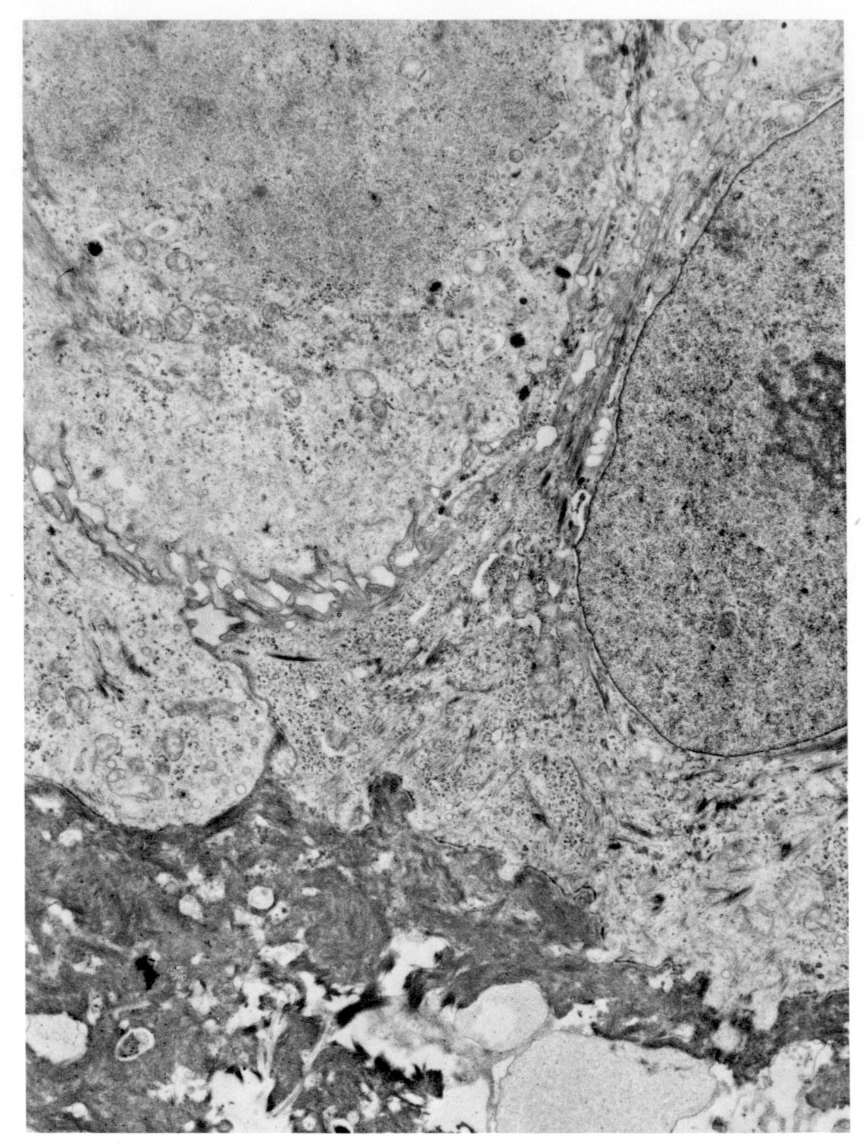

Fig. 4. Shows basal epidermal cell with organelles of synthesis and nucleus in the right-hand portion of the picture, and hemidesmosomes lying against a fibrin network. In the upper left quadrant of the picture is the nucleoplasm and organelles of an epidermal cell in mitosis. Electronmicrograph from a five-day-old wound. ×5,500.

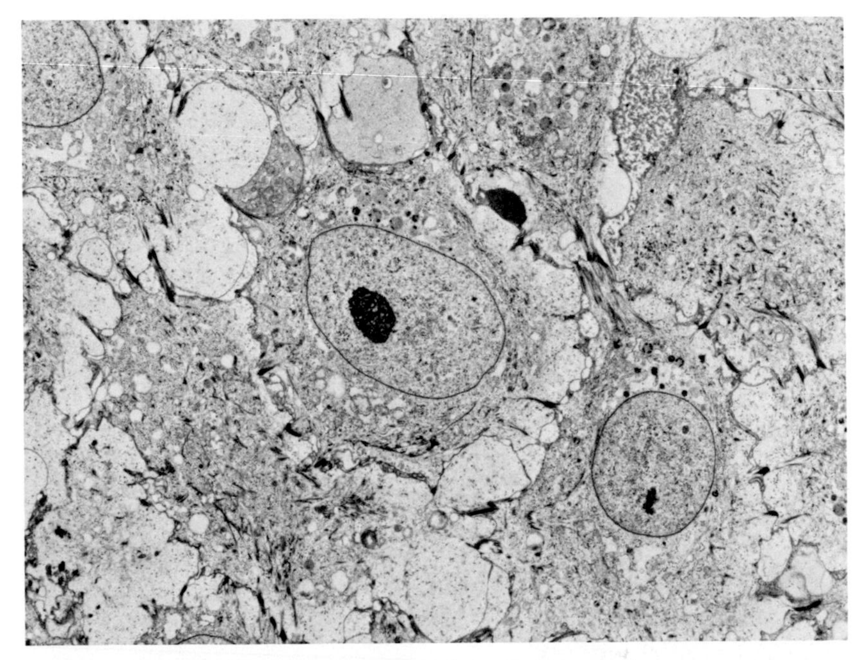

Fig. 5. Electronmicrograph of epidermal cells just above the basal layer in a three-day-old wound. They show extensive surface bleb formation along with phagocytic inclusions, but also organelles of synthesis. ×2,000.

characteristics are dominated by ribosomes, desmosomes, and filaments (Fig. 6). These same features are encountered in the flattened outer epidermal cells which transform to nucleated squames beneath the wound scab (Fig. 7). Hence, in hyperplastic epidermis of 3–5-day wounds, the central area beneath the scab frequently shows squamous transformation of uppermost cell layers without the occurrence of keratohyalin immediately subjacent to the wound crust (Fig. 8). Between such cells, intercellular lamellar granules may be observed (Fig. 9). Membrane thickening occurs in cells without detectable keratohyalin granules (Fig. 12).

On the other hand, more advanced differentiative changes appear to occur towards the wound margins (presumably an older area of regenerated epidermis) (cf. Fig. 3). There the transforming cells show scanty distribution of keratohyalin granules (Figs. 10 & 11).

Concomitantly, but closer yet to the margin of 3- to 5-day-old wounds, the formation of keratohyalin granules occurs in many layers of cells and in

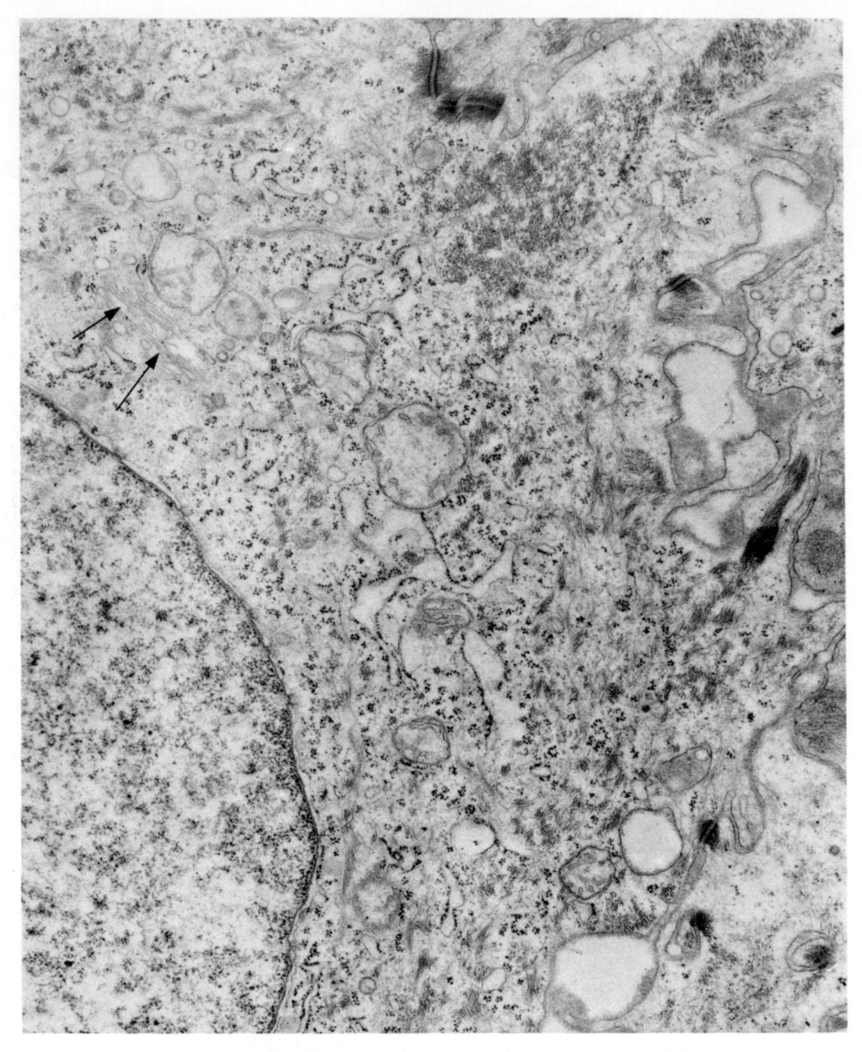

Fig. 6. An electronmicrograph of an epidermal cell in the midst of the hyperplastic epithelium of a three-day-old wound showing numerous organelles of synthesis, notably extensive rough endoplasmic reticulum, filaments, desmosomes at the margin of the cell in the right-hand aspect of the illustration, and a Golgi complex indicated by the arrows. This illustration exemplifies the morphologic markers of early differentiation, presumably associated with filament synthesis. $\times 13{,}000$.

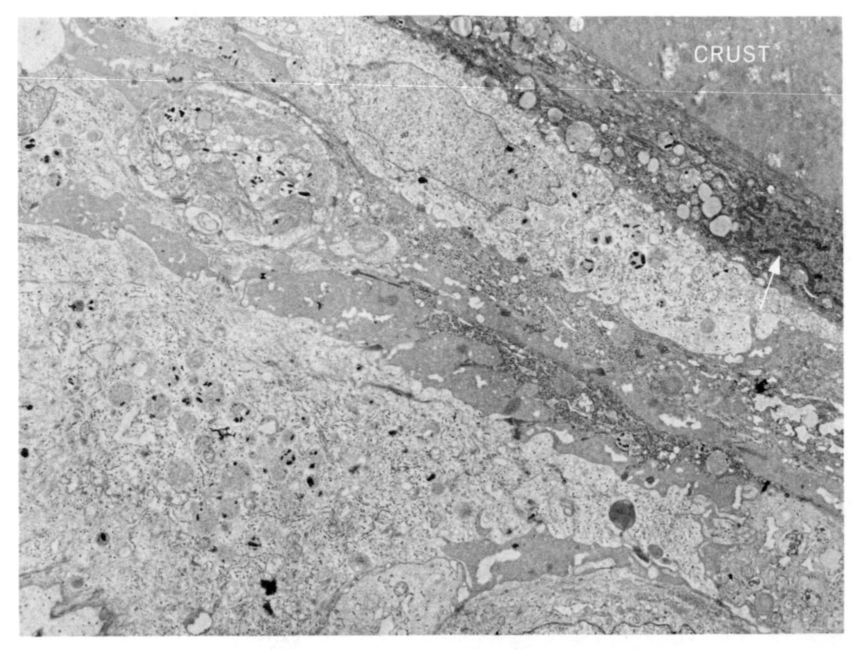

Fig. 7. Electronmicrograph from higher in the epidermis of a three-day-old wound showing epithelial squames flattened against the wound crust with some phagocytic inclusions and some cytoplasmic filaments. Note nucleated squames (arrows) at the base of the crust. ×4500.

great profusion (Fig. 13). The overlying cornified cells contain no nuclei, and reveal a consolidated filament/matrix pattern closely resembling the normal morphology (Fig. 14, cf. Figs. 11 & 12).

It is to be concluded that the early cytologic events of epidermal repair following experimentally-induced, superficial, linear, incised wounds do indeed spread out on a time scale a sequence of changes interpretable as stages in differentiation. To recapitulate, the earliest changes comprise appearance of numerous ribosomes, filaments, and formation of desmosomes, and hemidesmosomes in newly migrated epithelium covering a wound defect. This has been demonstrated before in amphibia (Singer and Salpeter[7]), and subsequently in mammals (Odland and Ross,[6] and Croft and Tarin[8]).

With wound closure, the locomotory behavior of the cells ceases (Abercrombie[9]), but with persistence of mitotic activity shown by Viziam and

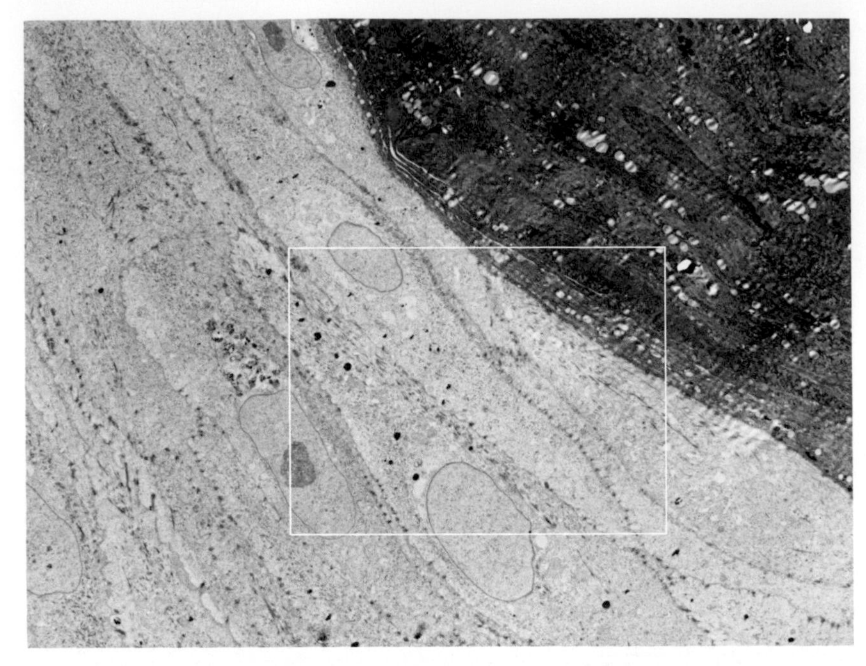

Fig. 8. Electronmicrograph in which the cells immediately beneath the wound crust in the upper right corner of the illustration show filaments and some cytoplasmic organelles, but no keratohyalin granules, whereas markers at this later stage of synthesis appear in some cells further toward the wound margin (upper left of this illustration) and in the deeper epidermal layers. The rectangle subtends an area depicted in Fig. 9. ×4,000.

Matoltsy,[10] surface ruffling prevails during the hyperplastic thickening of the newly repaired epithelium (Vaughan and Trinkaus[11]). Nonetheless, ribosomes and increments of filaments prevail in the newly regenerate cells, and only later do lamellar granules appear without immediately coinciding with the thickening of the cornified cell membrane. The organized internal structure of the normal stratum corneum does not develop until underlying cells have synthesized keratohyalin, whereas flattening and thickening of the cell membrane do occur without the preexisting evolution of keratohyalin.

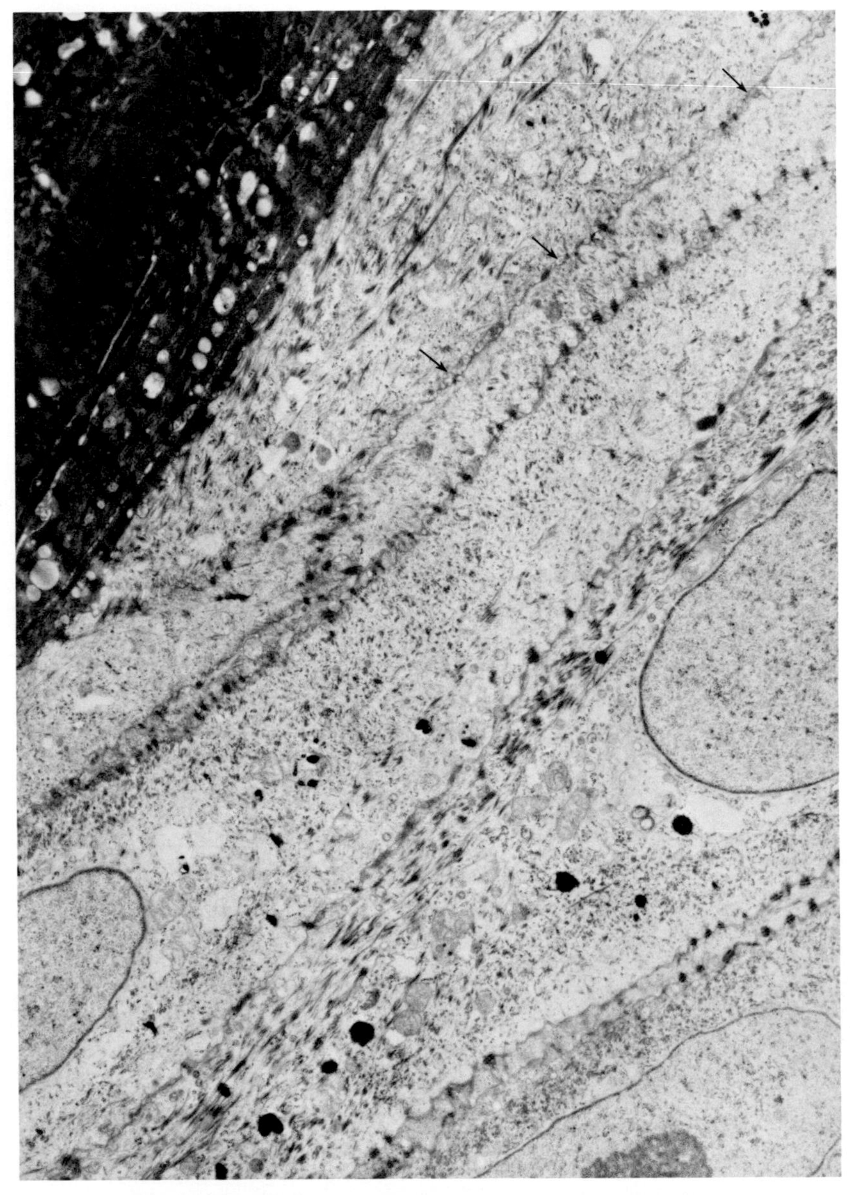

Fig. 9. Electronmicrograph of higher magnification from area seen in figure 8. This demonstrates cytoplasmic details of a cell just beneath dense squames of wound crust in upper right corner. At arrows, observe the intercellular disposition of lamellar granules without apparent thickening of the plasma membrane. ×7,000.

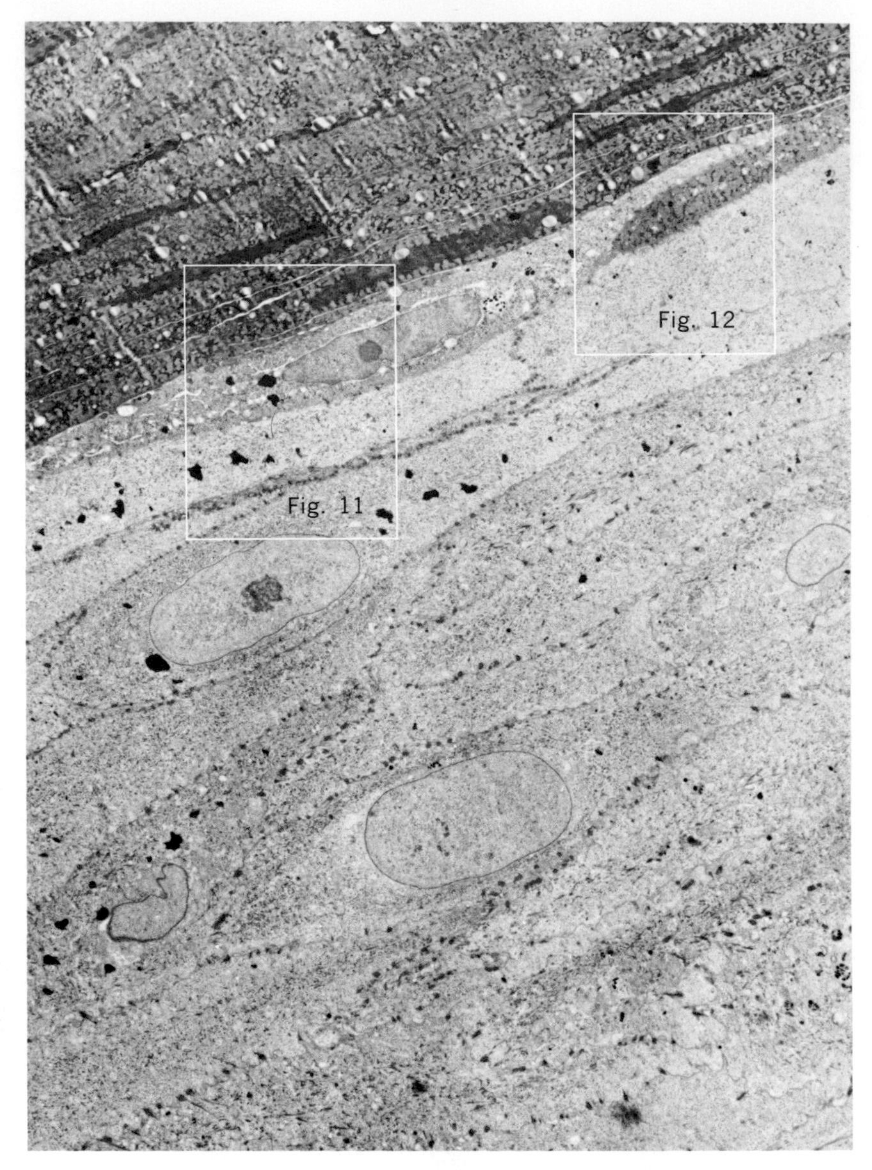

Fig. 10. A low-magnification electronmicrograph from the area depicted in figure 3 of a three-day-old wound just beneath the wound crust. The uppermost squames show a further degree of differentiation as manifested by keratohyalin granules on the left-hand side, but not among those cells on the right of the illustration. The two rectangles depict areas subtended by Figs. 11 and 12. ×3,250.

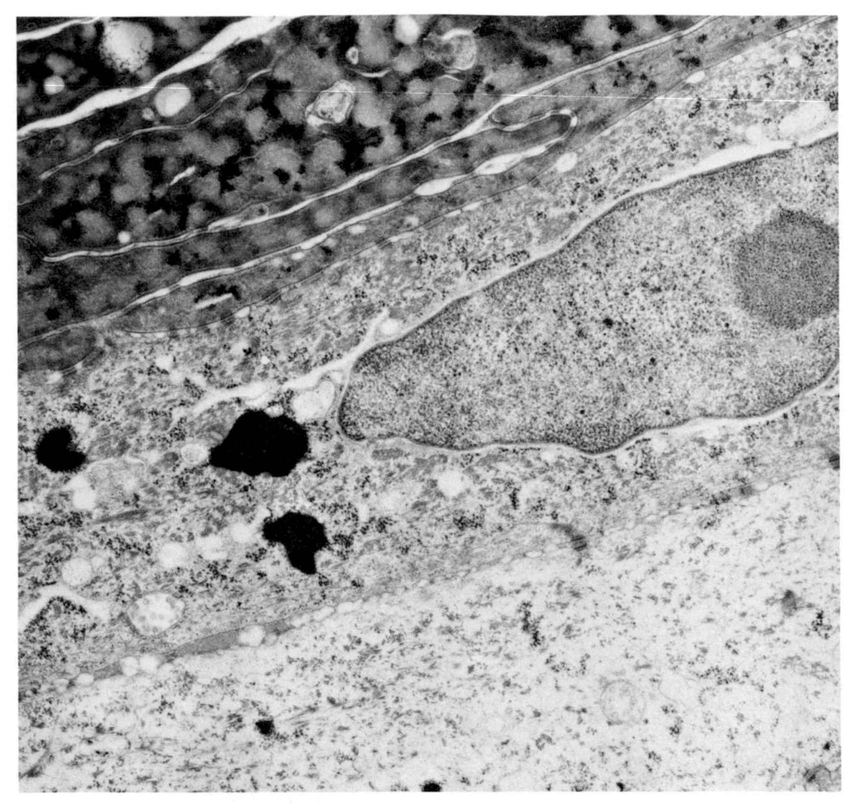

Fig. 11. An electronmicrograph showing a nucleated squamous cell in which keratohyalin granules are prominent, and immediately above it a cell with a thickened membrane whose cytoplasmic characteristics conform to some degree with those of a mature horny cell. ×7,000.

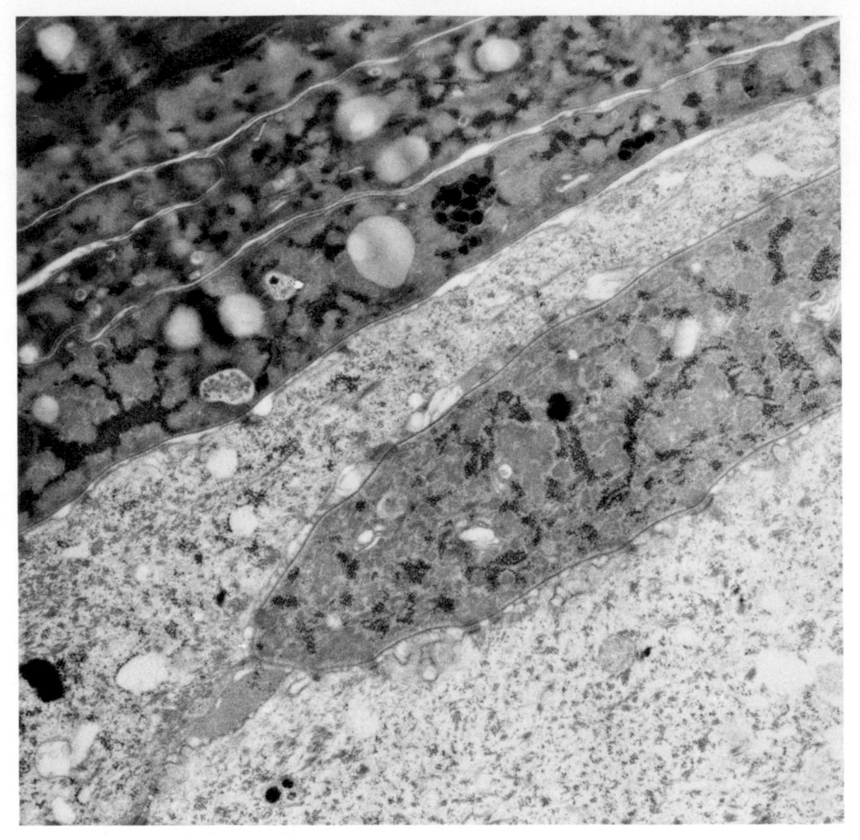

Fig. 12. A portion of a cell in which the subjacent cell shows no keratohyalin. About the margins of the glycogen-laden consolidated but immature squamous cell, a thickened membrane occurs. For location compare with Fig. 10. ×7,000.

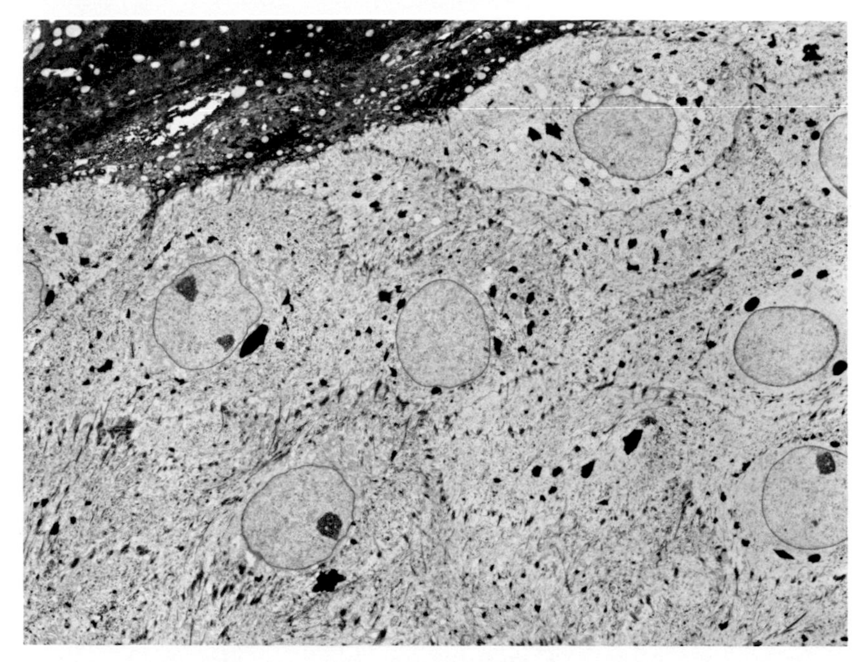

Fig. 13. Depicts the profusion of granular cells at the margin of a three-day-old wound, presumably a more advanced stage of differentiation in the hyperplastic wound epithelium. ×2,000.

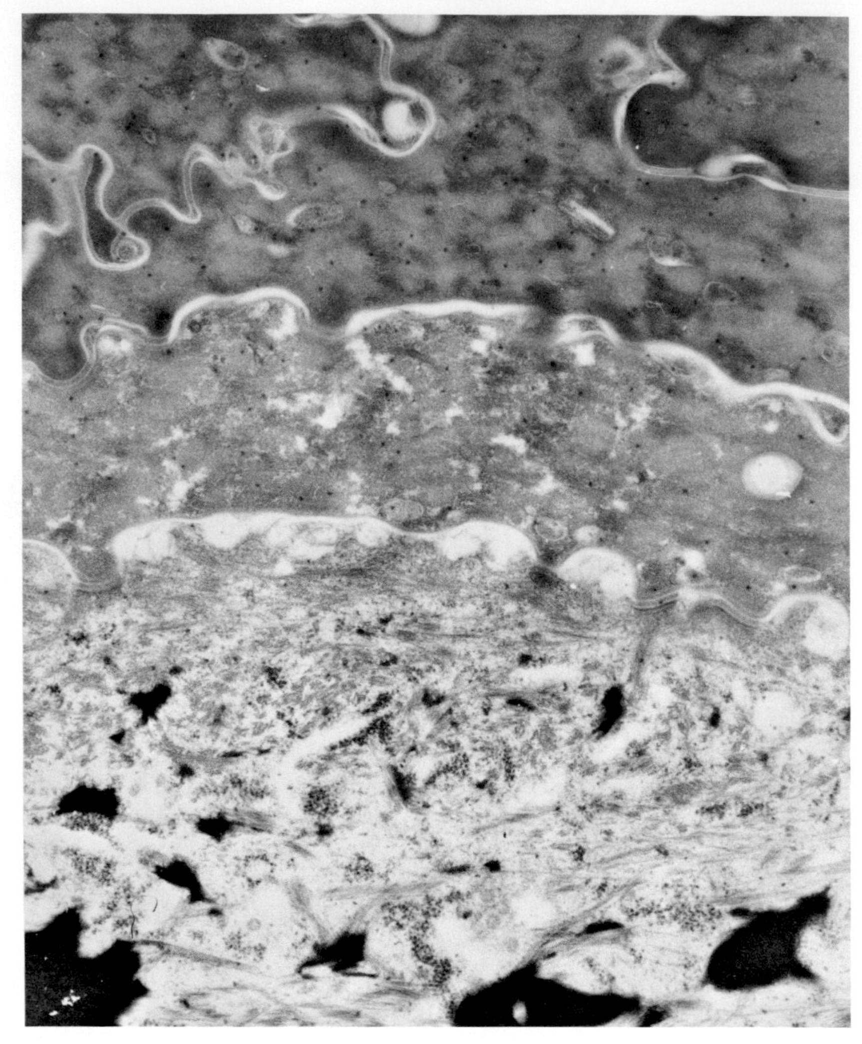

Fig. 14. Higher magnification electronmicrograph. In the lower half of the illustration is a portion of a stratum granulosum cell with keratohyalin granules, filaments, and some ribosomes, and immediately above it a consolidated cell with thickened membrane and internal structure resembling that seen in mature cells of the stratum corneum, presumably a normal end-stage differentiation. ×10,000.

REFERENCES

1. Odland, G. F. Histology and fine structure of the epidermis. In The Skin. International Academy of Pathology, Monograph No. 10. Williams & Wilkins, Baltimore, 1971, pp. 28–46.

2. Breathnach, A. S. An Atlas of the Ultrastructure of Human Skin. J & A Churchill, London, 1971.
3. Matoltsy, A. G. and Parakkal, P. F.: Keratinization. In Ultrastructure of Normal and Abnormal Skin (Zelickson A. S., ed.), Lea & Febiger, Philadelphia, 1967, pp. 76–104.
4. Hay, E. D. and Revel, J. P., Autoradiographic studies of the origin of the basement lamella in Ambystoma. *Develop. Biol.,* **7**: 152–168, 1963.
5. Briggaman, R. A., Dalldorf, F., and Wheeler, C. E., Jr. Formation and origin of basal lamina and anchoring fibrils in adult human skin. *J. Cell Biol.,* **51**: 384–395, 1971.
6. Odland, G. and Ross, R. Human wound repair. I. Epidermal regeneration. *J. Cell Biol.,* **39**: 135–151, 1968.
7. Singer, M. and Salpeter, M. M., Regeneration in vertebrates: The role of the wound epithelium. In Growth in Living Systems. Basic Books, Inc., New York, 1961, pp. 277–311.
8. Croft, C. and Tarin, D. Ultrastructural studies of wound healing in mouse skin. I. Epithelial behavior. *J. Anat.,* **106**: 63–77, 1970.
9. Abercrombie, M. The bases of locomotory behavior of fibroblasts. *Exptl. Cell. Res. Suppl.,* **8**: 188–198, 1961.
10. Viziam, C. B., Matoltsy, A. G., and Mescon, H. Epithelialization of small wounds. *J. Invest. Dermatol.,* **43**: 499–507, 1964.
11. Vaughan, R. B. and Trinkaus, J. P., Movements of epithelial sheets *in vitro.* *J. Cell. Sci.,* **1**: 407–413, 1966.

Discussion

Dr. Bernstein: Do you have any further information on the life history of the unique glycogen granules you see in the developing horny cells?

Dr. Odland: To be precise, no!

Dr. Bernstein: Is it true that you would not normally see glycogen in this condition or in the normal situation of keratinization?

Dr. Odland: Is there someone else who would like to answer that?

Dr. Matoltsy: Normally you don't see glycogen in the epidermis, but if you wound the epidermis, by incision or stripping, then mitotic activity increases and glycogen appears. About 20 years ago, I think Dr. Lobitz had a paper in The Journal of Investigative Dermatology on this point. It was a light microscopic study.

Dr. Bernstein: In that condition, is it found in the non-nucleated cell?

Dr. Matoltsy: Glycogen was seen in the living cells.

Dr. Odland: This material is present in the lowermost squames beneath the wound crust, cells which I have presumed to be those which are transformed with the passage of time but about which I can make no further comment.

DR. FUKUYAMA: I have two questions. Have you studied the effects of colchicine or cytochalasin B on wound healing? The second question is: Do you see any morphological differences between filaments in normal and in these migrating keratinocytes?

DR. ODLAND: To the first question: We have not used mitotic inhibitors in the wound study. As to the second question, we have not observed a difference in the filaments other than the way they are arranged in the cell. Did I interpret your questions correctly?

DR. FUKUYAMA: I was thinking more about the effect of colchicine on the locomotion of the cells. Do they migrate faster or slower under those conditions? How is the roughening of the cells affected? When you see the roughening, do you consider those fine filaments the same as tonofilaments, or are they some other type by morphological or chemical criteria?

DR. ODLAND: Well, I made my apologies early, Dr. Fukuyama. I can't make that type of interpretation and those of you who are present but who are not essentially interested in minutiae of techniques will recognize that a distinction between actin-type filaments, various microfilaments and microtubules in epithelial cells requires, if I'm not mistaken, fixation in formaldehyde for preservation. In view of the fact that we have not fixed our wounds and in view of societal pressures against doing human experimentation of this sort now as opposed to eight years ago, we have not repeated these studies. Dr. McGuire, would you care to make a comment on this question?

DR. MCGUIRE: This can be a comment to Dr. Odland's talk and to Dr. Fukuyama's question. If cells in tissue culture are treated with colchicine, motility continues. However, directed motility ceases. The cells appear to be incapable of determining in which direction they wish to go after something to eat or someone to meet or whatever cells want to do. The comment upon your paper is that at the time you interpreted the appearance of filaments as a concomitant of differentiation I think that you were correct. Now, however, there are enough examples of transformation of G actin to F actin, transformations of non-polymerized material to polymerized material in the absence of synthesis, that differentiation surely does take place in this wound reversion but the appearance of filaments can no longer be taken as a criterion of differentiation in the original sense.

DR. ODLAND: You mean that it cannot be taken as a marker?

DR. MCGUIRE: Yes, I think that initially the interpretation was that the filamentous material was probably a type-specific product of the keratinocyte, e.g. keratin, and I think that without further establishment of that fact, it is now a tenuous interpretation.

DR. FREEDBERG: I think we're getting into a problem that we should at least handle at the very beginning. Are cell division and differentiation in the epidermis mutually exclusive? It is clear that if one takes as a definition of a differentiated cell, an epidermal cell which has the morphologic products of epidermal differentiation and then that cell divides, the ground rules under which other cells, in other tissues, are said to be either dividing cells or differentiated cells, it seems to me, are not applicable to epidermis. The basic point being, that electron microscopically one can see cells which are in the process of mitosis which have filaments in them.

DR. ODLAND: And which persist through the process of mitosis.

DR. FREEDBERG: Yes. I have a question for which an answer is necessary. Where does mitosis really take place in these wounds? You did not want to discuss that as part of your paper, but it seems to me that it is important because if the mitosis is not in the group of cells that are migrating across the wound and if this migration is due to pressure from behind pushing these cells across the wound, then I think interpretations are different. Where do you think the mitoses are? Are they in the cells coming across or are they at the edge being pushed?

DR. ODLAND: I think they are in the migrating sheet but how far back from the advancing edge I do not know. Are you raising something more than the old time-honored question: Is wound coverage by migration or mitosis?

DR. FREEDBERG: Yes, because the interpretation which you put on a time-frame is that mitosis takes place and then cells differentiate. Here we can see, at least that's the way I interpreted your comments, in this process of wound healing a reduplication of maturation in the epidermis. Now, if the cells which divide are not the cells in the middle of the wound, then one has to go through a process of first dedifferentiation and then redifferentiation in the terms that you are using the word.

DR. ODLAND: Dedifferentiation is perhaps an old unacceptable term.

DR. FREEDBERG: Undifferentiation—I'm not sure what is unacceptable.

If you take a cell which is loaded with filaments, has keratohyalin granules and has lamellar bodies, and if that cell then starts to move across the wound, because it is pushed by its neighbors and if at that time, the state of the cell changes, e.g. by a G to F transformation as referred to by Dr. McGuire, then I can use the word dedifferentiate, you can use a different word, but that's a problem in semantics presuming that we are talking about the same thing.

DR. ODLAND: During a period of 18 to 24 hours, the morphologist is at a loss to say what is going on in these cells. Let us say the interpretation is that somehow something is going on at a level that is not demonstrable morphologically, but presumably is in preparation for the process of migration. If that is the kind of thing which you are referring to as dedifferentiation, i.e. a cessation of the ordinary programmed events of cell maturation, then I do not mind use of the term.

DR. FREEDBERG: It is not only cessation but it is moving, if you will, backward in the sense of what we call differentiation. However, the basic question still is, where are the mitoses?

DR. ODLAND: I took the material for mitoses from Dr. Matoltsy's work.

DR. MATOLTSY: I would like to clarify very briefly that it is the basal cells which migrate up from the cut edge. The granular cells for instance would not migrate. And now you have to use your imagination. A basal cell starts moving, 1 day, 2 days, 3 days and while it moves it divides. It moves over the wound and this is what Dr. Odland refers to as differentiation, and Dr. McGuire as dedifferentiation. These are cells which continually divide and then they have to replace material because only half remains after each division so they are synthesizing new material very actively. As they move forward, in front you find these basal cells very freshly dividing. A little bit farther back you see some of the cells settle and start to differentiate and, what Dr. Odland showed, all the way back, you see the complete process of differentiation. Is this clear? You have to use your imagination because it is related to time.

Fibrous Components in Abnormal Keratinization

Yasumasa Ishibashi

Department of Dermatology, University of Tokyo, Faculty of Medicine, Branch Hospital, Tokyo, Japan.

ABSTRACT

Fibrous components in the epidermis in various disorders of keratinization were ultrastructurally observed, and the following results were obtained.

Individual tonofilaments show morphologically no distinct abnormalities in these disorders. However, the aggregation process of filaments, the deposition of keratohyalin and the formation of keratin pattern with full dehydration seem to be more or less involved in these conditions. These findings seem to suggest the filament-interfilamentary substance relationship may play a significant role in keratinization, apart from the chemical changes of fibrous protein itself.

Besides tonofilaments the keratinocyte in the epidermis seems to synthesize other types of fibrous materials. These materials may also be "keratinized" in the horny cell as the keratin filament modified substance. However, they show no distinct aggregation process in the prickle cell.

The majority of the clinically abnormal features in the above mentioned conditions may eventually be caused by the faulty barrier formation in the horny cell, to say nothing of the thickening of horny layer.

INTRODUCTION

Keratinization is a process of cell differentiation in the epidermis from the basal cell to the horny cell. This process must be supported by consecutive intracellular as well as extracellular biochemical reactions, the details of which are not yet clarified satisfactorily. There are various abnormal keratinizations, most of which belong to the heritable disorders. Except for one[1] the pathogenesis of these disorders is not definitely known, though substantial[2] or metabolic defects[3,4] may presumably be responsible for the development of some of these conditions.

Fibrous components of the epidermis may play the most important role in the keratinization process. The majority of them is composed of tonofilaments which are mostly synthesized in the lower epidermis and alter finally into keratin filaments in the horny cells. Besides tonofilaments there are other types of fibrous components. "Actin" by McGuire[5] may be one of those filaments. "Amyloid-like filaments" and "wavy cytoplasmic filaments" in lichen amyloidosus by Hashimoto *et al.*[6] also seem to be other types of fibrous components in the epidermal cells, which are not seen in the normal epidermis.

In this paper ultrastructural changes of tonofilaments in various abnormal keratinizations and a group of filamentous structures in the epidermal cells in lichen amyloidosus will be shown in detail, and the significance of these findings in the keratinizing disorders will be discussed.

MATERIALS AND METHODS

The materials used for this investigation were obtained from patients with various keratotic disorders. They consisted of five cases of dominant ichthyosis vulgaris, one case of X-linked ichthyosis, one case of Sjögren-Larsson syndrome,[7] one case of Rud's syndrome,[8] three cases of epidermolytic hyperkeratosis,[9] one case of acrodermatitis enteropathica,[10] one case of erythrokeratodermia variabilis,[11] four cases of lichen amyloidosus and one case of lichen planus. As the control the materials from three normal adults were also used for this investigation.

The specimens were obtained from the skin lesions of the above mentioned patients by the banal or punch biopsy. For the controls they were taken from the extensor aspect of the upper arm, the abdomen and the buttock. They were immediately fixed in 2.5 % glutaraldehyde, and then 1 % OsO_4, both buffered with cacodylate at pH 7.4, or directly in 1% OsO_4 buffered with cacodylate at pH 7.4. Then they were dehydrated with ethanol series, and embedded in Epon 812. Ultrathin sections from these materials were obtained by LKB or Porter-Blum MT 2 microtome, and stained with uranyl acetate and lead citrate, or only with uranyl acetate. Observations and photographs were carried out by Hitachi HS-8, JEM-100 C and EM 9A Carl Zeiss electron microscopes.

RESULTS AND COMMENTS

Tonofilaments in the basal cell of dominant ichthyosis have well demarcated margins and a constant thickness of 50–75 Å in diameter (Fig. 1a), which is almost the same as in the normal.[12–15] The similar findings are also seen in those of variable erythrokeratoderma (Fig. 1b), and of

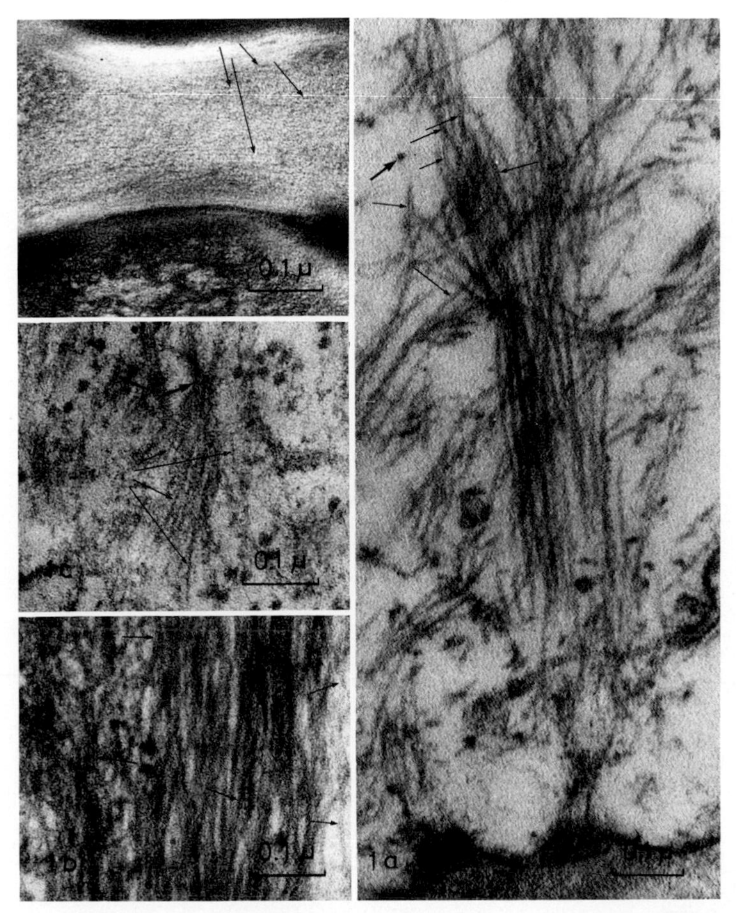

Fig. 1a. An electron micrograph of tonofilaments in the basal cell from a patient with dominant ichthyosis vularis. They show an apparently tube-like structure consisting of central electron translucent core and peripheral electron dense coat in vertical section (small arrows) as well as in cross section (large arrow). OsO_4 fixation. Stained with uranyl acetate (UA) and lead citrate (LC).

Fig. 1b. Tonofilaments in the basal cell from a patient with variable erythrokeratoderma. No distinct abnormalities. They reveal also a tube-like structure (small arrows). Glutaraldehyde (GA) and OsO_4 fixations. Stained with UA and LC.

Fig. 1c. Tonofilaments in a granular cell from the same patient as Fig. 1a. Besides the tube-like structure (large arrows) smaller filamentous elements are seen (small arrows). OsO_4 fixation. Stained with UA and LC.

Fig. 1d. Horny cells from the same patient as Fig. 1a reveal no distinct structural abnormalities. Arrows: keratin Filaments. OsO_4 fixation. Stained with UA and LC.

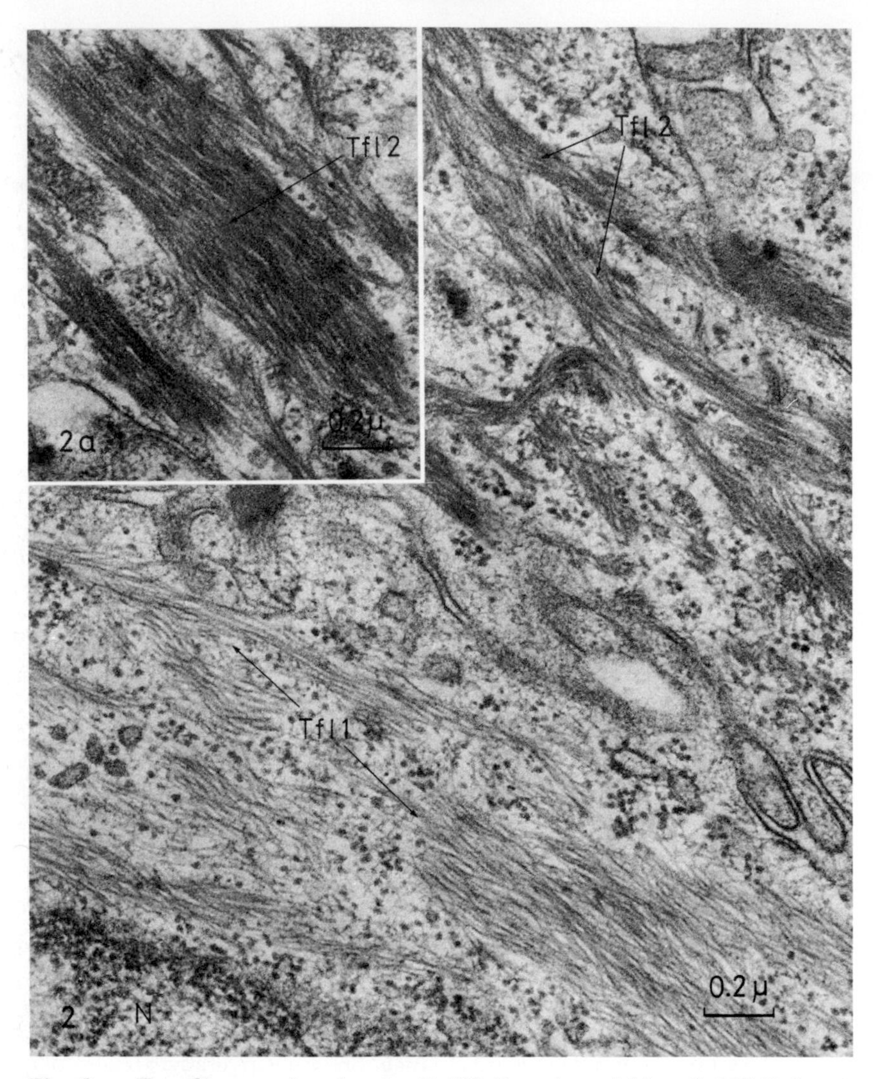

Fig. 2. Tonofilaments in a basal cell (Tfl 1) and a prickle cell (Tfl 2) from a patient with Sjögren-Larsson syndrome. Tfls 2 seem to form apparently more compact electron denser bundles than those of Tfls 1. GA and OsO_4 fixations. Stained with UA and LC. N: nucleus.

Fig. 2a. Tonofilaments in a prickle cell from the same patient as fig. 2. In these bundles of tonofilaments (tonofibrils), the electron opacity seems to be more prominent. Fixations: GA and OsO_4. Staining: UA and LC.

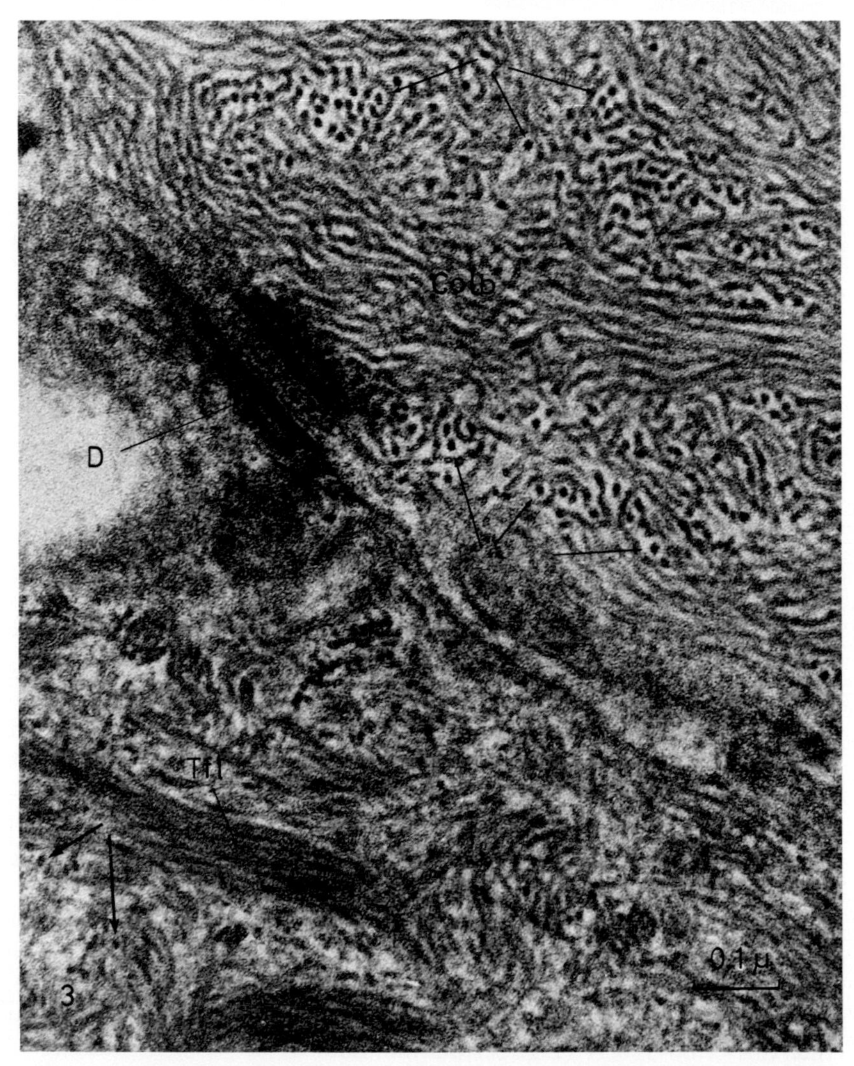

Fig. 3. So called "colloid body" (Colb) in the epidermis from a patient with lichen amyloidosus. In cross section filamentous structures (tonofilaments) in the colloid body (right-upper) seem to be separated from each other with almost constant distance. Therefore each filament seems to be coated with an electron transparent mantle (small arrows). This structure is not clear around tonofilaments (Tfl) in an unaffected keratinocyte (large arrows). D: desmosome. Fixation: GA and OsO_4. Staining: UA and LC.

other keratotic disorders[16,17,18] (Fig. 2, 3, 5).[19] These tonofilaments reveal distinctly a "tubular" structure (Fig. 1a, 1b) with central electron translucent core of 25–35 Å in diameter and electron denser peripheral part of 10–30 Å in thickness. This structure is also recognized in a cross section of filaments (Fig. 1a). The fine structure corresponding to the 9-plus 2-pattern[20] in hair filaments is not recognized in the tonofilaments of the basal cell. These tubular structures may exhibit the difference of stainability[13,21] to uranyl acetate and lead citrate exists between the peripheral part and the center of the tonofilament.

In various keratotic disorders those tonofilaments in the basal cells seem to reveal no essential structural abnormalities.[22–27]

In the prickle cell of dominant ichthyosis, however, far finer filamentous structures of 20–40 Å in diameter are often observed, besides normally structured tonofilaments (Fig. 1c). These filamentous structures may only reveal the above mentioned electron denser peripheral part of tonofilaments.

In the horny cell of dominant ichthyosis "keratin pattern"[12–28] shows no distinct abnormality, though some filamentous structures seem to be slightly thinner (Fig. 1c).

It is well known there are some differences between tonofilaments arranged in groups in the basal cell and those aggregated in the prickle cell forming tonofibrils in the normal epidermis.[21] This finding is more distinctly seen in Sjögren-Larsson syndrome (Fig. 2, 2a). Loosely arranged tonofilaments in the basal cell may be tightly packed in the prickle cells.[21] As a whole the electron density of tonofilament bundles (tonofibrils) apparently seem to be increased (Fig. 2, 2a). The nature and significance of this process is not yet clarified, though hypothetical elucidations were observed.[13,21]

In some conditions these processes may be completely disturbed. For instance, in the so called "colloid bodies", which are mainly composed of tonofilaments,[29] the aggregation of filaments never does occur (Fig. 3). Individual filaments in it do not attach to each other with constant intervals of 50–100 Å (Fig. 3). Similar findings are seen in bundles of probably newly synthesized tonofilaments in some conditions.[30] Between filaments there are electron translucent spaces. Therefore individual filaments are distinctly recognized (Fig. 3). In other words, each filament may be circularly surrounded by electron translucent substances.

On the contrary, in epidermolytic hyperkeratosis, normally structured tonofilaments in the basal cell clump in the prickle cell and form a well demarcated mass, in which individual filaments are no more easily recognized[18] (Fig. 4). This process never occurs in the basal cells in this disorder. The pathogenesis of this process is not yet known. It seems to be unlikely that synthesized fibrous protein itself in the basal cell changes the

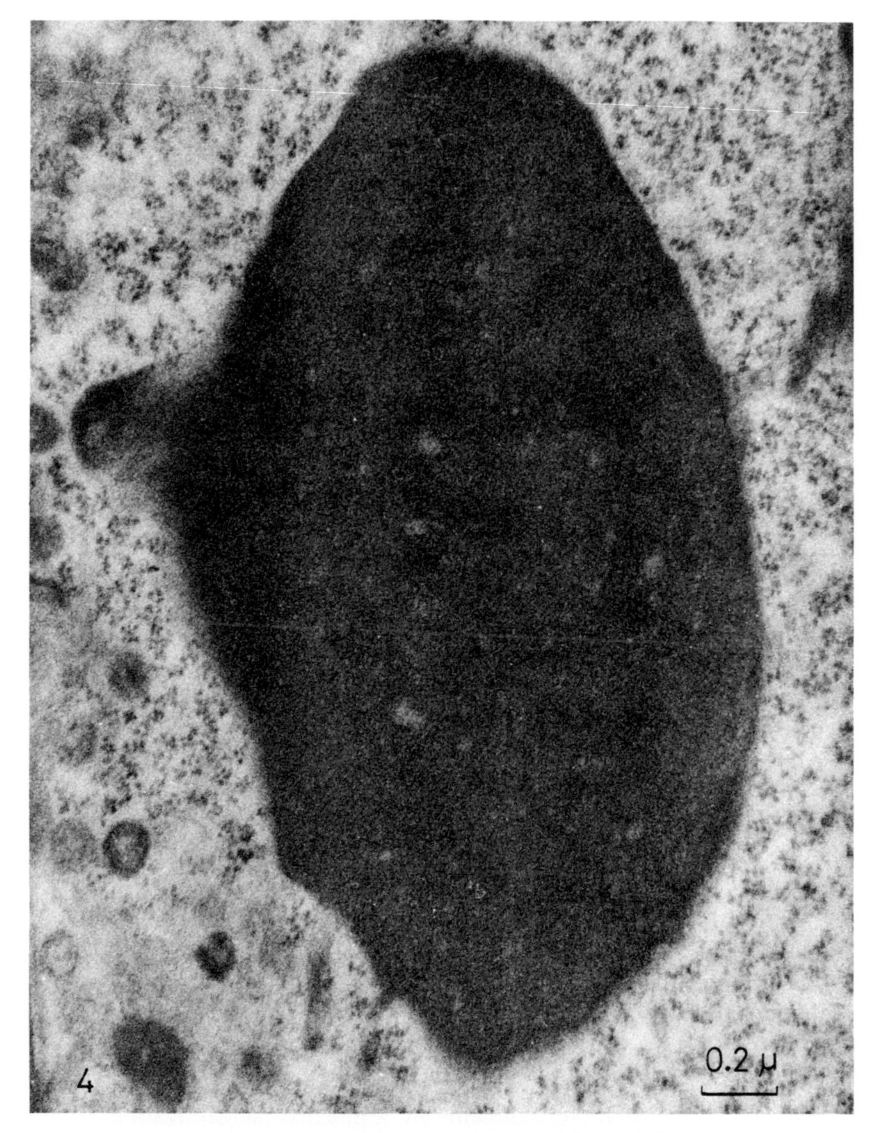

Fig. 4. A mass of clumped tonofilaments in a prickle cell from a patient with epidermolytic hyperkeratosis. In this mass individual filaments are no more distinguishable. OsO_4 fixation. Stained with UA.

chemical properties abruptly in the prickle cell just above it. The biochemical relationships betweeen filaments and interfilamentary substances seem to be rather responsible for this condition. In fact probably at the initial stage of clumping the interfilamentary space increases in its electron density.[30] The clumping may be an excessively advanced aggregation process of tonofilaments. Similar clumping processes of tonofilaments are seen in a variety of hyperkeratotic disorders.[31,32,33]

Acrodermatitis enteropathica is known as a zinc deficiency disease[2] or a disorder of lipid metabolism.[3] This inherited condition is characterized mainly by orificial and acral vesicobullous, or pustular and crusted eczematous skin lesions with partial or total alopecia and dyspeptic diarrhea. In a lesion of this disorder a characteristic ultrastructural finding is a marked hyperproduction of tonofilaments, which is usually seen in every layer of the epidermis (Fig. 5). However, individual tonofilaments are well recognized even in the granular layer (Fig. 5). In this disorder the aggregation process of tonofilaments in the prickle cell may be somewhat involved. Zinc deficiency or disturbed lipid metabolism may have some unfavorable effect on this process.

In dominant ichthyosis vulgaris tonofilaments seem to be rather sparse and the electron density of the peripheral part of filaments not so conspicuous (Fig. 5a). On the contrary, in some cases of lichen amyloidosus the electron density of bundles seem to be apparently increased, in addition to marked hyperproduction of tonofilaments, each of which is no more easily discernible (Fig. 5b). In this disorder the aggregation process of tonofilaments in the prickle cell may be emphasized. In Rud's syndrome this process seems to begin intensely in the basal cell.

From the morphological standpoint, it may be said that the different electron dense properties of fibrous components in the above mentioned various keratotic disorders seem to result not from the difference of the central electron translucent part of filaments but from the difference of the surface part, because there is no structural difference in newly synthesized individual tonofilaments in the various disorders.

Keratohyalins appear on bundles of tonofilaments in the granular layer. The morphological structures seem to be different in the various keratotic disorders. They show a peculiar spongy feature in some cases of variable erythrokeratoderma (Fig. 6). The similar, but somewhat different, structure may be seen in some cases of other keratotic disorders.[24] The significance of this peculiar keratohyalin in abnormal keratinization is not obscure. In acrodermatitis enteropathica, keratohyalin shows rather round shape with uniform electron density (Fig. 6b). In dominant ichthyosis, however, it reveals a somewhat reticular form with irregular striated electron translucent pattern (Fig. 6b). The meaning of these differences in the various disorders is not yet elucidated.

In the transitional zone of the granular layer and the horny layer, various structural abnormalities are seen in the keratotic disorders. In acrodermatitis enteropathica the dehydration process and formation of

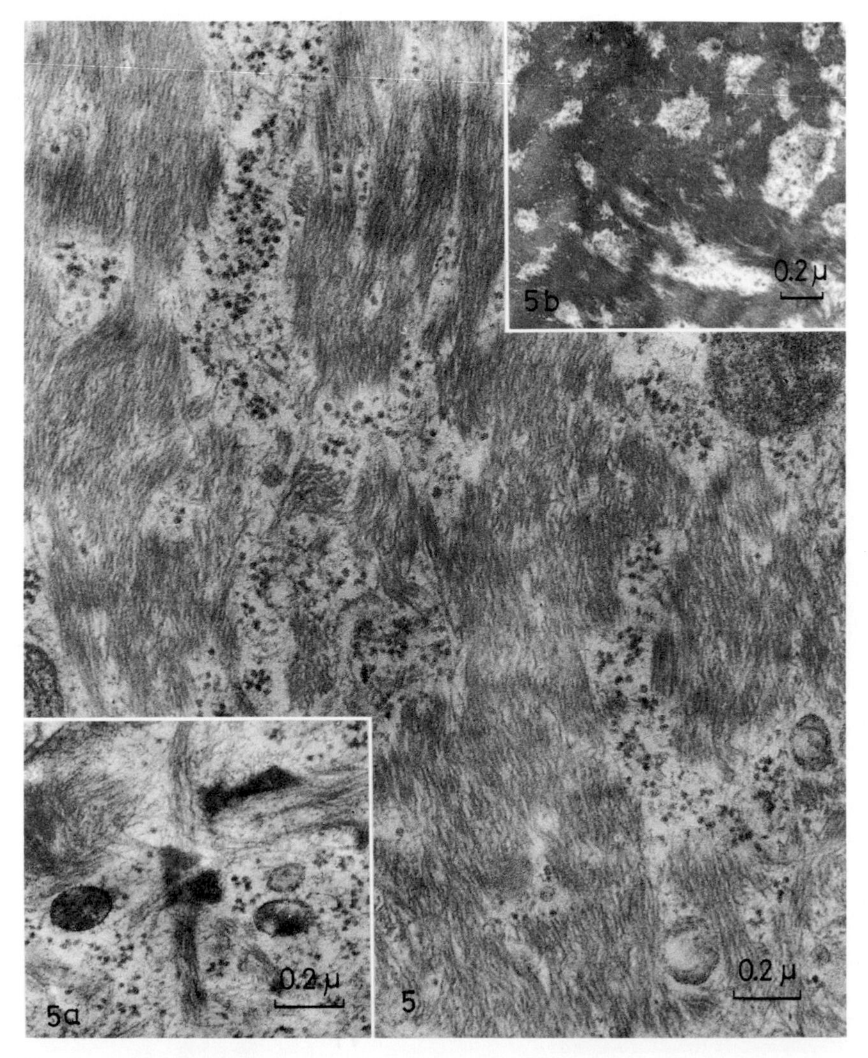

Fig. 5. Tonofilaments in a prickle cell from a patient with acrodermatitis enteropathica. Individual tonofilaments, which are easily recognized, reveal no apparent structural abnormality, though they seem to be distinctly increased in number. GA and OsO_4 fixations. Stained with UA and LC.

Fig. 5a. Tonofilaments in a granular cell from a patient with dominant ichthyosis vulgaris. They seem to be sparse and some of them apparently thinner. OsO_4 fixation. UA and LC staining.

Fig. 5b. Tonofilaments in a granular cell from a patient with lichen amyloidosus. Though individual filaments are not easily discernible, they seem to be apparently increased in number. GA and OsO_4 fixations. Stained with UA and LC.

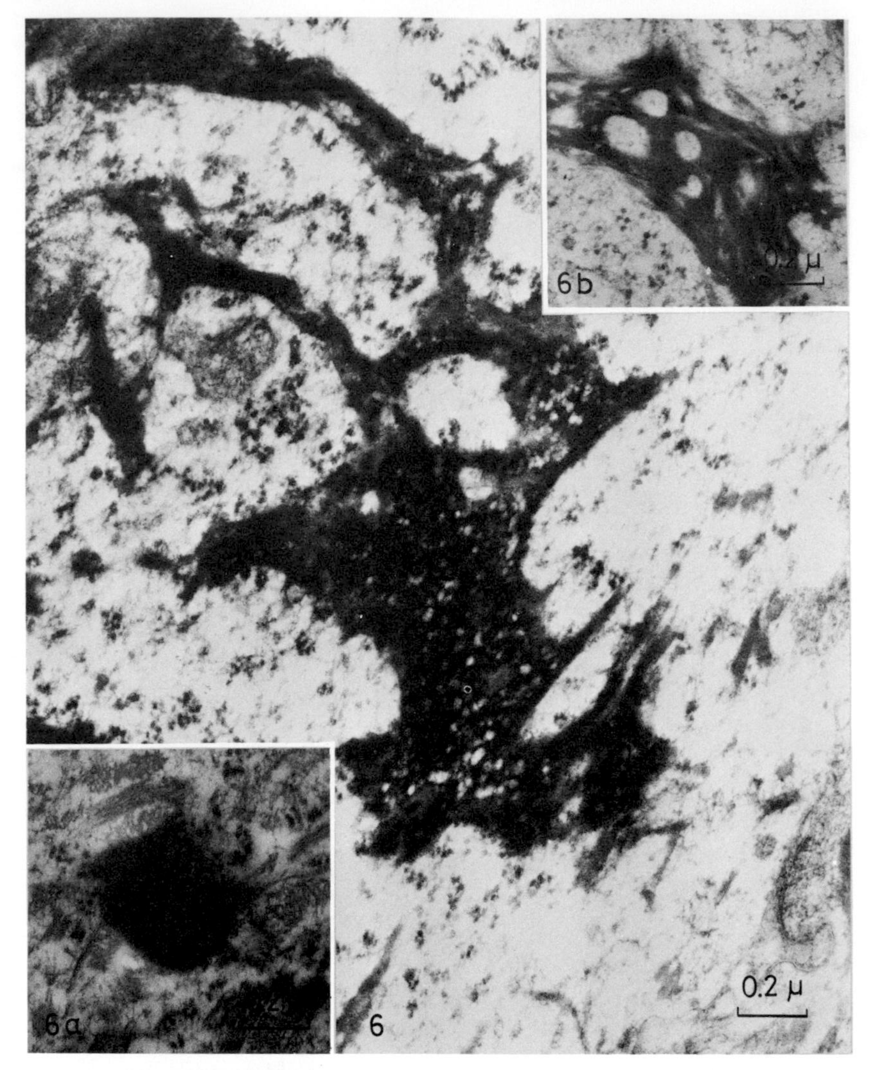

Fig. 6. Keratohyalin (Kh) in a granular cell from the same patient with variable erythrokeratoderma as fig. 1b. This Kh reveals a characteristic sponge-like structure. GA and OsO_4 fixations. Stained with UA and LC.

Fig. 6a. Round shaped Kh with smooth, almost uniform electron density in the same patient with acrodermatitis enteropathica as fig. 5. Fixation: GA and OsO_4. Staining: UA and LC.

Fig. 6b. Reticularly formed Kh in a patient with dominant ichthyosis vulgaris. OsO_4 fixation. Stained with UA and LC.

keratin filaments seem to be slow and incomplete, and granular components remain without enough degradation, in addition to marked hyperproduction of filamentous materials without the formal keratin pattern, while the distinct thickening of cell walls is already recognized (Fig. 7). This phenomenon may indicate not only a promoted turn over of cells, but also the results of disturbed aggregation process of tonofilaments, as well as inadequate keratohyalin formation.

A remarkable thickening of horny cells with incomplete degradation of granular components is seen in some cases of lichen amyloidosus (Fig. 7a). In dominant ichthyosis irregular shaped electron translucent spaces are seen over several layers of horny cell, though similar but not so prominent structures are also observed in the normal lowest horny cells[34] (Fig. 7b).

Besides tonofilaments there are other types of fibrous components in the epidermal cells in some specific conditions. In lichen amyloidosus two different types of fibrous structures were found by Hashimoto *et al.*[6] in the epidermal cells. They called the one "wavy cytoplasmic filaments" and the other "amyloid-like filaments".

In lichen amyloidosus fine filamentous structures are seen mainly between tonofilament bundles or directly attached to them from the basal to the granular cell (Fig. 8). These filaments are about 40–60 Å in diameter,[35] and are not capsulated with the membrane. Each filament is clearly recognized because of the electron transparency of interfilamentary spaces. From the different diameters they are not true amyloid fibrils, as observed in the dermis. These filaments may be in accordance with "wavy cytoplasmic filaments" (WCF) by Hashimoto. It is not obvious whether they will be "keratinized" in the horny cell or not. The aggregation process of filaments with increased electron density as in tonofibrils never occurs in WCF.

There are other types of fibrous structures in the epidermal cells in lichen amyloidosus (Fig. 9, 9a, 10a, 10b). These fibrous structures are usually seen in the cytoplasm as a large mass independent from tonofilaments. Frequently the filaments in the center of the mass are far thinner than those at the periphery[35] (Fig. 9). These masses are also not covered with any membraneous structures, and seen usually from the prickle to the horny cell, rarely in the basal cell. The fibers, which form a central core of the mass, show about 100–250 Å in diameter, whereas those in the periphery about 500–800 Å (Fig. 9, 9a). They seem to corespond well to "amyloid-like filaments" (ALF) found by Hashimoto *et al.*[6] In higher magnification these ALFs seem to be composed of far finer wavy filamentous materials about 40–60 Å in diameter. These finer filamentous materials are simultaneously seen in the spaces between the thicker fibers (ALF) (Fig. 9a). These ALFs may have a central core,[6] about 100 Å in diameter, which are recognized as electron dense round or elipsoid shaped dots in

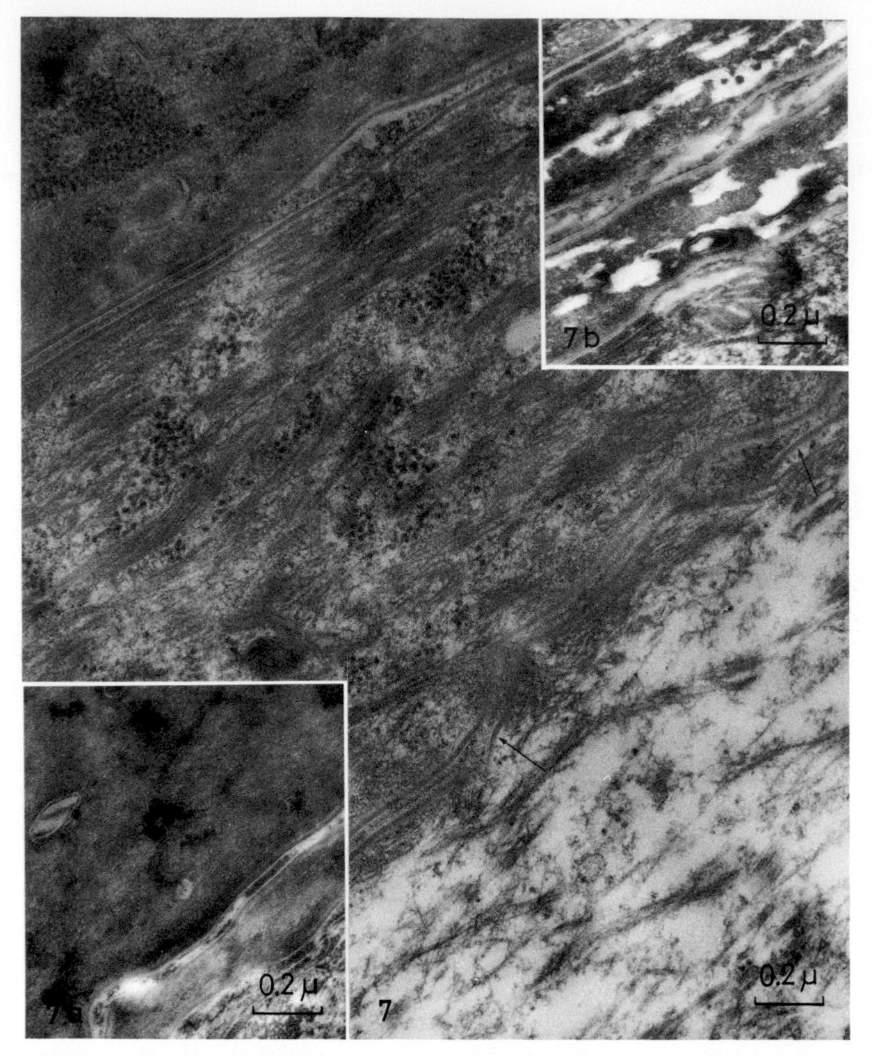

Fig. 7. Transitional zone of the granular layer and the horny layer in the patient with acrodermatitis enteropatica. In spite of distinct thickening of the cell wall (arrows) no apparent intracytoplasmic changes are observed (right-bottom). Even after considerable dehydration tonofilaments and granular components (ribosomes and glycogen particles) seem to remain unchanged (middle). No apparent keratin pattern is in the completely dehydrated horny cell with the granular components (left-upper). GA and OsO_4 fixations. UA and LC stainings.

Fig. 7a. Transitional zone of the granular layer and the horny layer from the patient with lichen amyloidosus. A remarkably thickened horny cell (left-top) with remnants of granular components. The thickening of the

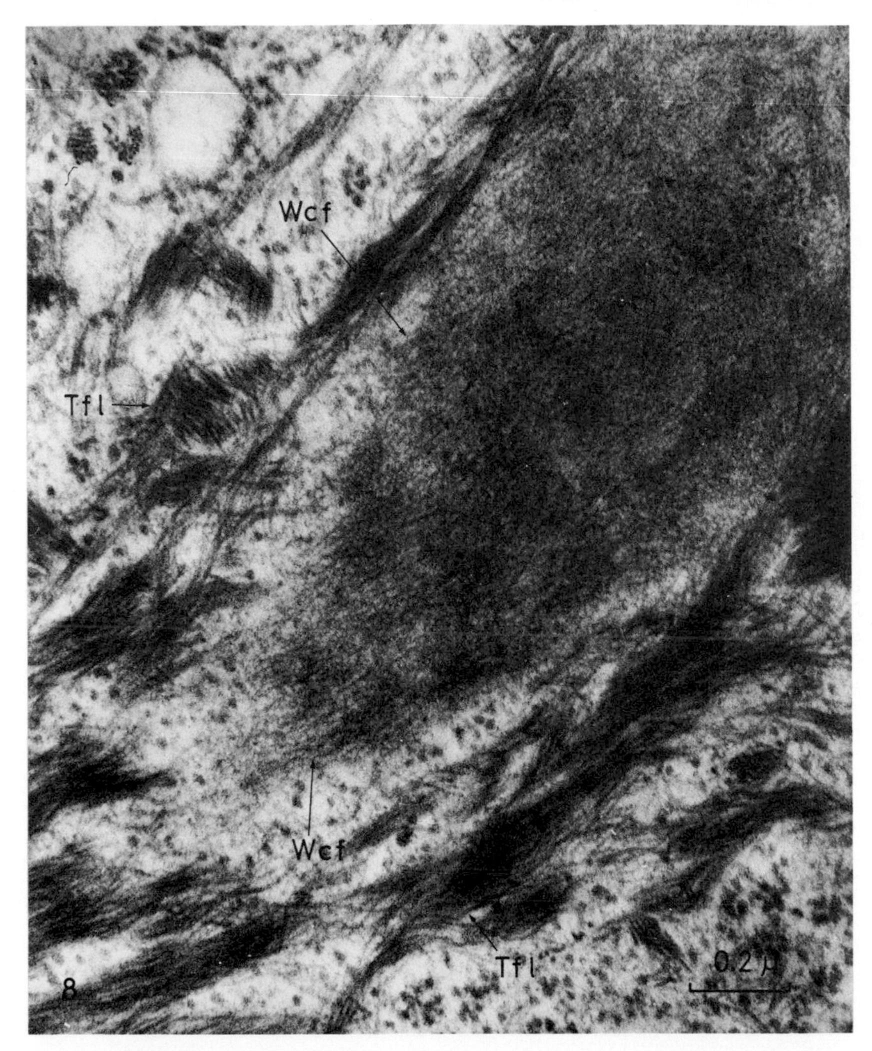

Fig. 8. "Wavy cytoplasmic filaments" (Wcf) in a prickle cell from a patient with lichen amyloidosus. Tfl: tonofilaments. Fixation: GA and OsO_4. Staining: UA and LC.

---

granular cell walls is not yet seen as well as in the normal (right-bottom). Fixations: GA and OsO_4. Staining: UA and LC.

Fig. 7b. Transitional zone of the granular layer and the horny layer from the patient with dominant ichthyosis. Irregularly shaped electron transparent spaces are seen in several horny cells. Partly observed "keratin pattern." The thickening of the granular cell walls just under the horny layer is not yet seen (right-bottom). Fixation: OsO_4. Staining: UA and LC.

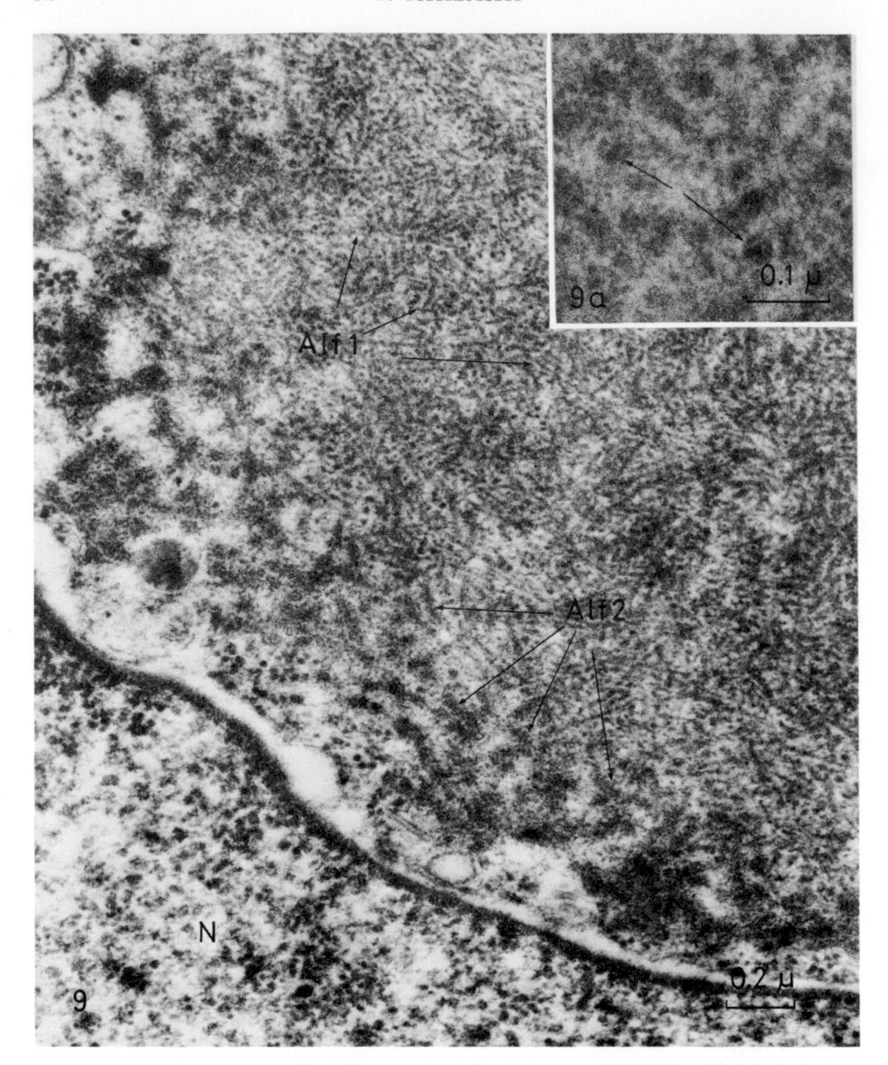

Fig. 9. "Amyloid-like filaments" (Alf 1, Alf 2) in a prickle cell from the same patient as in fig. 8 with lichen amyloidosus. Two different types of fibrous structures (Alf 1 and Alf 2) are distinctly observed. N: nucleus. Fixation: GA and OsO_4. Staining: UA and LC.

Fig. 9a. A high magnification of "amyloid-like filaments" (Alf 2) from the same patient as in fig. 9. In interfilamentary spaces wavy finer filamentous structures are seen. In cross section of Alf 2 an electron dense "core" is observed (arrows). Fixation: GA and OsO_4. Staining: UA and LC.

Fig. 10a. "Amyloid-like filaments" (Afl 3) in the granular cell from the patient with lichen amyloidosus. The diameter of these filaments is clearly different from those of Afl 1 and Afl 2. N: nucleus. Fixation: GA and OsO_4. Staining: UA and LC.

Fig. 10b. "Amyloid-like filaments" in the horny cell. The appearance of this substance seems to be rather "spongy" than "fibrous". Kfl: keratin filaments. D: desmosome. Fixation: GA and OsO_4. Staining: UA and LC.

the cross section (Fig. 9a). However, in the vertical section these central cores of fibers are not easily discernible, especially in far thicker fibers. In the other mass of ALFs the fibers reveal about 200–350 Å in diameter (Fig. 10a). Therefore it may be said ALFs are not uniform sized fibers, but consist of various sized fibers, which are composed of far finer filaments.[35] In the horny cells these ALFs show spongy, muddy rather than fibrous appearance as a kind of "keratinized material" (Fig. 10b). This change seems to be analogous to those in epidermolytic hyperkeratosis.[36] The nature and pathogenesis of these materials (WCF and ALF) are not yet known. Probably they are formed by keratinocyte[6] in loco, and may have a similar chemical structure to those of amyloids or tonofilaments. Further biochemical investigations are expected.

In the keratinization process newly synthesized tonofilaments in the lower epidermis alter finally into keratin filaments in the horny cell. In this process two morphologically noticeable steps are observed. The first is the so called aggregation of tonofilaments in the prickle cell and the second is the formation of keratin pattern with full dehydration in the horny cell. In some keratotic disorders the aggregation process is very conspicuous, but in others rather unnoticeable. In epidermolytic hyperkeratosis it seems to be excessive and extremely deviated from the normal, but in colloid bodies, filaments in Paget cells[37] and other fibrous components, such as wavy cytoplasmic filaments in lichen amyloidosus, in the epidermal cell, this process seems to be entirely lacking or disturbed. The significance and the detailed biochemical reactions of this process are not yet known. They may play some important role in second steps lately.

However, exceptionally, in a rare case of epidermolytic hyperkeratosis the disturbed first step does not always lead to defective barrier formation.[38] In this case abnormal hyperkeratotic lesions were seen only in some limited areas, i.e. the palms, soles, elbows, knees, etc., and the majority of the remaining skin revealed clinically absolutely normal appearance, except easily producing eroded lesions after ripping off adhesive tapes from the skin. This apparently intact skin showed ultrastructurally the typical clumpings of tonofilaments in the prickle cells of the epidermis, as well as in the hyperkeratotic lesions. In this case the aggregation process may be significant for protection against the tearing off of cells rather than formation of the barrier.

On the other hand, the second step seems to occur not only on tonofilaments, but also on other fibrous components, including amyloid-like filaments. This process must be significant to form a barrier to the outside. The majority of the above mentioned keratotic disorders show more or less disturbance of this process. The clinically abnormal features of the epidermis in the keratotic disorders may result mainly from a defect of this formal process.

REFERENCES

1. Mize, C. E., Herdon, J. H., Blass, J. P., Milne, G. W. A., Follansbee, C., Laudat, P. and Steinberg, D.: Localization of the oxidative defect in phytanic acid degradation in patients with Refsum's disease. *J. Clin. Invest.*, **48**: 1033–1040, 1969.
2. Moynahan, E. J.: Acrodermatitis enteropathica: A lethal inherited human zinc-deficiency disorder. *Lancet*, **2**: 399–400, 1974.
3. Cash, R. and Berger, C. K.: Acrodermatitis enteropathica: Defective metabolism of unsaturated fatty acids, *J. Pediatr.*, **74**: 717–729, 1969.
4. Hooft, C., Kriekemans, J., Devos, E., Traen, S. and Verdonk, G.: Sjögren-Larsson syndrome with exsudative enteropathy. Influence of medium-chain triclycerides on the symptomatology. *Helv. Paediat. Acta*, **22**: 447–458, 1967.
5. McGuire, J.: Actin is present in mammalian keratinocytes. Biochemistry of Cutaneous Epidermis Differentiation, 1977.
6. Hashimoto, K. and Leon Yoong Onn, L.: Lichen amyloidosus. Electron microscopic study of a typical case and a review, *Arch. Derm.*, **104**: 648–667, 1971.
7. Saida, T. and Ishibashi, Y.: Sjögren-Larsson syndrome, *Hifu-Rinsho*, **14**: 507–515, 1972 (in Japanese).
8. Ogawa, K., Saida, T., Ishibashi, Y. and Kukita, A.: Rud's syndrome, *Hifu-Rinsho*, **18**: 445–457, 1976 (in Japanese).
9. Ishibashi, Y. and Klingmüller, G.: Erythrodermia ichthyosiformis congenita bullosa Brocq. Über die sogenannte granulöse Degeneration. I., *Arch. Klin. Exp. Derm.*, **231**: 424–436, 1968.
10. Nakabayashi, K., Takizawa, K., Ishibashi, Y. and Kukita, A.: A case of acrodermatitis enteropathica. *Jap. J. Derm.*, **85**: 169, 1975.
11. Saida, T. and Ishibashi, Y.: A case of erythrokeratodermia variabilis, *Jap. J. Derm.*, **85**: 149, 1975 (in Japanese).
12. Brody, I.: The ultrastructure of the tonofibrils in the keratinization process of normal human epidermis. *J. Ultrastr. Res.*, **4**: 264–297, 1960.
13. Zelickson, A. S.: Normal human keratinization process as demonstrated by electron microscopy. *J. Invest. Derm.*, **37**: 369–379, 1961.
14. Odland, G. F.: The fine structure of the interrelationship of cells in the human epidermis. *J. Biophys. Biochem. Cytol.*, **4**: 529–538, 1958.
15. Roth, S. I. and Clark, W. H.: Ultrastructural evidence related to the mechanism of keratin synthesis. The Epidermis, edited by Montagna, W. and Lobitz, W. C., pp. 303–337, New York: Academic Press, 1964.
16. Wilgram, G., Caulfield, J. B. and Madgic, E. B.: A possible role of the desmosome in the process of keratinization. An electron-microscopic study of acantholysis and dyskeratosis. The Epidermis, edited by Montagna, W. and Lobitz, W. C., pp. 275–301, New York: Academic Press, 1964.
17. Lagerholm, B.: Cellular changes in the psoriatic epidermis. II. Submicroscopic organization in psoriatic lesions of different age, *Acta derm.-venereol.*, **45**: 99–122, 1965.
18. Ishibashi, Y. and Klingmüller, G.: Erythrodermia ichthyosiformis congenita

bullosa Brocq, Über die sogennante granulöse Degeneration. II., *Arch. Klin. Exp. Derm.,* **232**: 205–224, 1968.

20. Filshie, B. K. and Rogers, G. E.: The fine structure of alpha-keratin. *J. molec. biol.,* **3**: 784–786, 1961.
21. Brody, I.: Different staining methods for the electron-microscopic elucidation of the tonofibrillar differentiation in normal epidermis. The Epidermis, edited by Montagna, W. and Lobitz, W. C., pp 251–273, New York: Academic Press, 1964.
22. Anton-Lamprecht, I.: Zur Ultrastruktur hereditärer Verhornungsstörungen. I. Ichthyosis congenita. *Arch. Derm. Forsch.,* **243**: 88–100, 1972.
23. Anton- Lamprecht, I., Curth, H. O. and Schnyder, U. W.: Zur Ultrastruktur hereditärer Verhornungsstörungen. II. Ichthyosis hystrix Typ Curth-Macklin. *Arch. Derm. Forsch.,* **246**: 77–91, 1973.
24. Anton-Lamprecht, I.: Zur Ultrastruktur hereditärer Verhornungsstörungen. III. Autosomal-dominante Ichthyosis vulgaris. *Arch. Derm. Forsch.,* **248**: 149–172, 1973.
25. Anton-Lamprecht, I.: Zur Ultrastruktur hereditärer Verhornungsstörungen. IV. X-chromosomal-recessive Ichthyosis. *Arch. Derm. Forsch.,* **248**: 361–378, 1974.
26. Anton-Lamprecht, I.: Zur Ultrastruktur hereditärer Verhornungsstörungen. V. Ichthyosis beim Refsum-Syndrom. *Arch. Derm. Forsch.,* **250**: 185–206, 1974.
27. Ishibashi, Y., Tsuru, N. and Takizawa, K.: A case of Papillon-Lefèvre syndrome. *J. Derm.* (Tokyo) in preparation.
28. Charles, A.: An electron microscopic study of cornification in the human skin. *J. Invest. Derm.,* **33**: 65–74, 1959.
29. Ishibashi, Y., Harada, S. and Kukita, A.: An electron-microscopic investigation of colloid bodies. *Nishinihon J. Dermatol.,* **37**: 296, 1975 (in Japanese).
30. Ishibashi, Y. and Klingmüller, G.: Erythrodermia ichthyosiformis congenita bullosa Brocq. Über die sogenannte granulöse Degeneration. V., *Arch. Klin. Exp. Derm.,* **233**: 124–138, 1968.
31. Braun-Falco, O., Petzolat, D., Christophers, E. and Wolff, H. H.: Die granulöse Degeneration bei Naevus verrucosus bilateralis. Eine morphologische und funktionelle Studie. *Arch. Klin. Exp. Derm.* **235**: 115–137, 1969.
32. Hirone, T. and Fukushiro, R.: Disseminated epidermolytic acanthoma. *Acta Dermato.-venereol.,* **53**: 393–402, 1973.
33. Haustein, U. F. and Klug, H.: Ultrastrukturelle Befunde bei Keratosis palmoplantaris. *Dermatol. Mschr.,* **160**: 825–835, 1974.
34. Odland, G. F.: Tonofilaments and keratohyalin. The Epidermis, edited by Montagna, W. and Lobitz, W. C., pp 237–249, New York: Academic Press, 1964.
35. Ishibashi, Y. and Takizawa, K.: Ultrastructural changes of the epidermis in lichen amyloidosus. *Minophagen Medical Review.,* **21**: 157–158, 1976 (in Japanese).
36. Ishibashi, Y. and Klingmüller, G.: Erythrodermia ichthyosiformis congenita bullosa Brocq. Über die sogennante granulöse Degeneration. IV. *Arch. Klin. Exp. Derm.,* **233**: 107–123, 1968.

37. Ishibashi, Y., Niimura, M. and Klingmüller, G.: Elektronenmikroskopischer Beitrag zur Morphologie von Paget-Zellen. *Arch. Derm. Forsch.,* **245**: 402–416, 1972.
38. Saida, T., Ogawa, K. and Ishibashi, Y.: A case of erythrodermia ichthyosiformis congenita with peculiar localization of hyperkeratotic skin lesions. *Hifu-Rinsho,* in preparation (in Japanese).

Discussion

Dr. Goldsmith: Dr. Schnyder and his co-workers in Heidelberg have reported that in epidermalytic hyperkeratosis, there is an abnormal periodicity along the fibrous proteins. I think there is a periodicity of 170 angstrom units. In your studies did you find any abnormal periodicity on the fibrous proteins?

Dr. Ishibashi: I could not find those periodicities in my cases.

Dr. Fukuyama: Have you treated your patients having acrodermatitis enteropathica with zinc, since zinc may have some special relevance to keratinization?

Dr. Ishibashi: I only treated my patients with Kinoform but not with zinc. The zinc content in the blood of the patients was not low in the cases under my treatment.

Dr. Fukuyama: Were there any ultrastructural changes of the filaments during treatment?

Dr. Ishibashi: I did not study that.

Dr. Bernstein: Looking at the various tonofilamentous or rather, filamentous structures, that you have seen in these various diseases, I'm impressed with the advantage that you would have if you were able to investigate the differences utilizing immunologic probes. I would like to ask, is there sufficient material available, in your opinion, to isolate and make antibody to any of the filamentous structures that you have seen?

Second, if not, would it be possible to isolate tonofilaments or filamentous material from the normal epidermis and make antibody to this? If one had any antibody, especially antibody against normal filamentous structure, one could look for differences in chemical composition and/or structure in these diseases?

Dr. Ishibashi: It is probably possible, but I have no information on this point.

Dr. Baden: We have tried to see if there are differences in these proteins by isolating them and using immunological probes. It is too difficult. One major difficulty has to do with the considerable variability that one sees, i.e. variation in the polypeptide that one can isolate. Even from normal individuals there is significant variability in the ratio of the components that are present and it is not clear at this point whether that is a technical problem related to how you prepare them. However, to an antibody which we prepared to normal fibrous protein and which reacted only with fibrous protein, fibrous proteins in psoriasis and ichthyosis reacted with the antibody. That does not mean that there are no differences, but the immunologic reactivity to normal fibrous protein appears to be the same.

Dr. Odland: It seems to me that the kinds of questions that are being raised about identification of filaments presents a fairly important issue that at the moment is unresolved and is certainly not going to be resolved by those of us who have described for you, morphological events. It also seems to me inherent in the nature of this conference, if Dr. McGuire does not provide us with the answers we all seek, to raise questions and raise enough interest in those who are capable of approaching the answers to the problems.

Actin is Present in Mammalian Keratinocytes

Joseph McGuire,* Elias Lazarides** and Albert DiPasquale*

*Department of Dermatology, Yale University School of Medicine, New Haven, Connecticut, U.S.A. and **Department of Molecular, Cellular, and Developmental Biology, University of Colorado, Boulder, Colorado, U.S.A.

INTRODUCTION

Actin, a protein of molecular weight 45,000, forms the bulk of the thin filaments in muscle sarcomeres. The contractile force of muscle is thought to be related to the thin actin filament sliding over the thick myosin filament. The energy for this sliding motion and subsequent muscle contraction is derived from the hydrolysis of ATP, which is regulated by the concentration of calcium. This transduction of ATP hydrolysis to cellular movement, although well described in muscle, has only recently been described in nonmuscle cells.[1] Within the past three years the presence of actin and myosin has been described in many nonmuscle cells. The role of actin in these nonmuscle cells is related to pinocytosis, cytokinesis, secretion, and cell locomotion. In this report we will describe: (1) the presence of actin in keratinocytes, (2) the localization of actin within cultured keratinocytes, and (3) the differential distribution of actin throughout the epidermis. We will compare the actin content of cultured keratinocytes with epidermis and speculate on the function of actin within keratinocytes.

MATERIALS AND METHODS

Extraction of actin from calf hoof

Calf hoofs were obtained from the abbatoir, chilled, and dissected within two hours of death. The posterior aspect of the hoof just beneath the hairline was excised and sliced vertically into sections 2 to 3 mm thick. It was technically easy to dissect pigmented cow hoofs because of the sharp demarcation between the nonpigmented dermis and the pigmented basal area of the epidermis. Three fractions were dissected: stratum corneum, viable epidermis, and dermis. The viable epidermis was then cut into upper and lower (basal) fractions. These fractions were lyophilized, weighed, ground to powder in a Wiley mill, and weighed. They were homogenized in buffer A (tris hydrochloride 2 mM, ATP 0.2 mM, β-mercaptoethanol 0.5mM, $CaCl_2$ 0.2mM, pH 8). After stirring in this buffer for 1 hr at 4° the homogenate was filtered through sintered glass. The filtrate was centrifuged at 10,000 × *g* for 60 min at 4°. The supernatant from this centrifugation is referred to as the buffer A extract. The retentate on the sintered glass was stirred in urea buffer (urea 8M, Tris 0.5pm, mercaptoethanol 0.025m, pH 9). After stirring at 37° for 4 hrs, the mixture was centrifuged at 40,000 × *g* for 30 min, and the supernatant is referred to as urea extract.[2] Buffer A is identical to that used by Spudich and Watt.[3]

Extraction and purification of actin for amino acid sequencing

Viable calf hoof epidermis was extracted according to the procedure of Feuer *et al.*[4] Actin was purified by two polymerization steps. This calf hoof actin was used for amino acid sequencing as well as polymerization and reaction with heavy meromyosin.[5] Rabbit muscle actin which was used as a standard throughout these experiments was prepared and purified by repeated polymerization according to the technique of Feuer *et al.*[4]

Polyacrylamide gel electrophoresis

SDS polyacrylamide gels were electrophoresed at 2 mA per 7 cm gel. The gels contained 7% acrylamide and 0.1% sodium dodecyl sulfate (SDS). Tris-glycine, the running buffer, contained 1% SDS. The gels were fixed in isopropanol, 125:acetic acid, 50:water, 325 for 12 to 24 hrs. They were then stained in fixing solution containing 0.1% coomassie Blue for 1–4 hrs. Gels were destained with acetic acid, 70:methanol, 35:water, 395. This is a modification of the Weber and Osborn technique.[6] Quantitation of proteins on acrylamide gels was performed by a Beckman gel scanner. Gels were stained in 1% fast green *w/v* in 7% acetic acid (*v/v*) and destained in 7% acetic acid *v/v*.[7]

Cell Culture

Calf keratinocytes. Viable epidermis obtained by dissection of calf hoofs was suspended in 0.025% crystalline trypsin at 37°. After 15 min, one ml of fetal calf serum was added to this cell suspension. The cells were collected by centrifugation and placed in Corning plastic flasks containing F-10 medium with 10% horse serum, 2% fetal calf serum, and gentamycin 50 μg/ml.

Guinea pig keratinocytes. Pieces of skin from guinea pig ears were trypsinized in .025% Difco trypsin 1:250 at 37°. After 15 min the slices were transferred to MEM containing 10% fetal calf serum; the dermis was teased from the epidermis, and the epidermis was scraped with needles to release the keratinocytes. The cells settled and attached to the Falcon dishes. The identity of the keratinocytes was established by the presence of demosomes and by the extraction from these cultures of keratin peptides.

DNase affinity column. Actin was identified by its specific affinity for DNase I.[8] DNase type I (Sigma) was mixed with cyanogen bromide activated agarose A-1. 5M[9] overnight at 4°. The DNase-agarose was washed on a Buchler funnel with 2000 ml 0.1M $NaHCO_3$. Protein samples were added to a 3 ml DNase agarose column and eluted in 3 steps. The first elution was with buffer A (sodium acetate 0.5M, $CaCl_2$ 1mM and glycerol 30%, pH 6.5). The second elution was with buffer A containing 0.75M guanidine HCl. The third elution was with buffer A and 3M guanidine HCl. The three elutions are referred to as pools 1, 2, and 3.

Immunofluorescence

Actin was extracted from clone SV101 of mouse fibroblasts 3T3 cells and purified by electrophoresis on acrylamide slab gels. Antibodies to this material were raised in rabbits.[10] Cells growing on No. 1 Corning glass coverslips were washed with phosphate buffered saline (PBS) and fixed in 3.5% formaldehyde in PBS for 20 min. Following a second wash with PBS, they were treated with absolute acetone at −10°, air-dried, and then exposed to rabbit antibody diluted 1:20 37° for 1 hr in a humid atmosphere. The coverslips were washed three times with PBS and exposed for 1 hr to 1:10 dilution of goat antirabbit gammaglobulin-FITC (Miles). Following 3 washes with PBS and one wash with distilled water, the coverslips were examined by epi-fluorescence.

RESULTS

Epidermis of calf hoof contained actin (Table 1); however, no actin could be demonstrated in the dermis. Of the protein extracted from the epidermis by buffer A, 11.8% was actin. The buffer A extract of stratum cor-

Table 1. Comparative abundance of actin in various locations in calf epidermis.

				ACTIN		
	UX	BA	BA/UX	% TEP	% Dry Wt.	% BA
St. corneum	101.2	1.69	0.017	0.074	.028	4.5
Viable Epidermis	180.	8.1	0.045	1.39	.625	11.8
Dermis	122.5	3.2	0.026	0	0	0

UX, urea extract
BA, buffer A extract
TEP, total extracted protein (UX + BA)

neum contained 4.5% actin. On the basis of dry weight, 0.625% of the viable epidermis was actin while only 0.028% of the stratum corneum was actin. The percent of actin in the total extracted protein of viable epidermis was 20 times greater than stratum corneum, 1.39% versus 0.074%. There was virtually no actin in the urea extracted protein. Dermis contained no identifiable actin in the buffer A extract. Other tissues contained substantial amounts of actin. Four percent of the buffer A extract of newborn mouse epidermis was actin. Eighteen percent of buffer A extracted protein cultivated guinea pig keratinocytes was actin.

Table 2. Actin content of various tissues.

	Actin %		PROTEIN mg extracted	
TISSUE	TEP	BA	BA	UX
Mouse Epid.	2.41	4.1	2.3	1.6
Gunea pig keratinocytes	5.34	18	0.86	2.07
Cultivated Cloudman melanoma	7.94	12	10.8	13.5

Epidermal actin was identified by five techniques:

1. Coelectrophoresis with rabbit muscle actin
2. Affinity for DNase-agarose
3. Amino acid composition
4. Formation of complexes with HMM
5. Immunofluorescence

1. Proteins extracted from cultivated guinea pig keratinocytes with buffer A contain a species that coelectrophoreses with rabbit muscle actin as shown in Fig. 1a. The urea extracted protein from these cultivated cells does not contain actin (Fig. 1b). Buffer A extract of calf hoof epidermis contains a major protein that coelectrophoreses with actin (Fig. 2a). The urea extract contains the keratin peptides described by Steinert (2) but no actin (Fig. 2b). In cultivated guinea pig keratinocytes and calf epidermis all the actin is extracted by buffer A.
2. Epidermal actin was selectively retained by an agarose-DNase column. A mixture of 3 proteins including rabbit muscle actin was applied to the column; actin was selectively retained and was eluted by guanidine-HCl (Fig. 3). The elution of buffer A extract of calf hoof epidermis gave similar results. A protein retained by the agarose-DNase column was eluted by guanidine-HCl (Fig. 3b).

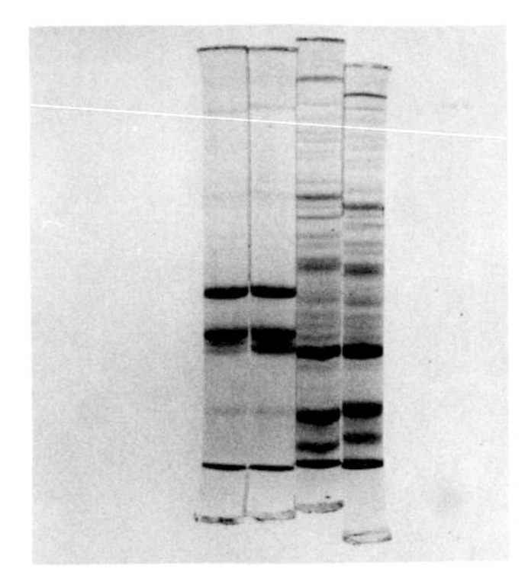

Fig. 1. Extraction of cultivated guinea pig keratinocytes.
gel 1 urea extract
gel 2 urea extract + rabbit actin
gel 3 buffer A extract
gel 4 buffer A extract + rabbit actin

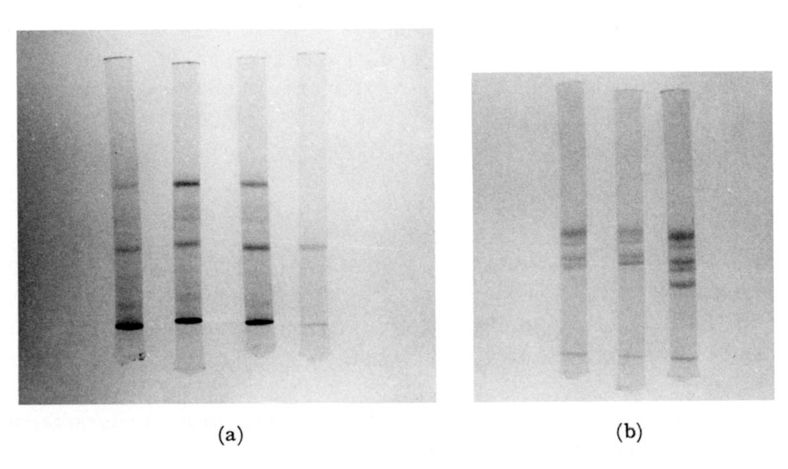

(a) (b)

Fig. 2. Gel electrophoresis of proteins of calf hoof epidermis.
a. Extraction with buffer A
gel 1 upper epidermis
gel 2 lower epidermis
gel 3 lower epidermis and added actin
gel 4 rabbit muscle actin
b. Extraction with urea
gel 1 upper epidermis
gel 2 lower epidermis
gel 3 upper epidermis and rabbit muscle actin

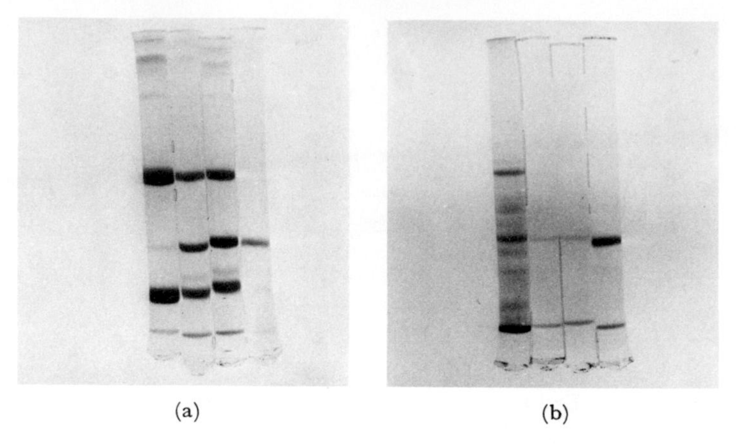

Fig. 3. Chromatography of actin on agarose-DNase column.
a. Chromatography of rabbit muscle actin, glyceraldehyde-3-phosphate dehydrogenase (MW 36000) and albumin (MW 65000)
gel 1 pool 1
gel 2 pool 2 (eluted with 0.75M guanidine HCl)
gel 3 pool 2 and added actin
gel 4 pool 3 (eluted with 3M guanidine HCl)
b. Chromatography of buffer A extract of calf hoof epidermis
gel 1 pool 1
gel 2 pool 2
gel 3 pool 3
gel 4 pool 3 and rabbit muscle actin

3. The amino acid composition of actin extracted from both mouse squamous cell carcinoma and cultivated mouse epidermal cells was nearly identical to amino acid analyses of rabbit and mouse muscle actin. 3-Methyl histidine was present in epidermal and muscle actin. (Table 3)[11]
4. Actin extracted from squamous cell carcinomas formed characteristic complexes with heavy meromyosin.
5. Keratinocytes cultivated from calf hoof contained fibrillar material which bound specifically antibody to 3T3 fibroblast actin.[12] The keratinocytes contained the characteristic keratin peptides and thus could be identified as keratinocytes. The filaments that reacted with the antibody coursed throughout the cells, and some of them correspond to stress lines seen by phase microscopy (Fig. 4).

SUMMARY AND DISCUSSION

The presence of actin within the keratinocytes has been verified by several criteria. Assuming that actin is functioning as a component of a

Table 3. Amino acid analyses of actin from muscle and mouse epidermis. (residues/100 residues)

Amino Acid	Muscle Actin		Epidermis	
	Rabbit*	Mouse	Tumor	Cultured cells
Aspartic acid	9.09	8.95	9.10	8.80
Threonine	7.22	7.30	7.35	7.10
Serine	6.15	5.95	6.10	6.15
Glutamic acid	10.16	10.25	10.15	10.40
Proline	5.08	5.20	4.90	5.05
Glycine	7.49	7.75	7.35	7.85
Alanine	7.75	7.60	7.70	7.60
Valine	5.61	5.75	5.85	5.35
Half-Cystine	1.34	1.05	1.10	1.05
Methionine	4.28	3.85	3.95	4.10
Isoleucine	8.02	8.25	7.85	8.10
Leucine	6.68	7.10	6.95	7.05
Phenylalanine	3.21	3.35	3.15	3.00
Tyrosine	4.28	4.15	4.10	3.90
3-Methylhistidine	0.27	0.25	0.25	0.20
Histidine	1.87	1.95	1.75	1.70
Lysine	5.61	5.70	5.80	5.95
Arginine	4.81	4.80	4.75	5.05

*From: Elzinga & Collins, *Cold Spring Harbor Symp. Quant. Biol.*, **37**, 1–7, 1972.

contractile system within the cell,[13] it seems likely that it is involved with such events as cytokinesis, phagocytosis, secretion, and cell locomotion.

Epithelial cells, under normal and abnormal conditions, undergo major translocations. They spread in wound repair and, as carcinomas, invade the dermis below the basement membrane. Under normal conditions, the epidermis is a dynamic tissue consisting of three populations: a stem cell population (basal cells), a differentiating population (stratum malphigii), and a terminally differentiated population (stratum corneum). Cells are constantly moving vertically away from the basement membrane to form the dead stratum corneum. Lateral translocation of cells within the normal epidermis (movement parallel to the basement membrane) must also occur. McKenzie[14] and Menton[15] showed that cells of the stratum corneum of thin epidermis are highly ordered into stacks. The large surface area of the squame makes it necessary that smaller cells of the malphigii layer move laterally to enter the procession of cells into the stacks.

Despite the belief that population pressure is responsible for the vertical movement of cells in normal epidermis,[16] considerable evidence exists to the contrary. Etoh *et al.*[17] have shown that, under conditions where no cell division takes place, cells move upward through the epidermis into the

Fig. 4. Indirect immunofluorescence using actin antibody. Calf hoof keratinocytes were cultured for six weeks, transferred to coverslips for 24 hours and then tested with actin antibody.

granular layer. A similar observation was made in 1960 by Sherman and Quastler,[18] who found that the upward movement of intestinal crypt cells continued following 1500 rad X-ray to C-57 mice. These findings suggest that actin plays an active role in the movement of cells within the normal epidermis in planes parallel and perpendicular to the basement membrane.

How might actin function to move cells through the epidermis? If the keratinocyte has sites of strong but transient contact with its neighboring keratinocytes, the cell could move through the epidermis by extending and adhering to the surface of a neighboring cell(s). It could then pull itself toward the neighboring cell(s). Subsequent breaking of this adhesion and extension to a more distal cell would enable the cell to move within the epidermis. Klaus has shown that desmosomes have a life span of less than 40 min in cultivated cells.[19] If the desmosomal complex were involved in the directed movement of a cell, then one might expect the contractile system to insert in or near the desmosome. It is clear from the immunofluorescent studies that bundles of filaments do extend to the cell margin.

Regions of the cell surface devoid of desmosomes may be involved with this type of movement. Heaysman and Pegrum[20] have shown that, in the area of contact between two fibroblasts in culture, there is a 200Å gap, without desmosomes. A dense filamentous mass materializes within two minutes in the cytoplasm of the two cells in the region of contact. Abercrombie has described retraction of cells away from one another following contact (contact retraction), presumably due to the contraction of musclelike filaments.[21] Epithelial cells making a strong intercellular adhesion would, in similar circumstances, be expected to draw themselves toward one another. Subsequent breaking of the adhesion would allow the cell to continue its migration. Whether any of the actin bundles observed by immunofluorescence will be found to correspond to tonofilaments is unknown and will require the use of a different probe. It is possible that more than one structure has been included under the general category tonofilament. Biochemical and immunologic studies will help establish the identity of the fibrillar elements and determine whether keratin or actin or both comprise the tonofilament.

Acknowledgements

Gratitude is expressed to Professor S. N. Klaus of Yale for providing cultures of guinea pig keratinocytes. The amino acid sequencing of actin and heavy meromyosin decoration were performed by Peter Steinert.[11] Supported by USPHS grant #AM 13929. Expert assistance was provided by Janet Shansky and Nancy Williams.

REFERENCES

1. Pollard, T. D. and Weiheng, R. R. Cytoplasmic actin and myosin and cell movement in Critical Reviews in Biochemistry. Vol. 2, 1–65, 1973.
2. Steinert, P. M. The extraction and characterization of bovine epidermal α-keratin. *Biochem. J.*, **149**: 39–48, 1975.
3. Spudich, J. A. and Watt, S. The regulation of rabbit skeletal muscle contraction. *J. Biol. Chem.*, **246**: 4866–4871, 1971.
4. Feuer, G., Molnar, F., Pettko, E., and Straub, F. B. The studies of the composition of the plemerization of actin. *Hung. Acta Physiol.*, **1**:150–163, 1948.
5. Ishikawa, H., Bischoff, R., and Holtzer, H. Formation of arrowhead complexes with heavy meromyosin in a variety of cell types. *J. Cell Biol.*, **43**: 312–328, 1969.
6. Weber, K. and Osborn, M. The reliability of molecular weight determination by dodecyl sulfate-polyacrylamide gel electrophoresis. *J. Biol. Chem.*, **244**: 4406–4412, 1969.
7. Gorovsky, M. A., Carlson, K., and Rosenbaum, J. L. Simple method for quantitative densitometry of polyacrylamide gels using fast green. *Anal. Biochem.*, **35**: 359–370, 1970.
8. Lazarides, E. and Lindberg, V. Actin is the naturally occurring inhibitor of deoxyribonuclease I. *Proc. Nat. Acad. Sci.*, **71**: 4742–4746, 1974.

9. Axen, R., Porath, J., and Arnback, S. Chemical coupling of peptides and proteins to polysaccharides by means of cyanogen halides. *Nature,* **214**: 1302–1304, 1967.
10. Lazarides, E. and Weber, K. Actin antibody: the specific visualization of actin filaments in non–muscle cells. *Proc. Nat. Acad. Sci.,* **71**: 2268–2272, 1974.
11. Steinert, P., Peck, G., Yuspa, S., McGuire, J., and DiPasquale, A. Isolation of an actin-like protein from epidermal tumors and cultured epidermal cells. *J. Invest. Derm.,* **66**: 276, 1976.
12. McGuire, J., Lazarides, E., and DiPasquale, A. Keratinocytes contain actin. *Clin. Res.,* **24**: 425A, 1976.
13. Weber, K. and Groeschel-Stewart, U. Antibody to myosin: the specific visualization of myosin-containing filaments in nonmuscle cells. *Proc. Nat. Acad. Sci.,* **71**: 4561–4564, 1974.
14. MacKenzie, I. C. Ordered structure of the epidermis. *J. Invest. Derm.,* **65**: 45–51, 1975.
15. Menton, D. N. and Eisen, A. Z. Structure and organization of mammalian stratum corneum. *J. Ultrastruc. Res.,* **35**: 247–264, 1971.
16. Leblond, C. P., Greulich, R. C., and Pereira, J. M. P. Relationship of the cell formation and cell migration in the renewal of stratified squames epithelia. Advances in Biology of Skin. Vol. V. Wound Healing. Ed. W. Montagna, R. E. Billingham. N. Y. MacMillan 39–67, 1964.
17. Etoh, H., Taguchi, Y. H., and Talachnick, J. Movement of β-irradiated epidermal basal cells to the spinous-granular layers in the absence of cell division. *J. Invest. Derm.,* **64**: 431–435, 1975.
18. Sherman, F. G. and Quastler, H. DNA synthesis in irradiated intestinal epithelium. *Exp. Cell Res.,* **19**: 343–360, 1960.
19. Klaus, S. N. On the lability of desmosomal attachments between guinea pig keratinocytes *in vitro.* Personal communication.
20. Heaysman, J. and Pegrum, S. Early contacts between normal fibroblasts and mouse sarcoma cells. *Expt. Cell Res.,* **48**: 479–481, 1973.
21. Abercrombie, M. The bases of the locomotory behavior of fibroblasts. *Exp. Cell Res. Suppl.* **8**: 188, 1961.

Discussion

Dr. Matoltsy: How did you determine the lifetime of the desmosome to be 40 minutes?

Dr. McGuire: This conclusion is based on observations of Dr. S. N. Klaus of Yale. He examined cultivated guinea pig keratinocytes by time-lapse photography and found that the lifetime of a single desmosome is shorter than 40 minutes. This is unpublished information.

Dr. Ogura: Muscle actin changes into a large molecular fibrous component in the presence of high concentrations of salt such as magnesium and calcium. Does your actin display a similar phenomenon?

DR. McGUIRE: The answer is yes. The G to F transition takes place and is in fact what we have done to purify the material when we have a large enough source of material. Ordinarily we don't do that. Instead we use chemical techniques. With a large amount of epidermis as a source, you can go through multiple cycles of polymerization to obtain a highly purified material. The decorated filaments that Dr. Steinert has, were prepared in such a way. They were polymerized with high salt and then decorated with meromyosin.

DR. ASO: We have been taught that in cell movement, microtubules play a significant role, though we don't know how they function. What is the difference between the function of microtubules and of actin in your system?

DR. McGUIRE: Actin is a part of the contractile machinery and is related to motility, cytokinesis and phagocytosis. Microtubules appear to be responsible for the directionality of these events, that is for their polarization, at least in regard to cell locomotion.

DR. TAKEBE: In your Table 2, I recall the amount of actin in epidermal cells of different animals varied from 4 to 18% of the total extractable protein. Do you have any explanation of that difference or can you correlate the amount of actin to the physiology of the cells? For example, do fast-moving cells have more actin?

DR. McGUIRE: This is a very important question for which I have no adequate answer at this time. We are analyzing and comparing actin in normal and psoriatic epidermis. Actin in confluent cultured keratinocytes is being compared in central and peripheral (spreading) areas. Preliminary evidence suggests in viable calf epidermis that there is more actin in the lower compared to the upper regions of the tissue. We don't have, as yet, a comparison between psoriatic and normal epidermis in this regard.

DR. FREEDBERG: Are there differences in the amount of actin in tissue culture cells during the time of culture? We have looked in a cell line which is of human epidermal origin and clearly there is actin but I felt that there should be more actin in younger cells if they are migrating across the dish. However, we did not find that. Have you found differences?

DR. McGUIRE: I would like to answer this in three parts. The first question: Can a cell retain its specific epidermal function at a time when it is making large amounts of actin? The answer to that is yes. Question No. 2: Do cells contain more or less actin depending on what they are asked to do? The only thing we can tell is that immediately after culture, that is, immediately after being cultured from viable epidermis, the amount of actin is increased over that in viable epidermis. The amount of actin continues increasing in culture up to about 2 months. So the amount of actin in cultivated cells increases. There is a third question and that is: If you wound the culture, i.e. if you have a monolayer and you wound the monolayer, do you find it easier to demonstrate actin nearer to the wound margin? The answer to that question is, yes, but are you looking at actin synthesis? Our data right now is not very good, but we believe that we are simply looking at the polymerization of previously synthesized actin present within the keratinocyte.

Cellular Envelope Changes during Differentiation of Human Epidermal Cells

Takae HIRONE and Yoshichika ERYU

Department of Dermatology, Kanazawa Medical University, Ishikawa, Japan

ABSTRACT

The ultrastructural changes of the cellular envelope during differentiation of epidermal cells were studied in disseminated epidermolytic acanthoma by electron microscopy. The cellular envelope of the basal and lower squamous cells consisted of only a trilaminar plasma membrane. In the upper squamous and granular cells, the outer leaflet of the plasma membrane was irregularly coated with the material derived from lamellar granules discharged into the intercellular space. In some of the uppermost 1–3 granular cells, the inner leaflet was thickened by the addition of the peripheral dense band without fusion from its inside. Consequently, the cellular envelope of the horny cells was composed of a plasma membrane, a peripheral dense band and a thin layer of dense material derived from lamellar granules, although the last one was a discontinuous layer. The evidence indicates that the peripheral dense band is formed not by the material from degraded organelles but by newly synthesized material in the cell, and that its development precedes the occurrence of other changes necessary for the transformation of the granular cell into the horny cell in some pathologic conditions, at least in disseminated epidermolytic acanthoma.

It is generally agreed that in mammalian epidermis cellular envelopes are thickened during transformation of epidermal cells into horny cells and become highly resistant to the action of kerationolytic agents.[1,2] Ultrastructural observations provided some evidence indicating that the transformation process is initiated in transition cells interposing between granular cells and horny cells.[3,4] However, the mechanism involved in the thickening has not yet been settled because transition cells are very few and usually appear as single cells and they are infrequently encountered under the electron microscope. In a previous study of disseminated epidermolytic acanthoma (DEA),[5] it was noted that cells with a thickened envelope were not infrequently encountered in the upper layer of the stratum granulosum. Consequently, this condition was considered to be suitable for studying the cellular envelope thickening during differentiation of epidermal cells.

The purpose of this study is to describe the ultrastructural changes of the cellular envelope which occur during differentiation of epidermal cells in DEA and to discuss the mechanism of the cellular envelope thickening.

MATERIALS AND METHODS

The biopsy specimens were obtained from a patient with DEA. Small blocks of the tissue specimen were fixed in ice-cold 2.5% glutaraldehyde in cacodylate buffer, pH 7.4, for 2 hours, washed several times in the same buffer, and then postfixed in ice-cold 2% osmium tetroxide in Veronal-acetate buffer, pH 7.4, for 2 hours. After dehydration in a graded series of acetone, all blocks were embedded in Epon 812. Thin sections were cut with glass knives on a Porter-Blum MT–2 or LKB III ultramicrotome, stained with uranyl acetate followed by lead citrate, and examined in a Hitachi HU–11E or HU–12A electron microscope. One micron sections were stained with Paragon reagent and examined by light microscopy to select the areas suitable for ultrastructural analysis.

RESULTS

General features of the epidermis in DEA were reported in a previous paper.[5] Striking changes were the development of abnormal aggregation and abnormal configuration of tonofilaments in the squamous and granular cells, the formation of extraordinarily large keratohyalin granules in the granular cells, and the occurrence of intracellular edema in the upper squamous and granular cells. The latter two changes were usually prominent in the interpapillary acanthotic regions of the epidermis. Furthermore, it was noted that the horny layer was composed of completely keratinized cells of flattened shape and incompletely keratinized cells of irregular shape. In this study, attention was mainly focused on the cellular envelopes of epidermal cells in the lesion of this condition.

The envelopes of all nonkeratinized cells, except for some cells in the uppermost 1–3 granular layers, were the same in structure. It was seen to be a trilaminar plasma membrane, 80–90 Å thick, consisting of dense inner and outer leaflets and a less dense central leaflet (Fig. 1).

In the uppermost 1–3 granular layers, the cells surrounded with a thickened envelope were not infrequently encountered (Figs. 2,3,7). The thick-

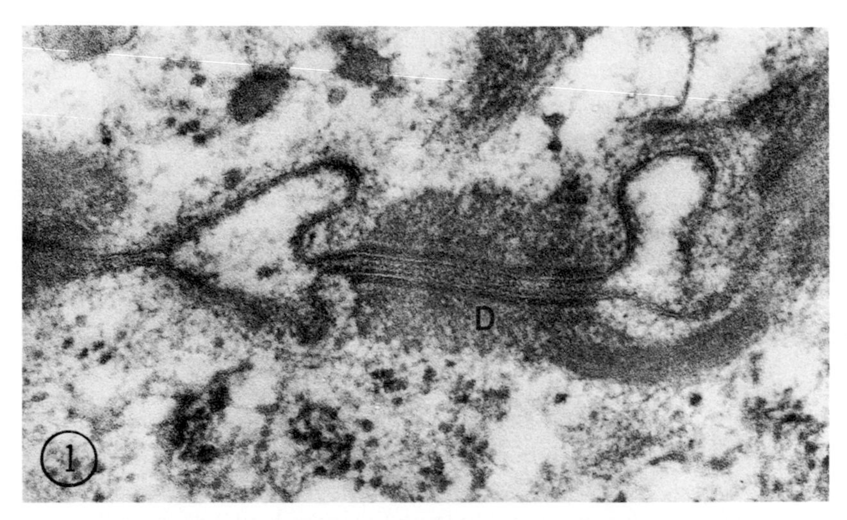

Fig. 1. Small area of two adjacent cells of the upper squamous layer. Trilaminar plasma membranes and a desmosome (D) are seen. ×53,600.

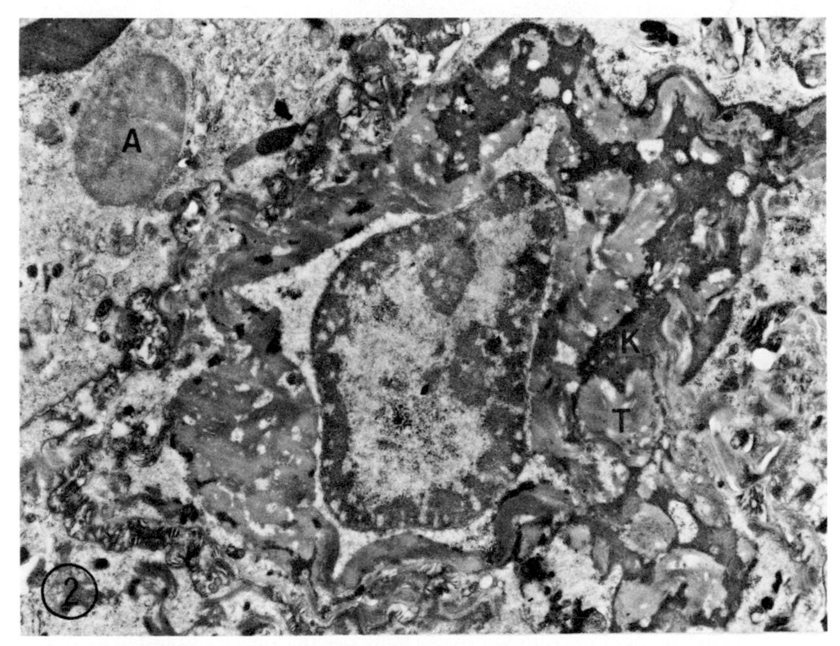

Fig. 2. Cell lying beneath the uppermost granular layer. Abnormal aggregates of tonofilaments (A) and perinuclear thickened bundles of tonofilaments (T) with partial deposition of keratohyalin substances (K) are seen in the cytoplasm. ×6,800.

ened cellular envelope consisted of a plasma membrane and a dense band running along the cytoplasmic face of the inner leaflet of the plasma membrane. The dense band, 100–160 Å thick, continued to the attachment plaques of desmosomes which had a similar thickness and electron-density, consequently forming a continuous dense layer in the entire cell periphery. This structure will be referred to as peripheral dense band according to Farbman,[6] who found a similar structure in tongue epithelium of rat embryo. The peripheral dense band was closely associated with the inner leaflet, so that the former was often indistinguishable from the latter.

In the areas where the thickened cellular envelope was apposed to the plasma membrane of adjacent cells, numerous lamellar granules were present in the interspace between them (Figs. 3,4). These granules showed varying degrees of dissociation into individual lamellae and some amorphous material. Dissociated lamellae lay perpendicular or parallel to the outer leaflet, which was separated from the inner leaflet or from the inner leaflet accompanied with the peripheral dense band, respectively, by the

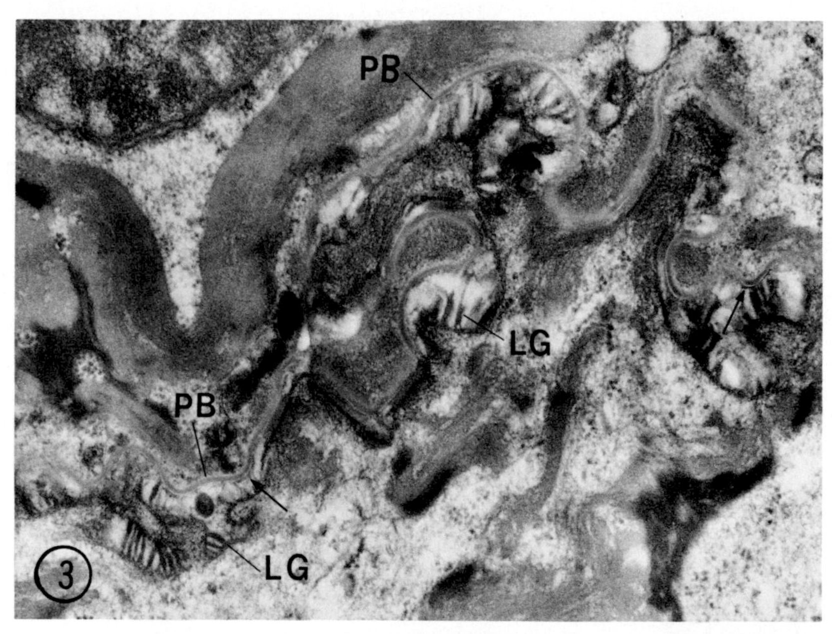

Fig. 3. Enlargement of part of Fig. 2. Dissociating lamellar granules (LG) are found in the intercellular space between the plasma membrane with a peripheral dense band (PB) of the upper cell and the plasma membrane of the lower cell. At arrows, the outer leaflet coated with some amorphous material is distinguished from the peripheral dense band (PB) by interposition of the less dense central leaflet. ×25,600.

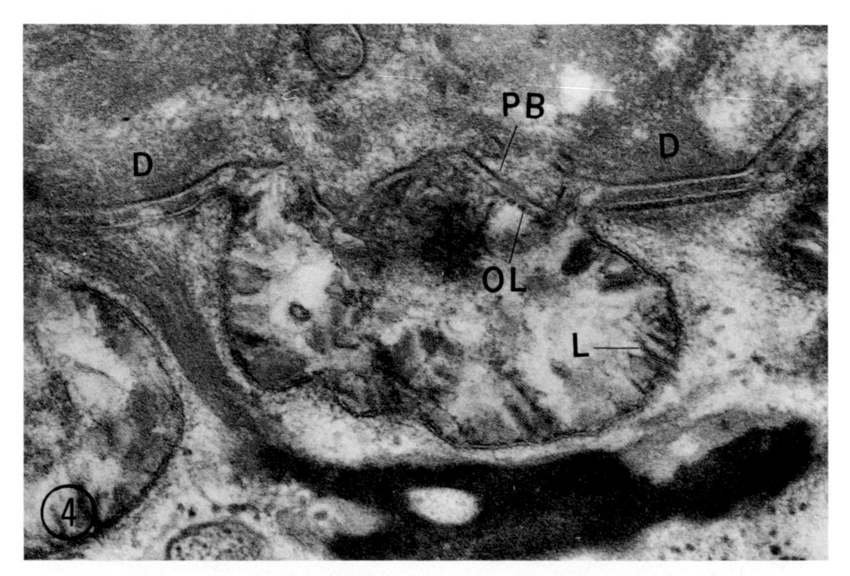

Fig. 4. Small area of two adjacent granular cells. The peripheral dense band (PB) is seen to be continuous with the attachment plaques of desmosomes (D). The outer leaflet (OL) coated with some amorphous material can be seen. Individual lamellae (L) lie perpendicular or parallel to the outer leaflet of the plasma membrane. ×48,000.

less dense central leaflet. Occasional images were encountered which suggested that the outer leaflet was irregularly coated by some lamellae and/or some amorphous material (Figs. 3,4). The same situation was seen in the areas where the thickened cellular envelopes of two adjacent granular cells were apposed each other (Figs. 7,8).

The peripheral dense band was usually seen in some granular cells without intracellular edema, but occasional images showed that it existed in other granular cells with marked intracellular edema (Figs. 5,6). In the latter cells, the cytoplasmic organelles were scanty and the cytoplasmic matrix was more electron-translucent; however, the peripheral dense band was persistent.

In the horny cell layer, the cellular envelopes became obscure. The peripheral dense band was usually indistinct because of the condensation of the filament-amorphous-matrix complex and because of the decrease of electron-density of the band itself (Fig. 9). In the uppermost 1–2 horny cells whose contents were disintegrated in varying degrees, however, the peripheral dense band was found to be a less dense zone running along the cytoplasmic face of the dense inner leaflet (Fig. 10). Although the outer leaflet was indiscernible, the discontinuous thin layer of dense material

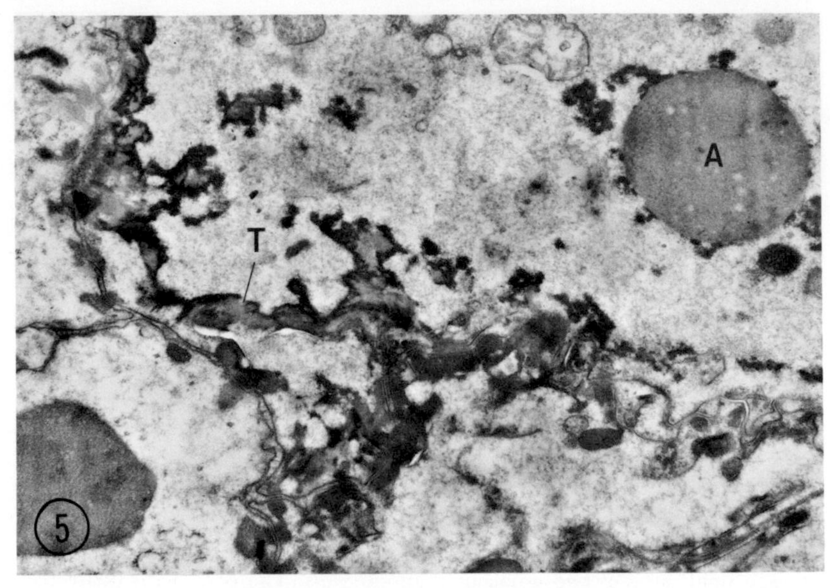

Fig. 5. Granular cells showing marked intracellular edema. Cell constituents are degraded except for abnormal aggregates of tonofilaments (A) and disarrayed bundles of tonofilaments (T). Cellular envelopes are intact. ×9,600.

existed, facing the dense inner leaflet across a less dense layer corresponding to the central leaflet.

DISCUSSION

The thickening of the cellular envelope in various keratinizing epithelia has been described and discussed. Matoltsy and Parakkal,[7,8] who studied mammalian epidermis and oral mucosa, suggested that lammellar granules are involved in the thickening and coating of the plasma membrane, a view that subsequently received support from Hashimoto *et al.*,[9] who studied human oral mucosa. In a study on rat tongue epithelium, Farbman[6] proposed that the plasma membrane thickening occurs primarily on the cytoplasmic side of the membrane and receives no substantial contribution from extracellular material derived from lamellar granules. He further stated that the thickening is due to the addition of the peripheral dense band from the inside of the membrane, a view that was favored by Brody[3,4] and Jessen.[10] In a recent study of isolated membranes of horny cells from various types of epidermis, Matoltsy[11] has noted the dense band of material consisting of 50 Å particles embedded in an amorphous substance and demonstrated their proteinaceous nature, which

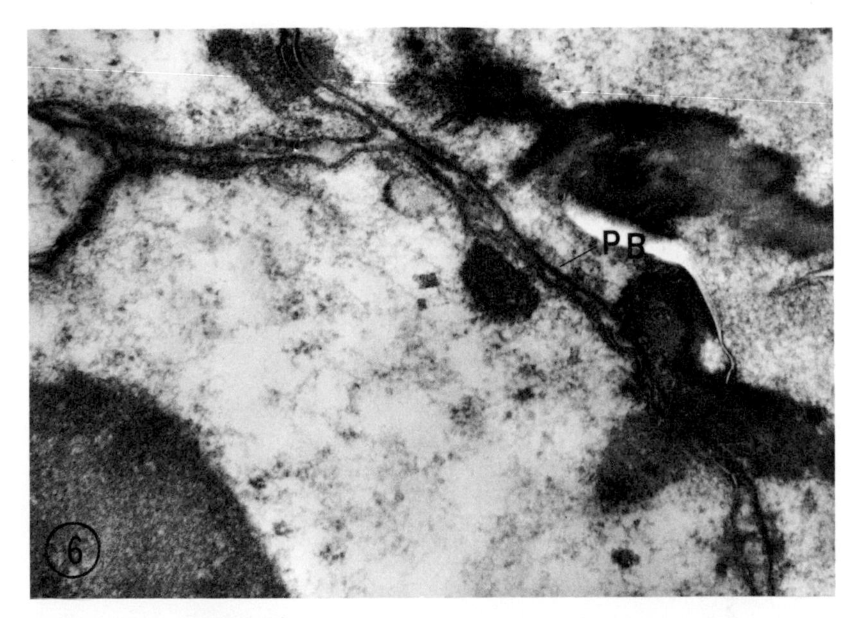

Fig. 6. Enlargement of part of Fig. 5. The peripheral dense band (PB) can be seen in the upper right cell. ×29,600.

had been suggested by Wolff and Schreiner[12] and Niebauer.[13] Frithiof,[14] who studied human oral epithelium, proposed that the plasma membrane is thickened by the addition of the dense band from its inside and the deposition of the intercellular dense material on its outside, together with the increased thickness of the central leaflet of the plasma membrane. The same view has been expressed by Martinez and Peters[15] and Raknerud,[16] although there are some differences among them.

Our findings substantially support the view of Frithiof.[14] The difference in our findings from Frithiof's findings is that there is no thickening of the central leaflet in the final stage of the transformation. It is also different from Martinez and Peters' and Raknerud's findings that the inner leaflet is thickened by the addition of the peripheral dense band without fusion, as indicated in the uppermost 1–2 horny cells, and that there is no formation of a dense layer delimiting the peripheral dense band from the interior of the cell in the final stage of the transformation. The sequential changes of the cellular envelope during differentiation of epidermal cells are summarized in Fig. 11. It should be remarked that the trilaminar plasma membrane of the epidermal cell is maintained throughout the epidermis, although the outer leaflet is indiscernible in the uppermost 1–2 cells of the horny layer. Our observations have shown that the thin layer of

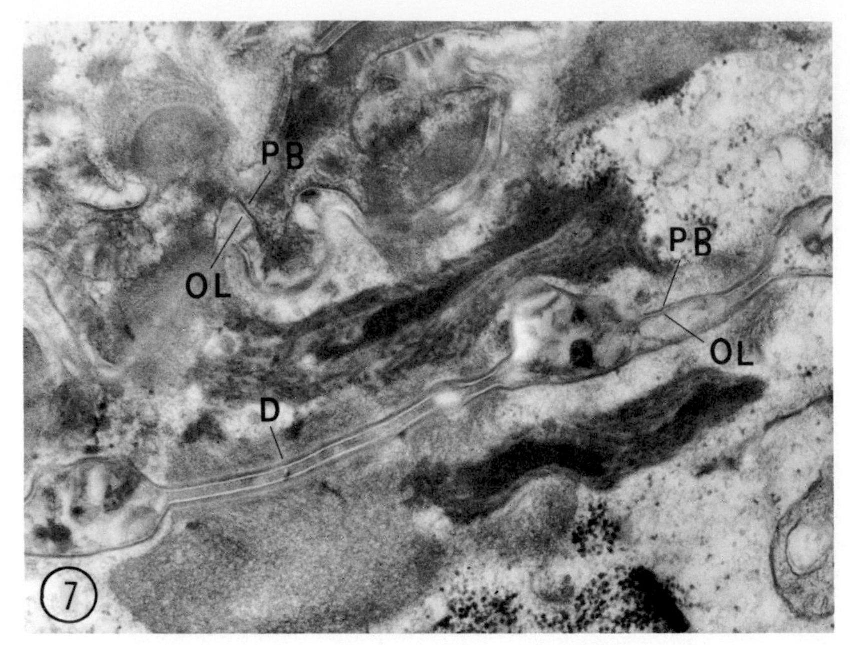

Fig. 7. Small area of the uppermost 2 and 3 cells of the granular layer. The peripheral dense band (PB) continuing to the attachment plaques of desmosomes (D) and the outer leaflet (OL) of the plasma membrane can be seen. Individual lamellae and some amorphous material are observed in the intercellular space. ×24,000.

dense material derived from lamellar granules is not continuous in the horny cell layer. The evidence suggests that lamellar granules are not sufficiently numerous to provide a complete coating to the plasma membranes of horny cells. It appears warranted that their main function is not to strengthen the plasma membrane but to modify the intercellular material. On the other hand, it has shown that the peripheral dense band remains unchanged in horny cells, even in desquamating cells whose contents have been disintegrated. The evidence indicates that the peripheral dense band is responsible for such toughness of the cellular envelope of the horny cell as suggested by Kligman.[17] Concerning the substructure of the dense band, we could not resolve it in this study.

The origin of the peripheral dense band has been discussed for some times. Snell,[18] who studied human epidermis, suggested that the thickened plasma membrane might be formed by the deposition of keratohyalin granules on the inside of the plasma membrane. In a study on rat tongue epithelium, Jessen[10] noted that small spherical keratohyalin granules were

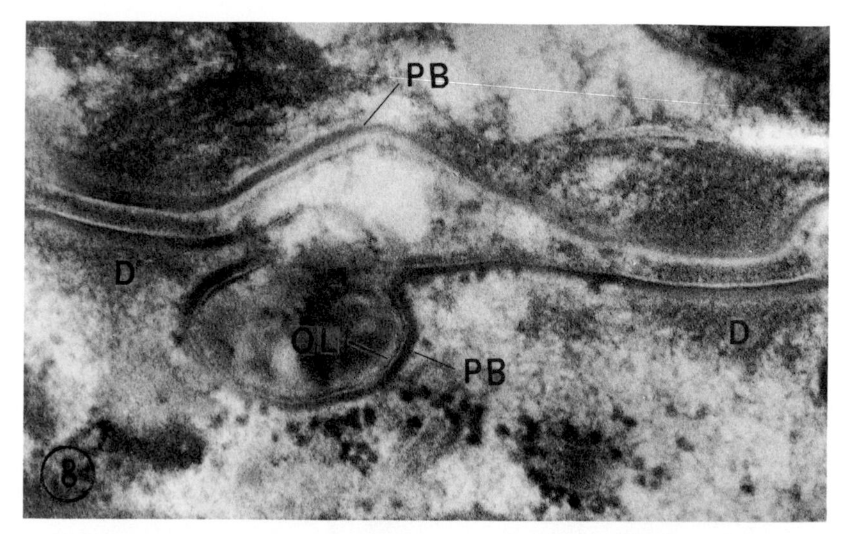

Fig. 8. High magnification of part of the uppermost 2 and 3 granular cells. The peripheral dense band (PB) is continuous with the attachment plaques of desmosomes (D). The outer leaflet (OL) coated with some amorphous material is separated from the peripheral dense band (PB) by the less dense central leaflet. ×62,400.

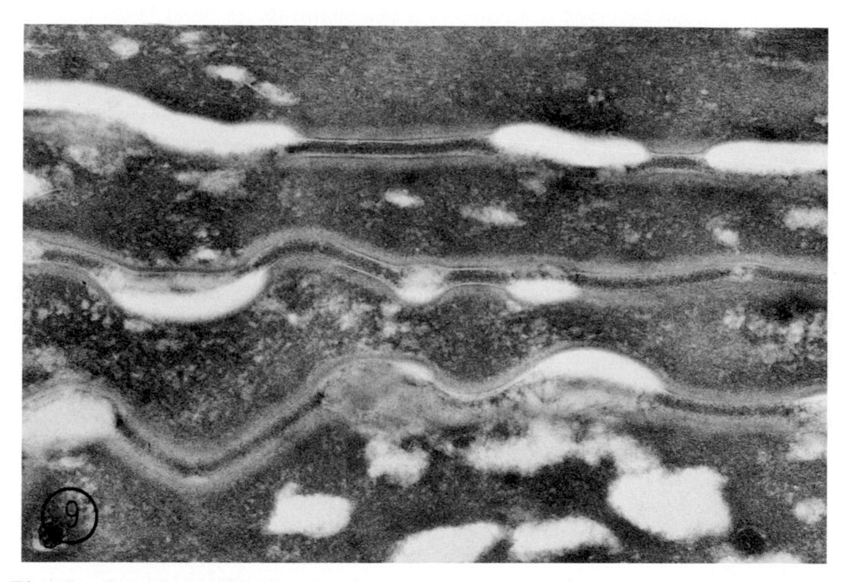

Fig. 9. Low magnification of completely keratinized cells. A few lamellae and some amorphous material are seen in the intercellular space. The peripheral dense band is indistinct because of the increased electron-density of the interior of the cell. ×42,400.

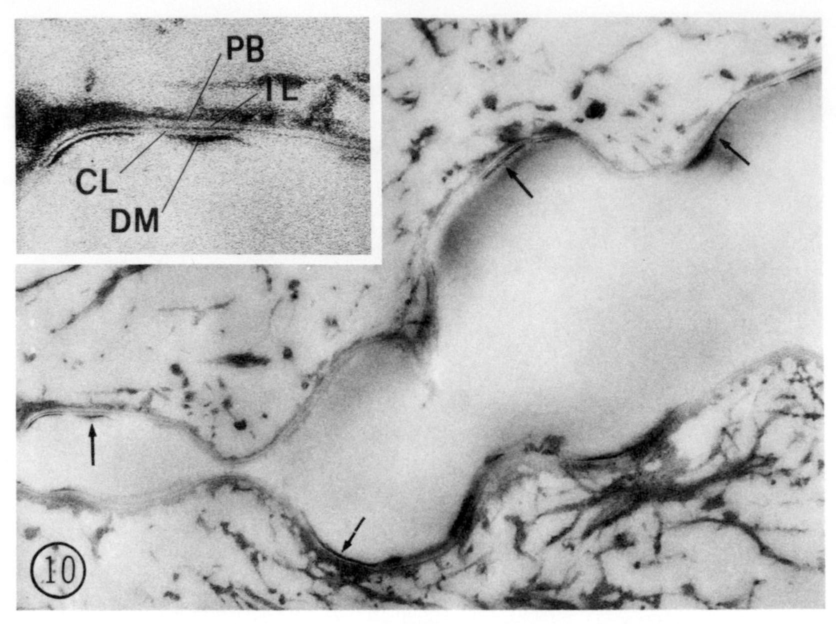

Fig. 10. Small area of two desquamating cells. A discontinuous thin layer of dense material is seen at arrows. ×38,400. Inset shows a part of the cellular envelope composed of a peripheral dense band (PB), a dense inner leaflet (IL), a less dense central leaflet (CL), and a thin layer of dense material (DM). ×86,400.

associated with the attachment plaques of desmosomes in the lowermost layer of the stratum granulosum, and suggested the relation of the granules with the thickening of the plasma membrane. This type of granule has not been observed in normal and pathologic human skin except for bullous congenital ichthyosiform erythroderma.[19] On the other hand, Hashimoto[20] has demonstrated that the peripheral dense band is formed in the absence of keratohyalin granules in the human toenail. Although no evidence is given in this study, it seems unlikely that the peripheral dense band is derived from keratohyalin substances.

Finally, it should be noted that the peripheral dense band was clearly demonstrated in the uppermost 1–3 granular cells with no degradation of cell organelles. The evidence indicates that the peripheral dense band is formed not by the material from degraded organelles but by newly synthesized material in the cell, and that its development precedes the occurrence of other changes necessary for the transformation of the granular cell into the horny cell in some pathologic conditions, at least in DEA. The presence of the peripheral dense band in the granular cells with marked

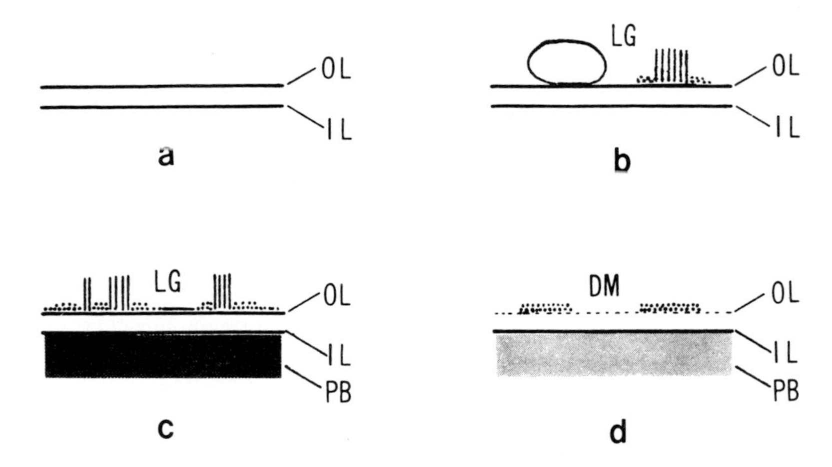

Fig. 11 a-d. Diagrammatic representation of cellular envelope changes during differentiation of the epidermal cells in DEA. (a) Cellular envelope, consisting of only a trilaminar plasma membrane, in basal or lower squamous cell. OL, outer leaflet of the plasma membrane. IL, inner leaflet of the plasma membrane. (b) Cellular envelope in upper squamous or granular cells except some cells in the uppermost 1–3 granular layer. Lamellar granules (LG) discharged from the cell are degraded in the extracellular space. (c) Cellular envelope, consisting of a trilaminar plasma membrane and a peripheral dense band (PB), in some of the uppermost 1–3 granular cells. The peripheral dense band is formed which is closely associated with the inner leaflet of the plasma membrane. Degraded lamellar granules are deposited on the outer leaflet of the plasma membrane. (d) Cellular envelope in completely keratinized horny cells. The peripheral dense band is retained, but becomes less dense. A discontinuous thin layer of dense material (DM) derived from lamellar granules is formed in the outside of the plasma membrane.

intracellular edema can be explained by suggesting that the development of the dense band precedes the occurrence of intracellular edema in the cells.

REFERENCES

1. Odland, G. F. and Reed, T. H., Epidermis. In Ultrastructure of Normal and Abnormal Skin (A. S. Zelickson, ed.), Lea & Febiger, Philadelphia, 1967, pp. 54–75.
2. Matoltsy, A. G. and Parakkal, P. F., Keratinization. In Ultrastructure of Normal and Abnormal Skin (A. S. Zelickson, ed.), Lea & Febiger, Philadelphia, 1967, pp. 76–104.

3. Brody, I., An ultrastructural study on the role of the keratohyalin granules in the keratinization process. *J. Ultrastruct. Res.,* **3**: 84–104, 1959.
4. Brody, I., The modified plasma membranes of the transition and horny cells in normal human epidermis as revealed by electron microscopy. *Acta Dermatovener.,* **49**: 128–138, 1969.
5. Hirone, T. and Fukushiro, R., Disseminated epidermolytic acanthoma: Non-systematized multiple verrucoid lesions showing granular degeneration. *Acta Dermatovener.,* **53**: 393–402, 1973.
6. Farbman, A. I., Plasma membrane changes during keratinization. *Anat. Rec.,* **156**: 269–282, 1966.
7. Matoltsy, A. G., and Parakkal, P. F., Membrane-coating granules of keratinizing epithelia. *J. Cell Biol.,* **24**: 297–307, 1965.
8. Matoltsy, A. G., Membrane-coating granules of the epidermis. *J. Ultrastruct. Res.,* **15**: 510–515, 1966.
9. Hashimoto, K., DiBella, R. J., and Shklar, G., Electron microscopic studies of the normal human buccal mucosa. *J. Invest. Derm.,* **47**: 512–525, 1966.
10. Jessen, H., Two types of keratohyalin granules. *J. Ultrastruct. Res.,* **33**: 95–115, 1970.
11. Matoltsy, A. G., The substructure and chemical nature of the horny cell membrane. *J. Invest. Derm.,* **62**: 343, 1974.
12. Wolff, K. and Schreiner, E., Differential enzymatic digestion of cytoplasmic components of keratinocytes: Electron microscopic observations. *J. Ultrastruct. Res.,* **36**: 437–454, 1971.
13. Niebauer, G., Enzymatic digestion of epon-embedded ultrathin sectioned tissue. In Proceedings of 14th International Congress of Dermatology, Excerpta Medica, 1974, pp. 457–460.
14. Frithiof, L., Ultrastructural changes in the plasma membrane in human oral epithelium. *J. Ultrastruct. Res.,* **32**: 1–17, 1970.
15. Martinez, I. R., Jr. and Peters, A., Membrane-coating granules and membrane modifications of keratinizing epithelia. *Am. J. Anat.,* **130**: 93–119, 1971.
16. Raknerud, N., The ultrastructure of the interfollicular epidermis of the hairless (hr/hr) mouse. II. Plasma membrane modifications during keratinization. *Virchows Arch. B Cell Path.,* **17**: 113–135, 1974.
17. Kligman, A. M., The biology of the stratum corneum. In The Epidermis (W. Montagna and W. C. Lobitz, Jr., eds.), Academic Press, New York, 1964, pp. 387–433.
18. Snell, R., An electron microscopic study of the human epidermal keratinocyte. *Z. Zellforsch. Mikroskop. Anat.,* **79**: 492–506, 1966.
19. Suzuki, H., Kurosumi, K., and Miyata, C., Electron microscopy of spherical keratohyalin granules. *J. Invest. Derm.,* **60**: 219–223, 1973.
20. Hashimoto, K., Ultrastructure of the human toenail. II. Keratinization and formation of the marginal band. *J. Ultrastruct. Res.,* **36**: 319–410, 1971.

The Membrane of Horny Cells*

A. Gedeon MATOLTSY

Department of Dermatology, Boston University School of Medicine, Boston, Massachusetts, U.S.A.

ABSTRACT

The membrane of epidermal horny cells was studied in an effort to advance our knowledge on the protective system of vertebrates. With the exception of teleost, representatives of vertebrate classes, such as the cow, chick, and frog, were found covered by horny cells with a triple-layered membrane consisting of a relatively thin outer and mid layer and a 70 to 150 Å thick inner layer. The inner layer contains 40 to 60 Å particles embedded into an amorphous substance and is anchored to the filament-matrix complex filling the horny cells. It can be separated from the outer and mid layers of the horny cell membrane and the filament-matrix complex by the use of 0.1 N NaOH without dissociation of its substructure. The isolated inner layer is elastic and highly resistant. It is not solubilized by denaturing, reducing, and chelating agents, acids, and organic solvents. Amino acid analyses indicate that in various vertebrate classes different proteins participate in its formation. It is proposed that the inner layer of the horny cell membrane together with the filament-matrix complex provide the structural basis of the protective system of both terrestrial and amphibious vertebrates.

INTRODUCTION

It is a characteristic feature of keratinizing epithelia that cornified cells are enveloped by a membrane that is about twice as thick as the plasma membrane of nonkeratinized cells. Thickening of the plasma membrane of differentiating epithelial cells has been studied in detail by Farbman[1] in the rat tongue epithelium. He found that at an advanced stage of differen-

* This investigation was supported by research grant AM05924 from the National Institute of Arthritis and Metabolic Diseases, United States Public Health Service.

tiation, when the epithelial cells still have an intact nucleus and complement of cell organelles, a moderately dense band of material appears in the peripheral cytoplasm, next to the plasma membrane. The density of this band increases and becomes identical to that of the inner leaflet. Subsequently the peripheral band and the inner leaflet are no longer distinguishable as they form a broad 100–150 Å thick layer. The mid and outer leaflets of the plasma membrane do not thicken: they remain unaltered. The outer leaflet of fully keratinized cells is fragmented and sheds into the intercellular spaces. Thickened horny cell membranes with a comparable structure have been noted in the epidermis of mammals,[2–5] aves,[6] reptiles,[7,8] and amphibia,[9,10] and the mammalian cervical,[11] esophageal,[12] and oral epithelium.[13–17]

While considerable information is available on structural features of vertebrate horny cell membranes, little is known about their chemical nature. Attention has been focused on the horny cell envelope after it was noted that it resists the action of keratinolytic agents such as 1% sodium sulfide,[18] alkaline protease subtilisin,[19] etc. We noted several years ago that the content of human epidermal horny cells is completely solubilized by 0.1 N sodium hydroxide.[20] The remaining material appears in the light microscope as ghosts of horny cells[20] and in the electron microscope as wrinkled membranes.[21] Chemical studies indicated that the isolated material consists of a protein-lipid-carbohydrate complex.[22] The protein component forms about two-thirds of the membranes and contains half-cytine and proline residues in a relatively large amount.[23]

Presently it is not known whether the avian, reptilian and amphibian horny cell membranes also resist the action of keratinolytic agents and are capable of providing protection for the horny cell content. In this study the structure and chemical nature of bovine, chick, and frog horny cell membranes were compared in an effort to learn more about the protective system of vertebrates.

MATERIAL AND METHODS

Cow's noses were obtained fresh and chilled from a local abbatoir. The upper part of the epidermis was removed in 0.5 mm slices with a Storz keratotome. The feathers were plucked from newborn chicks and subsequently the entire trunk skin was removed. The unpigmented area of the abdominal skin from frogs (*Rana pipiens*) was used.

For electron microscope investigation the skin samples were cut to small pieces under a dissecting microscope and fixed for 1 hour in chilled 1% osmium tetroxide buffered to pH 7.2 with veronal acetate. Subsequently the skin fragments were dehydrated in an ascending series of ethanol and embedded in Epon 812. Thin sections were cut with glass knives using the Porter-Blum ultramicrotome and stained with uranyl acetate and lead hydroxide. The specimens were studied in an RCA-EMU-3F or Philips 300 electron microscope.

For chemical studies bovine epidermis was obtained as described above. The epidermis of chick and frog was separated from the dermis as follows: The entire trunk skin of chick and abdominal skin of frogs was placed on a filter paper and immersed into a 2 M sodium bromide solution for 1 hour at room temperature. The epidermis was then separated in large sheets under a dissecting microscope. Isolated epidermis was washed three times with large amounts of distilled water and cut to small pieces with scissors.

Horny cell membranes were isolated by the "alkali" method established earlier for isolation of human horny cell membranes.[23] Epidermal fragments obtained from 10 cows' noses or 20 newborn chicken or 10 frogs were suspended in 100 ml of 0.1 N NaOH solution. The suspension was shaken with glass beads at room temperature for 24 hours. After the removal of glass beads the suspension was centrifuged at 3500 rpm for 10 minutes. The sediment was resuspended in fresh 0.1 N NaOH and shaken with glass beads for a second 24 hours. At the end of this period the remaining large tissue fragments were removed by centrifugation at 600 rpm for 5 minutes and discarded. The resulting supernatant was centrifuged at 3500 rpm for 10 minutes to sediment the horny cell membranes. To remove the remaining contaminants the membranes isolated from the frog epidermis were washed three times with alkali by resuspension and centrifugation. The same procedure was used for purification of cow membranes, except that the alkali treatment was extended 4–5 days and for chick membranes 8–10 days, changing the reagent every 24 hours. The final pellets were checked in the phase contrast and electron microscope, washed with large amounts of distilled water to remove alkali, and then stored in a cold room.

For electron microscope examination the membranes were spread on carbon coated grids and stained with 1% phosphotungstic acid (PTA) in phosphate buffer pH 7.2 or lead hydroxide. For thin sectioning preparations were pelletted by centrifugation at 20,000 rpm for 1 hour and processed as described above.

Solubilization of bovine membranes was attempted by treatment of samples with 6 M urea or 1–5% sodium dodecyl sulfate in TRIS buffers ranging from pH 7.2–9.8 with or without the addition of 1–2% 2-mercaptoethanol, 0.04 M dithiothreitol, or 1–10 mM ethylenediamine tetraacetic acid at 20–40°C for 24–72 hours. Some samples were also treated with concentrated acetic acid, or 0.25 M oxalic acid at 20°C for 24 hours and also at 100°C for 2 hours, or with 0.01 N HCl at 100°C for 24 hours. Concentrated pyridine and dimethylsulfoxide were also used at 20°C for 24 hours.

For amino acid analyses membrane preparations were dried to constant weight and 2 mg aliquots suspended in 3 ml concentrated HCl for 17 hours at room temperature. Subsequently HCl was diluted to 6 N with distilled water. The tubes were sealed under nitrogen and the samples hydrolyzed at 110°C for 48 or 72 hours. Amino acids were determined by automated column chromatography with the Technicon Auto Analyzer. Sulfur containing amino acids were separately estimated using performic acid oxidation prior to hydrolysis.

RESULTS

THE STRUCTURE OF HORNY CELL MEMBRANES *IN SITU*

The stratum corneum of bovine epidermis consists of loosely packed horny cells with abundant microvillous extensions protruding into relatively wide intercellular spaces. The membrane of these cells is highly convoluted and generally three distinct layers are seen. The inner layer is 100–150 Å thick and is electron dense. Both the electron-translucent mid layer and the electron-dense outer layer are 20–30 Å thick (Fig. 1).

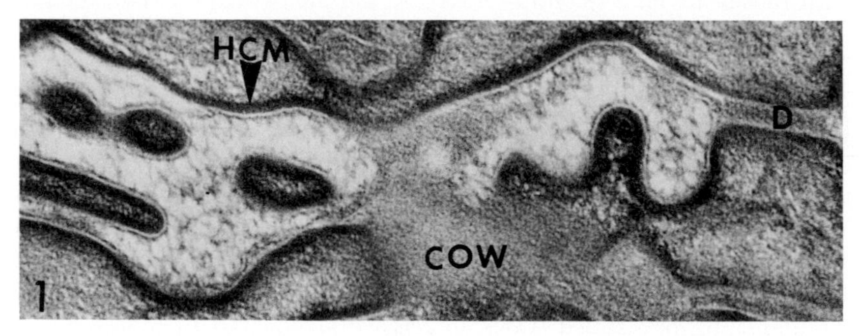

Fig. 1. Portion of the stratum corneum of the cow's nose epidermis. Convoluted membranes of horny cells consist of a thick inner layer and thin mid and outer layers (HCM). Filaments are anchored to the inner layer and desmosomes (D) connect adjacent cells. The intercellular spaces are relatively wide and filled with a fuzzy substance. ×83,600.

The horny layer of the chick epidermis is composed of strongly flattened and tightly packed cells. The intercellular spaces are very narrow. The membrane of the cells is smooth; microvillous extensions are very few or absent. The 70–100 Å thick inner layer is electron dense, the 15–25 Å mid layer is electron translucent, and the 15–20 Å outer layer is electron dense (Fig. 2).

In the stratum corneum of the frog epidermis, the intercellular spaces are wider than in the chick epidermis and the highly convoluted cell membranes interdigitate at several points. The 150–180 Å thick inner layer is electron dense, the about 30 Å mid layer is electron translucent, and the 30–35 Å outer layer is electron dense (Fig. 3).

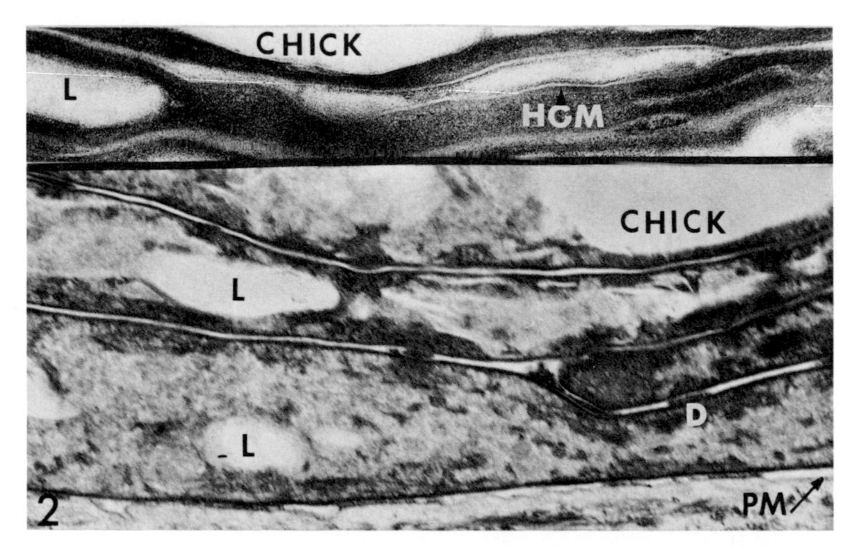

Fig. 2. Composite electron micrograph of the stratum corneum of the chick epidermis. Horny cells are enveloped by a thickened membrane and narrow intercellular spaces separate the strongly flattened cells. The upper micrograph ($\times 170{,}000$) shows well-preserved thick inner layer and thin mid and outer layers of the horny cell membrane (HCM). Relatively thin plasma membrane of a nonkeratinized cell (PM) is shown in the lower micrograph ($\times 38{,}600$). Large empty areas correspond to the site of lipid droplets (L) that were removed during tissue processing. Horny cells are connected by few desmosomes (D).

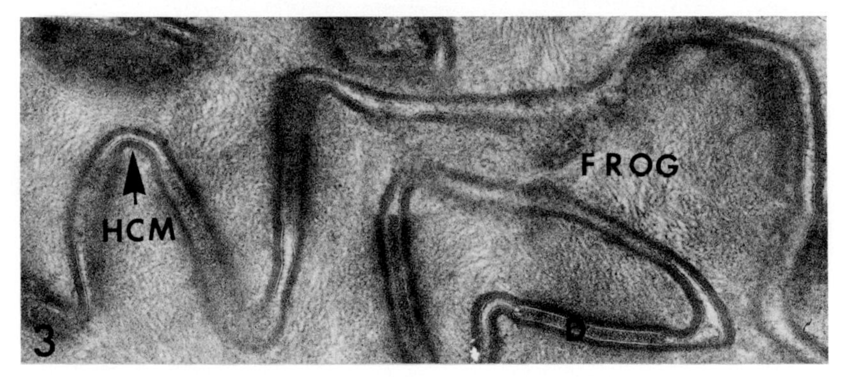

Fig. 3. Portion of the stratum corneum of the frog epidermis. The thick inner layer and thin mid and outer layers of the horny cell membranes (HCM) are well preserved in the marked areas. Filaments adjoin the convoluted horny cell membranes and numerous desmosomes (D) connect neighboring cells. The intercellular spaces are relatively narrow. $\times 52{,}800$.

It is a common feature of bovine, chick, and frog horny cell membranes that the outer and mid layers are continuous with the outer and mid leaflets of the plasma membrane seen in the desmosomal junction while the thickened inner layer fuses with the desmosomal plaque. The inner leaflet of the plasma membrane proper is only occasionally seen on the surface of the inner layer of the horny cell membrane (Fig. 4), or in the desmosomal junction. At high magnification the thickened inner layer of the horny cell membrane reveals a granular substructure (Figs. 4–7). Electron dense particles ranging in diameter from 40–60 Å can be seen

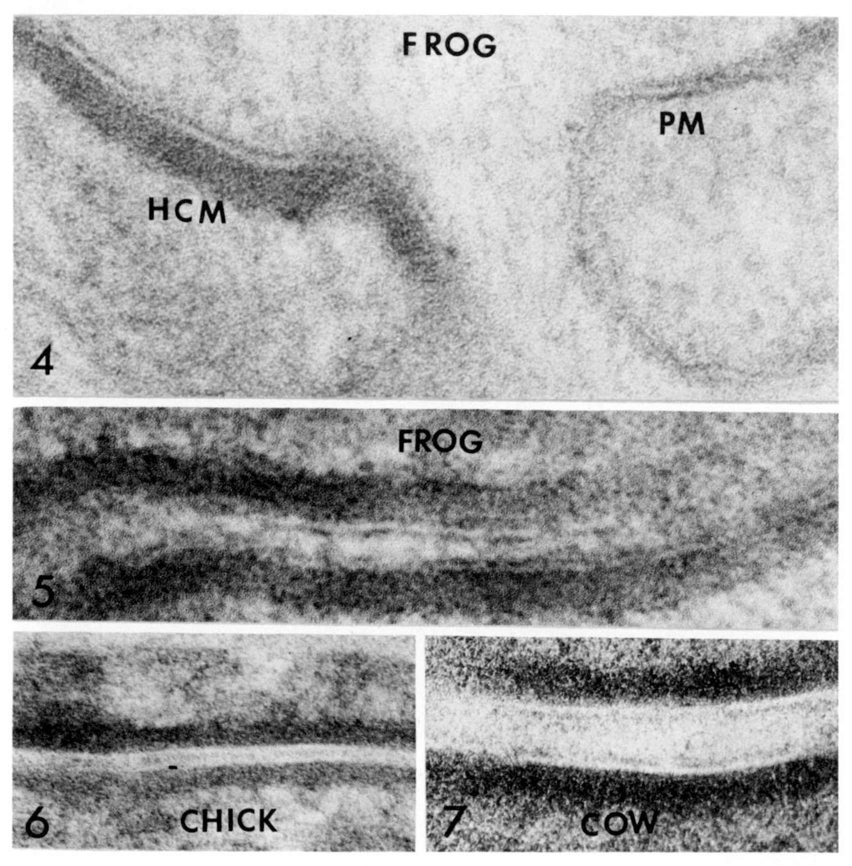

Fig. 4. The membrane of a horny cell (HCM) and the plasma membrane (PM) of an adjacent nonkeratinized cell of the frog epidermis are shown side by side. ×388,000.

Figs. 5, 6, and 7. A granular substructure is demonstrated in the inner layer of horny cell membranes of the frog, chick, and cow epidermis. Fig. 5 × 258,000, Fig. 6 ×180,000, and Fig. 7 ×240,000.

scattered and embedded into an electron-translucent amorphous substance. Cytoplasmic filaments (tonofilaments) are embedded into or terminate at the peripheral part of the thickened inner layer.

THE STRUCTURE OF ISOLATED HORNY CELL MEMBRANES

When isolated horny cell membranes of the cow or frog are suspended in distilled water and examined in the phase contrast microscope, the membranes appear flattened and assume various configurations when pressure is applied on the cover slip. On release of the pressure they regain their original shape. This procedure can be repeated several times without causing any damage or permanent change in the membranes. One other characteristic feature of these membranes is that they reveal a "dotted" surface under the phase contrast microscope. The "dots," that is, dense regions, show variable shapes and sizes, and are irregularly distributed (Figs. 8 and 9). Detailed electron microscope examination of the dense

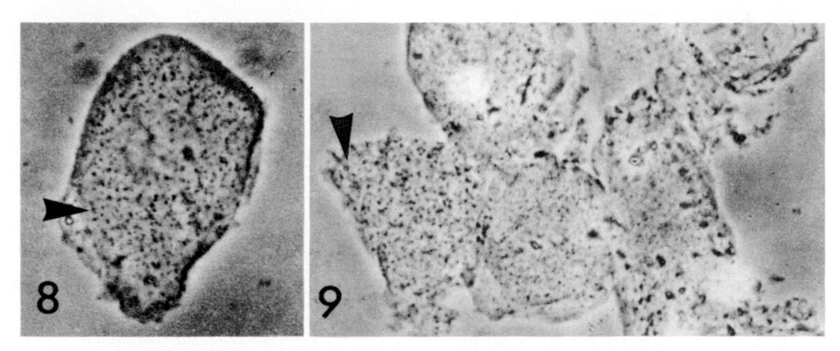

Fig. 8. Photomicrograph of a cell membrane appearing as a ghost of a cow horny cell. Note "dotted" surface structure (arrow). ×150.

Fig. 9. Photomicrograph of horny cell membrane fragments of the cow. Note "dotted" surface structure (arrow). ×140.

regions reveals fingerlike projections which are folded over flat membrane areas (Figs. 10,11,13). Chick membranes differ significantly since they do not show the dotted structure in the phase contrast microscope and have no fingerlike projections. They are seen as smooth sheets in the electron microscope (Fig. 12).

Bovine, chick, and frog membranes spread on carbon-coated grids reveal a granular substructure that can be seen only in specimens stained with lead hydroxide (Figs. 14, 15, 16). This structure is not seen in PTA stained preparations. The particles range in size from 40–60 Å and are embedded in a less electron-dense amorphous material.

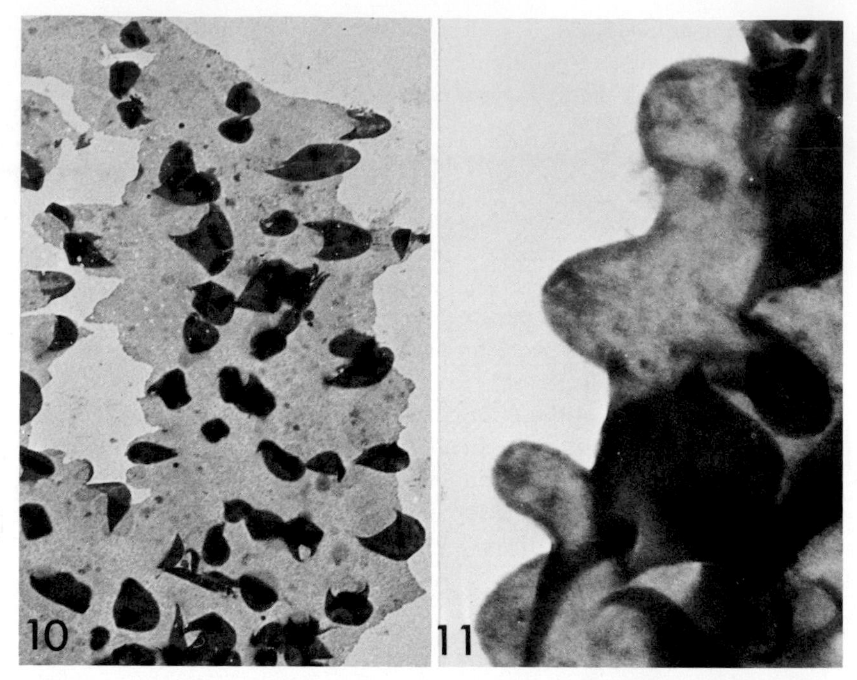

Fig. 10. Fragment of a horny cell membrane of the cow spread on carbon-coated grid and stained with PTA. Dense areas correspond to ''dots'' shown in Fig. 9. ×4,400.

Fig. 11. Fragment of a horny cell membrane of the cow is shown at a higher magnification. Note fingerlike projections at the edge of the fragment. Dense regions correspond to projections folded over the smooth parts of the membrane. The fragment was spread on carbon-coated grid and stained with PTA. ×10,600.

In thin sections, the membranes closely resemble their original configurations. Bovine and frog membranes are highly convoluted (Figs. 17 and 19); chick membranes are smooth (Fig. 18). The significant difference between isolated and *in situ* membranes is that the isolated membranes consist of a single layer. Apparently the outer and mid layers are solubilized and lost during the isolation procedure and only the inner, thickened layer resists the action of 0.1 N NaOH. The granular substructure of the inner layer remains intact (Fig. 20), and comparable to that seen *in situ* (Fig. 5).

THE CHEMICAL PROPERTIES OF ISOLATED HORNY CELL MEMBRANES

Samples of bovine membranes were selected for treatment with various denaturing, reducing, chelating agents, acids, and organic solvents in an

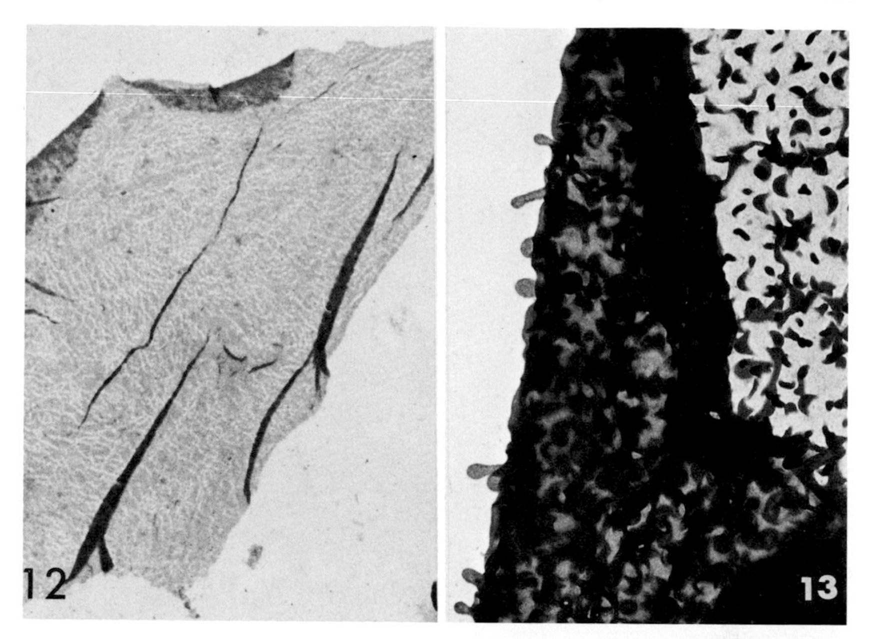

Fig. 12. Membrane fragment of chick horny cell spread on a carbon-coated grid and stained with PTA. Note that the membrane is smooth; projections are not present. ×10,400.

Fig. 13. Fragment of a rog horny cell membrane spread on a carbon-coated grid and stained with PTA. On the left side of the picture, membrane projections lie flat over the supporting film. On the right, the dense areas correspond to projections folded over the flat parts of the membrane. ×7,200.

effort to solubilize and characterize their components. The reagents used in these studies are listed in "Materials and Methods." Since none of them dissociate the membranes it has not been possible to isolate their components. Partial degradation takes place during a 24 hour hydrolysis in 6 N HCl. Complete degradation can be achieved only by pretreatment with concentrated HCl followed by hydrolysis in 6 N HCl for 48 to 72 hours.

The amino acid composition of the membrane proteins from cow, chick, and frog are shown in Table 1 together with data obtained in a previous study of human membrane protein.[23] It can be seen that the amino acid composition of mammalian membrane proteins (human and cow) is similar; each contains proline and half-cystine residues in a relatively large amount. However, the amino acid composition of chick and frog membrane proteins is significantly different. It is noteworthy that chick membranes are rich in glycine residues while relatively large amounts of aspartic acid and lysine residues are present in frog membranes.

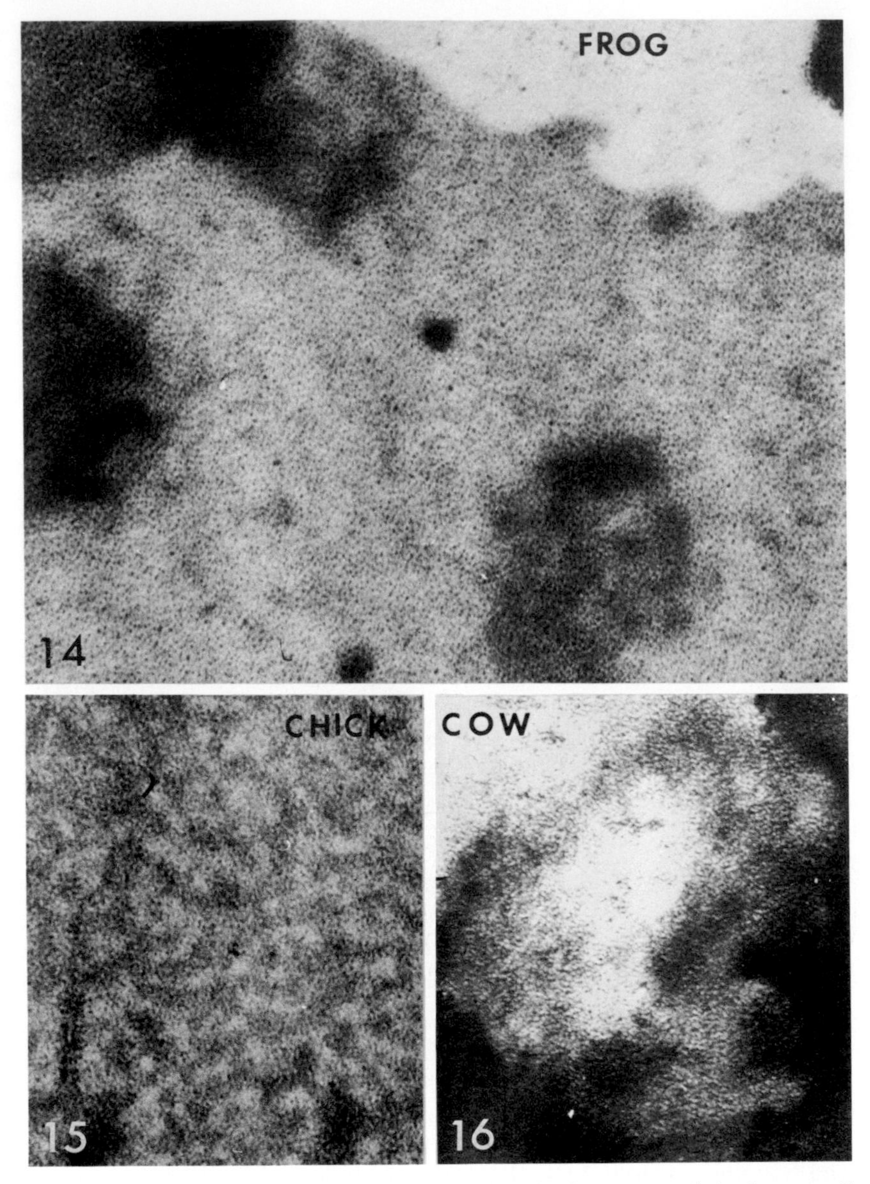

Figs. 14, 15, and 16. A granular substructure is demonstrated in horny cell membranes isolated from the frog, chick, and cow epidermis. The membranes were spread on carbon-coated grids and stained with lead hydroxide. Fig. 14 ×82,000, Fig. 15 ×85,000, and Fig. 16 ×80,000.

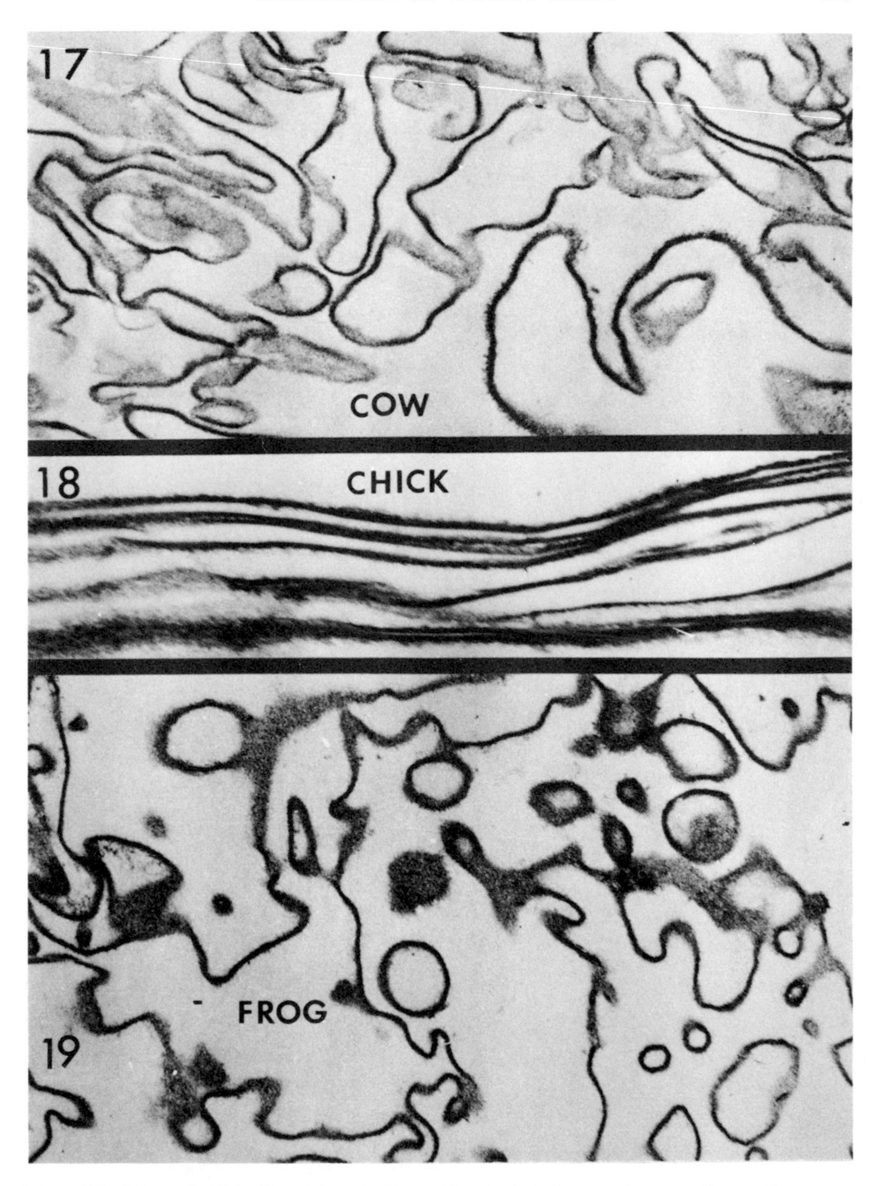

Figs. 17, 18, and 19. In thin section, the isolated membranes from the cow, chick, and frog epidermis closely resemble their original configurations as shown in Figs. 1, 2, and 3. Note that each membrane consists of a single layer corresponding to the thickened inner layer of the horny cell membrane. The mid and outer layers of the horny cell membranes, seen *in situ*, are absent. Fig. 17 ×40,200, Fig. 18 ×45,000, and Fig. 19 ×43,000.

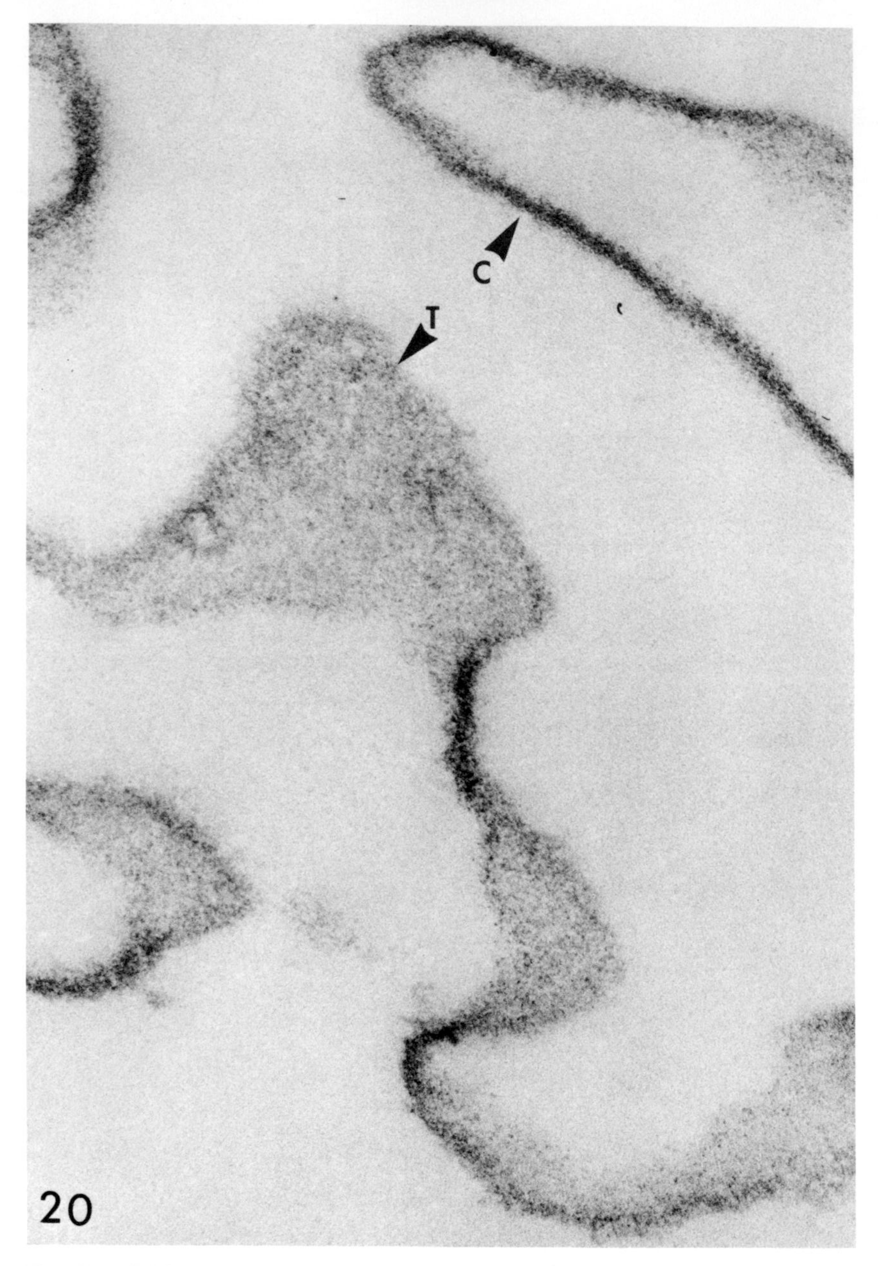

Fig. 20. Thin section is shown of the inner layer of the horny cell membrane isolated from the cow's epidermis. A granular substructure is seen in both cross (C) and tangential (T) sections of the inner layer. ×330,900.

Table 1. Amino acid composition of isolated horny cell membranes.

	Human (23)	Cow	Chick	Frog
Cys. acid	49	32	17	18
Asp.	58	59	44	104
Thr.	37	41	26	47
Ser.	74	67	92	44
Glu.	140	172	115	157
Pro.	137	125	49	83
Gly.	141	141	352	90
Ala.	52	43	32	64
Val.	52	59	38	68
Isoleu.	33	28	64	45
Leu.	58	53	38	70
Tyr.	11	13	9	11
Phe.	23	18	17	26
Lys.	66	75	50	106
His.	21	24	30	29
Arg.	46	50	31	41

48 h. hydr. Res. per 1000 res.

DISCUSSION

It is well known that the stratum corneum of the epidermis represents an important interface between the external and the internal environments of vertebrates and its integrity is essential for survival. The protective function of the stratum corneum, however, has not been satisfactorily explained since its components have not been characterized in detail. The protective system within the stratum corneum became better understood after the epidermal differentiation products were identified in various vertebrates and the transformation of differentiated cells into horny cells was elucidated. Presently it is thought that cell organelles are selectively degraded and partially or fully eliminated from the transforming cells and the remaining differentiation products associate into a fibrous-amorphous complex generally referred to as keratin.[24] Since the fibrous-amorphous complex is firmly stabilized and insoluble in usual solvents of proteins it has been held primarily responsible for protection of the vertebrate skin.

This study demonstrates that epidermal cells form another insoluble product besides keratin. The most characteristic feature of this differentiation product is that is surrounds the horny cell content as a 70–150 Å layer and has a much higher chemical resistance than keratin. Since this layer is anchored to the filament-matrix complex it is apparent that these three structural components of the horny cell form a functional unit. The matrix provides permanent orientation of filaments and the filaments assure flexibility and elastic recovery of the horny cell. Since the 70–150 Å layer is capable of reassuming its original shape after distortion it is reasonable to postulate that it consists of an elastic substance. This substance allows the filament-matrix complex to respond to applied force without disruption of the interconnecting junctions, and its unusally high chemical resistance assures integrity of keratin. Thus this three-component system protects the skin and inhibits passage of harmful agents into the organism. The outer and mid layers of the horny cell membrane do not seem to play an important role in protection as they have a much lower chemical resistance than the inner layer.

Electron microscopy has shown that the resistant layer that envelops keratin consists of two components: 40–60 Å particles and an amorphous substance. The origin of these components is not known. It is possible that they are formed during an advanced stage of differentiation and remain dispersed in the cytoplasm until the onset of the transformation process. The major role of the plasma membrane proper is presumably to provide a template in the transforming cell for formation of a thickened inner layer that ultimately encases keratin. Structural stability and chemical resistance of this layer most probably is related to extensive cross-linking of polypeptide chains by some unknown, highly stable bond. Hydrogen and disulfide bonds do not seem to play a primary role in stabilization of the proteins, as the isolated membraneous material cannot be dissociated by reagents that split these chemical bonds. The amino acid analyses suggest that the proteins participating in the formation of the resistant inner layer are similar in mammals and different in other vertebrates.

In an effort to include one representative of each vertebrate class in this study, horny cell membranes were also isolated from shed snake skin (Boa constrictor). So far it has not been possible to separate completely the content of the horny cell from the encasing membrane even after treatment with alkali for several weeks. Adequately purified membrane preparations were not obtained in sufficient quantity from reptilian skin; therefore the data of biochemical studies are not presented in this paper. Fish tail epidermis was separated from the dermis and treated with alkali. This tissue was completely solubilized in a very short time, thus indicating that resistant membranes are not formed by the teleost epidermis. This finding is in agreement with our previous studies which have shown that the teleost epidermis does not keratinize.[25]

The observations made in this study thus lead to the conclusion that, with the exception of teleost, a similar protective system is formed in vertebrates consisting of (1) filaments, (2) amorphous matrix, and (3) highly resistant and elastic membranes. The proteins that take part in the formation of this system presumably are different in various vertebrate classes.

Acknowledgement

The expert assistance of Ms. Margit N. Matoltsy and Judith A. Bednarz is greatly appreciated by the author.

REFERENCES

1. Farbman, A. I. Plasma membrane changes during keratinization. *Anat. Rec.* **156**: 269–282, 1966.
2. Rhodin, J. A. G. and Reith, E. J. Ultrastructure of keratin in oral mucosa, skin, esophagus, claw, and hair. In Fundamentals of Keratinization (E. O. Butcher & R. F. Sognnaes, eds.), Amer. Assoc. for the Advancement of Sci., Washington, D.C., 1962, pp. 61–94.
3. Odland, G. F. and Reed, Th. H. Epidermis. In Ultrastructure of Normal and Abnormal Skin (A. S. Zelickson, ed.), Lea & Febiger, Philadelphia, 1967, pp. 54–75.
4. Matoltsy, A. G. and Parakkal, P. F. Keratinization. In Ultrastructure of Normal and Abnormal Skin (A. S. Zelickson, ed.), Lea & Febiger, Philadelphia, 1967, pp. 76–104.
5. Brody, I. The modified plasma membranes of the transition and horny cells in normal human epidermis as revealed by electron microscopy. *Acta derm.-venereol.*, **49**: 128–138, 1969.
6. Matoltsy, A. G. Keratinization of the avian epidermis. *J. Ultrastr. Res.*, **29**: 438,–458, 1969.
7. Roth, S. I. and Maderson, P. F. A. The ultrastructure of the maturation processes of the epidermis of snakes and lizards. *Ann. Ital. Dermatol. Clin. & Sperim.*, **22**: 295–321, 1968.
8. Matoltsy, A. G. and Huszar, T. Keratinization of the reptilian epidermis: An ultrastructural study of the turtle skin. *J. Ultrastr. Res.*, **38**: 87–109, 1972.
9. Farquhar, M. G. and Palade, G. E. Cell junctions in amphibian skin. *J. Cell Biol.*, **26**: 263–291, 1965.
10. Lavker, R. M. Horny cell formation in the epidermis of Rana pipiens, *J. Morph.*, **142**: 365–378, 1974.
11. Grubb, C., Hackeman, M., and Hill, K. R. Small granules and plasma membrane thickening in human cervical squamous epithelium. *J. Ultrastr. Res.*, **22**: 458–468, 1968.
12. Parakkal, P. F. An electron microscopic study of esophageal epithelium in the newborn and adult mouse. *Amer. J. Anat.*, **121**: 175–196, 1967.
13. Schroeder, H. E. and Theilade, J. Electron microscopy of normal gingival epithelium. *J. periodont. Res.*, **1**: 95–110, 1966.
14. Osmanski, C. P. and Meyer, J. Differences in the fine structure of the mucosa of mouse cheek and palate. *J. Invest. Dermatol.*, 309–317, 1967.
15. Frithiof, L. Ultrastructural changes in the plasma membrane in human oral epithelium. *J. Ultrastr. Res.*, **32**: 1–17, 1970.

16. Hashimoto, K. Fine structure of horny cells of the vermilion border of the lip compared with skin. *Arch. oral Biol.*, **16**: 397–410, 1971.
17. Martinez, R. I. and Peters, A. Membrane-coating granules and membrane modification in keratinizing epithelia. *Amer. J. Anat.*, **130**: 93–120, 1971.
18. Lagermalm, G., Philip, B., and Lindberg, J. *Nature*, **168**: 1080, 1951.
19. Loomans, M. E. and Hannon, D. P. An electron microscopic study of the effects of subtilisin and detergents on human stratum corneum. *J. Invest. Dermatol.*, **55**: 101–114, 1970.
20. Matoltsy, A. G. and Balsamo, C. A. A study of the cornified epithelium of human skin. *J. Biophys. Biochem. Cytol.*, **1**: 339–360, 1955.
21. Matoltsy, A. G. and Parakkal, P. F. Membrane-coating granules of keratinizing epithelia. *J. Cell Biol.*, **24**: 297–307, 1965.
22. Matoltsy, A. G. The envelope of epidermal horny cells. 13 Congr. Internatl. Dermatol.—München, 1967, Springer-Verlag, Berlin, Heidelberg, New York, 1968, pp. 1014–1015.
23. Matoltsy, A. G. and Matoltsy, M. N. The membrane protein of horny cells. *J. Invest. Dermatol.*, **46**: 127–129, 1966.
24. Matoltsy, A. G. Desmosomes, filaments, and keratohyalin granules: Their role in the stabilization and keratinization of the epidermis. *J. Invest. Dermatol.*, **65**: 127–142, 1975.
25. Henrikson, R. C. and Matoltsy, A. G. The fine structure of teleost epidermis. I. Introduction and filament-containing cells. *J. Ultrastr. Res.*, **21**: 194–212, 1968.

Discussion

Dr. McGuire: Are the membranes resistant to sequential extraction first with an organic solvent such as acetone-benzene or acetone-methanol and then with 8 M urea-mercaptoethanol?

Dr. Matoltsy: Yes. We tried to solubilize bovine horny cell membranes after extraction with acetone for 24 hours. We used a mixture of 6 M urea, 2% SDS, 0.1 M 2-mercaptoethanol, 0.1 M EDTA and 0.1 M NaOH at 40° for 72 hours. The membranes appeared swollen in the phase contrast microscope. A significant amount of protein, however, was not recovered from the supernatant solution after centrifugation at 3,500 rpm.

Dr. Ohkido: What kind of lipids does your membrane contain? Does it contain sterols, phospholipids, glycerides or fatty acids?

Dr. Matoltsy: We did not study the nature of lipids which are contained in the thickened horny membranes since we have not had sufficient lipid to do such analyses.

Dr. Karasek: This is indeed a remarkable protein and I couldn't help thinking how happy I am that I have such a resistant covering. The content of aspartic acid and lysine was very interesting. In protein chemistry, we are always trying to look for the ester bond and the amide bond as linkages in protein structure. I was wondering if the lysine residues or the aspartic residues are indeed blocked in some way in these proteins indicating a coordinate covalent type of linkage.

DR. MATOLTSY: I think your question will be answered by Dr. Sugawara who will speak later. He believes that his fraction V, obtained with 0.5 M NaOH, is related to membranes and contains ester bonds.

DR. SUGAWARA: I obtained the γ-glutamyl-ε-lysine peptide from the alkaline-insoluble fraction and I have very much interest in your presentation. There is one question I would like to ask. Do you have any information about the residual fraction which sedimented at 600 rpm and which was disregarded during the preparation of your membranes? Does that fraction have the same chemical composition as the membrane fraction?

DR. MATOLTSY: It appears from microscopic observation that the sedimented residual fraction is primarily material which was not broken up during the process because of its very resistant nature. In this way we lose 10 to 15 percent of the original material. I think that shaking with glass beads would break up more of the material. I think that it has the same chemical composition, but I am not sure.

DR. TESUKA: Would you please tell us your concept of the synthetic mechanism for this membrane? Do you believe in the concept for this process described by Dr. Jessen about 5 years ago?

DR. MATOLTSY: Dr. Jessen proposed that the thickened part of the horny cell membrane is derived from keratohyalin because he saw keratohyalin near these membranes and both have similar electron densities. Of course, structural similarity does not necessarily mean chemical identity. Currently, we assume that the components of the thickened membrane are distributed in the cytoplasm and assembled during a late state of differentiation at the cell periphery to form a highly resistant and thickened inner layer of the plasma membrane.

If I have a moment of time, I would like to clarify a question which was asked previously. About 15 years ago, we published a paper in which we indicated that the membrane-coating granules (MCG) were involved in the thickening of the membranes. This relationship was derived from our observation that these granules were discharged into the inter-cellular spaces at a time when the membranes were thickened and when they appeared to become more resistant. But currently we are convinced that these are two different processes. The thickening of the membrane is not related to the discharge of the MCG.

Influence of Vitamin A on Differentiating Epithelia

Gary L. Peck,* Peter M. Elias ** and Bruce Wetzel*

Dermatology Branch, NCI, Bethesda, Maryland, U.S.A.* and *Department of Dermatology, University of California, School of Medicine, California, U.S.A.*

INTRODUCTION

Vitamin A induced mucous metaplasia of embryonic chick skin has served for over 20 years as a well-defined model system for the study of altered epithelial differentiation. Since 1953 when Fell and Mellanby[1] initially observed the inhibition of keratinization in chick embryo skin and its subsequent transformation into a mucous secreting structure by vitamin A, many studies have been concerned with the mode of action of vitamin A on a variety of epithelia. Results from these studies indicate that the response to vitamin A varies with the dose of vitamin A and with the species, age, and site of tissue. In addition to chick skin,[2–10] excess vitamin A inhibits keratinization in the stratified squamous epithelium of the adult hamster cheek pouch,[11–14] embryo mouse epidermis[15] and lip vibrissae,[16–17] embryonic rat esophagus,[18] tongue,[19] palate,[19] tail, and foot pad,[20] rat vagina,[12] adult rabbit keratoacanthoma,[21] mouse,[22–23] guinea pig[24–26] and human[27–28] epidermal cells, adult guinea pig ear,[29] chick esophagus and cornea,[30] as well as in embryonic,[31] normal adult,[32–33] and diseased[33] human skin. Mucin is produced in the chick skin, the cheek pouch, mouse vibrissae, cornea, esophagus, tongue, and keratocanthoma. Conversely, vitamin A deficiency leads to squamous metaplasia in calf parotid gland,[34] rat bladder,[35] rat trachea,[36,37] and cornea,[38] hamster trachea,[39] and mouse prostate.[40] There is a marked reduction in the number of goblet cells in the small intestine of vitamin A deficient rats, but squamous metaplasia does not occur.[41] In all of these models the initial effects were focal and reversible, selectively altering the differentiation of germinative layers of epithelia. In three tissues, the chick esophagus,[30] hamster cheek pouch,[11] and embryonic mouse lip vibrissae,[17] vitamin A induced intraepithelial mucus gland formation. Hydrocortisone and citral may reverse or inhibit completely the metaplastic effects of vitamin A.[42–44]

Most studies of excess vitamin A have used retinol in low to moderate dosage. However, it has become evident that retinoic acid is more potent than retinol in skin and other extracutaneous tissues,[45–47] possibly because of its greater solubility in aqueous media, or possibly because of the presence of specific retinoic acid-binding proteins.[48–50] In this report explants of chick embryonic skin were exposed to retinoic acid in low, moderate, and high doses and examined by light microscopic histochemistry, and transmission electron microscopy (TEM), and scanning electron microscopy (SEM).

METHODS

1. Tissue culture techniques

Tarsometatarsal shank skin was taken from white leghorn chick embryos ranging from 13 to 21 days of gestation, and up to 3 days after hatching. The shanks were immersed in Minimal Essential Medium (MEM), and after a longitudinal incision the metatarsal bones, cartilage, and loose areolar connective tissue were gently removed. The skin was cut into 2 × 2 mm pieces and placed on silicone-treated lens paper in Falcon organ culture dishes. The lens paper was supported by a stainless steel grid overlying a center well into which tissue culture media was placed up to the level of the supporting grid (0.6 to 0.7 cc). Several media have produced similar results including MEM, Trowell's medium, Earle's Basal Medium (BME), and Medium 199 with and without added retinol (0.1mg/L). In each case, the medium was supplemented with 5% to 15% fetal calf serum, glutamine, and antibiotics (penicillin, streptomycin, and fungizone). Experiments using serum-free medium failed to produce mucous metaplasia.[8,10] Cultures were kept in a Wedco model 2–17 incubator at 37°C, 100% humidity, under 5% CO_2 in air for periods varying from 3 hours to 8 days. Typically, skin from 14-day chick embryos (Stage 40)[51] was cultured for 3 days in Medium 199 with 5% fetal calf serum with and without 20 i.u. (6.7ugm)/ml retinoic acid. Vitamin A acid (all-trans-retinoic acid, Eastman Chemicals) was dissolved in absolute ethanol with sonication and then added to the complete medium immediately prior to use in final concentrations between 0.01 and 100 i.u./ml. The final ethanol concentration was 0.03% in all cultures including controls. Specimens were prepared for LM, SEM, and TEM after culture periods between 3 hours and 8 days, and from normal chicks at comparable periods of development *in ovo*.

2. Tissue preparation

For light microscopy tissue was fixed in 10% formalin in tap water with 2% sodium acetate for one hour at room temperature, then dehydrated in a graded series of methanol, embedded in paraffin, and sectioned at 6 um, then stained with periodic acid Schiff with and without diastase pretreatment, alcian blue at pH 0.4 and pH 2.4, toluidine blue, colloidal iron, or hematoxylin and eosin. Frozen sections of fixed tissue were stained with Oil Red 0.

Specimens for TEM were fixed in 6% glutaraldehyde in 0.1 M phosphate buffer for 2 hours at 4°C, rinsed in phosphate buffer with 0.1 M sucrose, postfixed in 1% osmium tetroxide for 90 minutes at 4°C, dehydrated through a graded ethanol series into propylene oxide, then infiltrated and embedded in Epon 812. Thin sections cut with an LKB microtome and a diamond knife were stained with lead citrate and uranyl acetate and examined in a Siemens 1–A Elmiskop TEM. Specimens for SEM were fixed by adding an equal volume of 6% glutaraldehyde in 0.1M phosphate buffer to the medium (3% final concentration of glutaraldehyde) for 90 minutes at room temperature. The specimens were washed in distilled water, placed in small stapled packets of lens paper, dehydrated through a graded ethanol series into amyl acetate, and dried by the critical point method with CO_2. They were then attached to aluminum stubs with 3M transfer tape, coated with carbon and gold-palladium, and viewed in an ETEC Autoscan at 20 KV.

RESULTS

Epidermal Differentiation in Control Cultures

Skin from 14-day chick embryos maintained in culture without added retinoic acid for periods up to 8 days show normal epidermal differentiation as compared with maturation *in ovo*. The characteristic tarsometatarsal shank skin scales develop normally. In short-term explants (1–2 days) the epidermis consists of stratified squamous epithelium covered by a two-layered periderm. The surface epithelial cells are flat with abundant tonofilaments and junctional complexes, but not yet cornified. Individual keratinocytes contain fine tonofilaments, many free ribosomes but very few polysomes, numerous mitochondria, and scattered profiles of rough endoplasmic reticulum and Golgi elements. No keratohyalin, glycogen, or mucin was observed in the epidermis of these specimens cultured for 2 days. Peridermal cells contained their characteristic granules and some glycogen. By SEM one can appreciate the continuous covering of peridermal cells over the developing skin scales. The peridermal surface appears smooth at low magnification, but on closer examination is uniformly covered with distinctive short microvilli and small anastomosing ridges.

In older explants (3–6 days) from 14-day embryos the epidermis has thickened and cornified, and the periderm is lost. Keratinocytes contain increased numbers of tonofilaments aggregated into tonofibrils, and keratohyalin granules are present. The distribution and appearance of Golgi elements, ribosomes, and mitochondria appear similar to earlier specimens. By SEM, some periderm is shed after 3 days in culture, exposing an even continuous layer of flat, smooth surfaced, polyhedral epidermal cells. Lipid droplets are seen both in the basal layer and the stratum corneum. PAS and other stains for mucopolysaccharides reveal no mucin and little glycogen.

Effects of Retinoic Acid on Epidermal Differentiation

The changes observed in treated cultures are dependent on both the dose of vitamin A acid and the time of exposure to this agent. Thus, with small amounts of retinoic acid (0.1 IU/ml) early metaplastic changes occur after 6 days *in vitro*; with larger doses (20 IU/ml) metaplastic changes occur within 18 hours after exposure. These metaplastic changes are reversible when retinoic acid treated skin is transferred to control medium. However, the mucin-containing epidermal cells do not redifferentiate to produce keratin. Instead they are gradually replaced by a new population of keratinizing cells.

Focal metaplastic changes are seen with small amounts of retinoic acid (e.g., 1 IU/ml for 3 days) or short exposure to larger doses (e.g., 20 IU/ml for 1 day); these foci are surrounded by normal, apparently nonresponding, epidermis. Accentuated metaplasia consistently appears at the cut edge of the explant and in the creases of the skin scales. With low doses skin scales flatten, and at higher doses (20 IU/ml/3 days) the explants decrease considerably in size as dermal elements are lost. By TEM most ribosomes in the keratinocytes appear as polyribosomes and Golgi elements increase in number. The epidermis becomes more hyperplastic leading to acanthosis and loss of horizontal stratification. The periderm begins to slough after only one day of retinoic acid treatment and is lost after two days (Fig. 1). In contrast to the control cultures keratohyalin granules do not form, and cornification does not occur. Even in the superficial layers epidermal cells are spherical rather than flat and may detach. Individual epidermal cells are variably affected (Fig. 2). Numerous microvilli appear on these cells with axial filaments and surface glycocalyx, but they become less prominent in the more rounded cells which appear to be sloughing from the epidermal surface. Cilia do not occur. Desmosomal cleavage is seen with widened intercellular spaces. Tight junctions appear between superficial epidermal cells and complex interdigitations occur below such junctions. Elongated gap junctions are seen three to four cells deep into the epidermis. Glycogen appears early in the cytoplasm of superficial epidermal cells and, occasionally, within phagocytic vacuoles. Filaments do not disappear completely from epidermal cells, but are finer and fewer in number than in control cultures. In-

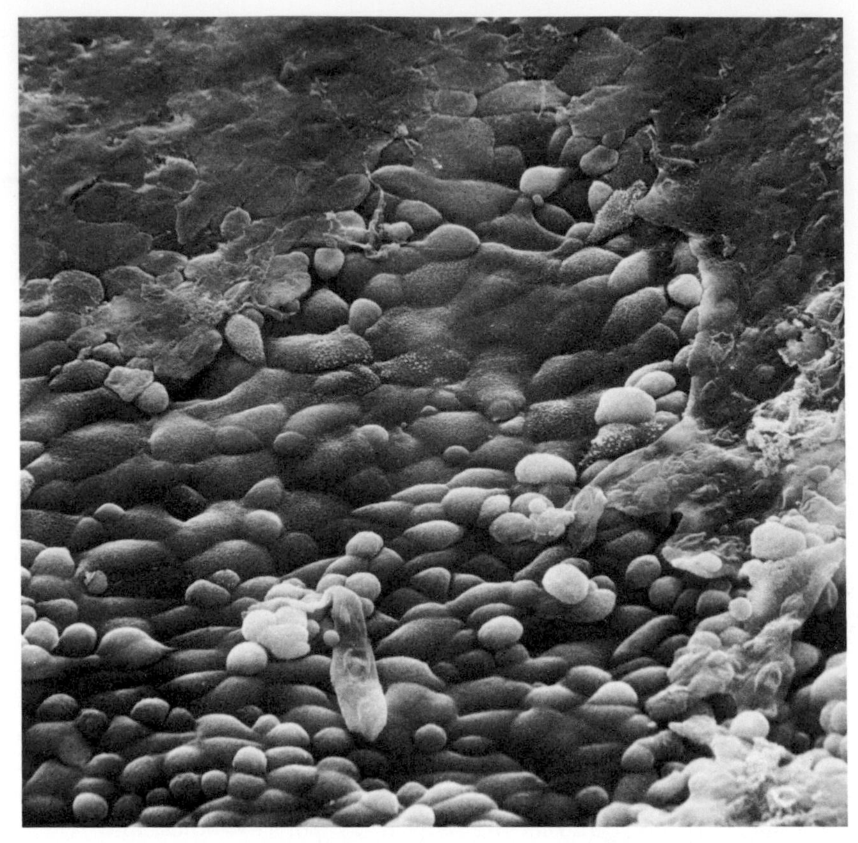

Fig. 1. Early retinoic acid effects include premature sloughing of the periderm, rounding and bulging upwards of underlying epidermal cells, and the development of surface villi and folds on the epidermal cells. 14 day chick skin cultured for two days. ×700.

creased numbers of large secondary lysosomes and presumptive lipid droplets are also seen. The basement membrane is discontinuous, and fingerlike projections of the basal cells extend through the gaps, occasionally contacting dermal fibroblasts. Collagen fibers sometimes appear above the basement membrane in contact with basal cells (Fig. 3).

Prolonged exposure to retinoic acid at high doses induces extensive elaboration of both inter- and intra-cellular mucin, and in most instances saccular glandlike dilatations develop between epidermal cells. Under these conditions vacuoles of varying sizes appear in the superficial layers and their contents stain intensely with colloidal iron, alcian blue, and PAS (with and without diastase pretreatment). A similarly staining material also fills the glandular structures and coats the luminal surfaces. The

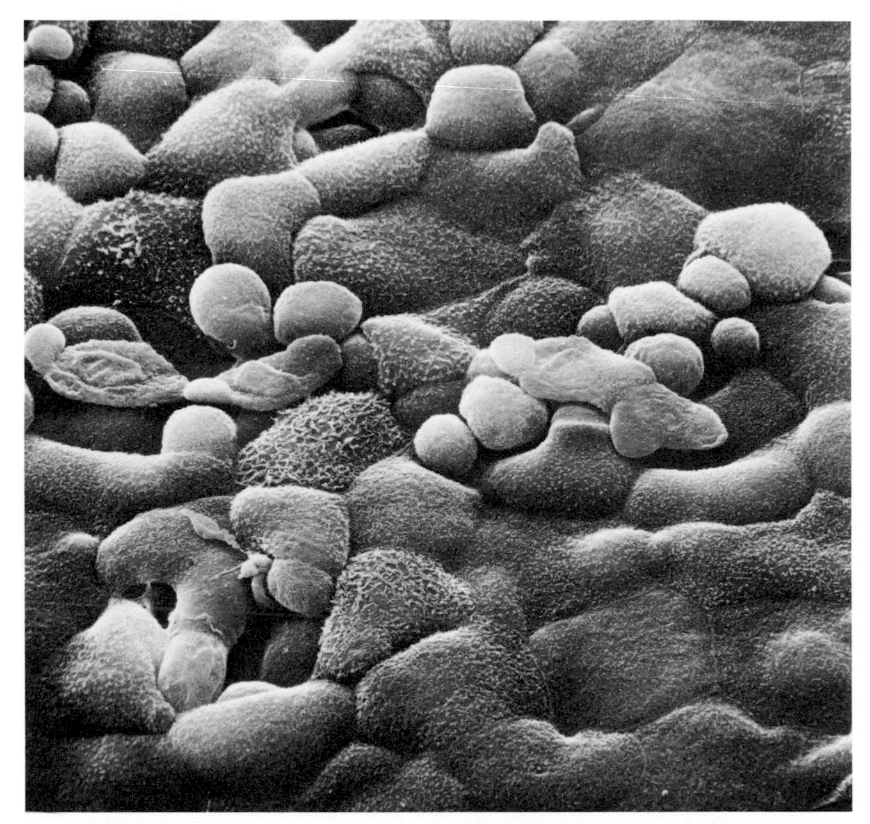

Fig. 2. A higher magnification of a two day retinoic acid treated specimen illustrates that the individual epidermal cells may vary in their response. ×1300.

saccular dilatations, possibly representing acini, are surrounded by epithelial cells with tight junctions, and microvilli which occasionally form a brush border and apical mucin granules. Ductlike structures often extend from these dilatations to the surface. Material similar in density to the mucin granules is seen within these acini, and coats the epidermal surface.

Retinoic acid is seen to specifically affect the further development of epidermal germinative cells in skin from older chicks (between 16 and 24 days of gestation; hatching occurs on day 21). Development initially proceeds normally in the more differentiated cells of the superficial epidermis; they appear to be unaffected by exposure to retinoic acid. However, a thickened, metaplastic germinative layer appears beneath the normal stratified squamous epithelium, producing irregular bumps and folds in

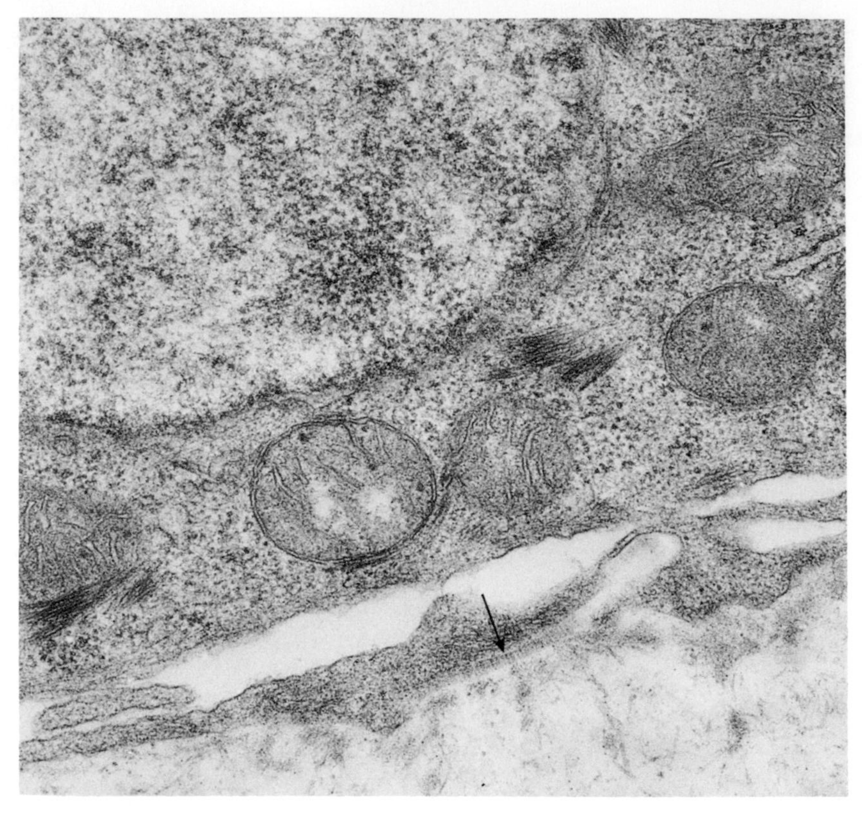

Fig. 3. Gaps in the basement membrane after retinoic acid exposure allow contact between a collagen fiber (↓) and an epidermal basal cell. ×5000.

the surface. After 3 days' exposure, however, the superficial epidermal cells become edematous and rupture, eventually resulting in a cleavage plane between the responsive and nonresponsive cells.

DISCUSSION

The mechanism of action of inductive agents in differentiation is thought to involve primarily the germinative cell population which is capable of DNA synthesis. Holtzer suggests that the inductive event occurs during the S-phase of a "quantal" cell cycle. The effects of the inductive stimulus are not seen, however, until after several "proliferative" cycles. In effect, cells that are already committed to the production of one "luxury" molecule, that is, keratin, myosin, or mucin, do not dedifferentiate and synthesize a new terminal product. Rather a new product must come from previously uncommitted cells, such as the epidermal basal cells.[52]

Most of the evidence in the field of vitamin A effects on epithelia tend to support this view. For instance, hyperplasia of the basal layer of the epidermis after vitamin A exposure has been noted repeatedly, including in this study. Metaplasia secondary to vitamin A deficiency also leads to basal cell hyperplasia. Basal cell effects are most easily observed in organ culture when vitamin A is discontinued and the tissue is exposed to control medium. In the reversal experiment in this study the mucous containing cells were pushed to the surface by a new generation of keratinizing cells. In addition, when older embryos are used, metaplasia is limited to the basal epidermis and the superficial epidermis remains keratinized. These data indicate that although the epidermis may modulate between mucin formation and keratin synthesis depending on the presence or absence of vitamin A acid, the individual epidermal cells, once committed to a specific course of differentiation, cannot reverse themselves. The enhanced metaplasia at the cut edges of explants and in the folds of the skin scales may be related to the increased mitotic index and labelling index in these areas as compared to other areas of the epidermis.[53,54] The foci of early mucous metaplasia may also indicate cell cycle specificity. Wong suggests that the focal mitotic activity seen in the hamster cheek pouch may account for the focal metaplasia induced by vitamin A.[13]

Studies on direct effects of vitamin A on epidermal cells in culture also favor the germinative cell theory in that these cells are capable of DNA synthesis and mitosis. In epidermal cell culture vitamin A acid stimulates the mitotic index and labelling index.[24,27,28] *In vivo*, vitamin A acid also stimulates mitotic and labelling indices and epidermal turnover, again indicating a germinative cell effect.[55] Lazarus suggests that the acceleration of the mitotic rate in skin treated with vitamin A may be a result of lysosomal proliferation with subsequent increased concentration and release of lysosomal hydrolases. He has demonstrated an increased concentration of the lysosomal proteinase, Cathepsin D, in cultured rabbit skin exposed to vitamin A acid.[56] Proteases in other systems have been shown to be potent mitogens.[57,58] These findings are in accord with Holtzer's concept of the need for proliferative cell cycles in the mechanism of differentiation.

In this report the germinative layer was hyperplastic and these cells contained prominent Golgi apparatus, rough endoplasmic reticulum, and polyribosomes as an early metaplastic event. The observed increase in Golgi elements, the subcellular site of glycoprotein synthesis, is consistent with De Luca's comment[59] that the primary action of vitamin A in mucous cell differentiation may be at the level of the biosynthesis of specific glycopeptides which in turn affect RNA synthesis[60,61,62] and cell differentiation.

Hardy suggests that the prime effect of vitamin A may be the production of basement membrane gaps which allow extensive dermal-epidermal cell to cell contact.[16] These contacts were also seen in our study but not as extensively as in Hardy's. She suggests that inappropriate signals from the dermis lead to metaplasia as opposed to a direct action of the vitamin on the epidermis. The dermal signal provided by cell to cell contacts could represent a metabolic cooperation[63] or a transfer of informational macromolecules.[64] In another system, cell contacts, as opposed to diffusible signals, have proven essential to the induction of metanephric tubules by several embryonic and neoplastic tissues.[65] Additional evidence in favor of a dermal mechanism comes from McLoughlin, who demonstrated that gizzard mesenchyme has the ability to induce mucous metaplasia in overlying isolated chick epidermis.[66] Similarly, Sweeny and coworkers found that full thickness canine oral mucous membrane when transplanted to the ear began to keratinize. Upon retransplantation to the tranchea the graft shed its stratum corneum and underwent a mucous metaplasia with production of a simple tubular mucous gland within its center.[67] However, this dermal induction hypothesis, at first glance, seems unable to explain mucous metaplasia in those studies of vitamin A effect on isolated epidermis or on isolated epidermal cells. This theory would presumably require the presence of contaminating fibroblasts or other dermal elements in the epidermal preparations to produce the dermal signal. Furthermore, in all the studies using isolated epidermal cells, frank mucous metaplasia was not seen. In one system, although retinyl acetate treated cells were stained with alcian blue[22] and produced a new glycopeptide,[23] they did not contain mucous granules by TEM. It would appear that, if there is a direct metaplastic effect of vitamin A on epidermal cells, then a dermal presence is necessary to fully amplify the inductive signal.

We also saw direct contact between collagen fibers and basal cells through the basement membrane gaps. Basement membrane gaps have been observed in neoplastic and hyperplastic conditions, such as carcinoma and psoriasis.[39] However, the significance of basement membrane gaps in this and other experiments using excess vitamin A[9,16,31] becomes less clear because they were not seen in vitamin A deficiency induced squamous metaplasia in tracheal epithelium, although carcinogen induced squamous metaplasia in the same study did produce basement membrane gaps.[39] In addition, normal columnar orientation and mitotic activity were observed in cultured epidermis lacking a basement membrane. Whether collagen itself plays a role in determining the direction of epidermal cell differentiation or whether it only serves to amplify the vitamin A metaplastic effect is unclear. Hay and Meier have demonstrated that collagen-cell surface interactions are necessary for corneal morphogenesis.[69] One possible specific role for collagen in this system could be to participate as a substrate in epidermal-cell surface glycosyltransferase reactions,[70] perhaps also involving a retinol-glycolipid intermediate. The possible roles of collagen in cell differentiation are discussed at length in recent reviews.[71–73]

In spite of this preponderance of evidence indicating a vitamin A action on the basal germinative cells, it is clear that there are, in addition, direct effect on the postmitotic, postsynthetic, superficial epidermal cells. Rothberg has demonstrated that in 13-day embryonic chick skin tritiated thymidine is taken up only by the basal layer of epidermal cells.[74] Therefore, the superficial cells, whose surface appearance is markedly altered by retinoic acid, are postsynthetic. Since direct effects of vitamin A on membranes have long been demonstrated,[75] it is conceivable that in this system vitamin A acid may bind to all the epidermal cell membranes with resultant alteration in cell surface morphology. Upon transmission of this surface signal to the genome, perhaps via membrane messengers,[76] new messenger RNA and glycoproteins may be formed only in those cells capable of DNA synthesis.

We cannot as yet determine whether the surface mucin cells seen after three days of incubation are derived from superficial postsynthetic epidermal cells or represent metaplastic basal cells that have migrated to the surface. If the source of the mucin-containing cells were indeed the superficial postsynthetic epidermal cells, then this would support the view that the action of vitamin A occurs at the posttranslational level as suggested by De Luca. In this instance vitamin A could serve as a carrier of monosaccharide units and glycosylate preexisting proteins to form a glycoprotein moiety.[59]

The development of a rapid assay for metaplasia should permit short term experiments with metabolic inhibitors which could serve to further define the mechanism of action of vitamin A in this organ culture system. In this regard, pretreatment but not posttreatment with actinomycin D prevents the formation of mucin in the rabbit keratoacanthoma exposed to vitamin A acid.[77] Pretreatment prevents the uptake of ^{3}H-vitamin A acid into the cell which suggests that actinomycin D may be inhibiting cell permeability to vitamin A acid rather than inhibiting it at the genetic level. Continuous applications of puromycin given concurrently with vitamin A acid also prevent mucin formation in the keratoacanthoma and may represent an inhibition of mucin granule development in the golgi apparatus rather than an inhibition of protein synthesis.[78]

The demonstration that the epidermis can develop into a branching tubuloalveolar mucous gland is an indication both of the sensitivity and potentiality of chick skin to an inductive stimulus and of the strength of the stimulus provided by retinoic acid. Glandular metaplasia has been noted in the hamster cheek pouch,[11] the chick esophagus,[30] and in embryonic mouse vibrissae[17] after exposure to vitamin A. Hardy suggests that the basal lamina gaps permit the glandular metaplasia seen in the vibrissae.

Clinically, vitamin A and its analogs have been shown to be effective in the prevention and treatment of several types of cancer.[79] Evidence indicating that the metaplastic action of vitamin A may be linked to its efficacy in cancer treatment was provided by Prutkin, who treated experimentally induced rabbit keratoacanthoma with topical vitamin A acid and found that prior to involution the keratoacanthomas underwent mucous metaplasia.[21]

Other clinical uses of vitamin A and its analogs have been in the treatment of acne and disorders of keratinization.[45–47] The clinical effectiveness of retinoic acid, in particular, in these conditions may result from its ability to induce desmosomal cleavage and subsequent cell separation and desquamation.[80] Retinoic acid may depend on a specific binding protein for its metaplastic activity. Indeed, a high affinity retinoic acid binding protein has been found in embryonic chick skin and other tissues,[48–50] The involvement of these binding proteins in the mechanism of action of vitamin A or vitamin A acid is supported by the finding that the ability of cis isomers of retinol to bind to cellular retinol binding protein parallels their potency in promoting growth in vitamin A deficient rats.

The sensitivity of the chick skin system should stimulate its use as a model for the continuing effort to elucidate the control mechanisms involved in differentiation, and in determining the mechanism of action of vitamin A in cancer treatment and prevention, and for testing vitamin A analogs.[81]

REFERENCES

1. Fell, H. B. and Mellanby, E. Metaplasia produced in cultures of chick ectoderm by high vitamin A. *J. Physiol.*, **119**: 470–488, 1953.
2. Jackson, S. F. and Fell, H. B. Epidermal fine structure in embryonic chicken skin during atypical differentiation induced by vitamin A in culture. *Developmental Biology*, **7**: 394–419, 1963.
3. Fell, H. B. The role of organ cultures in the study of vitamins and hormones. *Vitamins and Hormones*, **22**: 81–127, 1964.
4. Fell, H. B. The experimental study of keratinization in organ culture. The Epidermis, W. Montagna and W. C. Lobitz, Jr., eds. Academic Press, New York, Chapter 4, pp. 61–81, 1974.
5. Elias, P. M. and Friend, D. S. Vitamin-A-induced mucous metaplasia: An in vitro system for modulating tight and gap junction differentiation. *J. Cell Biol.*, **68**: 173–188, 1976.
6. McLoughlin, C. B. The importance of mesenchymal factors in the differentiation of chick epidermis. I. The differentiation in culture of the isolated

epidermis of the embryonic chick and its response to excess vitamin A. *J. Embryol. exp. Morph.*, **9**: 370–384, 1961.

7. Freinkel, R. K. Lipogenesis during cornification of chicken skin in organ culture. *J. Invest. Dermatol.*, **50**: 339–344, 1972.
8. Rothberg, S. The cultivation of embryonic chicken skin in a chemically defined medium, and the response of the epidermis to excess of vitamin A. *J. Invest. Dermatol.*, **49**: 35–38, 1967.
9. Sagami, S. and Kitano, Y. Electron microscopic study of the effect of vitamin A on the differentiation of reconstructed embryonic chick skin. *Br. J. Derm.*, **83**: 565–571, 1970.
10. Beckingham-Smith, K. The proteins of the embryonic chick epidermis. II. During culture in serum-containing medium with and without added vitamin A. *Developmental Biology*, **30**: 263–278, 1973.
11. Lawrence, D. J. and Bern, H. A. Vitamin A and mucous metaplasia. *Ann. NY Acad. Sci.*, 106 (Art. 2): 646–653, 1963.
12. Bern, H. A. and Lawrence, D. J. Influence of vitamin A on keratinization. Fundamentals of Keratinization. Ed. by E. O. Butcher, R. D. Sognnaes, Am. Assoc. for the Advancement of Science, No. 70, Washington, D. C., 1962.
13. Wong, Y. C. Mucous metaplasia of the hamster cheek pouch epithelium under hypervitaminosis A. *Exp. and Molec. Path.*, **23**: 132–143, 1975.
14. Cavalaris, C. J., Mutakas, V. J., and Krikos, G. A. Histochemistry of the mucins of vitamin A produced mucous metaplasia in hamsters. *Archs. oral Biol.*, vol. **14**: 1313–1322, 1969.
15. Hardy, M. H., Sweeny, P. R., and Bellows, C. Early ultrastructural changes in fetal mouse epidermis induced by vitamin A in organ culture. *In Vitro*, **9**: 358, 1974.
16. Hardy, M. H. Epithelial-mesenchymal interactions in vitro altered by vitamin A and some implications. *In Vitro*, **10**: 338, 1974.
17. Hardy, M. H. Glandular metaplasia of hair follicles and other responses to vitamin A excess in cultures of rodent skin. *J. Embryol. exp Morph.*, **19**: 157–180, 1958.
18. Lasnitzki, I. The effect of excess vitamin A on the embryonic rat oesophagus in culture. *J. Exp. Med.*, **118**: 1–6, 1963.
19. New DAT. Effects of excess vitamin A on cultures of skin and buccal epithelium of the embryonic rat and mouse. *Brit. J. Dermatol.*, **75**: 320–325, 1963.
20. New DAT. Effects of excess vitamin A on cultures of skin from the tail and pads of the embryonic rat, and from the trunk, tail and pads of the embryonic rabbit. *Exp. Cell Res.*, **39**: 178–183, 1965.
21. Prutkin, L. Mucous metaplasia and gap junctions in the vitamin A acid-treated skin tumor, keratoacanthoma. *Cancer Research*, **35**: 364–369, 1975.
22. Yuspa, S. H. and Harris, C. C. Altered differentiation of mouse epidermal cells treated with retinyl acetate in vitro. *Exp. Cell Res.*, **86**: 90–105, 1974.
23. DeLuca, L. and Yuspa, S. H. Altered glycoprotein synthesis in mouse epidermal cells treated with retinyl acetate in vitro. *Exp. Cell Res.*, **86**: 106–110, 1974.
24. Christophers, E. Growth stimulation of cultured postembryonic epidermal cells by vitamin A acid. *J. Invest. Dermatol.*, **63**: 450–455, 1974.

25. Wolff, H. H., Christophers, E., and Braun-Falco, O. Beeinflussung der epidermalen ausdifferenzierung durch vitamin A-saure. *Arch. klin. exp. Derm.,* **237**: 774–795, 1970.
26. Lukacs, I., Christophers, E., and Braun-Falco, O. Zur wirkungsweise der vitamin A-saure (Tretinoin). *Arch. Derm. Forsch.,* **240**: 375–382, 1971.
27. Karasek, M.A. Effect of all-trans-retinoic acid on human skin epithelial cells in vitro. *J. Soc. Cosmet. Chem.,* **21**: 925–932, 1970.
28. Chopra, D. P. and Flaxman, B. A. The effect of vitamin A on growth and differentiation of human keratinocytes in vitro. *J. Invest. Dermatol.,* **64**: 19–22, 1975.
29. Barnett, M. L. and Szabo, G. Effect of vitamin A on epithelial morphogenesis in vitro. *Exp. Cell Res.,* **76**: 118–126, 1973.
30. Aydelotte, M. B. The effects of vitamin A and citral on epithelial differentiation in vitro. 2. The chick oesophageal and corneal epithelia and epidermis. *J. Embryol. exp. Morph.,* **11**: 621–635, 1963.
31. Lasnitzki, I. The effect of carcinogens, hormones, and vitamins on organ cultures. *Internat. Rev. Cytol.,* **7**: 103–104, 1958.
32. Szabo, G. Cultivation of skin, pure epidermal sheets, and tooth germs in vitro. Fundamentals of Keratinization. Ed. by E. O. Butcher, R. D. Sognnaes, Am. Assoc. for the Advancement of Science, no. 70, Washington, D. C., 1960.
33. Plewig, G., Wolff, H. H., and Braun-Falco, O. Lokalbehandlung normaler and pathologischer menschlicher haut mit vitamin A-saure. *Arch. klin. exp. Derm.,* **239**: 390–413, 1971.
34. Hayes, K. C., McCombs, H. L., and Faherty, T. P. The fine structure of vitamin A deficiency. I. Parotid duct metaplasia. *Laboratory Investigation,* **22**: 81–89, 1970.
35. Hicks, R. M. Hyperplasia and cornification of the transitional epithelium in the vitamin A-deficient rat. *J. Ultrastr. Res.,* **22**: 206–230, 1968.
36. Wong, Y. C. and Buck, R. C. An electron microscopic study of metaplasia of the rat tracheal epithelium in vitamin A deficiency. *Laboratory Investigation,* **24**: 55–66, 1971.
37. Marchok, A. C., Cone, M. V. and Nettesheim, P. Induction of squamous metaplasia (Vitamin A deficiency) and hypersecretory activity in tracheal organ cultures. *Laboratory Investigation,* **33**: 451–460, 1975.
38. Beitch, I. The induction of keratinization in the corneal epithelium. *Investigative Ophthalmology,* **9**: 827–843, 1970.
39. Harris, C. C., Sporn, M. B., Kaufman, D. G., Smith, J. M., Frank, B. A., Jackson, F. E., and Saffiotti, U. Histogenesis of squamous metaplasia in the hamster tracheal epithelium caused by vitamin A deficiency or benzo[a] pyrene-ferric oxide, *J. Nat. Cancer Inst.,* **48**: 743–761, 1972.
40. Lasnitzki, I. Hypovitaminosis-A in the mouse prostate gland cultured in chemically defined medium. *Exp. Cell Res.,* **28**: 40–51, 1962.
41. DeLuca, L., Little, E. P., and Wolf, G. Vitamin A and protein synthesis by rat intestinal mucosa. *J. Biological Chem.,* **244**: 701–708, 1969.
42. Aydelotte, M. B. The effects of vitamin A and citral on epithelial differentiation in vitro 1. The chick tracheal epithelium. *J. Embryol. exp. Morph.,* **11**: 279–291, 1963.

43. Barker, S. A., Cruickshank, C. N. D., and Webb, T. The effect of vitamin A, hydrocortisone and citral upon sulphate metabolism in skin. *Exp. Cell Res.*, **35**: 255–261, 1964.
44. Fell, H. B. The influence of hydrocortisone on the metaplastic action of vitamin A on the epidermis of embryonic chicken skin in organ culture. *J. Embryol. exp. Morph.*, **10**: 389–409, 1962.
45. Logan, W. S. Vitamin A and keratinization. *Arch. Derm.*, **105**: 748–753, 1972.
46. Ashton, H., Stevenson, C. J., and Frenk, E. Therapeutics XVI. Retinoic Acid. *Br. J. Derm.*, **85**: 500, 1971.
47. Fulton, J. E. Vitamin A acid—The last five years. *J. Cutaneous Pathology,* **2**: 155–160, 1975.
48. Sani, B. P. and Hill, E. L. Retinoic acid: A binding protein in chick embryo metatarsal skin. Biochem. Biophys. *Res. Commun.*, **61**: 1276–1282, 1974.
49. Ong, D. E. Retinoic acid binding protein: Occurrence in human tumors. *Science,* **190**(4209): 560–561, 1975.
50. Chytil, F. and Ong, D. E. Mediation of retinoic acid-induced growth and antitumour activity. *Nature,* **260**: 49–51, 1976.
51. Hamburger, V. and Hamilton, H. L. A series of normal stages in the development of the chick embryo. *J. Morphology,* **88**: 49–92, 1951.
52. Holtzer, H., Weintraub, H., Mayne, R., and Mochan, B. The cell cycle, cell lineages, and cell differentiation. Current Topics in Developmental Biology, vol. 7 (A. A. Moscona and A. Monroy, eds.) Academic Press, NY, 1972.
53. Gelfant, S. A study of mitosis in mouse ear epidermis in vitro. 1. Cutting of the ear as mitotic stimulant. *Exp. Cell Res.*, **16**: 527–537, 1959.
54. Sawyer, R. H. Avian Scale Development. II. A study of cell proliferation. *J. Exp. Zool.*, **181**: 385–408, 1972.
55. Plewig, G. and Fulton, J. E. Autoradiographic analysis of epidermis and adnexa after topical vitamin A acid (retinoic acid). *Hautarzt,* **23**: 128–136, 1972.
56. Lazarus, G. S., Hatcher, V. B., and Levine, N. Lysosomes and the skin. *J. Invest. Dermatol.*, **65**: 259–271, 1975.
57. Noonan, K. D. Role of serum in protease-induced stimulation of 3T3 cell division past the monolayer stage. *Nature,* **259**: 573–676, 1976.
58. Allison, A. C. Lysosomes in cancer cells. *J. Clin. Path, 27, Suppl.* (Roy. Coll. Path), **7**: 43–50, 1974.
59. DeLuca, L. and Wolf, G. Mechanism of action of vitamin A in differentiation of mucus-secreting epithelia. *J. Agr. Food Chem.*, vol. **20**: no. 3:474, 1972.
60. DeLuca, L., Kleinman, H. K., Little, E. P., and Wolf, G. RNA metabolism in rat intestinal mucosa of normal and vitamin A-deficient rats. *Arch. Biochem. Biophys.*, **145**: 332–337, 1971.
61. Sporn, M. B., Dunlop, N. M., and Yuspa, S. H. Retinyl acetate: Effect on cellular content of RNA in epidermis in cell culture in chemically defined medium. *Science*, **182**: 722–723, 1973.
62. Lukacs, I. and Braun-Falco, O. RNS-Synthese der haut. *Arch. klin. exp. Derm.*, **238**: 187–196, 1970.
63. Cox, R. P., Krauss, M. R., Balis, M. E., and Dancis, J. Metabolic cooperation

in cell culture. Cell Communication, R. P. Cox, ed. John Wiley & Sons, NY, 1974, pp. 67–95.

64. Kolodny, G. M. Transfer of macromolecules between cells in contact. Cell Communication, R. P. Cox, ed. John Wiley & Sons, NY, 1974, pp. 97–111.
65. Saxen, L., Lehtonen, E., Karkinen-Jaaskelainen, M., Norling, S., and Wartiovaara, J. Are morphogenetic tissue interactions mediated by transmissible signal substances or through cell contacts? *Nature,* **259**: 662–663, 1976.
66. McLoughlin, C. G. The importance of mesenchymal factors in the differentiation of chick epidermis. II. Modification of epidermal differentiation by contact with different types of mesenchyme. *J. Embryol. exp. Morph.,* **9**: 385–409, 1961.
67. Sweeny, P. R., Farkas, L. G., Farmer, A. W., Wilton, W., and McCain, W. G. Metaplasia of adult oral mucous membrane. *Journal of Surgical Research,* **19**: 303–308, 1975.
68. Kallman, F., Evans, J., and Wessells, N. K. Normal epidermal basal cell behavior in the absence of basement membrane. *J. Cell Biol.,* **19**: 231–236, 1966.
69. Hay, E. D. and Meier, S. Role of collagen-cell surface interaction in corneal morphogenesis. *J. Cell Biol.,* **67**: 321a, 1975.
70. Dorsey, J. K. and Roth, S. The effect of polyprenols on cell surface galactosyltransferase activity. Control of proliferation in animal cells (B. Clarkson and R. Baserga, eds.), Cold Spring Harbor Conference on Cell Proliferation, Vol. 1, Cold Spring Harbor Laboratory, 1974.
71. Hay, E. D. Origin and role of collagen in the embryo. *Amer. Zool,* **13**: 1085–1107, 1973.
72. McMahon, D. Chemical messengers in development: A hypothesis. *Science,* **185**: 1012–1021, 1974.
73. Reddi, A. H. Collagen and cell differentiation. Biochemistry of Collagen (G. N. Ramerchandran and A. H. Reddi, eds.), Plenum Press, NYC, 1976.
74. Rothberg, S. and Ekel, T. M. Mitotic activity in the chick embryo epidermis. *Nature,* **216**(5122): 1352, 1967.
75. Roels, O. A., Anderson, O. R., Lui, N. S. T., Shah, D. O., and Trout, M. E. Vitamin A and membranes. *Am. J. Clin. Nut.,* **22**: 1020–1032, 1969.
76. Roseman, S. Complex carbohydrates and intercellular adhesion. The cell surface in development. (A. A. Moscona, ed.) John Wiley & Sons, NY, 1974.
77. Prutkin, L. The effect of actinomycin D on the incorporation of labeled vitamin A acid in normal and tumor epithelium. *J. Invest. Dermatol.,* **57**: 323–329, 1971.
78. Prutkin, L. Inhibition of mucous metaplasia in the skin tumor keratoacanthoma by continual applications of puromycin. *Experientia,* **31**: 491–493, 1975.
79. Sporn, M. B., Dunlop, N. M., Newton, D. L., and Smith, J. M. Prevention of chemical carcinogenesis by vitamin A and its synthetic analogs (retinoids). *Federation Proceedings,* **35**: 1332–1338, 1976.
80. Kaidbey, K. H., Kligman, A. M., and Yoshida, H. Effects of intensive application of retinoic acid on human skin. *Brit. J. Dermatol.,* **92**: 693–706, 1975.
81. Sporn, M. B., Clamon, G. H., Dunlop, N. M., Newton, D. L., Smith, J. M., and Saffiotti, U. Activity of vitamin A analogues in cell cultures of mouse epidermis and organ cultures of hamster trachea. *Nature,* **253**: 47–50, 1975.

Discussion

Dr. Kurosumi: Are the keratinizing cells of the hair follicles or sweat ducts, which are already differentiated, capable of undergoing mucous metaplasia in either the case of Vitamin A acid treatment or in the case of transplantation?

Dr. Peck: There is work reported by New which he has interpreted as indicating that areas of the skin which have a large number of hair follicles are more resistant to the metaplastic effects of Vitamin A than areas which have no hair. However, in Vitamin A deficiency you see increased keratinization of the hair follicles and keratinization of the sweat ducts. Excess Vitamin A has not been shown to produce mucous metaplasia in these structures.

Dr. Aso: I am confused by the physiological action of Vitamin A because we know that Vitamin A acid is a precursor of Vitamin A and Vitamin A acid is very quickly metabolized in the skin. How can we distinguish between the physiological and pharmacological actions of Vitamin A acid in epidermal cell differentiation? I have applied Vitamin A acid to the skin of the guinea pig ear and during the first three or four days epithelial cells proliferated, many mitoses were seen, but after continued application of up to 12 days this mitotic activity stopped. If we were applying Vitamin A acid in excessive amounts, we may have been observing pharmacological rather than physiological effects. What is your view on this question of the physiological versus the pharmacological effect?

Dr. Peck: Certainly, the level of Vitamin A acid which we used in our skin experiments should be considered to have been pharmacological although not a toxic dose. Our goal was to establish a model system of altered epidermal differentiation induced by a single agent and to identify the mechanisms involved. We were not primarily involved in elucidating the physiological role of Vitamin A in the skin. Perhaps a more physiological reproach would be to simulate DeLuca's experiments studying Vitamin A deficiency in the intestine. Effects of physiological doses of Vitamin A after the establishment of the Vitamin A deficiency state would be more suitable for study of the physiological role of Vitamin A.

Dr. Adachi: In response to Dr. Aso's comment, I think it would be very nice if you could demonstrate a Vitamin A acid receptor in your system then everything would be resolved. Although Vitamin A acid is extensively metabolized, the receptor can bind the material and preserve its action. A Vitamin A receptor has been shown in other tissues and I think this would be quite interesting.

DR. PECK: Other workers have demonstrated specific Vitamin A acid binding proteins in embryonic chick skin as well as other tissues. This may be one of the reasons why Vitamin A acid is more potent than Vitamin A.

DR. OGURA: Dr. DeLuca at the National Cancer Institute has reported that one of the biochemical functions of Vitamin A is to behave as a transmitter of sugars to glycoproteins in the plasma membrane of the cell. Vitamin A-mannose phosphate is such a transmitter. Has a Vitamin A-sugar phosphate, similar to the substance which has been found in the epithelium of the lung, been identified in epidermis?

DR. PECK: This work has not been done. We do intend to carry out such experiments in the near future.

II. CONTROL MECHANISMS OF SKIN METABOLISM

Biochemistry of Nucleases in Relation to Epidermal DNA Catabolism

R. Ogura, T. Hidaka and H. Koga

Department of Medical Biochemistry, Kurume University, School of Medicine, Kurume, Japan

ABSTRACT

In order to clarify the regulation mechanism of nucleic acid metabolism during cellular differentiation, the enzyme activities related to DNA degradation process were taken up in this paper. Acid DNase, phosphodiesterase, acid phosphatase, purine nucleoside phosphorylase and phosphoribosyl transferase have been demonstrated and purified from cow snout epidermis. Highly polymerized DNA was degraded to nucleoside and purine base along a similar pathway as in other mammalian tissues. The salvage pathway enzymes for purine nucleotide biosynthesis were also involved in keratinization process. The DNA-nucleolytic enzyme activity was high in the upper layer of epidermis, and the salvage pathway enzyme was relatively high in the lower layer. The DNA nucleolytic enzymes were discussed for a possible role in nuclear decomposition during keratinization and cellular differentiation.

Epidermis is one of the tissues typified by rapid proliferation and differentiation. In normal epidermis, no nuclei are present in the stratum corneum, but imperfect keratinization is accompanied with the retention of nuclei in the stratum corneum. During normal keratinization, nucleoproteins are first broken down to DNA and RNA. Further catabolism of nucleic acids probably proceeds along the same pathways as in other mammalian tissues. The endogenous levels of DNA are under the control of both synthetic and catobolic enzymes in the epidermis. DNA-nucleolytic enzymes which effect this degradation may be of great significance in the overall biological control processes in the epidermis. For these reasons, we have attempted to purify the DNA-nucleolytic enzymes from cow snout epidermis, and to extend the study of their biological characteristics in cellular differentiation.

MATERIALS AND METHODS

Material

Cow snout epidermis was used as a source of nucleolytic enzymes. The cow snout was obtained in the slaughter house. The skin was excised, and the subcutaneous tissue was removed. The epidermis was sliced by a laboratory cryostat-microtome (Yamatokoki, Japan Model 1118) with the cutting blade adjusted to 10 micron thickness. About 60 slices were obtained from snout epidermis. The slices obtained from whole epidermis were pooled, and homogenized in 10 vols of distilled water with a glass homogenizer. The tissue preparation was centrifuged at 3,000 rpm for 15 minutes to remove cell debris. The resulting supernatant fluids are hereafter referred to as epidermal homogenate.

Overall assay of DNA breakdown

DNA-nucleolytic activity was quantitated by measuring spectophotometrically the release of low molecular weight products from highly polymerized salmon DNA.[1] The reaction mixture contained the following in a final volume of 2.0 ml: 1 mg of highly polymerized salmon sperm DNA, 0.12 mole of acetate buffer (pH 5.0), and the enzyme solution to be assayed. After 30 minutes of incubation the reaction was stopped by addition of 2.0 ml of 5% perchloric acid (PCA). The reaction tube was chilled in an ice-cold water bath for 20 minutes, and clarified by centrifugation for 30 minutes at 10,000 × *g*. The optical absorption at 260 nm (E_{260}) of supernatant was determined (diluted, if necessary, with water). A DNA-blank and sample-blank were incubated simultanously with each reaction tube, and these blanks were subtracted from the reading of each mixture. The extent of DNA hydrolysis was expressed as an increase of optical density at 260 nm.

Determination of protein

The amount of protein was determind by the method of Lowry.[2] Enzyme activity was expressed per mg protein.

DNA content in the epidermis

Determination of DNA content of the epidermis was made by a micro method using ultraviolet absorption.[3]

Enzyme assay

1. DNase

The assay procedure is the same as used in the method of overall assay. The reaction mixture for the assay of acid DNase contained the following components in a final volume of 2.0 ml: 1 mg of salmon DNA, 0.12 mole of acetate buffer (pH 5.0), 0.006 mole of EDTA, and the enzyme solution to be assayed. For the assay of neutral DNase, 0.12 mole of Tris buffer (pH 7.0) and 0.006 mole of $MgSO_4$ instead of EDTA were used in the reaction mixture.[1] One unit of enzyme activity is defined as an increase of 1.0 optical density at 260 nm for one minute under the above conditions. Specific activity was calculated by dividing the activity by the amount of protein. 2. Phosphodiesterase activity (Exonuclease, PDase)

Two different types of enzyme have been reported. Phosphodiesterase (exonuclease) catalyzes the hydrolysis of oligonucleotide at the end of chain. The enzyme proceeds as an exonuclease with the stepwise release of nucleoside 5′-monophosphate (5′-mononucleotide) or nucleoside 3′-monophosphate (3′-mononucleotide).

(1) 5′-mononucleotide forming phosphodiesterase (PDase I)

In this case, 0.5 ml of 0.1 M Tris-HCl buffer (pH 7.5) was mixed with 0.2 ml of 0.2 M $MgCl_2$, 0.1 ml of 2 mM thymidine 5′-phosphate nitrophenyl-phosphate and 0.2 ml of enzyme solution. After incubation for 60 minutes at 37°C, the reaction was stopped by addition of 1.0 ml of 0.5 M NaOH. Precipitated protein was removed by centrifugation, and the E_{410} of the supernatant was measured spectrophotometrically.[4] The concentration of liberated nitrophenol was calculated from the standard calibration curve. One unit of enzyme is defined as the amount of enzyme that liberates one μ mole of nitrophenol per one minute.

(2) 3′-mononucleotide forming phosphodiesterase (PDase II)

For the purpose of this measurement, 0.1 M acetate buffer (pH 5.0), 0.2 M EDTA instead of $MgCl_2$ and 2 mM thymidine 3′-phosphate nitrophenyl phosphate as substrate were used in reaction mixture.[4]

3. Acid phosphatase activity (ACPase)

Assay mixture contained 1.0 ml of 0.2 M acetate buffer (pH 5.0), 0.8 ml of 16 mM *p*-nitrophenyl phosphate, and the enzyme solution in a total volume of 2.0 ml. After incubation at 37°C for 30 minutes, 0.1 ml portions of the reaction mixture were mixed with 4.0 ml of 0.25 M NaOH. The absorbance of this solution was measured at 410 nm.[5] One unit of enzyme activity was defined as the amount of enzyme liberating one μ mole of nitrophenol per minute.

4. Purine nucleoside phosphorylase

The enzyme assay is the method coupled with xanthine oxidase reaction,[6] which is based on the measurement of the increase in absorbance at 293 nm due to the formation of uric acid. One unit of enzyme activity is defined as the amount of enzyme which gives an optical density increase of 12.5 per minute (μ mole of uric acid formation) at 293 nm under standard assay conditions. The reaction mixture for the standard assay contained the following components: inosine 0.5 μ mol, potassium phosphate buffer (pH 7.5) 50 μ mole, xanthine oxidase 0.02 μ M unit, and the enzyme solution to be assayed in a final volume of 1 ml at room temperature. The mixture of all components, except the enzyme solution, was preincubated for about 1 minute, to remove any trace of hypoxanthine or xanthine introduced as a contaminant in the inosine. After a preincubation, the reaction was started by the addition of enzyme solution.

5. Phosphoribosyl transferase (PRTase)

The reaction mixture contained the following components in a total volume of 45 μl: 30 μl of 150 mM Tris-HCl buffer (pH 7.4) containing 7.5 mM $MgCl_2$ and 7.5 mM phosphoribosylpyrophosphate (PRPP), 5 μl of isotope labeled substrate (guanine-8-^{14}C, hypoxanthine-8-^{14}C, adenine-8-^{14}C, respectively), and 10 μl of enzyme solution to be assayed. After incubation for 15 minutes at 37°C, the reaction was stopped by the addition of 5 μl 4 N formic acid. The authentic inner standards (purine base, nucleoside, and nucleotide) were added to the reaction mixture, and 10 μl aliquot of the solution was applied on cellulose acetate membrane to separate the reaction products electrophoretically. An ultraviolet lamp (258 nm) was used to detect the areas corresponding to the known products, which were cut out from the cellulose strip, and applied on a liquid scintillation counter to count radioactivity.[7]

RESULTS

1. Effect of pH on the DNA hydrolysis

To study the effect of pH on DNA hydrolysis, a pH dependent activity curve ranging from pH 3.5 to 10 was determined. In the reaction mixture without any additional metal or EDTA, DNA hydrolysis had pH optimum at 4.5, as shown in Fig. 1, and a significant increase of hydrolysis was

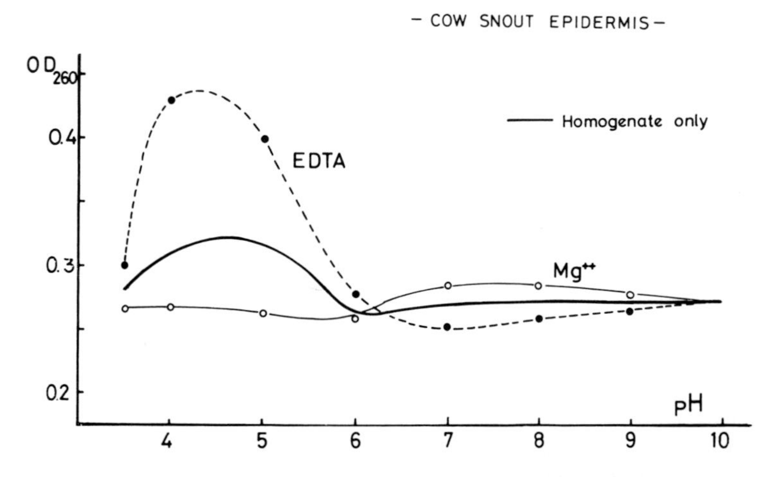

Fig. 1. Effect of pH on the DNA hydrolysis.

observed in the presence of EDTA in a final concentration of 0.024 mol. The presence of Mg^{++} led to an extensive hydrolysis of DNA in pH ranging from 7.0 to 8.0, but the activity was extremely less in the alkaline range than in the acidic area. From these results, the DNA hydrolysis is to be discussed at pH 5.0 in this paper.

2. Time course of DNA hydrolysis

One ml of 5 mg/ml salmon DNA was incubated with 5.0 ml of 0.25 M acetate buffer (pH 5.0), 1.5 ml of epidermal homogenate (0.1 g wet weight) in total volume of 10.0 ml. Two ml portions of the incubation mixture were mixed with 2 ml of 5% PCA in the following intervals: 1, 4, 14, and 20 hours. The optical density at 260 nm (E_{260}) of the PCA soluble product of the hydrolysis of DNA was determined. The optical density caused by the release of oligonucleotides gradually increased with the time of incubation, as shown in Figure 2. In order to clarify the DNA breakdown process, Sephadex G-75 chromatography was used. Five-tenth ml portions

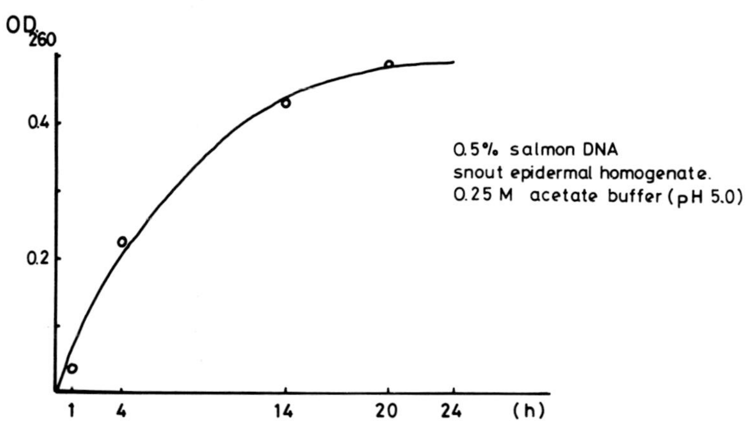

Fig. 2. Time course of DNA hydrolysis. Acid-soluble product of hydrolysis of DNA by water extract of cow snout epidermis.

of the above incubation mixture were applied to the column of sephadex G-75 (0.9 × 20 cm) at every 2, 5, and 20 hours of incubation time. The elution profile depicted in Fig. 3 has two UV-absorbing areas (E_{260}) corresponding to the molecular weight of products. The first peak eluated in void volume was shown to be referable to intact DNA, while the second elution peak was derived from the products of the low molecular weight caused by DNA hydrolysis. The first UV absorbing area showed the prominent peak at the short time of incubation, but further incubation produced an inversion in the elution profile. After incubation for 24 hours, most of the eluated hydrolysates were recovered in the second area.

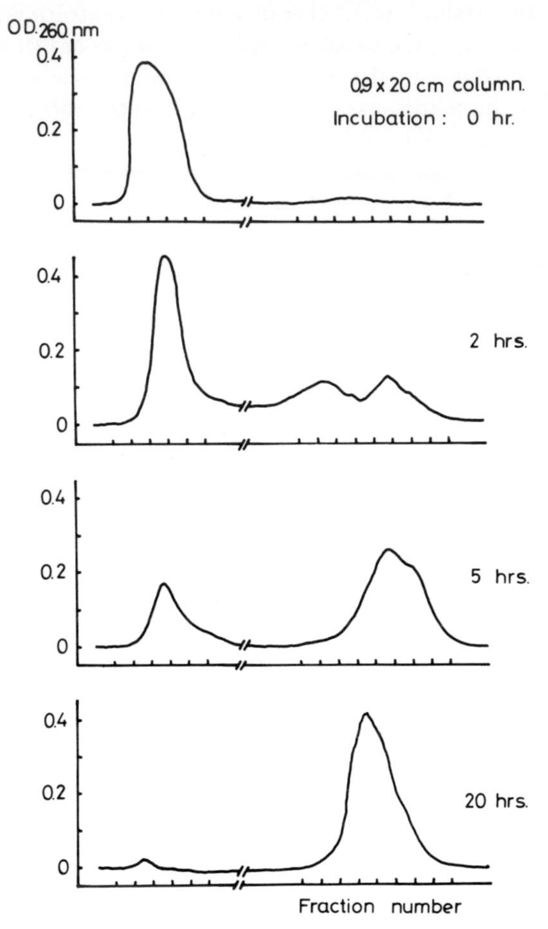

Fig. 3. Sephadex G-75 chromatography of degradation product of hydrolysis of DNA.

Appearance of these results means that enzymes in the cow snout epidermis hydrolyze the highly polymerized DNA to the low molecular weight products.

3. Degradation product of DNA

Salmon DNA was incubated with epidermal homogenate for one hour under the conditions described above. Five percent PCA solution was added to the incubation mixture, and the clear supernatant was obtained by centrifugation. The supernatant was treated with potassium hydroxide to remove an excess of PCA. The resulting acid-soluble supernatant was applied to a column (1.5 × 17 cm) of DEAE-cellulose.[8] The column was eluated with 0.02 M Tris-HCl buffer (pH 7.6)-7 M urea containing a linear gradient concentration of NaCl (0–1.0 M). The elution profile shown in Fig. 4 indicated a prominent peak in a void volume. Another small peak appeared in the latter eluation area. In a separate experiment, an

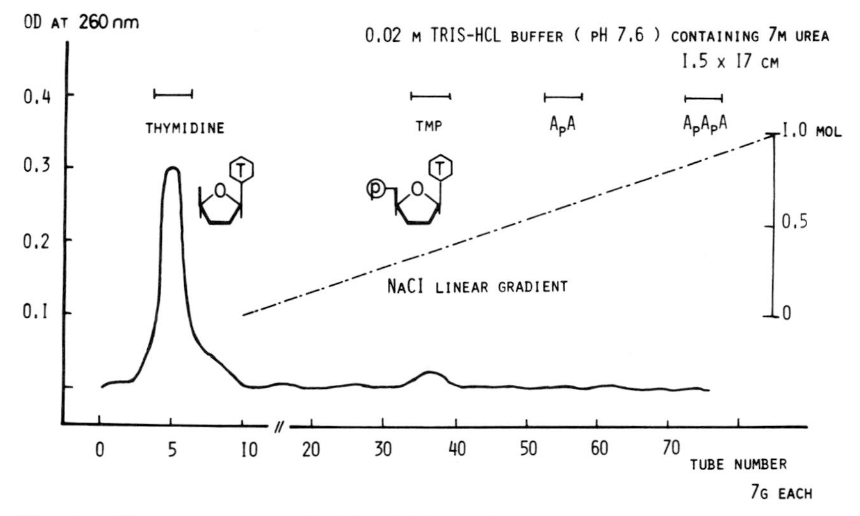

Fig. 4. Chromatography of acid soluble product of hydrolysis of DNA on DEAE cellulose-urea column.

authentic oligonucleotide was used as internal standard: thymidine, TMP, ApA and ApApA. The first peak eluated in void volume is referred to nucleoside, and the second small peak is referred to the TMP eluated in position. According to the above results, it is suggested that DNA was degraded along the same pathway as in other mammalian tissues (Fig. 5).

4. Purification and significance of nucleolytic enzymes in cow snout epidermis

(1) Acid DNase

DNase was purified approximately 200-fold in specific activity by DEAE-cellulose and Sephadex G-100 column.[9] The enzyme had a pH optimum at 5.5, and the temperature optimum at 50°C (Table 1). Any physiologic cation was not required on enzyme activity. Mg^{2+} was slightly inhibitory (98% inhibition in 0.05 M), and EDTA enhanced the enzymatic activity. Sulfate ion was a very strong inhibition. By the results obtained from the average chain length (not less than 3), the absence of a lag phase and alteration of viscosity, the DNase was characterized as the endonuclease. The native DNA was hydrolyzed to approximately 70% hydrolysed

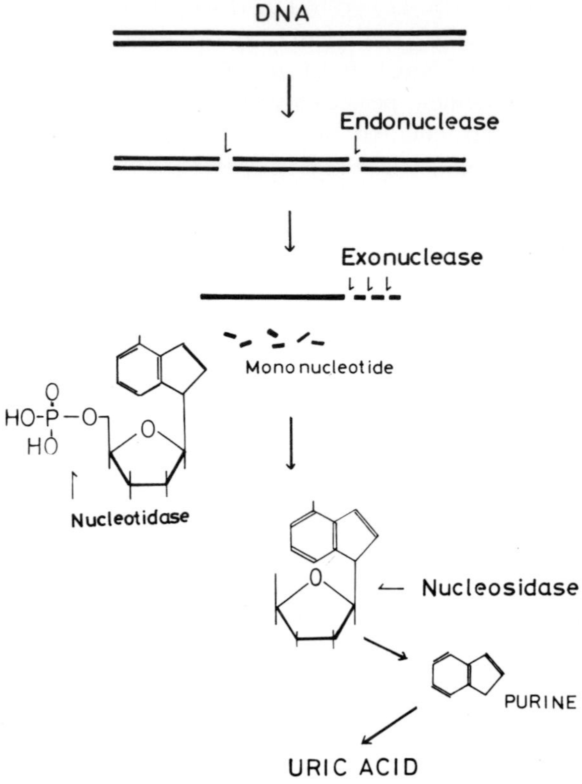

DNA
Endonuclease
Exonuclease
Mono nucleotide
HO-P-O
HO
O
Nucleotidase
Nucleosidase
PURINE
URIC ACID

Fig. 5. DNA catabolism.

Table 1. Biochemical properties of deoxyribonuclease isolated from mammalian skin.

	Santoianni, Rothman 1961	J. Tabachnick 1964		Miyagawa, Ogura 1971
Material	Human, Rat	Guinea pig	Guinea pig	Cow snout
Classification	DNase II (ACID)	DNase I (NEUTRAL)	DNase II (ACID)	DNase II (ACID)
Mode of action	Endonuclease	Endonuclease	Endonuclease	Endonuclease
Optimum pH	5.0	7.2	5.0	5.5
Optimum temperature		(LABILE)	(STABLE)	50°C
Inhibition	Mg^{++}	Na^{-}	Mg^{++}	sulfate, Mg^{++}
Activation	Na^{+}, K^{+}	Mg^{++}	Na^{+}, K^{+}	EDTA
DNA substrate	—	—	—	double strand DNA

product in acetate buffer (pH 5.5). Heat-denatured DNA was less readily degraded than native double-strand DNA.

(2) Phosphodiesterase (Exonuclease)

Phosphodiesterase hydrolyzed polynucleotide at the end of DNA-chain. The two different types of phosphodiesterase (I and II) were recognized in cow snout epidermis (Table 2). 3′-mononucleotide forming phosphodiestarase (type II) was about 6 times higher in activity. Phosphodiesterase was partially purified, approximately 100-fold in specific activity from the 20,000 × g supernatant by ammonium sulfate precipitation and DEAE-cellulose chromatography. The physiologic properties of phosphodiesterase were summarized in Table 2.[10]

Table 2. Enzymological properties of phosphodiesterase.

	Phosphodiesterase I	Phosphodiesterase II
Mode of action	Exonuclease	Exonuclease
Product	Nucleoside 5′-monophosphate	Nucleoside 3′-monophosphate
Optimum pH	7.5	5.0
Optimum temperature	45°C	50°C
Km (*p*-nitrophenyl-pT)	1.32×10^{-7} M	1.7×10^{-7} M
EDTA	not required	required
Metal ion	Mg^{++}	not required
Phosphate ion	competitive inhibition	competitive inhibition

(3) Acid phosphatase (ACPase)

The cow snout epidermis contained two kinds of acid phosphatase, termed ACPase-I and ACPase-II, which were separated by gel chromatography on Sephadex G-100. Although both of ACPase had hydrolyses *p*-nitrophenyl phosphate, the nucleotidase activity was found only in ACPase-I fraction. ACPase-I fraction was further purified approximately 80-fold in specific activity by the ammonium sulfate preparation (30%–65%), pH 4.0 treatment and DEAE-cellulose chromatography.[11] The partially purified acid phosphatase had a pH optimum at 5.0, Km value of 6.7×10^{-4} M for p-nitrophenyl phosphate hydrolysis, and relative heat stability. This enzyme was insensitive to formaldehyde, but was completely inhibited by fluoride and tartrate. No inhibitory effect by p-chloromercuribenzonate was observed. The inactivation was not observed by Mg^{2+}, Ca^{2+}, Na^{+}, K^{+}, and NH_4^{+}. The enzyme activity toward a number of phosphoric esters was investigated, and the results were summarised in Tables 3 and 4. Although the enzyme appeared to be most active toward p-nitrophenyl phosphate, a marked activity was observed with α-naphthyl phosphate and nucleoside monophosphate as substrate. Of the five nucleo-

Table 3. Relative rate of hydrolysis of several substrates by acid phosphatases purified from cow snout epidermis.

Substrate*		Acid phosphomonoesterase activity	
		I	II
p-Nitrophenyl phosphate		100% activity (37.74)**	100% (0.50)**
α-Naphtyl phosphate,		80.4	0
β-Glycerophosphate		27.0	—
Glucose-6-phosphate		2.7	—
Adenosine-5-monophosphate	(AMP)	34.1	0.5
Adenosine diphosphate	(ADP)	7.0	0
Adenosine triphosphate	(ATP)	4.0	0
Cyclic AMP		0	0

* Substrate concentration was 0.0008 M in all cases.
**( ): activity, units per ml sample.

Table 4. Relative rate of hydrolysis to several nucleotides by acid phosphomonoesterase in cow snout epidermis.

Substrate*	I
5′-AMP	100%
5′-ADP	21
5′-ATP	11
2′ or 3′-AMP	121
Cyclic AMP	0
5′-dGMP	144
5′- dCMP	86
5′-TMP	129
5′-dUMP	130

*in 2 M of final concentration.

side 5′-monophosphates, a relatively low activity was observed with CMP. Nucleoside 2′ or 3′-monophosphate was hydrolyzed as well as nucleoside 5′-monophosphate. No activity toward cyclic AMP was detected.

To find the biological significance of acid phosphatase, nucleotidase activity was determined in acidic and alkaline conditions. AMP and TMP were used as substrate. Acid or alkaline nucleotidase activity was measured at pH 5.0 or pH 9.0, in the presence of EDTA (10 mM) or Mg^{2+} instead of EDTA, respectively. β-glycerophosphate was also used as substrate to detect a so-called nonspecific alkaline phosphomonoesterase. The amount of inorganic phosphate released after incubation for 30 minutes at 37°C was determined.[12] As shown in Table 5, the purified enzyme has a highly acid phosphomonoesterase activity. The partially purified ACPase-I was

Table 5. Acid and alkaline phosphomonoesterase activities in cow snout epidermis.

Condition	Substrate		Activity
Acid	*p*-Nitropheny phosphate		267.41
Acid	Deoxyadenosine-5′-Monophosphate	(5′-dAMP)	86.24
Acid	Thymidine-5′-monophosphate	(5′-TMP)	144.52
Alkaline	β-Glycerophosphate		10.00
Alkaline	Deoxyadenosine-5′-monophosphate	(5′dAMP)	11.28
Alkaline	Thymidine-5′-monophosphate	(5′-TMP)	17.52

Activity: Micromole P*i* per gm fresh tissue per 30 min incubation at 37°C
Sample: Homogenate.

considered to be a typical nonspecific acid phosphatase playing the role of nucleotidase.

(4) Metabolism of nucleosides.

Purine nucleoside phosphorylase activity was assayed with the method of coupled xanthine oxidase reaction. The activity was as low as n mole units per mg protein. Xanthine oxidase activity in the cow snout epidermis was hardly detected by the present method. Purine nucleoside phosphorylase had a pH optimum at 7.5, optimum temperature at 37°C. Mg^{2+} was not required for enzyme activity.[13] Hypoxanthine-guanine phosphoribosyl transferase (HGPRTase) and adenine-phosphoribosyl transferase (APRTase) activity were observed in the cow snout epidermis (Table 6). HGPRTase appeared to be more active toward guanosine than hypoxanthine as substrate. APRTase was lower in activity than HGPRTase. HGPRTase was partially purified by pH 4.5 treatment, 65°C heat treatment, 40–70% ammonium sulfate preparation and sephadex G-200 chromatography.[14] The partially purified HGPRTase had a pH optimum at 7.0; optimum temperature at 56°C. The enzyme was stabilized by PRPP, and Mg^{2+} was required for the activation of enzyme. The molecular weight determined by gel filtration was about 80,000.

Table 6. Purine phosphoribosyl transferase activity in cow snout epidermis 20,000 × *g* Supernatant.

Hypoxanthion-Guanine Phosphoribosyl Transferase		
substrate:	Hypoxanthione	2.17 ± 0.53
	Guanine	4.39 ± 0.80
Adenine Phosphoribosyl Transferase		
substrate:	Adenine	0.75 ± 0.29

Enzyme activity: N mole/min./mg protein.

5. Distribution of nucleolytic enzymes in the epidermis

The DNA and RNA contents of cow snout epidermis were 18.5 ± 0.7 μg and 44.1 ± 4.4 μg per mg protein, respectively. The RNA/DNA ratio was 2.38. The DNA content of successive layers of snout epidermis is shown in Table 7. The cow snout epidermis was divided into three successive layers by laboratory microtome: the upper, middle, and lower layers. DNA content progressively decreases in successive layers. DNA content in the upper layer is low, amounting at most to only about 30% of the DNA content in the lower layer. The existence of a lower DNA level in the upper layer may reflect the DNA breakdown in the epidermis. Table 7 shows the distribution of nucleolytic enzyme activities in the three successive layers. Nucleolytic enzyme activities related with DNA degradation are found to be high in the upper layer of epidermis. It is of interest that the enzyme activity of the salvage pathway is relatively high in the lower layer.

In Table 8 are listed the nucleolytic enzymes from various mammalian tissue extracts, bovine liver, kidney and snout epidermis. As compared

Table 7. Distribution of nucleolytic enzyme activities in successive layers of cow snout epidermis.

DNA Content (μg)	Upper layer 12.9 ± 1.1	Middle layer 21.4 ± 0.8	Lower layer 40.9 ± 1.9
Acid DNase	3.56 unit	2.71 unit	2.70 unit
Phophodiesterase I	2.39	2.42	2.49
II	6.13	5.64	5.24
Acid phosphatase	144.2	89.5	77.3
Purine nucleotide phosphorylase(+)	372.1	327.1	758.3
Hypoxanthine-guanine phosphoribosyl transferase(+)	1.40	1.25	1.81

Enzyme activity: μ mole/min/mg potein
(+):n mole/min/mg protein

Table 8. Nucleolytic enzyme activities in bovine tissues.

	Epidermis	Liver	Kidney
Acid DNase	8.96 unit	6.96 unit	5.03 unit
Phosphodiesterase I	3.08	3.21	6.13
II	9.71	2.63	4.38
Acid phosphatase	150.3	44.8	75.4
Purine nucleotide phosphorylase(+)	416.2	86.6	129.7
Hypoxanthine-guanine phosphoribosyl tranferase(+)	2.17	1.37	1.59

Enzyme activity: μ mole/min/mg protein
(+): n mole/min/mg protein

with other tissues, the snout epidermis contains relatively high activity of DNA hydrolytic enzymes.

DISCUSSION

The presence of only a small amount of DNA and RNA in a normal stratum corneum suggests that during normal keratinization almost complete degradation of nucleoproteins occurs.[15] Catabolism of nucleic acid is related to the process of keratinization. Santoianni *et al.*[16] and Tabachnick[1] have studied the nucleic acid splitting enzymes in the animal epidermis. The salmon DNA incubated with the epidermal extract rapidly released the acid-soluble oligonucleotides. DNA was less degraded at alkaline pH than at acid pH.

In our separate experiment, nucleus fraction prepared from bovine liver was incubated with the epidermal homogenate (700 × *g* supernatant) in the acetate buffer (pH 5.0). The optical density at 260 nm caused by the release of oligonucleotides into the acid-soluble fraction gradually increased with the time of incubation (Fig. 6). The results of this experiment indicate that the degradation of nucleoprotein associates with lysosomal enzymes. Several lysosomal enzymes are involved in the decomposition of nucleic acid.[17] The various steps leading to a complete breakdown are shown in Fig. 5. The presence of these related enzymes could be proved in cow snout epidermis, and the enzymes were partially purified.

The primary attack on DNA occurs in accordance with an endonucleolytic manner. Endonuclease is known to split DNA-strands simultaneously at about the same level. Under the endonucleolytic action, DNA is split into oligonucleotides with an average size of 10 to 12 nucleotide units. DNase purified from cow snout epidermis has endonucleolytic properties and is

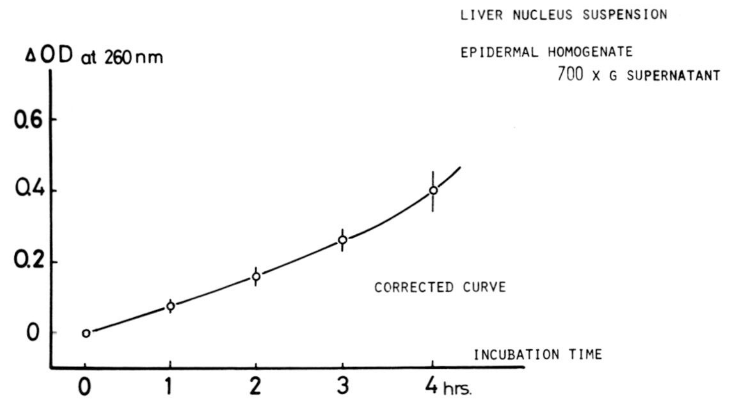

Fig. 6. Degradation of bovine liver nucleus DNA by snout epidermal homogenate.

classified as DNase II. Neutral DNase activity is also present in the guinea pig epidermis, but is weak or absent in the cow snout, rat, and human epidermis. The biological significance of neutral DNase is not clear.

The oligonucleotides formed are further hydrolyzed by exonuclease (phosphodiesterase) that releases free phosphomononucleotides stepwise from OH-end of an oligonucleotide chain. Cow snout epidermis contains two different types of phosphodiesterase (I & II). The hydrolysis rate was higher in 0.2 M acetate buffer-0.2 M EDTA (pH 5.0) than in 0.1 M Tris buffer-0.2 M $MgCl_2$ (pH 7.5). The hydrolyzed product results in nucleotide with 3′-phosphate. Nucleotide with 5′-phosphate may be less.

Finally the phosphomononucleotides are themselves split into phosphates and nucleosides by nucleotidase. In cow snout epidermis, at least two different types of acid phosphtase are present, and one of them appears to be acid nucleotidase of broad specificity, acting equally well on most 3′- and 5′-nucleotides.

A series of DNA breakdown enzymes, acid DNase—phosphodiesterase—acid phosphatase, could hydrolyze DNA to nucleosides. It is reasonable to assume that these nucleases play a role in degrading the nucleic acid during keratinization.

The decomposition products of nucleic acid are to be detected in water extracts of normal stratum corneum. However, only small amounts of breakdown products are found in the keratinized layer. The presence of purine derivatives such as xanthine and hypoxanthine has been reported by Wheatley,[18,19] Hodson,[20] De Bersaques,[21] and Miyagawa.[22] The decomposition of nucleic acid results in the formation of nucleosides and nucleotides, which in turn are broken down to pentose, phosphate, purine and pyrimidine base. Of these breakdown products, pentose can be readily detected in the water extract of the stratum corneum.[20] Xanthine and hypoxanthine are present, but no other nucleic acid bases have been detected. The purine bases are present in both DNA and RNA, and then are released during nucleic acid decomposition. It is of interest that the bases detected in normal stratum corneum are the purine derivates: xanthine and hypoxanthine. Uric acid is also derived from purine base through xanthine oxidase. Xanthine oxidase activity has been shown in guinea pig epidermis,[21] but similar experiments with cow snout epidermis[13] and human epidermis[21] proved it to be almost negative.

There remain unsolved problems regarding nuclease-mediated breakdown of nucleic acid. In 1963 Schwarz[24] reported that adenylic and guanylic nucleotides were detected after incubation of epidermal homogenate with adenine-^{14}C. Enzymes directly yielding nucleotides from purine and pyrimidin bases are known to play an important part in the salvage pathway of nucleic acid. Imura[23] has reported that the incorporation of labeled precursors into DNA through the *de novo* pathway is very low, while that of labeled precursors or the salvage pathway is quite high in rat epidermis. It is suggested that the salvage pathway is the way of DNA synthesis in the epidermis. One of the most interesting observations is that adenine phosphoribosyl transferase (APRTase) activity is found in the epidermis, although adenine is not detected in cow snout epidermis and the inversion activity from adenosine to adenine is very low.[14] The biological role of APRTase in the epidermis still remains unclear.

It was considered that DNA was degraded along a similar pathway as in the other mammalian tissues. The activity pathway in the cow snout epidermis may be accepted as illustrated in Figs. 7 and 8. The results of

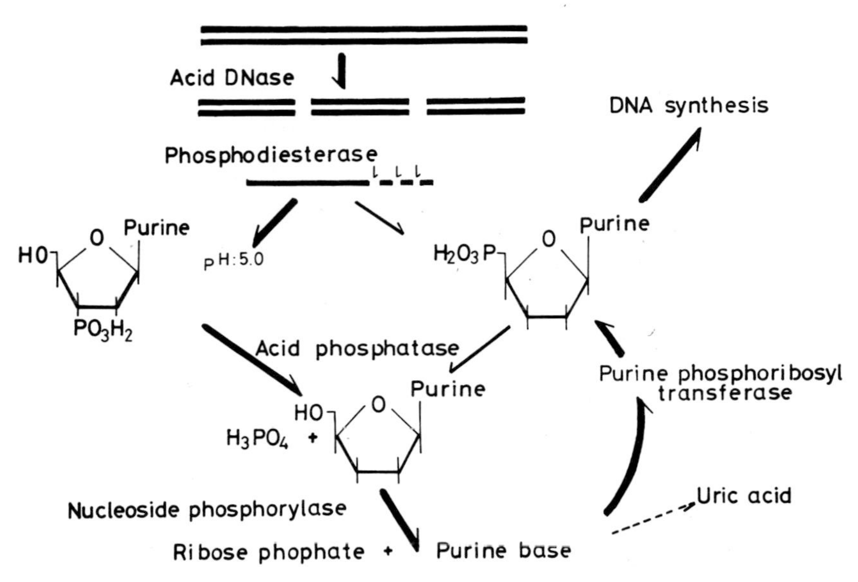

Fig. 7. Activity pathway of DNA catabolism in cow snout epidermis.

our present experiment do not directly answer the original problem as to whether the epidermal nucleases play a role in the biological control mechanism during cellular differentiation. However, the purification of enzymes will be applicable to fluorescent antibody technique. The antibody technique may improve the determination of cellular and subcellular localization of enzymes. Further work is in progress to establish the biological characteristics of nucleolytic enzymes in cellular differentiation.

Acknowledgements

This work was partially supported by the Basic Research Grant (1975) from the Japanese Dermatological Society. We wish also to express gratitude to Miss Y. Takashima and Miss A. Fujii, laboratory technicians in our department, for their generous help.

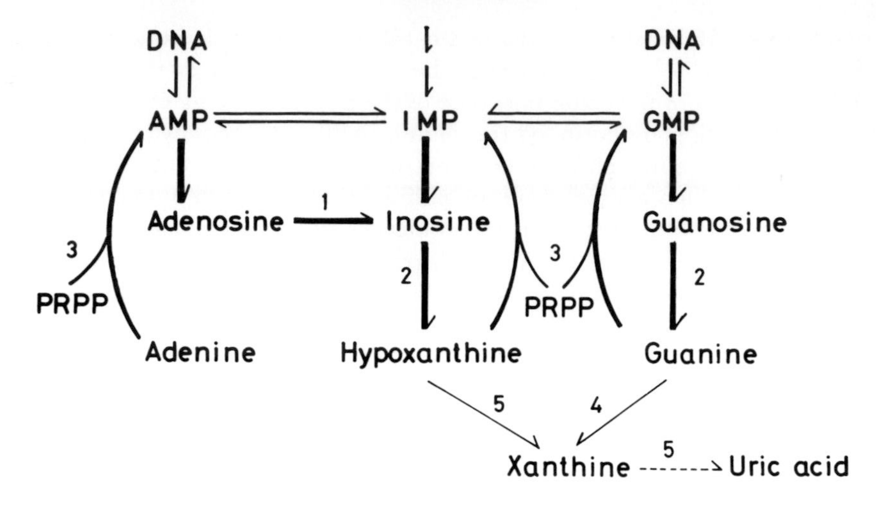

Fig. 8. Activity pathway of purine metabolism in cow snout epidermis
1. Adenosine deaminase; 2. Purine nucleoside phosphorylase; 3. Purine phosphoribosyl transferase; 4. Guanase; 5. Xanthine oxidase.

REFERENCES

1. Tabachnick, J. Studies of the biochemistry of epidermis (II). *J. Invest. Derm.*, **42**: 471–478, 1964.
2. Lowry, O. H., Rosebrough, N. J., Farr, A. L., and Randall, R. J. Protein measurement with Folin Phenol reagent. *J. Biol. Chem.*, **193**: 256–275, 1951.
3. Ogura, R., Freeman, R. G., and Knox, J. M. Morphological and biochemical changes in epidermal nuclei and nucleic acid following ultraviolet exposure. *Dermatologica*, **136**: 11–17, 1968.
4. Laskowski, M. Exonuclease (phosphodiesterase) and other nucleolytic enzymes from venom. In Procedures in Nucleic Acid Research (Cantoni, G. L., and Davies, D. R.) (G. L. Cantoni & D. R. Davies, eds.) Harper International Edition, New York, 1967, 154–187.
5. Lowry, O. H. Micromethods for the assay of enzymes. In Methods in Enzymology, vol. 4 (S. P. Colowick et. al., eds.), Academic Press, New York, 1957, 371–372.
6. Kim, B. K., Cha, S., and Parks, R. E. J. Purine nucleotide phosphorylase from human erythrocytes. *J. Biol. Chem.*, **243**: 1763–1770, 1968.
7. Kizaki, H. and Sakurada, T. A method for determination of hypoxanthine-guanine and adenine phosphoribosyltransferase based on the isolation of the reaction products by means of electrophoresis on cellulose acetate membrane. Proceeding of Jap. J. Clin. Chem., **14**, 110, 1974.
8. Uozumi, T., Hino, T., Tamura, G., and Arima, K. Studies on the autolysis of Aspergillus oryzal (part II). *Agr. Biol. Chem.*, **36**: 434–441, 1972.
9. Miyagawa, T. and Ogura, R. M. The biochemical property and physiologic

significance of deoxyribonuclease in the epidermis. *J. Invest. Derm.,* **57**: 330–336, 1971.
10. Maruta, T. Purification and characterization of phosphodiesterase from cow snout epidermis. Being prepared for publication.
11. Hidaka, T. Purification and characterization of acid phosphatase from cow snout epidermis. *Kurume Med. J.,* **39**: 34–53, 1976.
12. Sanui, H. Measurement of inorganic orthophosphate in biological materials: Extraction properties of butyl acetate. *Anal. Biochem.,* **60**: 489–504, 1974.
13. Kinoshita, M. Purification and characterization of purine nucleoside phosphorylase from cow snout epidermis. Being prepared for publication.
14. Koga, H. Purification and characterization of purine phosphoribosyl transferase from cow snout epidermis. Being prepared for publication.
15. Santoianni, P. and Ayala, M. Sc. D., Fluorometric ultramicroanalysis of deoxyribonucleic acid in human skin. *J. Invest. Derm.,* **45**: 99–103, 1965.
16. Santoianni, P. and Rothman, S. Nucleic acid-splitting enzymes in human epidermis and their possible role in keratinization. *J. Invest. Derm.,* **37**: 489–495, 1961.
17. Vaes, G. Digestive capacity of lysosomes. In Lysosomes and Storage Diseases (H. G. Hers & F. Van Hoof, eds.) Academic Press, New York, 1973, 43–77.
18. Wheatly, V. R. and Farber, E. M. Studies on the chemical composition of psoriatic scale. *J. Invest. Derm.,* **36**: 199–211, 1959.
19. Wheatly, V. R. and Farber, E. M. Chemistry of psoritic scale (II). *J. Invest. Derm.,* **39**: 79–89, 1962.
20. Hodyson, C. Nucleic acids and their decomposition products in normal and pathological horny layer. *J. Invest. Derm.,* **39**: 69–78, 1962.
21. De Bersaques, J. Purine and pyrimidine metabolism in human epidermis. *J. Invest. Derm.,* **48**: 169–173, 1967.
22. Miyagawa, T. and Urabe, H. Nucleic acid catabolism in epidermis. *Nishinihon J. Derm.,* **36**: 783–786, 1974.
23. Imura, H. Studies on nucleic acid metabolism in the skin (I). *Nishinihon J. Derm.,* **36**: 670–676, 1974.
24. Schwarz, E. Untersuchungen zum Schicksal der aus Nucleinsäuren beim Zellabbäuwahrend der epidermalen keratinisation freiwerdenden Purine. *Arch. Klin. Exp. Derm.,* **216**: 427–445, 1963.

Discussion

DR. GOLDSMITH: As you purify these enzymes, do you find any natural inhibitors which keep these hydrolases from degrading DNA *in vivo* at the wrong time?

DR. OGURA: Do you mean a natural biological inhibitor?

DR. GOLDSMITH: Yes.

DR. OGURA: Until now, we have not looked for such inhibitors, but in the future we hope to do so.

DR. TAKEBE: Does the endonuclease in your study break double-stranded DNA as you showed in the schematic figure?

DR. OGURA: I'm not sure right now but probably it does break double-stranded DNA. Like DNase II, our enzyme does not require any metal ion for activity so the activity is probably on double-stranded DNA.

DR. MIYAGAWA: Do you have any additional evidence on this point?

DR. OGURA: In general, acid DNase (DNase II) has the property of hydrolyzing double-stranding DNA. I purified a DNase from cow snout epidermis and found it to have activity on double-stranded DNA.

DR. STEINERT: The cow snout contains a large flora of microorganisms, many of which produce hydrolytic enzymes. What precautions were taken to avoid contamination of your enzyme preparation from this source?

DR. OGURA: So far, we haven't taken any precautions to avoid contamination with bacterial enzymes. I will keep this problem in mind for future work.

DR. BERNSTEIN: It seems to me that the amount of activity recovered is too great for the enzyme to be present only as a result of bacterial contamination.

DR. MCGUIRE: I wonder if it wouldn't be useful to compare these nucleases with those of the intestinal epithelium and bone marrow rather than with the enzymes in liver? The high levels of nucleases in epidermis may reflect salvage pathways rather than *de novo* pathways such as take place in the liver.

DR. OGURA: That is an interesting suggestion. I will determine the nucleolytic activities in intestinal epithelium and compare them with the activities in the epidermis. Thank you for your suggestion.

DR. TEZUKA: Have you ever done the same type of experiments using psoriatic skin?

DR. OGURA: No, I have never determined the level of enzyme activity in pathological skin.

DR. BERNSTEIN: Extending Dr. Goldsmith's question, did you at any time note a major increase in activity at a particular step during the course of the purification? If you had lost an inhibitor, you might have gained activity at one step in the purification.

DR. OGURA: We used DEAE-cellulose columns and Sephadex G-100 columns for purification of the enzymes. The total activity increased remarkably at the stage of DEAE-cellulose chromatography. Dr. Miyagawa, who purified the enzyme from cow snout epidermis, will be able to answer in detail. Dr. Miyagawa, would you please give us the information?

DR. MIYAGAWA: The total activity of DNase II from guinea pig epidermis increased significantly after the DEAE-cellulose chromatographic step. When crude extracts of guinea pig epidermis were dialyzed against buffer at pH 5.0 and centrifuged, the supernatant solution contained about 150 percent of the original activity. The addition of the resuspended pellet to the supernatant solution significantly decreased the total activity of the preparation providing evidence for the presence of an inhibitor of DNase in the guinea pig epidermis.

Purine Metabolism in Epidermis

Itsuro MATSUO,* Muneo OHKIDO,* Harutoshi KIZAKI ** and Tomomi SAKURADA**

*Department of Dermatology, Tokai University School of Medicine, Kanagawa, Japan and
**Department of Biochemistry, Keio University School of Medicine, Tokyo, Japan

ABSTRACT

The enzymes concerning the purine nucleotide metabolism were measured quantitatively by microprocedure with a homogenate of guinea pig and human epidermis.

The degradation enzymes of purine nucleotides, such as AMP deaminase, 5′-nucleotidase, adenosine deaminase and purine nucleoside phosphorylase, were demonstrated in guinea pig epidermis; and AMP deaminase, purine nucleoside phosphorylase, guanine deaminase, and xanthine oxidase were demonstrated in human epidermis.

Purine phosphoribosyltransferases, salvage enzymes for purine nucleotide synthesis, were also demonstrated in guinea pig and human epidermis.

It may be suggested from these results that purine nucleotides in guinea pig and human epidermis are degraded to purines by these degradation enzymes and are also synthesized by the salvage pathways utilizing the purines produced by the breakdown of nucleic acids of nucleotides.

INTRODUCTION

It is well known that nuclear and cytoplasmic nucleic acids are actively synthesized in epidermal cells during their keratinazation process.[1,2,3] However, no DNA and RNA or degradation products of nucleotides are detected in normal stratum corneum. The degradation enzymes of DNA and RNA are demonstrated in animal and human epidermis by biochemical methods,[4,5,6,7,8,9] and their roles in the disappearance of nuclei in the stratum corneum are discussed.

The present study was undertaken to clarify the mechanisms of degradation and synthesis of purine nucleotides in epidermis.

The enzymes concerning these pathways were measured quantitatively with a homogenate of guinea pig and human epidermis. As degradation, enzymes of purine nucleotides, AMP deaminase, 5′-nucleotidase, adenosine deaminase, purine nucleoside phosphorylase, guanine deaminase and xanthine oxidase were measured, and as salvage pathway enzymes for purine nucleotide synthesis, purine phosphoribosyltransferases were measured.

MATERIALS AND METHODS

The skin of five male guinea pigs, weighing 250–300g, was removed by peeling. Normal human skin was obtained from the abdomens of seven surgical patients aged 30 to 72, whose serum uric acid values were within a normal level. The epidermis was separated from the dermis by the application of physical stretch method.[10] Guinea pig epidermis, weighing ca 200 mg, was homogenized in a Potter-Elvehjem all glass homogenizer at 4°C with nine volumes of 50 mM Tris-HCl buffer, pH 7.4, and the homogenate was filtrated through a nylon net. The filtrated homogenate was centrifuged at 800 × *g* for 10 minutes and then 8,000 × *g* for 30 minutes. The supernatant at 800 × *g* centrifugation was used as a sample for 5′-nucleotidase assay, and the supernatant at 8,000 × *g* was used as a sample for the other enzymes. Human epidermis, weighing ca 70 mg, was homogenized with ten volumes of 50 mM Tris-HCl buffer, pH 7.4. The supernatant at 800 × *g* centrifugation for 15 minutes was used as an enzyme sample.

AMP deaminase assay. The solution (0.5 ml final volume), consisting of 20 mM AMP, 2 mM ATP, 250 mM sodium citrate buffer, pH 6.0, and enzyme sample, was incubated for 30 minutes at 37°C. The enzyme activity was calculated from the amount of released ammonium.[11]

5′-Nucleotidase assay. The solution (1 ml final volume), consisting of 3 mM AMP, 5 mM $MgCl_2$, 0.125 M Tris-HCl buffer, pH 8.0, and enzyme sample, was incubated for 30 minutes at 37°C. The released phosphorus was analyzed according to a modification of the procedure of Fiske and Subbarow.[12]

Adenosine deaminase assay. The reaction mixture consisted of 0.04 mM adenosine, 50 mM Tris-HCl buffer, pH 7.4, and enzyme sample in a final volume of 1 ml. The decrease of adenosine was directly measured by spectrophotometry at 265 nm.

Purine nucleoside phosphorylase assay. The solution (1 ml final volume), consisting of 4 mM inosine, 45 mM Tris-HCl buffer, pH 7.4, and enzyme sample, was incubated at 37°C for 10 minutes, and then was hydrolysed with 0.1 ml of 3N HCl at 37°C for 15 minutes. The reaction mixture was neutralized by adding 0.1 ml of 3N KOH, and the released ribose was measured by the method of Dygert *et al.*[13]

Guanine deaminase assay. An enzyme sample was incubated for 30 minutes at 37°C in a micro tube containing 0.35 mM guanine-8-^{14}C (specific activity 7.8 mCi/mmol, New England Nuclear Co.), 80 mM Tris-HCl buffer, pH 8.0, in a final volume of 35 μl. The reaction was stopped by placing the tube in boiling water for 3 minutes. After centrifugation of the micro tube, an aliquot of the supernatant was separated by cellulose acetate membrane (Cellogel, Chemetron Italy) electrophoresis. Electrophoresis was carried out in 0.1 M borate buffer, pH 9.0, at 150 V for 40 minutes. The radioactivity of xanthine was measured with Nuclear Chicago, Counter Model 6725, and enzyme activity was calculated from the radioactivity of xanthine.

Xanthine oxidase assay. An enzyme sample was incubated for 30 minutes at 37°C in a micro tube containing 0.1 mM hypoxanthine-8-^{14}C (specific activity 58 mCi/mmol, Radiochemical Center, Amersham), 100 mM Tris-HCl buffer, pH 7.4, in a final volume of 25 μl. The reaction was stopped by placing the tube in boiling water for 3 minutes. After centrifugation of the micro tube, an aliquot of the supernatant was separated by electrophoresis under the same conditions of guanine deaminase assay. Enzyme activity was calculated from the radioactivity of xanthine and uric acid.

Purine phosphoribosyltransferase assay. Hypoxanthine, guanine, and adenine phosphoribosyltransferase (HPRTase, GPRTase and APRTase) were assayed according to the procedure of Kizaki and Sakurada.[14] Enzyme sample was incubated for 15 minutes at 37°C in a series of micro tubes containing 0.1 M Tris-HCl buffer, pH 7.4, 5 mM $MgCl_2$, 2 mM 5-phosphoribosyl-1-pyrophosphate tetra sodium (Kyowa Research Biochemicals) and substrate [0.11 mM hypoxanthine-8-^{14}C, 0.51 mM guanine-8-^{14}C, or 0.25 mM adenine-8-^{14}C (specific activity 17.8 mCi/mmol, New England Nuclear Co.)] in a final volume of 45 μl. The reaction was stopped by the addition of 5 μl of 4N formic acid. The reaction products were separated by electrophoresis and the enzyme activity was calculated from the sum of the radioactivity of nucleotide and nucleoside.

Protein concentration was measured by the method of Lowry *et al.*[15] using bovine serum albumin as a standard.

RESULTS

The specific activities of the enzymes in guinea pig epidermis are shown in Table 1 and those in human epidermis in Table 2.

Microprocedures used in our experiments made it possible to measure quantitatively some of the enzymes concerning purine metabolism with a very small amount of tissue homogenate.

Table 1. Specific activities of the enzymes in guniea pig epidermis.

Enzyme	Specific Activity (Mean ± standard error)
	nmols/mg. protein/min.
AMP deaminase	108.757 ± 14.538
5′-Nucleotidase	74.160 ± 4.910
Adenosine deaminase	9.289 ± 0.267
Purine nucleoside phosphorylase	116.726 ± 2.840
Purine phosphoribosyltransferase	
HPRTase	1.514 ± 0.120
GPRTase	3.665 ± 0.389
APRTase	1.605 ± 0.160

Table 2. Specific activities of the enzymes in human epidermis.

Enzyme	Specific Activity (Mean ± standard error)
	nmols/mg.frotein/min.
AMP deaminase	40.461 ± 2.518
Purine nucleoside phosphorylase	29.748 ± 2.709
Purine phosphoribosyltransferase	
HPRTase	0.894 ± 0.094
GPRTase	1.407 ± 0.201
APRTase	3.470 ± 0.242
	pmols/mg.protein/min.
Guanine deaminase	51.3 ± 7.7
Xanthine oxidase	12.7 ± 3.1

AMP deaminase could be demonstrated in both guinea pig and human epidermis. AMP deaminase is one of the enzymes of purine nucleotide interconversion and catalyzes the deamination of AMP to IMP.

Catabolic enzymes of purine nucleotide, 5′-nucleotidase, which catalyzes the dephosphorylation of nucleotides to nucleoside, and adenosine deaminase which catalyzes the hydrolytic deamination of adenosine to inosine, could be demonstrated in guinea pig epidermis. No activities could be demonstrated by the same procedures in human epidermis.

Purine nucleoside phosphorylase is known to catalyze reversible interconversion between purine nucleoside and its base. This enzyme could be demonstrated in both guinea pig and human epidermis and was found very active in epidermal homogenate.

Guanine deaminase is a specific deaminase and catalyzes the hydrolytic deamination of guanine to xanthine. The final pathway of degradation of purine nucleotide is the formation of uric acid from xanthine by xanthine oxidase. Xanthine oxidase catalyzes the oxidation of hypoxanthine and xanthine. In our microprocedures, xanthine oxidase and guanine deaminase could be demonstrated as pmols/mg protein/min unit in human epidermis.

Purine nucleotide, IMP, was formed from 5-phosphoribosyl-1-pyrophosphate and simple precursors by the pathway called *de novo* synthesis. Purine nucleotides can be formed reutilizing the degradation products of purine nucleotides by the pathways which may be regarded as salvage pathways. Purine phosphoribosyltransferases catalyze the formation of purine nucleotide from free purine with 5-phosphoribosyl-1-pyrophosphate. HPRTase, GPRTase, and APRTase were demonstrated in both guinea pig and human epidermis.

DISCUSSION

In the epidermis, DNA and RNA are actively synthesized for cell division and protein synthesis, but in normal stratum corneum, no DNA and RNA, and the degradation products of nucleotides are demonstrated. However, in pathological psoriatic stratum corneum which contains nuclei, DNA and RNA, and the degradation products of nucleic acids, such as purine base and uric acid were demonstrated,[16,17,18] and it was suggested that the degradation mechanism of nucleotide is present in the epidermis together with the synthesis of nucleotides.

In this paper, the enzymes concerning the degradation of purine nucleotides to purine bases and uric acid, and the synthesis of purine nucleotides by utilizing the degradation products of purine nucleotides in epidermis are discussed. There are numerous reports on chemical properties of some of the degradation enzymes of DNA and RNA in animal and human epidermis.[4,5,6,7,8,9] Endonuclease and exonuclease, which catalyze the degradation of DNA to oligonucleotide, and oligonucleotide to nucleotide, were also demonstrated in guinea pig epidermis by Miyagawa *et al.*[19,20]

The interconversion of purine derivatives and the major pathway of purine nucleotide catabolism and the salvage pathway of purine nucleotide are shown in Fig. 1.

Enzymes concerning these pathways demonstrated in animal and human epidermis qualitatively or indirectly, but not quantitatively, are purine nucleoside phosphorylase,[21,22] adenosine deaminase,[21,22,23] guanosine deaminase,[23] guanine deaminase,[22,24] xanthine oxidase,[25,26] and purine phosphoribosyltransferase.[22,27] Although adenine deaminase was also demonstrated in human epidermis,[22] hypoxanthine was not formed in our experiment from ^{14}C-adenine in guinea pig epidermis.

Purine nucleoside phosphorylase was demonstrated in psoriatic scales by Bersaques,[21] and the disturbed metabolism of nucleic acid derivatives

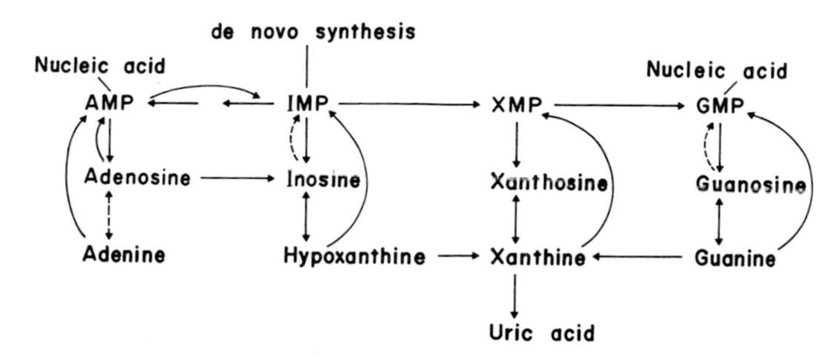

Fig. 1. The interconversions of purine derivatives.

in psoriasis was suggested by the high activity of this enzyme in psoriatic epidermis. The accumulation of the degradation products of nucleic acids in psoriatic scales might be related to this high enzyme activity.

The final pathway of degradation of purine nucleotides is the oxidation of xanthine to uric acid by xanthine oxidase activity. This enzyme was demonstrated in animal skin but not at all in human epidermis. But, recently Santoianni *et al.*[28] demonstrated this enzyme activity in human hair roots. Our results may suggest that uric acid can be formed in human epidermis, and the presence of uric acid in psoriatic stratum corneum is due to this enzyme activity.

Purine phosphoribosyltransferases have a significant role as purine salvage and these enzymes have been found and characterized in a number of human and mammalian tissues.[29,30,31,32,33,34] In the epidermis of human and rat skin, the presence of the salvage pathway of purines has been suggested,[22,27] but quantitatively these enzymes in the epidermis have not yet been studied. By the method of Kizaki and Sakurada,[14] which is applicable to crude extracts, the activities of these enzymes could be demonstrated in guinea pig and human epidermis and assayed accurately even in the presence of 5′-nucleotidase, adenosine deaminase, guanine deaminase and xanthine oxidase.

It has already been reported that pyrimidine nucleotides were synthesized both by the *de novo* and the salvage pathways in the mouse and rat epidermis,[35,36,37] and in the rat skin, the salvage pathway was considered to be more important than the *de novo* pathway for pyrimidine nucleotide synthesis.[37] And, in fact, glutamine phosphoribosyl pyrophosphate amidotransferase activity, the rate limiting enzyme of the *de novo* synthesis of purines, in guinea pig epidermis was very low (unpublished data).

It may be suggested from these results that purine nucleotides in guinea pig and human epidermis are synthesized mainly by the salvage pathways utilizing the purines produced by the breakdown of nucleic acids of nucleotides.

REFERENCES

1. Bernstein, I. A. and Foster, P. The metabolism of the nucleic acids in the skin of young rats. *J. Invest. Dermatol.*, **29**: 415–422, 1957.
2. Fukuyama, K. and Bernstein, I. A. Autoradiographic studies of the incorporation of thymidine- H^3 into deoxyribonucleic acid in the skin of young rats. *J. Invest. Dermatol.*, **36**: 321–326, 1961.
3. Fukuyama, K. and Epstein, W. L. Synthesis of RNA and protein during epidermal cell differentiation in man. *Arch. Dermatol.*, **98**: 75–79, 1968.
4. Santoianni, P. and Rothman, S. Nucleic acid-splitting enzymes in human epidermis and their possible role in keratinization. *J. Invest. Dermatol.*, **37**: 489–495, 1961.
5. Liss, M. and Lever, W. F. Purification and characterization of ribonuclease from psoriatic scales. *J. Invest. Dermatol.*, **39**: 529–535, 1962.
6. Tabachnick, J. Studies on the biochemistry of epidermis. II. Some characteristics of deoxyribonucleases I and II of albino guinea pig epidermis and saline extracts of hair. *J. Invest. Dermatol.*, **42**: 471–478, 1964.
7. Miyagawa, T. and Ogura, R. M. The biochemical property and physiologic significance of deoxyribonuclease in the epidermis. I. Partial purification and properties of DNase isolated from cow snout epidermis. *J. Invest. Dermatol.*, **57**: 330–336, 1971.
8. Miyagawa, T. Nucleolytic enzymes of epidermis and hair root. I. Deoxyribonuclease. *Nishinihon J. Derm.*, **36**: 379–381, 1974.
9. Miyagawa, T. Nucleolytic enzymes of epidermis and hair root. II. Ribonuclease. *Nishinihon J. Derm.*, **36**: 552–554, 1974.
10. Van Scott, E. J. Mechanical separation of the epidermis from the corium. *J. Invest. Dermatol.*, **18**: 377–379, 1952.
11. Tamura, Y., Niinobe, M., Arima, T., Okuda, H. and Fujii, S. Studies on aminopeptidases in rat liver and plasma. *Biochim. Biophys. Acta*, **327**: 437–445, 1973.
12. Fiske, C. H. and Subbarow, Y. The colorimetric determination of phosphorus. *J. Biol. Chem*, **66**: 375–400, 1925.
13. Dygert, S., Li, L. H., Floride, D. and Thoma, J. A. Determination of reducing suger with improved precision. *Anal. Biochem.*, **13**: 367–374, 1965.
14. Kizaki, H. and Sakurada, T. A micro assay method for hypoxanthine-guanine and adenine phosphoribosyltransferases. *Anal. Biochem.* **72**: 49–56, 1976.
15. Lowry, O. H., Rosebrough, N. J., Farr, A. L. and Randall, R. Protein measurement with the Folin phenol reagent. *J. Biol. Chem.* **193**: 265–275, 1951.
16. Wheatley, V. R. and Farber, E. M. Studies on the chemical composition of psoriatic scales. *J. Invest. Dermatol.*, **36**: 199–211, 1961.
17. Wheatley, V. R. and Farber, E. M. Chemistry of psoriatic scales. II. Further studies of the nucleic acids and their catabolites. *J. Invest. Dermatol.*, **39**: 79–89, 1962.
18. Hodgson, C. Nucleic acids and their decomposition products in normal and pathological horny layers. *J. Invest. Dermatol.*, **39**: 69–78, 1962.

19. Miyagawa, T. Epidermal phosphodiesterase I and II. *Nishinihon J. Derm.*, **36**: 677–679, 1974.
20. Miyagawa, T. and Urabe, H. Nucleic acid catabolism in epidermis. *Nishinihon J. Derm.*, **36**: 783–786, 1974.
21. Bersaques, J. Nucleosidase in human epidermis and in normal and abnormal scales. *J. Invest. Dermatol.*, **38**: 133–135, 1962.
22. Bersaques, J. Purine and pyrimidine metabolism in human epidermis. *J. Invest. Dermatol.*, **48**: 169–173, 1967.
23. Block, W. D. and Johnson, D. V. Studies of the enzymes of purine metabolism in skin. II. Nucleoside deaminases of rat skin. *J. Invest. Dermatol.*, **23**: 471–478, 1954.
24. Block, W. D. and Johnson, D. V. Studies of the enzymes of purine metabolism in skin. I. Guanase activity of rat skin. *J. Biol. Chem.*, **217**: 43–48, 1955.
25. Westerfeld, W. W. and Richert, D. A. The xanthine oxidase activity of rat tissues. *Proc. Soc. Exp. Biol. Med.*, **71**: 181–184, 1949.
26. Block, W. D. and Johnson, D. V. Factors influencing xanthine oxidase activity in rat skin. *Arch. Biochem. Biophys.*, **56**: 137–142, 1955.
27. Schwarz, E. Zur Genese von Imidazol-Derivaten der wasserlöslichen Substanzen epidermaler Hornschicht. In-vitro-Untersuchungen mit [2-^{14}C] Adenin. *Z. Physiol. Chem.*, **339**: 110–114, 1964.
28. Santoianni, P. and Ayala, F. Nucleic acids enzymes in human hair roots. *Acta Dermatovener.*, **53**: 449–452, 1973.
29. Gartler, S. M., Scott, R. C., Goldstein, J. L. and Campbell, B. Lesch-Nyhan syndrome: Rapid detection of heterozygotes by use of hair follicles. *Science*, **172**: 572–574, 1971.
30. Rosenbloom, F. M., Kelly, W. N., Miller, J., Henderson, J. F. and Seegmiller, J. E. Inherited disorder of purine metabolism. *J. Amer. Med. Ass.*, **202**: 175–177, 1967.
31. Krenitsky, T. A., Neil, S. M., Elion, G. B. and Hitchings, G. H. Adenine phosphoribosyltransferase from monkey liver. *J. Biol. Chem.*, **244**: 4779–4784, 1969.
32. Krenitsky, T. A. Tissue distribution of purine ribosyl- and phosphoribosyltransferases in the Rhesus monkey. *Biochim. Biophys. Acta*, **179**: 506–509, 1969.
33. Murray, A. W. The activities and kinetic properties of purine phosphoribosyltransferases in developing mouse liver. *Biochem. J.*, **104**: 675–678, 1967.
34. Murray, A. W. Purine-phosphoribosyltransferase activities in rat and mouse tissues and in Ehrlich ascites tumor cells. *Biochem. J.*, **100**: 664–670, 1966.
35. Pillarisetty, R. J. and Karasek, M. A. Pyrimidine biosynthesis de novo in skin. *J. Biol. Chem.*, **245**: 358–362, 1970.
36. Lewis, R. A. and Karasek, M. Studies of pyrimidine deoxynucleoside metabolism in newborn mouse skin. *J. Invest. Dermatol.*, **59**: 317–322, 1972.
37. Imura, H. Studies on nucleic acid metabolism in the skin. I. On deoxyribonucleic acid synthesis and pyrimidine nucleotide synthesis in rat skin. *Nishinihon J. Derm.*, **36**: 670–676, 1974.

Discussion

Dr. Freedberg: Do you have any idea as to where the enzymes are? This is the problem that has been discussed after the previous paper and was discussed this morning. Are they localized in a particular place in the epidermis? Are they held in an inactive state and then activated? Do you have any information on these matters?

Dr. Matsuo: No, I haven't. It is very hard to prepare subcellular fractions using biopsy specimens.

Dr. Ogura: Adenine phosphoribosyl transferase is active in guinea pig epidermis and in cow snout epidermis. But adenine is not detectable in the epidermis. How do you explain this biochemical oddity?

Dr. Matsuo: I also am concerned about the same point. But the presence of this enzyme in the epidermis was indirectly shown by the assay for adenyl cyclase in the epidermis using adenine as a substrate.

Molecular Mechanisms of DNA Repair in Normal and Xeroderma Pigmentosum Skin Fibroblast Cells

Hiraku TAKEBE and Mituo IKENAGA

Department of Fundamental Radiology, Faculty of Medicine, Osaka University, Osaka, Japan

ABSTRACT

When epidermal cells are exposed to ultraviolet light, the most efficient type of repair of UV damage may be excision, which has been extensively investigated in this work. Epidermal cells of patients of xeroderma pigmentosum (XP), having decreased excision repair activity, cannot undergo normal differentiation and will develop skin cancers in almost all cases. Direct measurement of excision of pyrimidine dimers from UV-irradiated cells in culture during post-UV incubation, and other repair tests, revealed that all six known types of XP (5 complementation groups and a "variant") had decreased activity of excision repair. Such decrease in excision repair is not only affecting the repair of UV damage, but appears to be common to some other types of DNA damage. We found that the damage caused by a potent carcinogen as well as a mutagen 4-nitroquinoline 1-oxide (4NQO) was also susceptible to excision repair in human cells. XP cells were more sensitive to 4NQO than normal cells for inactivation of colony forming ability, and showed lower activity of excising 4NQO-DNA adducts than normal cells. Therefore, any agent which causes DNA damage susceptible to excision repair may enhance the development of XP symptoms and may interfere with normal differentiation of epidermal cells. Correlation between keratosis and repair deficiency is also suggested. We propose that enhancement of "error-prone" repair (probably postreplication repair), due to decreased activity of "error-free" excision repair which is otherwise predominant, may be the major cause of high incidence of skin cancers in XP patients.

INTRODUCTION

There are three types of DNA repair in human cells, namely, excision repair, postreplication repair, and photoreactivation. Epidermal as well as fibroblast cells of the xeroderma pigmentosum (XP) patients have been shown to have decreased activities in these repair systems. Cleaver[1] first reported that the XP cells were deficient in excision repair. Investigation on many XP cells revealed that there were six different types of XP with different levels of DNA repair activity. XP cells belonging to five complementation groups, A, B, C, D, and E, have been reported to have decreased excision repair activity,[2,3,4] while another group, "variant," has been shown to have decreased postreplication repair.[5] The finding by Sutherland[6,7] that all of these XP cells had decreased activity of photoreactivation suggested the whole repair system may be coordinated by the cellular regulatory mechanism.

High incidence of skin cancers among XP patients prompted the investigators to correlate the repair deficiency and carcinogenesis through some molecular mechanisms which have not been clearly understood yet. This paper intends to provide some clues for elucidating the relationship between the repair deficiency and the carcinogenesis based on the survey of XP patients in Japan and on the molecular studies on the repair in their cells. Possible correlation between abnormal epidermal differentiation and the development of skin cancers will also be discussed.

XP PATIENTS IN JAPAN AND THE REPAIR TESTS

Table 1 summarizes the age distribution of XP patients in Japan and DNA repair of their cells. The repair activities were classified into four groups depending on the relative amount of unscheduled DNA synthesis (UDS) as measured by autoradiography of ^{3}H-thymidine incorporation after UV irradiation of the cultured fibroblast cells. More detailed information on most of the patients and the methods of repair tests were described previously.[8,9] The youngest patient ever identified as XP was 6 months old and the oldest surviving patient in our survey was 75. Table 1 shows that nearly half of the patients are children under 10 years, and all of them had

Table 1. Age distribution and DNA repair of XP patients in Japan.

Group	UDS after UV (% of normal)	Number of Patients	Ages[a] 0 – 9	10 – 19	20 – 29	30 – 39	40<
I	<5	28(9)	21(3)	6(5)	1(1)	0	0
II	5 – 10	2(1)	1	1(1)	0	0	0
III	25 – 50	7(7)	0	1(1)	2(2)	2(2)	2(2)
IV	70<	10(5)	0	1(0)	4(2)	2(1)	3(2)
Total		47(22)	22(3)	9(7)	7(5)	4(3)	5(4)

a. Numbers in parentheses are number of patients with skin cancers.

extremely low DNA repair activity. This is probably because the lower the repair activity is, the more rapidly the abnormal development of skin takes place so that the XP symptoms may be easily diagnosed at early ages. Although only 9 out of 28 patients with the lowest repair activity, or group 1, have developed skin cancers, almost all cases in this group had very severe skin symptoms and may eventually develop skin cancers later. The reasoning for this prediction will be discussed in the following sections.

There are 10 patients whose repair activities were measured as nearly normal or as indistinguishable from a normal level. To have a decisive diagnosis, host-cell reactivation of UV-irradiated herpes simplex virus was compared. The UV-irradiated viral DNA, entering into the nucleus of the cell, is expected to be repaired by the cellular excision repair capacity or host-cell reactivation. Cells with reduced excision repair activity may repair the UV damage of viral DNA to a less extent than the normal level. As shown in Fig. 1, all XP cells tested showed reduced host-cell reactivation levels even though some of them had UDS levels nearly or indistinguish-

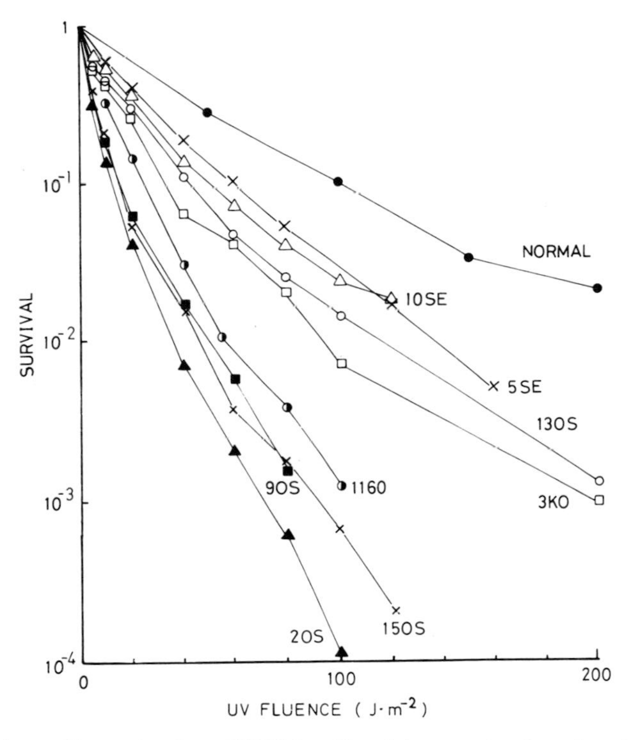

Fig. 1. Host-cell reactivation of UV-irradiated herpes simplex virus in normal and XP cells. Virus particles were irradiated in suspension followed by infection to the monolayer of the cells. Detailed methods were described previously.[8]

able from a normal level as listed in Table 2. Among 10 patients listed as group IV in Table 1, 8 patients were tested by this method for repair activity in their cells and all of them showed a reduced level of host-cell reactivation. One of them, XP3KO, has been shown to belong to "variant" by Fujiwara (personal communication), suggesting that variants which have been reported to have decreased postreplication repair may also have decreased excision repair.

Table 2. D_0 values of UV survival curves of herpes simplex virus.

Cells	Comp. Group	UDS	D_0(J/m^2)
Normal		100	38
XP5SE	E	70	21
XP10SE		100	18
XP13OS		80	13
XP3KO	variant	100	12
1160	D	25 — 55	7.5
XP9OS	A	<2	5.5
XP15OS		<2	5
1166	C	15 — 25	4 — 6
1170	C	15 — 25	
1199	B	3 — 7	
1200	D	25 — 55	
1223	A	<2	
XP2OS	A	<2	4

HIGH SENSITIVITY OF XP CELLS TO SOME CARCINOGENS AND ITS MOLECULAR MECHANISM

XP cells have been shown to be more sensitive than normal cells to a potent carcinogen 4-nitroquinoline 1-oxide (4NQO) for the inactivation of colony forming ability.[10] Treatment of normal and XP cells by ^{3}H-4NQO followed by extraction of DNA and subsequent acid hydrolysis and paper chromatography revealed the presence of 4NQO-DNA adducts, as shown in Fig. 2. The amount of the adducts in normal cells (FL cells in Fig. 2A) decreased by incubating the treated cells for 24 hours. On the other hand, XP cells similarly treated and incubated did not show such decrease of the adduct.[11] Figure 3 indicates that the normal cells had more efficient excision capacity for both 4NQO-purine adducts (Q-Pu) and 4AQO, 4-aminoquinolinc 1-oxide, released from the unstable fraction of 4NQO-purine adduct during hydrolysis. The XP cell line in Figs. 2 and 3, XP2OS(SV) has been shown to lack the excision of thymine-containing dimers after UV irradiation. From these observations, we conclude that the XP cells are deficient in excision repair of both UV

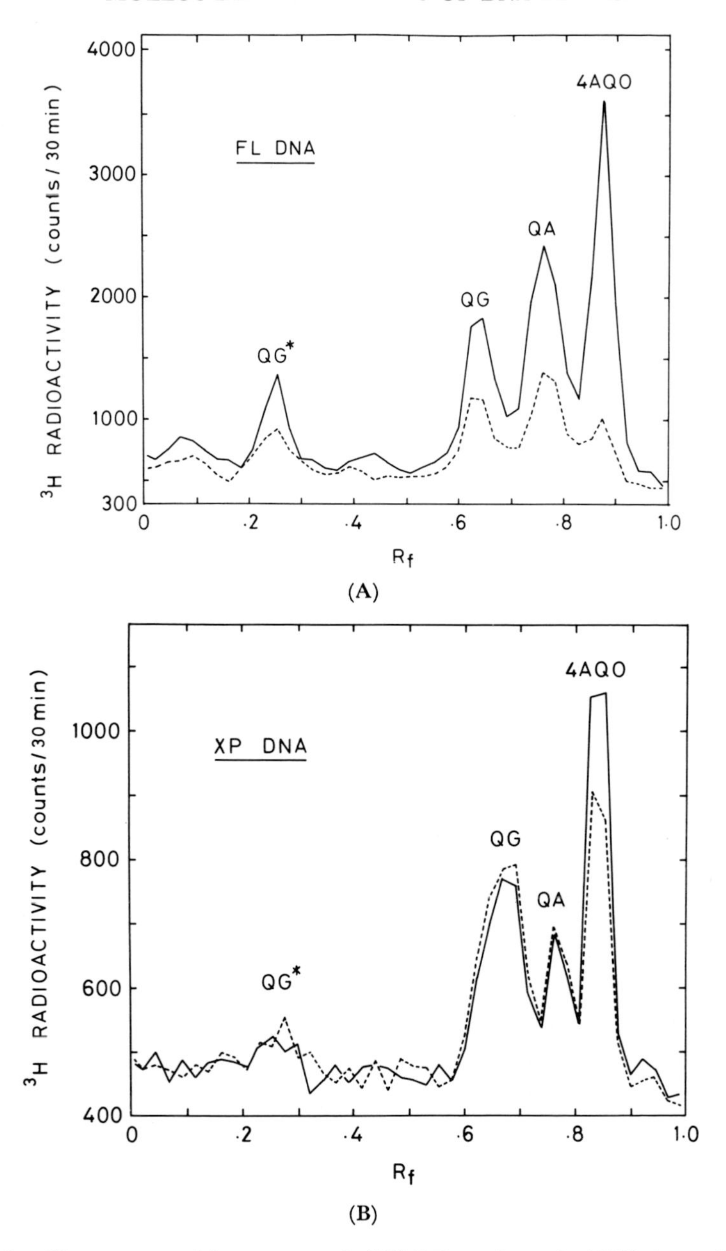

Fig. 2. Chromatographic patterns of 4HAQO-products in DNA immediately after the treatment (straight line) and after 24-hour postincubation (dashed line). A: FL (wild type) cells, B: xeroderma pigmentosum cells. Details of the experiment are to be published elsewhere (Ikenaga, Takebe, and Ishii, in preparation).

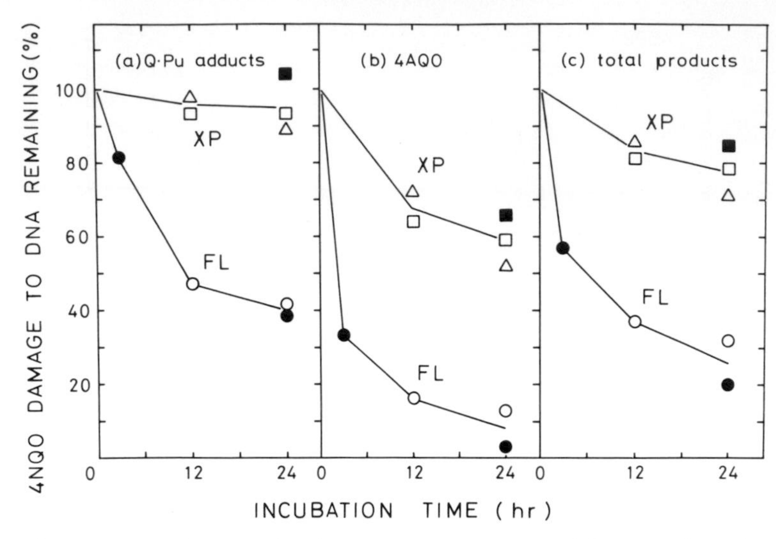

Fig. 3. Relative decrease in the amounts of stable 4NQO-purine adducts, 4AQO and total products in DNA of FL and XP cells after different periods of incubation following the treatment of the cells by ^{3}H-4NQO. A part of the data was published previously.[11]

damage and 4NQO damage in DNA. Similar mechanisms of repair may exist for DNA damage caused by some other carcinogenic chemicals which are known to have more effect on XP cells than normal cells, such as acetylaminofluorene compounds.[12] It should be noted that these results suggest XP patients may have higher susceptibility to some of the carcinogenic chemicals which cause excisable damage in DNA. Proper care should be taken during the treatment of the skin lesions of XP patients not to apply chemicals with potential action on DNA.

SKIN CANCERS AND DNA REPAIR IN XP PATIENTS

As shown in Table 1, there are 6 patients without skin cancers at the age of 10 or over. Five of them belonged to group IV, or nearly normal level of repair. This suggests that most of the XP patients will develop skin cancers unless they have nearly a normal level of DNA repair. Figure 4 gives the age distribution of XP patients bearing skin cancers in Japan as expressed in the fraction of cases not yet having developed skin cancers at age indicated, or the expression as used by Knudson.[13] The curve indicates that it is necessary to have some accumulation of UV exposure, at least for 5 years, for the development of the skin cancers. Assuming that the curve may be expressed as $S=1-(1-e^{-kD})^{m}$, where S is ordinate and the D is abscissa of Fig. 4, and k is a constant and m is number of targets, the event which

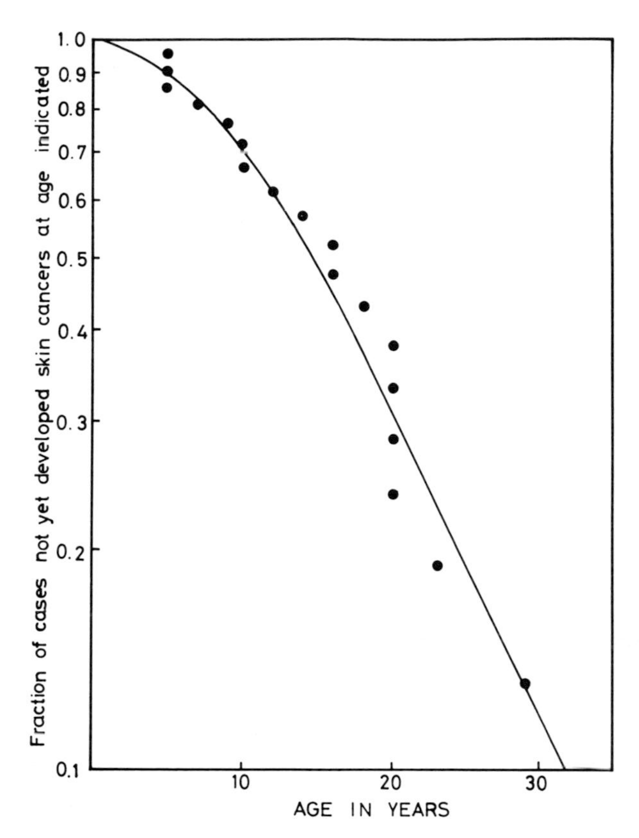

Fig. 4. Semilogarithmic plot of fraction of cancer-bearing xeroderma pigmentosum patients who have not yet developed skin cancers *versus* age in years.

causes cancer, the years (D) required to cause an average of 1 cancer-causing event per target is 10 years. Therefore if the considerable presence of repair activity is necessary for avoiding carcinogenesis in XP, all patients under 10 years of age without cancers at present should develop skin cancers soon, possibly by the age of 20.

POSSIBLE CORRELATION BETWEEN KERATINIZATION AND DNA REPAIR DEFICIENCY

Although XP is not known to have any correlation with keratinization at present, there are a few reports suggesting possible involvement of DNA repair deficiency in the diseases with abnormal keratinization. Lambert *et al.*,[14] at the Second International Workshop on DNA Repair in Mammalian Cells, 1976, presented the facts that patients with actinic keratosis exhibited decreased capacity for UV-induced DNA repair synthesis in peripheral leukocytes. A genetic disease, dyskeratosis congenita, has been known to develop leukemia. This could be due to Fanconi's anemia, known to often accompany the disease. Fanconi's anemia is another well-known DNA repair deficient genetic disease, possibly with reduced repair activity on DNA cross links. Urbach *et al.*[15] showed there is good correlation between increased prevalance of skin cancers and keratoses with age of the populations from different latitudes and of different genotypes. We assume that the error-prone type of repair, probably post-replication repair, should be the major mechanism of carcinogenesis. Since most of the skin cancers are believed to be caused by solar UV irradiation, both keratosis and skin cancers may have same molecular process(es) during abnormal differentiation.

Acknowledgements

The authors wish to thank Drs. Makoto Seiji, Yoshiharu Miki, and Hiroshi Akiba for the information and biopsy materials on XP patients as well as valuable suggestions. Encouragement and advice given by Drs. Sohei Kondo, Yoshisada Fujiwara, Mitsuhiko Tada, and Masao S. Sasaki on DNA repair studies are greatly appreciated. This work was supported by Grants in Aid from the Ministry of Education, Science and Culture, and by the Subsidy for Cancer Research from the Ministry of Health and Welfare.

REFERENCES

1. Cleaver, J. E. Defective repair replication of DNA in xeroderma pigmentosum. *Nature*, **218**: 652–656, 1968.
2. Robbins, J. H., Kraemer, K. H., Lutzner, M. A., Festoff, B. W., and Coon, H. G. Xeroderma pigmentosum. An inherited disease with sun sensitivity, multiple cutaneous neoplasms, and abnormal DNA repair. *Ann. Intern. Med.*, **80**: 221–248, 1974.
3. Kraemer, K. H., Coon, H. G., Petinga, R. A., Barrett, S. F., Rahe, A. E., and Robbins, J. H. Genetic heterogenecity in xeroderma pigmentosum: Complementation groups and their relationship to DNA repair rates. *Proc. Natl. Acad. Sci. U.S.*, **72**: 59–63, 1975.
4. Kraemer, K. H., de Weerd-Kastelein, E. A., Robbins, J. H., Keijzer, W., Barrett, S. F., Petinga, R. A., and Bootsma, D. Five complementation groups in xeroderma pigmentosum. *Mutation Rrs.*, **33**: 327–340, 1975.
5. Lehmann, A. R., Kirk-Bell, S., Arlett, C. F., Paterson, M. C., Lohman, P. H., de Weerd-Kastelein, E. A., and Bootsma, D. Xeroderma pigmentosum cells with normal levels of excision repair have a defect in DNA synthesis after UV-irradiation. *Proc. Natl. Acad. Sci. U.S.*, **72**: 219–223, 1975.
6. Sutherland, B. M., Rice, M., and Wagner, E. K. Xeroderma pigmentosum cells contain low levels of photoreactivating enzyme. *Proc. Natl. Acad. Sci. U.S.*, **72**: 103–107, 1975.

7. Sutherland, B. M., and Oliver, R. Low levels of photoreactivating enzymes in xeroderma pigmentosum variants. *Nature*, **257**: 132–134, 1975.
8. Takebe, H., Nii, S., Ishii, M. I., and Utsumi, H. Comparative studies of host-cell reactivation, colony forming ability and excision repair after UV irradiation of xeroderma pigmentosum, normal human and some other mammalian cells. *Mutation Res.*, **25**: 383–390, 1974.
9. Takebe, H., Miki, Y., Kozuka, T., Furuyama, J., Tanaka, K., Sasaki, M. S., Fujiwara, Y., and Akiba, H. DNA repair characteristics and skin cancers of xeroderma pigmentosum patients in Japan. *Cancer Res.* In press.
10. Takebe, H., Furuyama, J., Miki, Y., and Kondo, S. High sensitivity of xeroderma pigmentosum cells to the carcinogen 4-nitroquinoline-1-oxide. *Mutation Res.*, **15**: 98–100, 1972.
11. Ikenaga, M., Ishii, Y., Tada, M., Kakunaga, T., Takebe, H., and Kondo, S. Excision repair of 4-nitroquinoline-1-oxide damage responsible for killing, mutation and cancer. In Molecular Mechanisms for the Repair of DNA, (Hanawalt, P. C. and Setlow, R. B., eds.), Plenum Press, New York, 1975, pp. 763–771.
12. Cleaver, J. E. DNA repair with purines and pyrimidines in radiation and carcinogen-damaged normal and xeroderma pigmentosum human cells. *Cancer Res.*, **33**: 362–369, 1973.
13. Knudson, A. G., Jr. Mutation and cancer: Statistical study of retinoblastoma. *Proc. Natl. Acad. Sci. U.S.*, **68**: 820–823, 1971.
14. Lambert, B., Ringborg, U., and Swanbeck, G. Repair of UV-induced DNA lesions in peripheral lymphocytes from healthy subjects of various ages, *Mutation Res*. In press (abstract).
15. Urbach, F., Rose, D. B., and Bonnem, M. Genetic and environmental interactions in skin. In Environmental Cancer, Williams and Wilkins, Baltimore, 1972.

Discussion

Dr. Baden: Is the variant that shows normal repair by standard testing seen in Japan?

Dr. Takebe: Yes, with the help of Dr. Y. Fujiwara of Kobe University, we have seen that there are three variants which show the xeroderma pigmentosum trait, but whose cells show normal amounts of unscheduled DNA synthesis after exposure to ultraviolet light. They have reduced activities of post-replicative repair as shown by Lehmann and one of them has reduced host-cell reactivation (the other two were not tested) suggesting that they may be also reduced in excision repair.

Dr. Kurosumi: What are the pathological types of skin cancer which develop in patients with xeroderma pigmentosum?

DR. TAKEBE: They are either basal cell carcinoma or squamous cell carcinoma or both. We have no cases with malignant melanoma which itself is seen less in Japan than in other countries.

DR. BERNSTEIN: Do you know why some—but not all—of your normal cell lines show an initial shoulder in the curve of ultraviolet dose plotted against survival of cells?

DR. TAKEBE: We don't see shoulders in all cases, particularly in xeroderma pigmentosum cells. After X-ray exposure, shoulders appear almost all the time. The loss of shoulders in ultraviolet inactivation may be due to experimental error, since the plating efficiency of primary cultures is very low.

DR. FUKUYAMA: Did the second case show a decrease in MED (minimal erythema dose)? Is it related to the phenomenon of DNA repair?

DR. TAKEBE: Yes. MED is definitely very small in many xeroderma pigmentosum cases and as shown by Dr. Nakayama of the Shiseido Institute, we don't know how DNA damage is related to erythema, directly or indirectly.

DR. FUKUYAMA: What is your view of light repair which might occur in humans *in vivo*?

DR. TAKEBE: You may be referring to the finding by Dr. B. Sutherland that photoreactivation exists in the human. Although the presence of the photoreactivating enzyme was confirmed, biological repair in cultured cells has not been successful so far. It is very likely that all repair systems are coordinated *in vivo* as I previously discussed.

DR. ASO: It is understood that xeroderma pigmentosum cells can repair X-ray damaged DNA. You have mentioned that a part of the damage resulting from exposure to 4NQO (4-nitroquinoline 1-oxide) in DNA is similar to that of X-ray damage. Could you explain how 4NQO damages DNA?

DR. TAKEBE: The major type of damage is of the ultraviolet type which is susceptible to excision repair. As shown in Figs. 2 and 3, 4NQO was released after purification of DNA. We assume that one possible origin of the 4NQO might have been the damage which may be repaired in both normal and xeroderma pigmentosum cells as with X-ray damage, although the relative amount of excisable damage may be small.

Role of DNA Repair in the Susceptibility to Chromosome Breakage and Cell Killing in Cultured Human Fibroblasts

M. S. SASAKI*, K. TODA** and A. OZAWA*

* *Medical Research Institute, Tokyo Medical and Dental University, Tokyo, Japan* and ** *Department of Dermatology, Tokyo Teishin Hospital, Tokyo, Japan*

ABSTRACT

In view of the close association between DNA repair, cell killing, and malignant transformation, some human hereditary diseases of high cancer susceptibility associated with diverse skin disorders have been cytologically tested for the possible involvement of defective DNA repair. In addition to xeroderma pigmentosum (XP), the cytological responses highly suggestive of the defective DNA repair were found in Fanconi's anemia (FA). The foregoing experiments were performed to gain further insight into the defect in the FA cells and the results were compared with those for the XP cells. The FA cells were highly susceptible to chromosome breakage and cell killing specifically by bifunctional alkylating agents, while the XP cells were only slightly sensitive to these bifunctional agents. These findings suggest that in human cells the repair of DNA cross-links is indifferent to the excision ability responsible for the repair of UV-induced pyrimidine dimers and thus it is normally functioning in the XP cells but is impaired in the FA cells.

DNA with its metabolism plays an important role in the organization and function of cells, and in man its structural and functional alterations are causally related to many pathological changes. Among many other control systems residing in DNA metabolism, we have now a wealth of evidence that human cells possess excellent repair mechanisms to tolerate the damage produced in their DNA. The excess amount of repair load or the impairment of the repair systems may also relate to many later unpleasant outcomes. Stimulated by the first discovery of James Cleaver[1] that a recessively transmited hereditary disease, xeroderma pigmentosum, is analogous to *uvr*⁻ strain of bacteria and lacks the ability to excise UV-induced DNA damage, many investigators have directed their attention to the study of DNA repair in human hereditary diseases of unknown etiology. Such researches are now increasingly important in the rapidly unfolding area of cell differentiation, cell killing, aging, mutation induction, and malignant transformation in man. Although there is now evidence that mechanisms analogous to those present in bacteria may also exist, the DNA repair system responsible for a variety of DNA damage in human cells has not yet been fully understood.

Here we are going to present a brief review of the results of our search for the possible defective DNA repair in the causation of some human hereditary diseases of high cancer susceptibility and, further, to present some peculiar responses of XP and FA cells at chromosomal and cellular levels, which will be interpreted as a possible reflection of the presence of a unique tolerating system responsible for the repair of DNA cross-links.

Defective DNA repair and chromosomal instability

Table 1 summarizes the results of our attempt made cytologically to detect possible defective DNA repair in some rare human hereditary diseases. The table also includes some disorders which have not yet been studied in this respect and also some disease states in which genetic background is suspected for their cancer susceptibility. Of particular interest is that these inherited traits are manifested as a variety of skin disorders, such as abnormal keratosis, atrophic dermatosis, pigmentation anomalies, and telangiectasia. Some are characterized by abnormally sun-sensitive skin. The cytogenetic approach has several advantages in the search of human hereditary disorders for possible defective DNA repair. First, although mechanisms remain to be elucidated, the association between DNA damage and chromosome aberration formation has been clearly demonstrated and is now widely accepted,[2-7] and the defective DNA repair may be manifested as an abnormal susceptibility to chromosome breakage. Second, chromosomal instability of an intrinsic nature and in response to some DNA damaging agents can easily be tested by simple peripheral blood culture technique. Third, the development of a new technique to differentiate between sister chromatids by a simple staining technique[8,9] facilitated the quantitative analysis of sister chromatid exchanges (SCEs). The SCEs are now considered as cytological manifestations of a type of DNA repair process, possibly of the recombination character.[10-14]

The spontaneous levels of chromosome aberrations and SCEs, and chromosomal susceptibility to γ-ray, mitomycin C (MMC), methyl methanesulfonate (MMS) and 4-nitroquinoline 1-oxide (4NQO) tested in cultured skin fibroblasts or peripheral blood lymphocytes are given in this table by stating extremely high (+++), high (++), slightly, but significantly high (+), and normal (−) levels. In view of the diverse mechanisms relating to the development of cancers, many of the diseases listed in this table may be categorized into other groups which are not directly related to the defective DNA repair, as already suggested by the present results. Yet, some are highly suggestive of the presence of some defects in a process essential to the repair or replication of DNA. Xeroderma pigmentosum (XP) is now well documented to be defective in the excision repair of DNA damage induced by UV and certain chemicals. The XP cells are highly sensitive to the induction of chromosome aberrations specifically by UV[18] and 4NQO.[2,3] Abnormal chromosome responses suggestive of possible defective DNA repair are in Fanconi's anemia (FA) cells after treatment with MMC and also other bifunctional lakylating agents[19] and in Louis-Bar syndrome or ataxia telangiectasia (AT) cells after γ-ray irradiation. Skin fibroblasts from AT patients have been demonstrated to be highly sensitive to killing by X-ray,[20] and more recent study[21] suggests that the AT cells are defective in the repair of γ-ray-induced DNA base damage. Bloom syndrome is characterized by the manifold increase in the spontaneous SCEs[17] and is considered to have some problems in DNA replication or replication-mediated repair process.[22] Hutchinson-Gilford syndrome or progeria has been reported to be defective in the rejoining of X-ray-induced DNA single strand breaks,[23,24] but we failed to find any abnormal response to γ-ray at the chromosomal level. Four of the dominantly transmitted diseases, i.e., neurofibromatosis, familial polyposis coli, Peutz-Jeghers syndrome, and Gardner syndrome, are characterized by the frequent occurrence of karyotypically abnormal clones in skin fibroblasts, but show no indication of defective DNA repair. These diseases may be categorized into another class of cancer proneness, possibly due to the presence of premutational or precancerous alteration.

Susceptibility of FA and XP cells to killing and chromosome breakage

In order to obtain further insight into the possible defective DNA repair in the FA, we studied the susceptibility of the FA cells to killing and chromosome breakage by some mutagens. The experiments were made in parallel with those with XP cells. The susceptibility to killing was assayed by the inactivation of colony-forming ability of cultured skin fibroblasts. The FA cells were obtained from typical FA patients, FA5TO, FA7TO, FA9TO, and FA12TO. The XP cells, XP6TO and XP7TO, are cells belonging to complementation group A as determined by complementation test.[25] Mutagens used in this experiment were ^{60}Co γ-ray, 254nm UV, 4NQO, MMS, ethyl methanesulfonate (EMS), tetramethylene dimethanesulfonate (Busulfan, a bifunctional derivative of MMS and EMS), nitrogen mustard (HN2), MMC, decarbamoyl mitomycin C (DCMMC,

Table 1. Some rare hereditary disorders and disease states predisposing to ma-

Hereditary disorders	Heredity	Skin lesions
Xeroderma pigmentosum (De Sanctis-Cacchione syndrome or standard type)	a. rec.	Roughness and dryness, scaly desquamation, atrophy, telangiectasia, pigmentation.
Bloom syndrome	a. rec.	Telangiectasia, café-au-lait spots, ichthyotic skin.
Louis-Bar syndrome (Ataxia telangiectasia)	a. rec.	Telangiectasia, café-au-lait spots.
Fanconi's anemia	a. rec.	Dark skin pigmentation, café-au-lait spots.
Cockayne syndrome	a. rec.	Telangiectasia, pigmentation, thin skin.
Rothmund-Thomson syndrome	a. rec.	Erythema, telangiectasia, scarring, irregular pigmentation.
Werner's syndrome	a. rec.	Atrophic dermatosis, scleropoikioderma.
Hutchinson-Gilford syndrome (Progeria)	a. rec.?	Thin skin, alopecia, scleroderma, pigmentation.
Incontinentia pigmenti	X-linked	Vesiculation, verrucous change, atrophy, irregular pigmentation.
Dyskeratosis congenita	X-linked	Hyperkeratosis, hyperhidrosis, bullae, telangiectasia, pigmentation.
Acanthosis nigricans	?	Hyperkeratosis, pigmentation.
Maffucci's syndrome	?	Vitiligo, hyperpigmentation.
Pachyonychia congenita	a. dom.	Hyperkeratosis, callosities, keratosis pilaris, epidermal cysts.
Neurofibromatosis	a. dom.	Hyper- or hypopigmentation, café-au-lait spots, freckling.
Familial polyposis coli	a. dom.	
Peutz-Jeghers syndrome	a. dom.	Epidermal pigmentation.
Gardner's syndrome	a. dom.	Epidermal inclusion and/or sebaceous cysts, fibroma, invasive desmoid.
Bowen's disease (multiple)	?	Scaly or crusted lesion, hyperkeratosis.

* Ref. 15. ** American patient (Ref. 16). *** Ref. 17. **** Normal level
pically clones in cultured skin fibroblasts.

lignant transformation.

Sun-sensitivity	Cancer	No. of cases studied	Spontaneous		Chromosomal susceptibility to			
			Chrom. aberr	SCE	γ-ray	MMC	MMS	4NQO
+	+	3	±	(−)*	−	−	−	⧻
+	+	1**	⧻	(⧻)***				
−	+	2	+	−	⧻	−	−	−
−	+	5	⧻	−	−	⧻	−	−
+	?	0						
+	+	0						
−	+	1	−		−	−	−	−
−	?	1	−		−	−	−	−
−	+	2	+	−	−	−	+?	−
−	+	0						
−	(+)	1	−					
−	+	0						
−	?	0						
−	+	1	−****	−	−	−	−	−
−	+	14	−****	−	−	−	−	−
−	+	6	−****	−	−	−	−	−
−	+	3	−****	−	−	−	−	−
−	+	1	+	±	−	−	−	−

of spontaneous chromosome breakage but high frequencies of cells with karyoty-

a monofunctional derivative of MMC) and 8-methoxypsoralen (8MOP) followed by exposure to 355nm light.

The survival characteristics of these cells are presented in Figs. 1 and 2. The FA and XP cells were as sensitive as normal cells to γ-ray and MMS. The XP cells were highly sensitive to UV and 4NQO, while the FA cells were only slightly sensitive to UV. The FA cells showed an intermediate sensitivity to 4NQO, between XP and normal cells. Peculiar responses of these cells were the susceptibility to killing by bifunctional agents, from which we are going to proceed to ask to what extent the repair systems proposed for bacteria can be carried over to human cells. The FA cells were highly sensitive to bifunctional agents such as HN2, MMC, Busulfan, and 8MOP plus light reaction, while they were only slightly sensitive to the monofunctional agents including EMS and DCMMC. However, whereas the XP cells were highly sensitive to DCMMC, as they were, to UV and 4NQO, they were only slightly sensitive to those bifunctional agents. An equal sensitivity of XP cells to MMS, EMS, and bifunctional Busulfan may not be surprising since it has been proved that the MMS type damage can be repaired in the XP cells.[26] However, the insensitiveness of the XP cells to the other bifunctional agents cannot be explained simply by the DNA repair model proposed for bacteria. The amount of repair synthesis as determined by the unscheduled DNA synthesis in the XP cells was approximately 50% for HN2 treatment as compared with normal cells, only a residual amount for 8MOP treatment[27] and also very low, if any, for the treatment with MMC and DCMMC (Fig. 3). The FA cells showed the unscheduled DNA synthesis of an amount comparable to normal cells to UV, 4NQO, MMS, MMS and DCMMC. In bacteria, excision defective *uvr*⁻ mutant cells are cross sensitive to UV, HN2, MMC, and 8MOP plus light reaction, and repair of DNA cross-links has been proposed to be mediated by a sequential process, that is, *uvr*⁻ gene dependent incision followed by *rec* A gene dependent recombination of homologous duplexes.[28,29] Assuming that the sensitivity to killing is a cellular manifestation of DNA repair capacity, we are now inclined to believe that in human cells, unlike in bacteria, the repair of DNA cross-links is indifferent to the incision ability responsible for the excision repair of pyrimidine dimers and this process is normally functioning in the XP cells, and that although unidentified, a step needed for the repair of DNA cross-links is impaired in the FA cells. Recently, in alkaline sucrose sedimentation analysis, Fujiwara and Tatsumi[30] demonstrated that the fast sedimenting DNA from MMC-treated FA cells, possibly DNA containing cross-links, did not regain the normal sedimentation rate during the posttreatment incubation.

Further supporting evidence has been obtained from the experiment on the cell stage susceptibility of FA and XP lymphocytes to chromosome breakage by MMC and DCMMC,[31] which will be briefly presented here

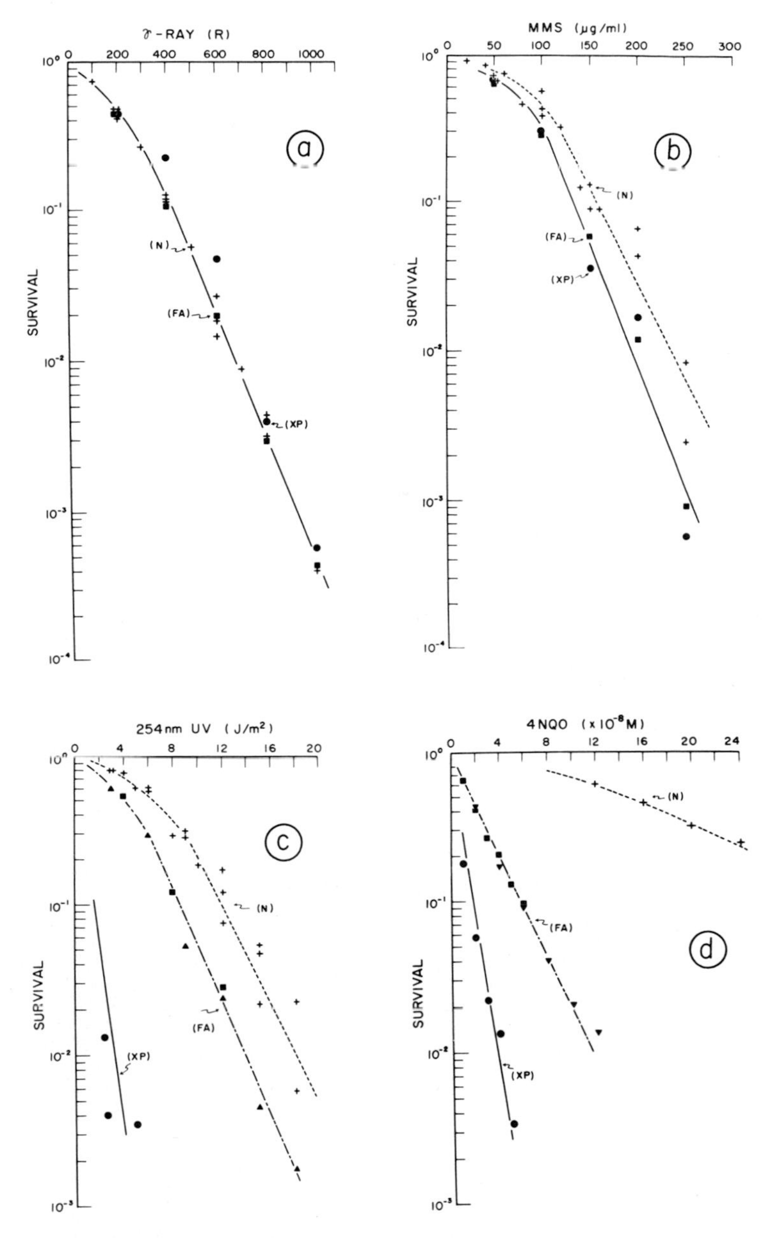

Fig. 1. Dose-response curves of survival for γ-ray (a), MMS (b), 254nm UV (c) and 4NQO (d). For γ-ray irradiation, cells suspended in culture medium were exposed to ^{60}Co γ-ray. For MMS, UV, and 4NQO treatments, cells incubated for 8–10 hrs were either exposed to UV or treated for 1 hr with MMS or 4NQO. Colonies were counted after 2-week incubation. (+) Normal cells. (■) FA5TO. (▲, ▼) FA9TO. (●) XP6TO in Figs. 1c, d and XP7TO in Figs. 1a, b.

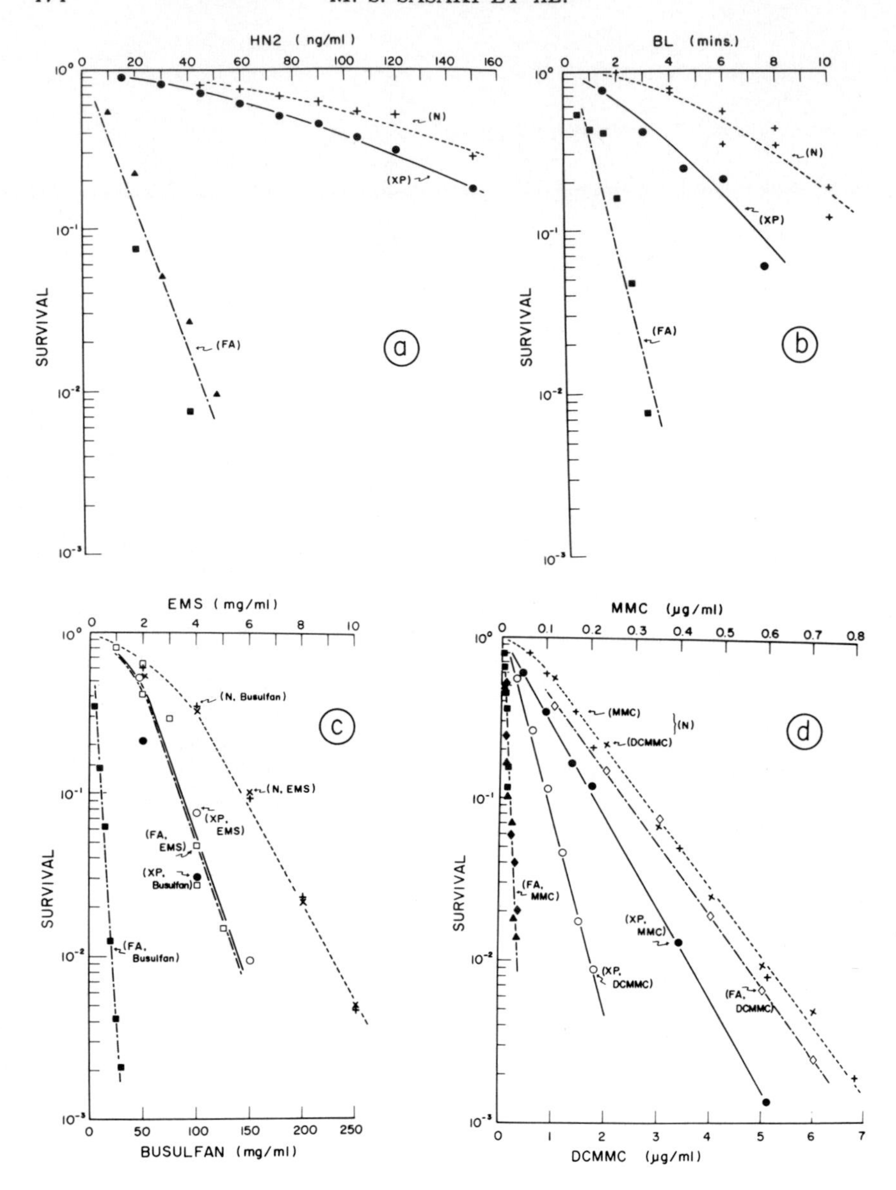

Fig. 2. Dose-response curves of survival for HN2 (a), 8MOP plus light reaction (b), EMS (c), Busulfan (c), MMC (d) and DCMMC (e). After incubation for 8–10 hrs, the cells were treated for 1 hr with HN2, EMS, Busulfan, MMC or DCMMC, and incubated for 2 weeks to count colonies. For 8MOP plus light reaction, cells were incubated for 1h with 1μg/ml 8MOP prior to exposure to black light (BL) with wavelength of maximum emission with 355nm (FL20S.BLB black-light fluorescent lamp, Toshiba Electric Co., at a distance of 10 cm). After exposure the cells were washed and reincubated for 2 weeks in the dark. (+, ×) Normal cells. (■, □) FA5TO. (▲) FA9TO. (◆, ◇) FA12TO. (●,○) XP6TO in Figs. 2a, d and XP7TO in Figs. 2b,c.

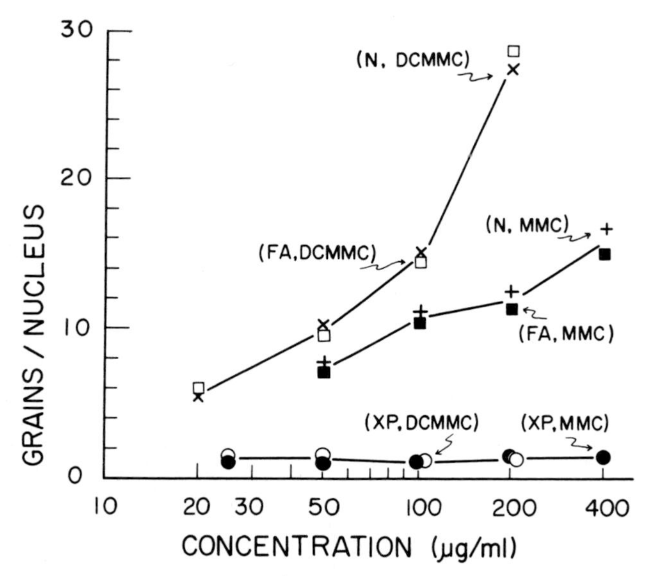

Fig. 3. Unscheduled DNA synthesis of cultured skin fibroblasts after treatment with MMC and DCMMC. The cells were incubated for 24 hrs in argine-isolecuine deficient medium, exposed to MMC or DCMMC for 30 mins, washed and then incubated for 3 hrs in the presence of 10μCi/ml ^{3}H-thymidine. The slides were processed for autoradiography and grains were counted in non-S phasc nuclei after a 1-week exposure. Experiments include FA5TO, XP7TO, and normal cells (N).

together with some additional data. In this cxperiment, peripheral blood lymphocytes were stimulated to progress mitosis by phytohemagglutinin. During culture, the cells were pulse treated with MMC or DCMMC with simultaneous labeling with ^{3}H-thymidine. The cells were fixed at varying time from treatment, and chromosome aberrations were scored in labeled and unlabeled mitoses. Since repair of DNA damage is a time-limited process and chromosome aberrations are assumed to arise when the DNA damage passes through the S phase, we would expect that in the repair proficient cells the treatment at the beginning of S phase results in the highest aberration yield, and in the treatment in G_1 phase, the further away from S phase the fewer aberrations, since in the treatment at the part of G_1 phase, the further away from S phase, the more time for damage to be repaired. In the repair deficient cells, however, we would expect constant aberration yield at whatever part of G_1 phase the cells are treated, away from or close to the S phase.

The results are shown in Figure 4. The normal cells followed the response pattern to MMC and DCMMC as expected for the repair pro-

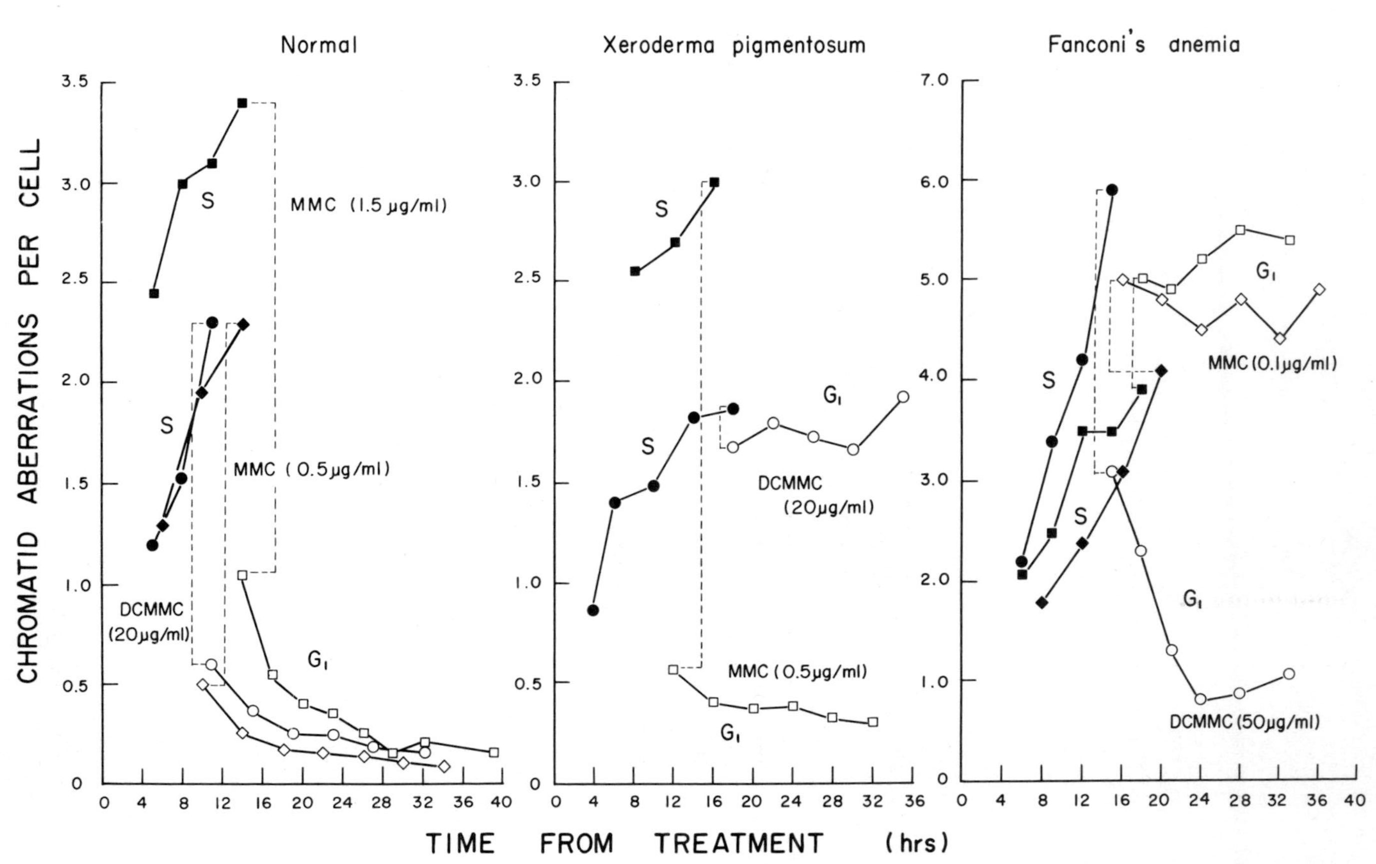

Fig. 4. Changes in the frequency of chromatid aberrations in labeled (solid symbols) and unlabeled (open symbols) mitoses as a function of time from treatment with MMC or DCMMC. The aberration frequencies are expressed as the net amounts

ficient cells. The response of the XP cells was repair deficient type to monofunctional DCMMC, while they showed repair proficient type response to bifunctional MMC. The response of the FA cells was exactly the other way around; they showed repair deficient type response to MMC, while response to DCMMC was repair proficient type. These differential chromosome responeses of the FA and XP cells to mono- and di-functional mitomycins were in agreement with the differential susceptibility to killing as assayed for cultured skin fibroblasts.

At this moment, we cannot either pinpoint the location of the primary defect in the FA cells or infer the mechanism to tolerate DNA cross-links in the XP cells, yet in view of the close association between DNA repair, cell killing, induction of mutation, and malignant transformation, these characteristic responses to cross-linking agents may be of particular importance in the medical care of these hereditary diseases, provided that they can be carried over to the *in vivo* state. Many of the hereditary diseases are obviously left to be studied, yet the elucidation of the primary defect may provide valuable clues to the understanding of genetic control of many pathological changes, as is already in progress in xeroderma pigmentosum.

We thank Professor A. Tonomura for his continuous interest in this research and Drs. S. Matsubara and J. Utsunomiya of Tokyo Medical and Dental University and Dr. Ikuo Ikemura of Tokyo Teishin Hospital for their cooperation. This work was supported by grants from the Ministry of Education, Culture and Science and from the Ministry of Health and Welfare, Japan.

REFERENCES

1. Cleaver, J. E. Defective repair replication of DNA in xeroderma pigmentosum. *Nature,* **218**: 652–656, 1968.
2. Stich, H. F. and San, R. H. C. DNA repair and chromatid anomalies in mammalian cells exposed to 4-nitroquinoline 1-oxide. *Mutation Res.,* **10**: 398–404, 1970.
3. Sasaki, M. S. DNA repair capacity and susceptibility to chromosome breakage in xeroderma pigmentosum cells. *Mutation Res.,* **20**: 291–293, 1973.
4. Bender, M. A, Griggs, H. G., and Walker, P. L. Mechanisms of chromosomal aberration production. I. Aberration production by ultraviolet light. *Mutation Res.,* **20**: 387–402, 1973.
5. Bender, M. A, Bedford, J. S., and Mitchell, J. B. Mechanisms of chromosomal aberration production. II. Aberrations induced by 5-bromodeoxyuridine and visible light. *Mutation Res.,* **20**: 403–416, 1973.
6. Comings, D. E. What is a chromosome break? In "*Chromosomes and Cancer*" (German, J., ed.), John Wiley & Sons, New York, 1974, pp. 95–133.
7. Griggs, H. G. and Bender, M. A. Photoreactivation of ultraviolet-induced chromosomal aberrations. *Science,* **179**: 86–88, 1973.

8. Latt, S. A. Microfluorometric detection of deoxyribonucleic acid replication in human metaphase chromosomes. *Proc. Nat. Acad. Sci. U. S. A.*, **70**: 3395–3399, 1973.
9. Perry, P. and Wolff, S. New Giemsa method for the differential staining of sister chromatids. *Nature*, **251**: 156–158, 1974.
10. Kato, H. Induction of sister chromatid exchanges by UV light and its inhibition by caffeine. *Exptl. Cell Res.*, **82**: 383–390, 1973.
11. Kato, H. Induction of sister chromatid exchanges by chemical mutagens and its possible relevance to DNA repair. *Exptl. Cell Res.*, **85**: 239–247, 1974.
12. Latt, S. A. Sister chromatid exchanges, indices of human chromosome damage and repair: Detection by fluorescence and induction by mitomycin C. *Proc. Nat. Acad. Sci. U. S. A.*, **71**: 3162–3166, 1974.
13. Wolff, S., Bodycote, J. and Painter, R. B. Sister chromatid exchanges induced in Chinese hamster cells by UV irradiation of different stages of the cell cycle: The necessity for cell to pass through S. *Mutation Res.*, **25**: 73–81, 1974.
14. Perry, P. and Evans, H. J. Cytological detection of mutagen-carcinogen exposure by sister chromatid exchange. *Nature*, **258**: 121–125, 1975.
15. Wolff, S., Bodycote, J., Thomas, G. H., and Cleaver, J. E. Sister chromatid exchange in xeroderma pigmentosum cells that are defective in DNA excision repair or postreplication repair. *Genetics*, **81**: 349–355, 1975.
16. Landau, J. W., Sasaki, M. S., Newcomer, V. D., and Norman, A. Bloom's syndrome. *Arch. Derm.*, **94**: 687–694, 1966.
17. Changanti, R. S. K., Schonberg, S., and German, J. A manifold increase in sister chromatid exchanges in Bloom's syndrome lymphocytes. *Proc. Nat. Acad. Sci. U. S. A.*, **71**: 4508–4512, 1974.
18. Parrington, J. M., Delhanty, J. D. A., and Baden, H. P. Unscheduled DNA synthesis, u.v.-induced chromosome aberrations and SV_{40} transformation in cultured cells from xeroderma pigmentosum. *Ann. Hum. Genet.*, **35**: 149–160, 1971.
19. Sasaki, M. S. and Tonomura, A. A high susceptibility of Fanconi's anemia to chromosome breakage by DNA cross-linking agents. *Cancer Res.*, **33**: 1829–1836, 1973.
20. Taylor, A. M. R., Harnden, D. G., Arlet, C. F., Harcourt, S. A., Lehmann, A. R., Stevens, S., and Bridges, B. A. Ataxia telangiectasia: a human mutation with abnormal radiation sensitivity. *Nature*, **258**: 427–429, 1975.
21. Paterson, M. C. and Smith, B. P. Personal communication.
22. Hand, R. and German, J. A retarded rate of DNA chain growth in Bloom's syndrome. *Proc. Nat. Acad. Sci. U. S. A.*, **72**: 758–762, 1975.
23. Epstein, J., Williams, J. R., and Little, J. B. Deficient DNA repair in human progeroid cells. *Proc. Nat. Acad. Sci. U. S. A.*, **70**: 977–981, 1973.
24. Epstein, J., Williams, J. R., and Little, J. B. Rate of DNA repair in progeroid and normal fibroblasts. *Biochem. Biophys. Res. Comm.*, **59**: 850–857, 1974.
25. Takebe, H. In preparation.
26. Cleaver, J. E. Repair of alkylation damage in ultraviolet-sensitive (xeroderma pigmentosum) human cells. *Mutation Res.*, **12**: 453–462, 1971.
27. Baden, H. P., Parrington, J. M., Delhanty, J. D. A., and Pathak, M. A. DNA synthesis in normal and xeroderma pigmentosum fibroblasts following treat-

ment with 8-methoxypsoralen and long wave ultraviolet light. *Biochim. Biophys. Acta*, **262**: 247–255, 1972.

28. Cole, R. S. Repair of DNA containing interstrand cross-links in *Escherichia coli*: Sequential excision and recombination. *Proc. Nat. Acad. Sci. U. S. A.*, **70**: 1064–1068, 1973.

20. Howard-Flanders, P. and Lin, P. F. Genetic recombination induced by DNA cross-links in repressed phage λ. *Genetics* (suppl.), **73**: 85–90, 1973.

30. Fujiwara, Y. and Tatsumi, M. Repair of mitomycin damage to DNA in mammalian cells and its impairment in Fanconi's anemia cells. *Biochem. Biophys. Res. Comm.*, **66**: 592–598, 1975.

31. Sasaki, M. S. Is Fanconi's anemia defective in a process essential to the repair of DNA cross links? *Nature*, **257**: 501–503, 1975.

Discussion

Dr. Aso: Do you have any idea how DNA damage from bifunctional agents, such as 4NQO and methoxypsoralen, can be repaired?

Dr. Sasaki: At this moment, I have no idea about the repair of DNA which is cross-linked by such bifunctional reagents. I can only say that it is catalyzed by enzymes which are different from those which are responsible for excision repair of thymidine dimers.

Dr. Takebe: The only comment I can make is that human beings are more complicated than *E. coli*. Even in *E. coli*, repair systems are not simple. For example, endonuclease V, a UV-damage specific endonuclease purified from T phage-infected *E. coli* cells, works only on UV-damage and not on 4NQO-damaged DNA although all known UV-sensitive mutants of *E. coli* are 4NQO sensitive. There are some indications that the repair enzymes in human cells may be controlled by additional factors.

Dr. Baden: Did you say that in xeroderma pigmentosum, damage from psoralen and long rays could be repaired?

Dr. Sasaki: As far as cross-linking, the answer is yes. But a monofunctional attack may not be repaired.

Dr. Baden: I'm not sure how you distinguish that in your experiments. You take cells and you radiate them and give them psoralen and you give them UV light. What you are saying is that they behave as though they were normal cells.

Dr. Sasaki: They are only slightly sensitive.

DR. Baden: "Slight" meaning that they can be distinguished from normal cells or they cannot be distinguished from normal cells?

DR. SASAKI: They are distinguishable.

DR. BADEN: Is this measured by trying to grow the cells or by observing repair mechanisms?

DR. SASAKI: By measuring colony-forming ability.

DR. ASO: Is it true that bifunctional DNA damage is much safer than monofunctional damage in terms of carcinogenicity?

DR. SASAKI: I am not sure, but it is very likely. Bifunctional agents are more effective than monofunctional agents in inducing chromosome aberrations and will link to cell-lethality more effectively, but we are not sure about the efficiency of DNA cross-links in terms of induction of mutations or malignancy.

DR. BADEN: In general, cross-linking correlates with carcinogenicity.

Dr. McGUIRE: Does the bifunctional error result in cell death more often than the monofunctional error? That's the issue. In other words, does the bifunctional lesion result in cell death more often than repair, giving an abnormal chromosome with altered growth and cancer? It seems to me that is the riddle regarding "psoralen" excision. Dr. Cole showed some years ago that in *E. coli,* the psoralen-adduct was much more difficult to excise than were thymidine dimers. It is an order of magnitude more difficult in terms of time. That is how *E. coli* deal with it, I don't know how eukaryotes deal with the problem.

DR. BADEN: But in actual fact, in studies of carcinogenesis with light of 3,000 nm, carcinogenesis does not correlate directly with the dose of light. That is, you don't get more carcinogenesis with higher doses because of the killing effect and you have less carcinogenesis with lower doses because of the repair. I'm not sure that the argument about killing is valid.

The Effect of Chalone on the Cell Kinetics in the Epidermis during Wound Healing or Organ Culture

Takeo YAMAGUCHI and Tomohisa HIROBE*

Division of Biology, National Institute of Radiological Sciences, Chiba, Japan

ABSTRACT

1) To assay chalones sensitively, the test system should be released from the chalone control which exists naturally within tissues. Thus, the proliferation of mouse ear epidermis was stimulated by wound or organ culture and used as an assay system for the effects of epidermal chalones on different phases of the cell cycle.

2) When an epidermal extract was injected subcutaneously into the mouse 2 days after wound formation on its ear, grain counts on ^{3}H-thymidine autoradiography decreased quickly after injection. A decrease in the number of labeled cells was evident at 6 hrs when the grain counts reverted to a control level. The number of mitotic figures reduced quickly and temporarily. The result suggests that the epidermal chalones inhibit 3 processes of the cell cycle: the DNA synthesis in S phase (S-chalone), the transition from G_1 to S phase (G_1-chalone), and the transition from G_2 to M phase (G_2-chalone).

3) Aqueous extract from liver or kidney affected neither the DNA synthetic nor the mitotic activities.

4) The cell cycle analysis revealed that the shortening of the generation time during wound healing was mainly the result of the shortening of the duration of G_1 phase, indicating that the G_1-chalone might have the most important role in the physiological regulation of cell proliferation.

5) When the chalone was given at an early period of wound healing (12–24 hrs after wound formation), a direct evidence for G_1-chalone was demonstrated by a reduction in the increase of the number of labeled cells which were otherwise increasing rapidly during this period. The chalone must be given more than 2 hrs prior to the beginning of S phase in order to inhibit the flow of cells into S phase.

6) Using an improved method of organ culture, the cell cycle phase

* Present address: Biological Institute, Faculty of Sciences, Tohoku University, Aramaki, Sendai 980, Japan

specificities of G_1- and G_2-chalones, which were separated by a Sepahadex G-100 gel chromatography of the precipitate by 80% ethanol, were demonstrated; one fraction with a molecular weight of more than 100,000 had only G_1-chalone activity and another fraction with a molecular weight of 10,000–100,000 had only G_2-chalone activity.

7) Existence of S-chalone activity was again indicated by a decrease in grain counts when the skin extract or its alcoholic precipitate was applied to the culture of wounded ear epidermis at its highest proliferative activity. No such an action on the epidermis was found in the liver extract.

INTRODUCTION

The chalone hypothesis presupposes that chalones act primarily by inducing cell differentiation: the cell is prevented from entering a new mitotic cycle.[1] The shortening of both G_1 and G_2 phases in the cell cycle was observed in the regenerating epidermis after a local irradiation, compared with the phases in normal epidermis.[2] This suggests that in normal epidermis such blocks occur at both G_1 and G_2 phases. Functionalization from G_2 phase, however, seems to be a rather rare event. Thus Bullough[1] assumed that the cell in the late G_1 phase (dichophase) decides between entering a new phase of DNA replication or taking the path toward functional differentiation.

The evidence has accumulated that crude extract of epidermis produces two tissue-specific responses when injected into mice. These are a delayed inhibition of ^{3}H-thymidine incorporation into epidermal DNA and a rapid reduction in the number of epidermal cells entering mitosis.[3–9] G_1-chalone[7,9] and G_2-chanlone[10] corresponding to these two responses have been partially purified. Purified G_2-chalone did not inhibit DNA synthesis or influx of S phase cells and is therefore cell cycle phase specific, inhibiting only the flow of cells into M phase.[11] On the other hand, G_1-chalone has yet to be analyzed to establish its exact point of action in the cell cycle. Strictly speaking, no direct evidence has been shown that it inhibits the cells in G_1 phase from entering S phase.

Experimental difficulty in the chalone study, especially in that for G_1-chalone, comes mainly from insensitivity of the assay system commonly used.[11] For assaying the chalone activity, a system released from the chalone action is, at least theoretically, more advantageous than the normal epidermis which is already under chalone control. Thus we added chalones to the epidermis in which proliferation had been stimulated by wounding[12] or transferring into organ culture, and examined their inhibitory activities on the different phases of the cell cycle.

MATERIAL AND METHODS

A. Chalone preparation

Procedures for extraction of epidermal chalone were described in detail previously.[12] The epidermis was scraped from the body skin of C57BL/6J mouse (4-to 6-month-old male) or Hartley albino guinea pig (10- to 12-month-old male). A cold Hanks' balanced salt solution (BSS) was added to it. The epidermis was then homogenized with a glass homogenizer and the resulting suspension was centrifuged at 3,000 rpm for 20 min. The supernatant (crude extract) was frozen at $-20°C$ and thawed just before use. The time interval between the extraction and the application to the assay system did not exceed 2 days. When chalones were used for injection in mouse (*in vivo*), the final solution was diluted with Hanks' BSS so that a 0.5 ml solution contained the required amount of chalone. When chalones were used for organ culture (*in vitro*), the required amount of chalone was dissolved in the culture medium. The amount of epidermal chalone was expressed as the area of skin from which the chalone derived. Approximately 1.2 mg of epidermis in dry weight was obtained by scraping an area of 1.0 cm^2 of skin. Aqueous extracts of other tissues were similarly prepared and the equivalent concentrations were calculated on a dry weight basis.

Partial purification of chalones was carried out by an ethanol precipitation and a Sephadex G–100 gel chromatography. To a $18{,}000 \times g$-supernatant of the crude extract, cold pure ethanol was added in drops with continuous stirring up to a concentration of 80% (V/V). The precipitate was collected by centrifugation at 3,000 rpm for 30 min, dissolved in 1.0 ml of 0.9% NaCl solution, and applied to a Sephadex G–100 column of 1.6 cm in diameter and 30 cm in length. The elution was performed with a 0.9% NaCl solution at a rate of 0.045 ml/min. Portions of 1.76 ml were collected. The standard substances with known molecular weight (MW) were blue dextran ($MW > 200{,}000$), hemoglobin ($MW = 68{,}000$), cytochrome C (12,000) and 2,4-dinitrophenol (184). The samples were then pooled into two fractions; fraction A ($MW > 100{,}000$) and fraction B ($10{,}000 < MW < 100{,}000$). The fractions were concentrated by ultrafiltration through an Ulvac Diafilter G–10T ($MW = 10{,}000$) and applied to the epidermal culture.

B. Animals used for assay of chalone activity

The mice used for the assay of chalone activity were 4- to 6-month-old males of the C57BL/6J strain. They were given water, fed *ad libitum* on a commercial diet and maintained at $24 \pm 1°C$. Special care was taken to handle them as gently as possible. In addition to noting the general fitness of the animals, their ears were closely inspected for visible signs of injury and only normal healthy ears were used.

C. *In vivo* assay system

Details were described previously.[12] A full-thickness cut of *ca.* 5 mm in length was made on the ear conch. Chalones were injected subcutaneously into the back neck of the mouse. Proliferation of epidermal cells within 1.0 mm from the wound edge was the subject in this system. Colchicine, if required, was injected in a dose of 4 μg/g body wt intraperitoneally 4 hrs before the histological fixation of the ear. Thymidine-6-^{3}H (^{3}H-TdR) in a dose of 2.0 μCi/g body wt was injected intravenously or intraperitoneally 60 or 30 min before the ear sampling. Owing to the presence of the diurnal variation of proliferative activities, the wound was made at 11–12 A.M. and all the procedures were scheduled so that all the histological sampling of ears in the same experimental series were carried out at the same time.

D. *In vitro* assay system

A piece of mouse ear conch (*ca.* 3 × 7 mm) was split into two epidermal sides. One side having cartilage tissue was laid, epidermal surface up, on a disk of glass filter (15 mm in diameter and 5 mm in height) which was immersed into TC–199 medium within a Petri dish, so that the medium reached the epidermis by capillary phenomenon. The Petri dishes were placed in a container through which a humidified gas mixture of 95% O_2 and 5% CO_2 was flowing constantly to maintain the active proliferation of epidermal cells. The whole apparatus was kept at 37°C in an incubator.

Epidermal chalones were added in a dose of 1-cm^2 skin equivalent per ml medium at the time required by the experiment. Simultaneously, adrenalin (0.0025 μg/ml) was added to the medium, except as otherwise stated. Colchicine (4 μg/ml), if necessary, was added at the same time and the ear piece was fixed 4 hrs later. ^{3}H-TdR was added in a dose of 3 μCi/ml 30 or 60 min before the histological fixation.

E. Autoradiography

Immediately after sampling, ears or ear pieces were fixed in Bouin's solution and embedded in Paraplast for sectioning. A section thickness of 6 or 8 μm was chosen for technical convenience since we counted not the labeling index, but the number of labeled cells within a unit length, a parameter independent of the variation in thickness above the range of β-rays from ^{3}H in the tissue. When the labeling index was required, the sections were cut at 4 μm in thickness. Autoradiographs (ARG) were made by dipping slides in Sakura NR–M2 emulsion (Konishiroku, Tokyo). They were exposed with a desiccant at 4°C for a length of time required by the experiment and developed with a D–19 developer for 5 min at 20°C. The slides were stained with Delafield's hematoxylin.

F. Counting procedures

Each time or dose group consisted of 12 different skin sites, except as otherwise stated. In the wounded epidermis, all the mitotic figures or the labeled cells in the epidermis within 1.0 mm from the wound edge were counted on each section. In the normal epidermis, including the cultured one, the number of mitoses or labeled cells along the whole length of epidermis or section was counted and the value was divided by the length measured. The average on 10–20 sections was expressed as the measurement on each skin site. Each datum in the following results represents the mean of 12 measurements with the standard error of mean thus obtained.

RESULTS

A. Change in the cell kinetics after wounding

Both ^{3}H-TdR labeled cells and mitotic figures within the epidermis surrounding the wound were found most abundantly between 2 and 3 days after wound formation (Fig. 1). Therefore, the epidermis during this period

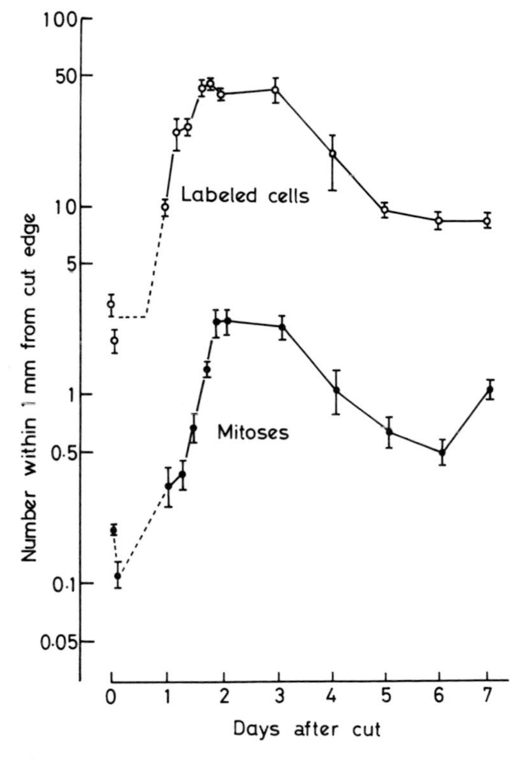

Fig. 1. Change in the proliferative parameters in the ear epidermis during wound healing. Dotted lines were drawn from the data for early changes.

of wound healing was subjected to the cell cycle analysis. Procedures for determination of cell cycle phases were the same as those used previously,[2] except the first order approximation was used here for normal epidermis instead of the second order approximation.

Percent labeled mitoses (PLM) curves for the normal epidermis and the epidermis during wound healing are shown together in Fig. 2. Generation time (T_c) of the cells during wound healing was obtained from the intersection between the halfway mark of the ascending limb in the first PLM wave and that in the second wave. The second wave, however, did not appear in normal epidermis.[2] Thus T_c of normal epidermis was calculated by the following first order approximation:[13]

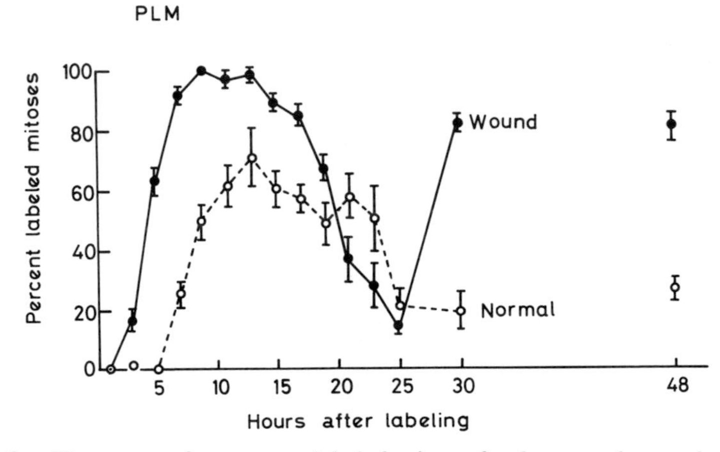

Fig. 2. The curves of percentage labeled mitoses for the normal control and for the epidermis between 2 and 3 days of wound healing.

$$T_c = \frac{T_s}{\mathrm{LI}} \ln 2 \qquad (1)$$

where T_s is the duration of S phase and LI is the labeling index of basal cells in which growth fraction is l^2.

Duration of mitosis (T_m) was also calculated for both normal and wounded epidermis by the formula:[14]

$$T_m = \frac{T_c \cdot \mathrm{MI}}{\ln 2} \qquad (2)$$

where MI is the mitotic index of basal cells.

T_s was estimated from the intersection of the halfway marks on the ascending and descending limbs in the first wave of PLM curve. Owing to the presence of diurnal rhythm of LI or MI, an average index throughout a day was used for the calculation. Both LI and MI were determined at 7 different time points throughout a 24-hr duration. Average indices were 0.00245 ± 0.00035 for MI, 0.0299 ± 0.0024 for LI in the basal cells of normal epidermis, and 0.0176 ± 0.0026 for MI in the basal cells during wound healing. Duration of G_2 (T_{g2}) was obtained from $T_2 - 1/2\ T_m$, where T_2 was the time from the beginning of labeling to halfway on the ascending limb.

The result obtained is summarized in Table 1. The greater portion of the shortening in T_c from 370 hrs in normal epidermis to 26 hrs during wound healing is attributed to the drastic decrease in T_{g1}. T_{g2} and T_m also underwent a considerable shortening. T_s remained unchanged. These changes were comparable to those found in the regenerating epidermis of guinea pig body skin after radiation damage[2] (Table 1).

Table 1. Cell cycle phases of epidermal basal cells.

	Duration (hrs)				
Epidermis	T_c	T_{g1}	T_s	T_{g2}	T_m
Mouse ear skin					
Normal	370	346	16	7	1.3
Wound healing	26	5	16	4	0.7
Guinea pig body skin[2]					
Normal	92	77	9.5	5.5	0.4
Regenerating after irrad.	16	3	9.5	2.5	0.4

B. Chalone effects on the epidermis during wound healing[12]

At first, the epidermis between 2 and 3 days after wound formation was selected as an assay system for the effect of chalone on the proliferation of epidermal cells.

Epidermal chalone from either syngeneic mouse or guinea pig inhibited the mitotic activity within the epidermis surrounding the cut when estimated as the number of mitoses accumulated by colchicine for 4 hrs (Fig. 3). Liver and kidney extracts had no effect on the mitotic activity (Fig. 3). The number of mitotic figures per unit length in this assay system was 20 times higher than in the assay system used commonly.[13]

In order to determine the locus of action of the chalone in the cell cycle of the regenerating epidermal cells, the changes with time after chalone application in the number of labeled cells, the number of grains on the labeled nuclei and the number of mitoses were investigated simultaneously on the same ARG which was developed after an exposure of 2 weeks (short exposure). Grain counts decreased temporarily in the early phase (0–2 hrs) after chalone injection (Fig. 4). This decrease in grain counts resulted in a decrease in the number of labeled cells on the ARG of a short exposure

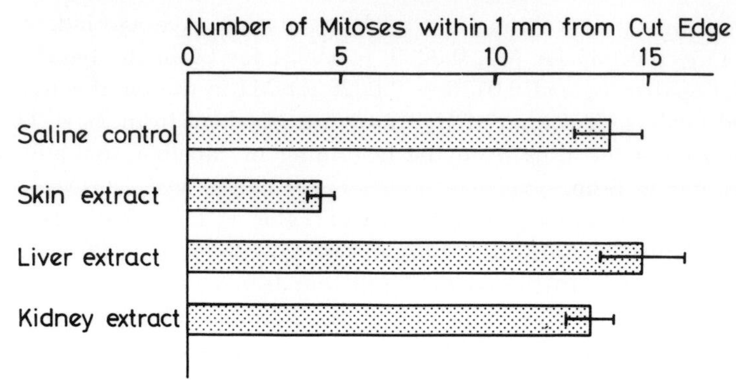

Fig. 3. Tissue-specific inhibition by chalone to mitotic activity in the regenerating epidermis during wound healing. The chalones and cochicine were injected 4 hours before sacrifice.

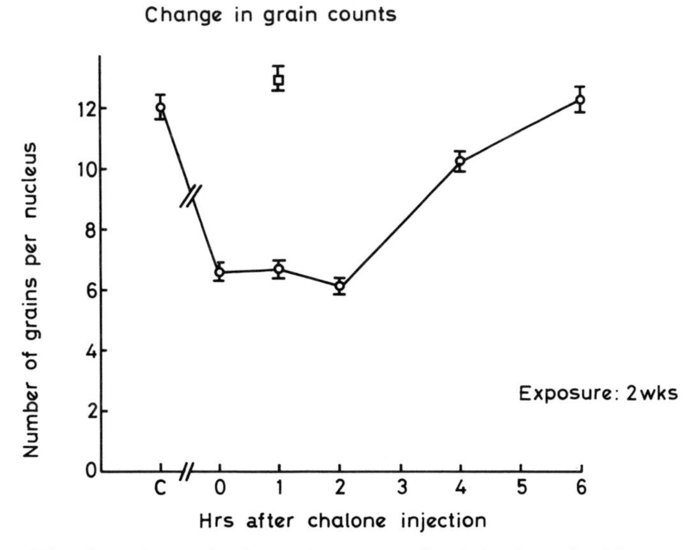

Fig. 4. The time change in the grain counts after injection of epidermal chalone (circles) or Hanks BSS only (square) in the regenerating epidermis. C, the control not injected except ^{3}H-TdR.

(circles in Fig. 5) but not in that on the ARG of a long exposure (6 weeks, triangles in Fig. 5). A decrease in the number of labeled cells on the ARG of a long exposure was evident at 6 hrs when the grain counts reverted to a level similar to the control without chalone. The number of mitoses reached a minimum at 2 hrs and then recovered quickly (Fig. 6), indicating a rapid disappearance of the inhibition of cells in G_2 from entering M phase. Mitoses decreased again thereafter (Fig. 6), presum-

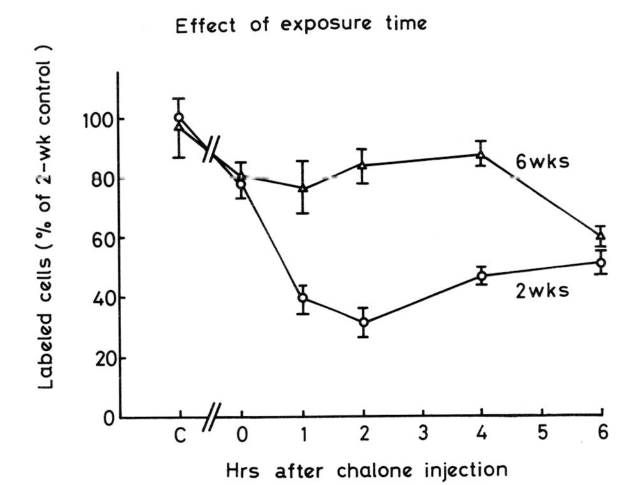

Fig. 5. The time change after injection of epidermal chalone in the number of labeled cells in the regenerating epidermis on ARG of 2 weeks or 6 weeks exposure.

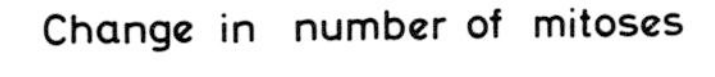

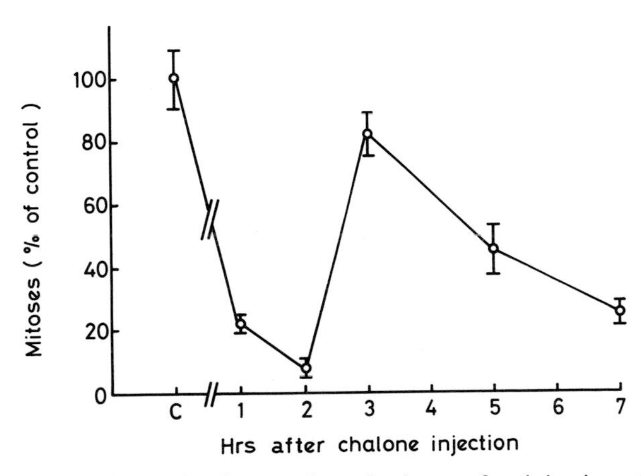

Fig. 6. The time change in the number of mitoses after injection of epidermal chalone in the regenerating epidermis.

ably a result caused by inhibition of cells in the preceding S phase from completing DNA synthesis, which was indicated above by a decrease in grain counts.

Again, injection of liver or kidney extract had no effect on the DNA synthetic activity of the regenerating epidermis judged as the change in the number of labeled cells on the ARG of a short exposure.[12]

These results indicate that the epidermal chalones inhibit the epidermal cell proliferation in, at least, 3 different processes of the cell cycle: the DNA synthesis in S phase, the transition from G_1 to S phase, and the transition from G_2 to M phase. From now on, chalones inhibiting these processes are called S-, G_1- and G_2-chalones, respectively. Among them the physiological significance of S-chalone is not well understood, since T_s was not changed after the stimulation of proliferation by wounding. This will be discussed later.

As mentioned above, the shortened T_c was largely the result of a drastic decrease in T_{g1}. This suggests that the most important regulation occurs at the transition from G_1 to S phase. To detect directly the activity of G_1-chalone, the chalone must be given at the onset of DNA synthesis. We found that an abrupt increase in DNA synthesis started between 12 and 14 hrs after the wound formation (dotted line in Fig. 7). The ear epidermis at this early stage of the wound healing was used as the assay system for G_1-chalone. A water soluble extract of the epidermis scraped from a 5-cm^2 area of mouse skin was injected in the test mouse 12 (A in Fig. 7), 14 (B), or 16 (C) hrs after wound formation. ^{3}H-TdR was injected 2, 4 or 6 hrs thereafter. The ear was fixed 30 min later for the histological sectioning. ARGs were exposed for 6 weeks and the number of labeled cells within 1.0 mm from the wound edge was counted.

The result is illustrated in Fig. 7. It shows clearly that the number of labeled cells continued to increase for 2 hrs after chalone treatment. The number of cells in S phase then stopped increasing, indicating the direct evidence for chalone inhibition of cells from entering S phase. It also indicated that to inhibit the epidermal cells from initiating DNA synthesis the chalone must act 2 hrs prior to the beginning of S phase.

C. Chalone effects on the epidermis in organ culture

Similar to the early stage of wound healing (Fig. 7), a rise in the number of labeled cells in epidermis was found to continue between 12 and 24 hrs after the initiation of organ culture (hollow circles in Fig. 8). The same idea described above was applied to detect G_1-chalone activity *in vitro* using this period of culture. Crude extract containing epidermal chalone (1 cm^2-area skin equivalent per ml medium) was added to the culture 14 hrs after the initiation of culture. ^{3}H-TdR was given 1, 2, 4, and 6 hrs later and the ear pieces were then fixed 30 min later. The result is shown as solid circles in Fig. 8. It indicates that chalone inhibited the cells from entering S phase. Adrenalin alone (hollow triangle in Fig. 8) did not inhibit the initiation of DNA synthesis. Chalone alone, however, inhibited the cells from entering S phase (solid triangle in Fig. 8). It is concluded therefore that G_1-chalone does not necessarily require adrenalin as a cofactor, in contrast to G_2-chalone which requires adrenalin as an essential cofactor.[14]

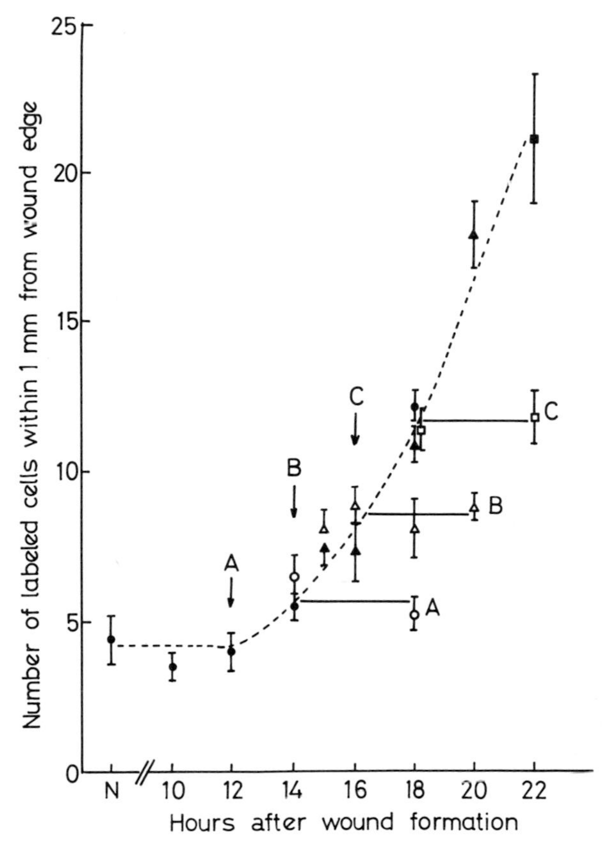

Fig. 7. Inhibition of the flow into S phase (G_1-chalone action) by epidermal extract (solid lines) injected at early periods of wound healing. During this period the number of labeled cells was increasing rapidly if chalone was not injected (dotted line). Arrows indicate the times of chalone injection. N, normal epidermis without wound.

Adrenalin requirement of G_2-chalone was also demonstrated in the present assay system, in which the crude extract and/or adrenalin were added to the medium simultaneously with colchicine at 16 hrs of culture and the ear pieces were fixed 4 hrs later (Table 2).

An attempt was made to separate G_1- and G_2-chalones by the difference between their molecular weights,[7,9,10] and the cell cycle phase specificity of them was examined. The elution pattern through a Sephadex G–100 chromatography of the precipitate of epidermal extract by 80% ethanol was shown in Fig. 9. Two fractions (A; MW $> 100{,}000$ and B; $10{,}000 <$ MW $< 100{,}000$) were then pooled and applied to a 16-hr culture. ^{3}H-

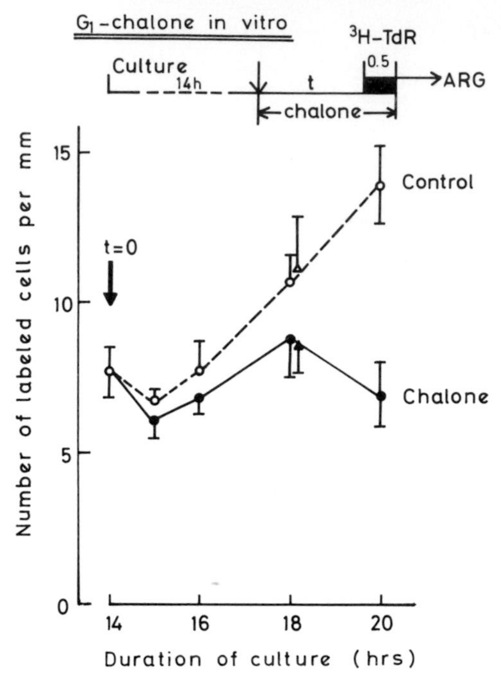

Fig. 8. Inhibition by chalone of the flow into S phase *in vitro* (G_1-chalone action) During this period of organ culture the number of labeled cells was increasing if chalone was not added (control). The addition of adrenalin alone (hollow triangle) was not effective, but that of chalone alone (solid triangle) was effective.

Table 2. The effect of epidermal chalone and adrenalin on the mitotic activity of the epidermis in organ culture. Experimental details are described in the text.

Treatment	No. of mitoses accumulated/cm
Control	13.0 ± 2.16
Skin extract	15.2 ± 2.48
Adrenalin	8.7 ± 1.09
Skin extract + Adrenalin	5.1 ± 0.88

TdR was added 4 hrs later and the ear pieces were fixed 60 min thereafter for ARG.

The precipitate of epidermal extract by 80% ethanol had both G_2- (Fig. 10) and G_1-chalone (Fig. 11) activities. After being heated at 100°C for 10 min, the ethanol precipitate lost its G_2-chalone activity, but not G_1-chalone activity.

The fraction B with MW less than 100,000 had only G_2-chalone activity

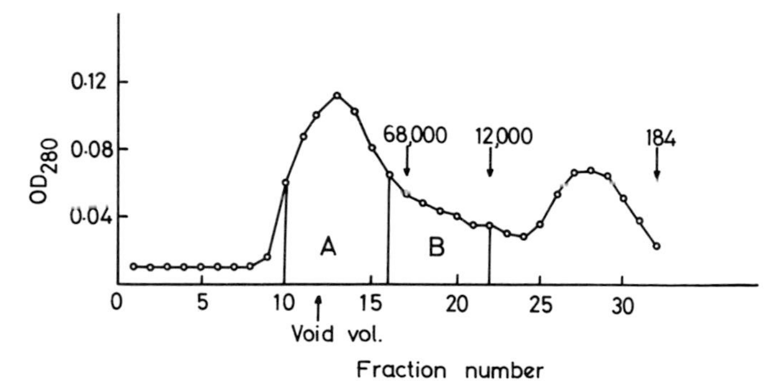

Fig. 9. Elution pattern of the alcoholic precipitate of skin extract through a Sephadex G-100 gel chromatography. Arrows indicate the molecular weight standards. Fractions A (MW more than 100,000) and B (MW between 10,000 and 100,000) were used in Figs. 10 and 11.

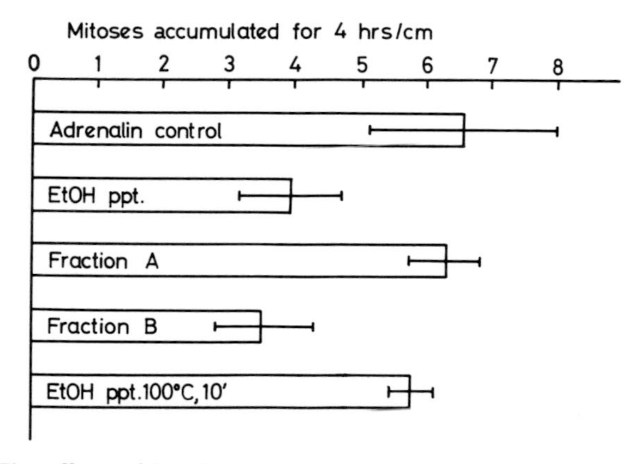

Fig. 10. The effects of fractions A and B (Fig. 9) on the flow into M phase of the epidermal cells in organ culture.

(Fig. 10) and the fraction A with MW more than 100,000 had only G_1-chalone activity (Fig. 11). Thus both the chalones were cell cycle phase specific in this meaning.

D. Chalone effects on the culture of wounded epidermis

If the above mentioned idea is accepted, that G_1-chalone is detectable by applying it to the rising phase of the number of labeled cells, the epidermis during wound healing may be used for the detection of G_1-chalone *in vitro* by transferring it into an organ culture at the time, in the course of increasing the number of labeled cells, that is, before 2 days of wound healing, as shown in Fig. 1. A similar idea is applicable to the *in vitro*

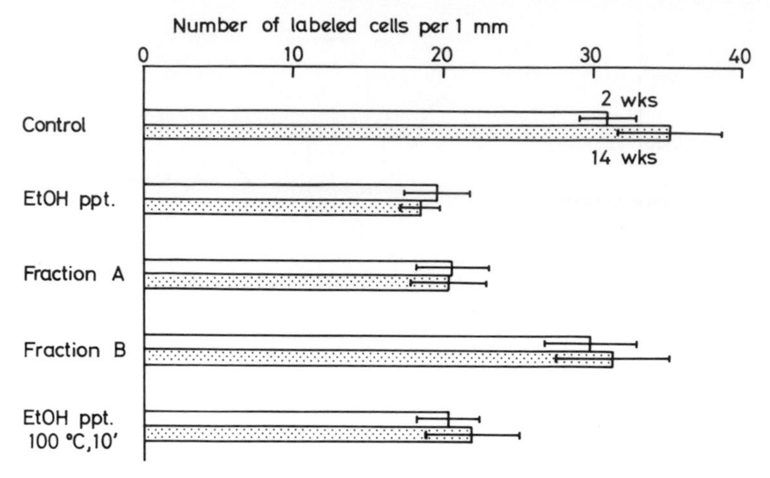

Fig. 11. The effects of fractions A and B (Fig. 9) on the flow into S phase of the epidermal cells in organ culture. Autoradiogram was exposed for 2 weeks (blank column) or for 14 weeks (dotted column).

assay of S-chalone by transferring the epidermis at the plateau phase of the number of labeled cells, that is, 2–3 days of wound healing (Fig. 1), and by counting the number of grains on labeled nuclei. The results of the experiments on this line were illustrated in Figs. 12 and 13, respectively.

The result shown in Fig. 12 is essentially the same as that shown already (Fig. 8) and indicates this convenient method can be used for assay of G_1-chalone *in vitro*.

When the epidermis at 3 days of wound healing was transferred into organ culture, the number of labeled cells did not differ significantly between the 4-hr cultures with or without chalone, even when the exposure time of ARG was lengthened up to 6 weeks (^{3}H-TdR of 3 μCi/ml was enough to saturate the increase in the number of labeled cells with the time of exposure of ARG at 2 weeks). If G_1-chalone inhibited the cells from entering S phase, a duration of 4 hrs was too short to cause a significant decrease in the number of the cells in S phase, because T_s was 16 hrs (Table 1) and the actual duration of the inhibition to the flow into S was $4 - 2 = 2$ hrs on the basis obtained in Fig. 7. On the other hand, the average number of grains on the labeled cells decreased markedly during this 4 hr-culture with chalone (Fig. 13). This must be ascribed to S-chalone activity which inhibits the cells in S phase from progressing to DNA synthesis. The result also indicated that S-chalone did not require adrenalin as a cofactor. The precipitate of skin extract by 80% ethanol had S-chalone activity, and liver extract had no such activity. Differing from G_1-chalone, S-chalone lost its activity when heated at 100°C for 10 min (Fig. 13).

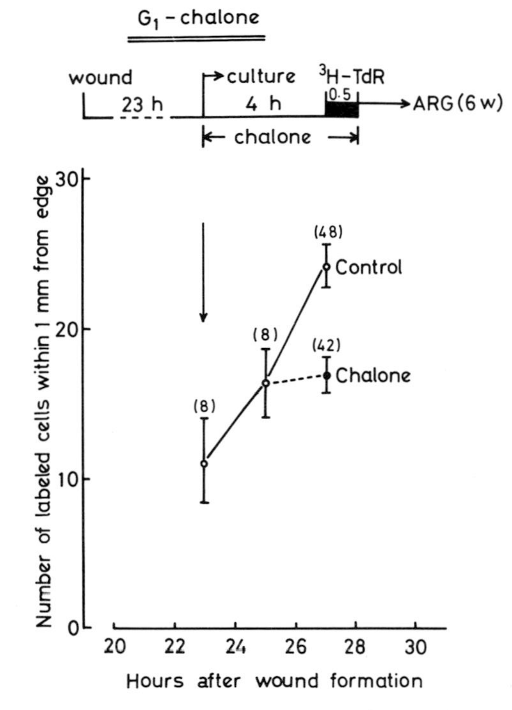

Fig. 12. The effect of chalone on the flow into S phase in a short culture of the wounded epidermis. The number of the cells in S phase was still increasing at this period of wound healing (24 hours) and G_1-chalone action was demonstrated.

DISCUSSION

Cell cycle analyses indicated the crucial role of G_1 phase in determining the duration of cell cycle (Table 1). Thus, in order to show that chalone is of greatest importance to the mechanism of growth control, we must present evidence that it inhibits cells in G_1 from entering S phase. To separate the action on the flow into S phase from that on the rate of DNA synthesis during S phase, it is necessary to detect the effect on the number of labeled cells in an ARG exposed satisfactorily. Such studies on the intact epidermis of hairless mouse *in vitro*[4,5,8] demonstrated the inhibition by epidermal chalone of the flow into S phase. This evidence for G_1-chalone, however, was rather indirect; the time delay required for obtaining a detectable decrease in the number of labeled cells after chalone injection obscured the exact locus of action on the cell cycle phases.[11] The method used by Marks[7,9] for detecting G_1-chalone was essentially the same as this indirect method; the chalone was injected into mouse and the *in vitro*

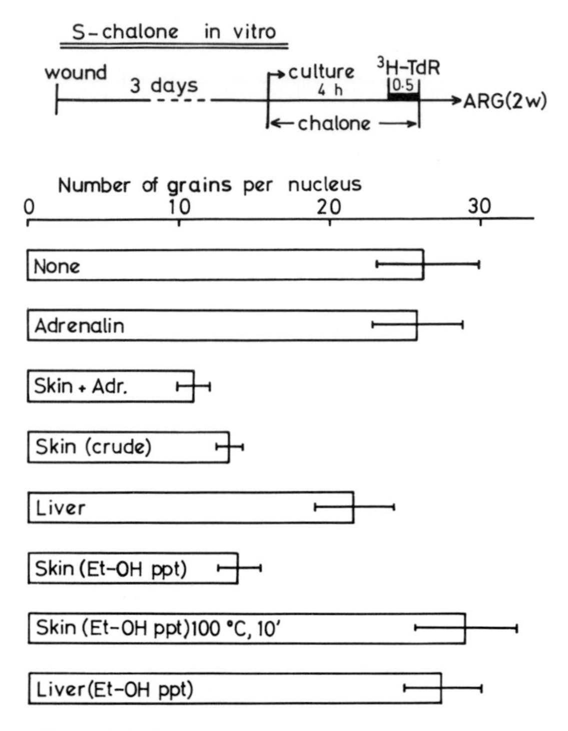

Fig. 13. The effect of chalone on grain counts in a short culture of the wounded epidermis transferred at a period of the highest proliferation during wound healing. Number of labeled cells was not affected by chalone. S-chalone action was demonstrated.

incorporation of ^{3}H-TdR into DNA in the skin piece excised after the time lag required was determined biochemically. On the other hand, our assay systems described in Figs. 7,8, and 12 supported direct evidence for the existence of the G_1-chalone which acted exactly on the flow into S phase by demonstraing a reduction in the increasing rate of the number of labeled cells at a definite period during wound healing or organ culture.

Differences between epidermal G_2- and G_1-chalones were suggested as follows:[8] G_2-chalone has a relatively short turnover time[5] and is present in or produced by the epidermal basal cells,[16] while G_1-chalone has a long turnover time[5] and is produced by the keratinizing cells.[17] They were also shown to be chemically different: G_2-chalone is a glucoprotein with a MW of 30,000-40,000,[10] heat-labile, and requires adrenalin as an essential cofactor,[15] while G_1-chalone is a glycoprotein with an apparent MW of 100,000–300,000, consisting of a large protein "coat" and a small active "core," probably a glycopeptide with a MW of 10,000–20,000,[9] heat-stable and adrenalin independent.[7] Our present data confirmed that the fraction with MW more than 100,000 included G_1-chalone and another fraction with MW of 10,000–100,000 included G_2-chalone. Adrenalin requirements and heat labilities of these also coincided with those reported by others. Cell cycle phase specificity of these chalones has been investigated only for G_2-chalone.[11] Whether G_1-chalone separated from G_2-chalone does inhibit the flow into M phase directly or not was an important question, because there had been a possibility that the delayed inhibition of DNA synthesis by G_1-chalone might be ascribed to an indirect inhibition through a block of the flow into the preceding M phase.[11] This possibility has now been denied by our present results, shown in Figs. 10 and 11: Fractions containing G_1-chalone inhibited only the flow into S phase, but not the flow into M phase, and that containing G_2-chalone inhibited the flow into M phase, but not the flow into S phase.

A question may arise on the existence of S-chalone. The present experiment showed an immediate decrease in the grain counts after chalone application (Figs. 4 and 13). On the other hand, scintillation countings of ^{3}H-TdR incorporated into DNA in normal epidermis have failed to detect the reduction of DNA synthesis during a few hours after chalone treatment both *in vivo*[4] and *in vitro*.[7,19] The reasons for this discrepancy were considered to be as follows: Normal DNA synthesis is, logically, already under chalone control and further addition of chalone should bring about only a minimal effect, which may be too small to detect biochemically. In fact, the experiments using the culture of single cells dissociated from epidermis[20,21] or a skin culture[22] demonstrated a rapid decrease after chalone application in the rate of DNA synthesis measured biochemically. Our *in vivo* (Fig. 4) and *in vitro* (Fig. 13) systems were regarded to be satisfactorily sensitive for detecting S-chalone activity because they were synthesizing DNA very actively. Cell cycle analysis, however, indicated that T_S remained unchanged after stimulation of proliferation by wounding (Table 1); if the inhibition by S-chalone had existed in the normal epidermis and if the inhibition would be released during wound healing, there should be a shortening of T_S. This fact obscures the physiological meaning of the S-chalone activity. A possible explanation may be that an excess amount of chalone might cause an unphysiological action on epidermal proliferation. Elgjo[23] found that after a single i.p. injection of 5 mg lyophilized skin extract, some of the affected epidermal cells in G_2 were permanently prevented from continuing their division. Wiley *et al.*[21] showed a large amount (78 mg protein/ml) of skin extract decreased ^{3}H-TdR incorporation into DNA measured in cpm during a short culture of isolated epidermal cells, but it had no such effect within a dose range used commonly for detecting G_1-chalone. Our assay system may be sensitive enough to detect S-chalone activity with a dose not excessive as such. It should be mentioned that according to our unpublished data the guinea pig skin used in the present experiment (Fig. 4) contained much more chalone activity per unit area than the mouse skin did. The S-chalone activity was shown to be tissue-specific and remained in the precipitate by 80% ethanol (Fig. 13), both indicating that it was not due to nonspecific inhibitors of the ^{3}H-TdR incorporation into DNA. The physiological meaning of this S-chalone activity is still to be understood.

Acknowledgements

We are deeply indebted to Mr. K. Manaka, Mr. Y. Kinjo, Mr. T. Yama-ai, Miss Y. Shiobara, Miss K. Nurishi, and Miss K. Eguchi, who were all coauthors in our original papers, published already or to be published. Thanks are also due to Mrs. J. Suzuki for her valuable assistance.

REFERENCES

1. Bullough, W. S. Mitotic and functional homeostasis: A speculative review. *Cancer Res.*, **25**: 1683–1727, 1965.
2. Yamaguchi, T. and Tabachnick, J. Cell kinetics of epidermal repopulation and persistent hyperplasia in locally β-irradiated guinea pig skin. *Rad. Res.*, **50**: 158–180, 1972.
3. Hall, R. G., Jr. DNA synthesis in organ cultures of hamster cheek pouch: Inhibition by a homologous extract. *Exp. Cell Res.*, **58**: 429–431, 1969.
4. Hennings, H., Eljo, K., and Iversen, O. H. Delayed inhibition of epidermal DNA synthesis after injection of an aqueous skin extract (chalone). *Virchows Arch. Abt. B Zellpath.*, **4**: 45–53, 1969.
5. Elgjo, K., Hennings, H., and Edgehill, W. Epidermal mitotic rate and DNA synthesis after injection of water extracts made from mouse skin treated with actinomycin D. Two or more growth-regulating substances? *Virchows Arch. Abt. B Zellpath.*, **7**: 342–347, 1971.
6. Frankfurt, O. S. Epidermal chalone. Effect on cell cycle and on development of hyperplasia. *Exp. Cell Res.*, **64**: 140–145, 1971.
7. Marks, F. Direct evidence of two tissue-specific chalone-like factors regulating mitosis and DNA synthesis in mouse epidermis. *Hoppe-Seyler's Z. Physiol. Chem.*, **352**: 1273–1274, 1971.
8. Eljo, K. Epidermal chalone: Cell cycle specificity of two epidermal growth inhibitors. In Chalones: Concepts and Current Researches (B. K. Forscher and J. C. Houck, eds.), Natl. Cancer Inst. Monogr., Vol. 38, pp. 71–76, 1973.
9. Marks, F., A tissue-specific factor inhibiting DNA synthesis in mouse epidermis. In Chalones: Concepts and Current Researches (B.K. Forscher and J.C. Houck, eds.) Natl. Cancer Inst. Monogr., Vol. 38, pp. 71–76, 1973.
10. Hondius Boldingh, W. and Laurence, E. B. Extraction, purification and preliminary characterisation of the epidermal chalone: A tissue specific mitotic inhibitor obtained from vertebrate skin. *Eur. J. Biochem.*, **5**: 191–198, 1968.
11. Thornley, A. L. and Laurence, E. B. Chalone regulation of the epidermal cell cycle. *Experientia*, **31**: 1024–1025, 1975.

12. Yamaguchi, T., Hirobe, T., Kinjo, Y., and Manaka, K. The effect of chalone on the cell cycle in the epidermis during wound healing. *Exp. Cell Res.*, **89**: 247–254, 1974.
13. Cleaver, J. E. Thymidine Metabolism and Cell Kinetics. North-Holland Publ., Amsterdam, 1967.
14. Smith, C. L. and Dendy, P. P. Relation between mitotic index, duration of mitosis, generation time and fraction of dividing cells in a cell population. *Nature*, **193**: 555–556, 1962.
15. Bullough, W. S. and Laurence, E. B. Mitotic control by internal secretion: The role of the chalone-adrenalin complex. *Exp. Cell Res.*, **33**: 176–194, 1964.
16. Elgjo, K., Laerum, O. D., and Edgehill, W. Growth regulation in mouse epidermis I. G_2-inhibitor present in the basal cell layer. *Virchows Arch. B Zellpath.*, **8**: 277–283, 1971.
17. Elgjo, K., Laerum, O. D., and Edgehill, W. Growth regulation in mouse epidermis II. G_1-inhibitor present in the differentiating cell layer. *Virchows Arch. Abt. B Zellpath.*, **10**: 229–236, 1972.
18. Bullough, W. S., Hewett, C. L., and Laurence, E. B. The epidermal chalone: A preliminary attempt at isolation. *Exp. Cell Res.*, **36**: 192–200, 1964.
19. Baden, H. P. and Sviokla, S. Effects of chalone on epidermal DNA synthesis. *Exp. Cell Res.*, **50**: 644–646, 1968.
20. Delescluse, C., Regnier, M., and Prunieras, M. Specific effect of epidermal extracts on DNA synthesis in tissue culture. *J. Invest. Dermatol.*, **58**: 253–254, 1972.
21. Wiley, C. L., Williams, W. W., and McDonald, C. J. The effects of the epidermal chalone on DNA synthesis in mammalian epidermal cells. *J. Invest. Dermatol.*, **60**: 160–165, 1973.
22. Rothberg, S. and Arp, B.C. Epidermal chalone and inhibition of DNA synthesis. In Chalones: Concepts and Current Researches (B. K. Forscher and J.C. Houck, eds.), Natl. Cancer Inst. Monogr., vol. 38, pp. 93–98, 1973.

Discussion

DR. BADEN: I would like to ask what tests you have done for cytotoxicity of the extracts?

DR. YAMAGUCHI: When the material is injected into the mouse, the action is almost entirely reversible so it must not be cytotoxic. We have not repeated the cytotoxicity tests which many investigators have already done. The inhibition of mitosis was effective only when adrenalin was present in the culture and this effect was also reversible.

DR. BADEN: I bring this up because we ran into trouble with this several years ago. It seems of value when you make a preparation to test it for cytotoxicity in some way since there can be variations in what you extract at different times.

DR. FUKUYAMA: Have you determined the effect of cutaneous chalone fractions on cell lines other than skin cells? For example, on fibroblasts?

DR. YAMAGUCHI: No, we have not examined the material's effect on cells other than epidermal because previous investigators have shown that the skin extract, prepared as we do, did not affect proliferative activity of intestinal crypt cells and other cultured cells. I feel, however, that the tests of our own skin extract or its fractions should be done since different fractions have specific effects at different phases of the cell cycle.

DR. ASO: I also began working on the chalone in 1972 and extracted chalone from pig epidermis. From one kilogram of tissue, we extracted about 100 mg of chalone by dry weight. After treatment by column chromatography and gel electrophoresis, we found about 12 bands of protein. Have you ever analyzed the purity of your chalone preparation by gel electrophoresis?

DR. YAMAGUCHI: No, our main effort has been focused on developing an effective assay system which separates the effects seen on different phases of the cell cycle. Therefore, biochemical analyses have not yet been done. Purification of G-1 and G-2 chalones, as performed by others, was discussed in my presentation. They are regarded as being heterogeneous.

DR. ASO: You inject these chalone preparations into the skin, do you not?

DR. YAMAGUCHI: Yes.

DR. ASO: I think it is a little dangerous for you to decide what you have. By injecting the chalone intradermally, you will find a great deal of release of prostaglandin and that will increase the cyclic AMP level and, in that case, we do not know what the effect of the cyclic AMP is and what the effect of prostaglandin is. So I think that the organ culture experiments are satisfactory, but you have to be careful of your interpretation of the injection experiments.

DR. MCGUIRE: It seems to me that many of the chalone effects that have been described, not only by you but by many other people who have studied chalone over the years, could be related to a change in pool size of thymidine within the cell or to an inhibition of the entry of labeled thymidine into the cells which would result in a reduced incorporation of labeled thymidine. This is an impossible question to consider in an animal experiment. However, it is possible to ask this question in cell culture or in organ culture since entry of exogenous thymidine into the pool can be measured. I wonder if you have done that. If the entry of tritiated thymidine is inhibited, then there would be a reduced number of grains.

DR. YAMAGUCHI: Some answer may be found in the fact that by 6 hours after application of chalone, the grain counts had reverted to a control level but the number of labeled cells were decreased. That is, during this period of 6 hours, the label was incorporated into the DNA within the cells at S phase to the normal extent but the number of cells passing through the S phase decreased presumably by being blocked in their entry into the S phase.

Inhibition of *in vitro* Protein Biosynthesis by the Partially Purified Fractions of the Melanoma Extract

M. Seiji,* H. Nakano,* S. Moro,* K. Ogata,** K. Ishikawa** and S. Odani**

* *Department of Dermatology, Tohoku University School of Medicine, Sendai, Japan* and
** *Department of Biochemistry, Niigata University School of Medicine, Niigata, Japan*

ABSTRACT

Additional studies on the chemical characteristics of fraction II_1 and IV_2 were carried out. In fraction II_1, there was a single absorption peak at 278 nm and the ratio of A260 to A280 was 0.83. And from the profile of the SDS-polyacrylamide gel electrophoretogram, fraction II_1 is assumed to consist mainly of proteins which are composed of two major and eleven minor components. In fraction IV_2, a single absorption peak was shown at around 260 nm but the spectrum varied slightly depending on the preparations, accordingly the ratio of A260 to A280 was between 2.0 and 3.95. Fraction IV_2 was assumed to consist of a large amount of small RNA and some proteins. Polysomes prepared from melanoma and liver and used in this series of experiments were found to be quite similar in their natures. Sufficient amount of polysomes larger than pentamer were found to be contained in total and free polysome preparations.

In the case of melanoma, it was clearly shown that both fraction II_1 and fraction IV_2 inhibited the incorporation of ^{14}C-L-leucine into proteins of various cell particles *in vitro*. That 100% inhibition was obtained with fraction IV_2, and 60% at the most with fraction II_1 might indicate that the inhibitory mechanisms of these two fractions might be quite different from each other. The inhibitory effect of fraction IV_2 on the melanoma system was dependent on the tumor microsome. In order to examine the specificities of fraction II_1 and fraction IV_2, nonspecific bovine serum albumin and yeast RNA were added to the C^{14}-L-leucine incorporation system of melanoma total polysome. No significant inhibitions were observed by albumin nor by yeast RNA.

INTRODUCTION

In 1968 Bullough[1] and Mohr[2] reported that the tissue specific inhibitor of cell division (chalone) was found to be present in melanoma extracts. A single injection of melanoma extract into the peritoneal cavity of mice bearing Harding-Passey mouse melanoma caused a 35–65 percent reduction of ^{14}C-thymidine incorporation into the DNA of melanoma melanocytes.[3] It was further shown that a single addition of melanoma extract to the tissue culture medium in which B-16 mouse melanoma melanocytes were cultured caused a reduction of mitotic index as well as the incorporation of ^{14}C-thymidine into DNA and that of ^{3}H-leucine into the protein of melanocytes. The inhibition of DNA and protein biosynthesis due to the melanocyte extract was well correlated.

The previous results showed that the biologically active substances of the melanoma extract appear to consist of two different kinds of fractions: fraction II_1, which is heat-labile and tissue nonspecific and consists mainly of protein, and fraction IV_2, which is heat-labile and tissue specific and consists of protein and RNA.[3]

In an attempt to clarify the inhibitory mechanisms, the effects of fraction II_1 and IV_2 were studied in the *in vitro* incorporation of ^{14}C-L-leucine by various cell particles of Harding-Passey mouse melanoma and mouse liver.

MATERIALS AND METHODS

Preparations of fraction II_1 and IV_2

Harding-Passey mouse melanomas used in the experiments were serially transplanted into a strain of Swiss and DD mice. Actively growing tumors were excised and promptly homogenized in 4 volumes of ice-cold distilled water at about 0°. All subsequent processes took place in a cold environment (about 3°C). The homogenate was centrifuged at 105,000 × *g* for 60 minutes in a Hitachi ultracentrifuge. The supernatant fraction was subjected to successive ethanol fractionation according to the method of Bullough *et al.*[4] by the slow addition of precooled 100% ethanol (0–4°C) to concentrations of 50, 70, and 95%. After each step the mixture was stirred for a short time and centrifuged after satisfactory flocculation of the precipitate occurred. The fraction precipitating between 70–95% ethanol was collected, dissolved in water, and lyophilized.

One hundred mg of the ethanol precipitate thus obtained were dissolved in 3 ml of 0.01M potassium phosphate buffer, pH 7.4, and subjected to gel filtration through a Sephadex G-100 column (2.5 × 90 cm), equilibrated with 0.01M phosphate buffer, pH 7.4; elution was conducted with the same buffer. Five-milliliter fractions were collected and the absorbances at 280 and 260 nm were measured. Four peaks were obtained and they were designated I, II, III, and IV according to the sequence of elution. The fractions II and IV were collected, frozen, and lyophilized individually.

Fractions II and IV of the Sephadex eluate were then chromatographed on a 3.5 × 30 cm column of diethylaminoethyl cellulose-Sephadex A-25, equilibrated with 0.01M potassium phosphate buffer, pH 7.4. Continuous gradient elution was then performed; a linear gradient from 0 to 0.2 M NaCl in 0.01M potassium phosphate buffer, pH 7.4 ,was used. Four-milliliter fractions were collected and the absorbances at 280 nm and 260 nm were measured. Two peaks, II_1 and II_2, were obtained at 280 nm peak and no peaks at 260 nm were observed. Four peaks, IV_1, IV_2, IV_3 and IV_4, were obtained in the elution pattern of fraction IV. The ratio of A260 nm to A280 nm above one was found in IV_2, IV_3, and IV_4. Among these subfractions, II_1 and IV_2 were collected, frozen and lyophilized individually.[3] Both fractions were dissolved in cold medium A (0.25M sucrose, 5mM $MgCl_2$, 25mM KCl, 50 mM Tris-HCl buffer (pH 7.6)) and dialysed in a cellophane tube at 4°C against the medium A solution for 8 hours to eliminate the sodium chloride contained and then immediately used for the incorporation experiments.

Chemicals ^{14}C-L-leucine and uniformly labeled 331 mCi/m mole was obtained from Radiochemical Centre, Amersham, England. Adenosine triphosphate (disodium salt, ATP), guanine triphosphate (sodium salt, GTP), glutathione (GSH), creatine phosphate sodium salt, creatine kinase (EC 2, 7, 3, 2), sodium dodecylsulfate (SDS), and bovine serum albumin were obtained from the Sigma Chemical Corporation. Sodium deoxycholate came from Difco Co.

Preparation of the various cell components

Harding-Passey mouse melanomas were excised when they reached 1.0 to 1.5 cm in diameter and were promptly homogenized in 3 volumes of medium A at about 0°C. All subsequent processes took place in a cold environment. Microsomes and cell saps were prepared by the method of Keller and Zamecnik.[5] The total polysomes were prepared by the method of Wettstein *et al.*[6] and the free and membrane-bound polysomes with the modified method of Sugano *et al.*[7], in which the medium A containing rat liver cell sap was used at the first homogenization step in order to inhibit the RNase activity. The rough surface membrane was prepared by Rothschild's method.[8] The polysomes were suspended in the medium A and centrifuged at 3000 rpm for 5 minutes and the supernatant thus obtained was used as polysomes. The cell sap was taken from the supernatant obtained by the centrifugation of homogenate of the melanoma and liver at 105,000 × *g* for 90 minutes. Only the upper one-third of the supernatant in the centrifuge tubes after centrifugation was pipetted out, and it was designated as a cell sap. Various cell particles and cell sap were also obtained from the liver of DD or Swiss strain mice in the same way as were the melanoma tissues, except that the rat liver cell sap was not used in the preparation of total and free polysomes.

Chemical analyses The protein content was determined according to Lowry *et al.*[9] using bovine serum albumin as a standard. The RNA content was determined by the method of Dische[10] using yeast RNA as a standard.

Inhibitory effects of fraction II_1 and IV_2 on the incorporation of ^{14}C-leucine into proteins of various cell particles *in vitro*

The reaction mixture for studying the incorporation of amino acid into protein contained in a total volume of 0.3 ml: 0.25 M sucrose, 25 mM KCl, 5 mM $MgCl_2$, 50 mM Tris-HCl buffer (pH 7.6), 1 mM ATP, 0.25 mM GTP, 10 mM creatine phosphate, 3 μg protein of creatine kinase, 5mM GSH, 0.2 μCi ^{14}C-L-leucine, about 0.5 mg protein of microsomes and rough surface membrane and/or 0.15 mg protein of various polysomes of the melanoma and liver, and 1.5 mg and/or 0.3 mg protein of the melanoma and liver cell saps. Various amounts of fraction II_1 and IV_2 were added to the complete reaction system described above in order to determine their inhibitory effects. After incubation at 30°C with shaking, the reaction was stopped by adding equal volume of cold 10% trichloroacetic acid.

Determination of radioactivity of various cell particles incubated

The mixture of cell particles and 5% trichloroacetic acid was centrifuged after overnight storage at 2°C. The precipitate thus obtained was treated with 5% hot trichloroacetic acid and then washed three times with 5% trichloroacetic acid, two times with absolute ethanol, and two times with an alcohol-ether mixture (3:1 *v/v*) at 60°C. The final precipitate was dissolved in formic acid and then transferred in a stainless steel planchet, dried and weighed, and its radioactivity was measured by a windowless-gasflow counter. Samples were corrected for variations in the protein content from self-absorption curves prepared from labeled rat liver proteins. The radioactivities incorporated into protein are given as counts per minute at infinite thinness. The inhibitory effects are expressed as percentages of the control.

Sodium dodecyl sulfate-acrylamide gel electrophoresis

Fraction II_1 and IV_2 were dialyzed against 0.01M phosphate buffer pH 7.0, and then incubated with 1% sodium dodecyl sulfate and 50 mM 2-mercaptoethanol at 37°C for 2 hours. Proteins in Fraction II_1 and IV_2 were electrophoresed in sodium dodecyl sulfate-acrylamide columns (6 × 80 mm) for 5–8 hours at 8 mA per tube using Bromphenol blue as a marker. Then, the gel was stained with 2.5% coomassie brilliant blue. The molecular weights of the stained materials were calculated by the method of Weber and Osborn.[11]

Sucrose density gradient centrifugation

The various polysome preparations were subjected to centrifugation in the 27 ml of 15–35% sucrose gradient solution containing TKM buffer (50 mM Tris-HCl (pH 7.6), 25 mM KCl, 5 mM $MgCl_2$) at 24,300 rpm for 3 hours at 2°C. Then the sedimentation profiles were obtained with an automatic gradient analyser of ISCO Instrument Co.

RESULTS

Absorption spectra of fraction II_1 and IV_2

The absorption spectra of fraction II_1 and IV_2 in 20 mM tris-HCl (pH 7.9) solution are shown in Fig. 1. There was a single absorption peak at 278 nm in fraction II_1 and the ratio of A260 to A280 was 0.83. In the case of fraction IV_2, the absorption spectrum differed slightly, depending on the individual preparations, with the maximum peak between 253–262 nm. The ratios of A260–A280 were 2.00–3.95.

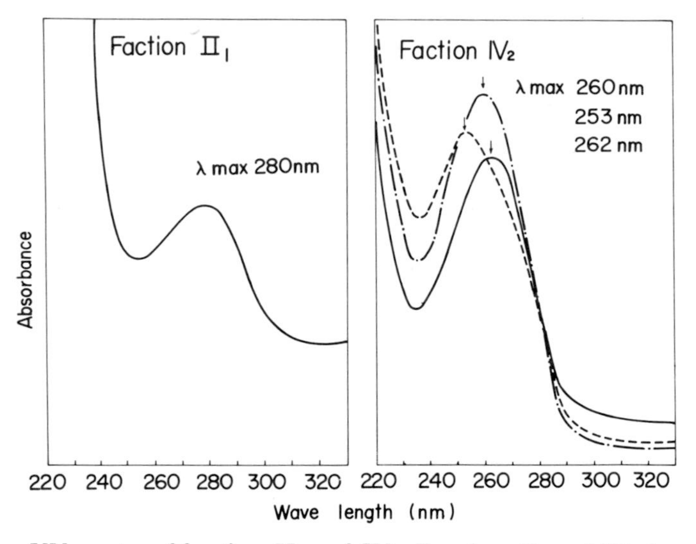

Fig. 1. UV spectra of fractions II_1 and IV_2. Fractions II_1 and IV_2 in medium A were dialysed as described in Materials and Methods and scanned by Hitachi 124 double beam spectrophotometer. The absorption spectra of three different fraction IV_2 preparations are shown.

Determination of molecular weights of various components contained in fraction II_1 and IV_2 by SDS-polyacrylamide gel electrophoresis

The SDS-polyacrylamide gel electrophoretograms of fraction II_1 and fraction IV_2 are shown in Fig. 2. The scanning profile at 610 nm of the coomassie brilliant blue stained SDS-polyacrylamide gel electrophoretogram of fraction II_1 is shown in Fig. 3. There were eleven minor protein components and two major absorption peaks, of which molecular weights were briefly estimated from the reference line as 50,000 and 26,000, respectively. Figure 4 shows the scanning profile of the stained SDS-polyacrylamide gel electrophoretogram of fraction IV_2. Fraction IV_2 was assumed to consist of a large amount of RNA and a small amount of proteins, molecular weight of which may possibly be very small, less than 10,000, because only one-tenth of the total protein which was subjected to the gel electrophoresis, remained in the gel after electrophoresis and no proteins were detected at the origin. Among the proteins remaining in the gel, there were three dominant fractions, molecular weights of which were briefly estimated as 53,000, 47,500 and 24,500, respectively. The nature of RNA contained in the fraction IV_2 was not investigated yet.

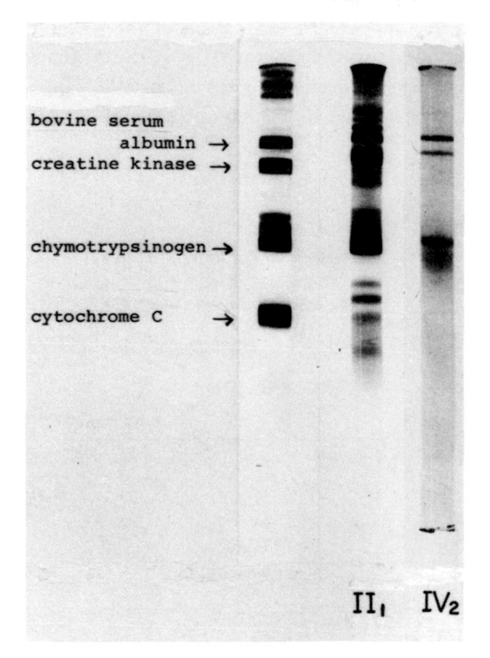

Fig. 2. Sodium dodecyl sulfate polyacrylamide gel electrophoretograms of fractions II_1 and IV_2. Fraction II_1, which contained Folin-reaction positive material that corresponds to about 30 μg of protein, and fraction IV_2, which contained the same material that corresponds to about 50 μg of protein, were incubated with 1% SDS and 50 mM 2-mercaptoethanol at 37°C for 2 hours and electrophoresed in SDS-polyacrylamide gel as described in Materials and Methods. Gels were stained with coomassie brilliant blue.

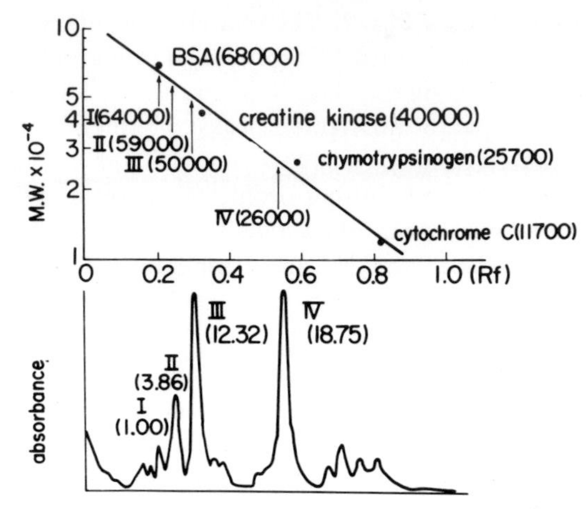

Fig. 3. Scanning of SDS-polyacrylamide gel electrophoretogram of fraction II_1 and the molecular weight estimation of the stained bands. Scanning of the stained bands was performed by a Joyce microdensitometer at 610 nm. Standard proteins were run in parallel. Bovine serum albumin, 68000; creatine kinase, 40000; chymotrypsinogen, 25700; cytochrome C, 11700.

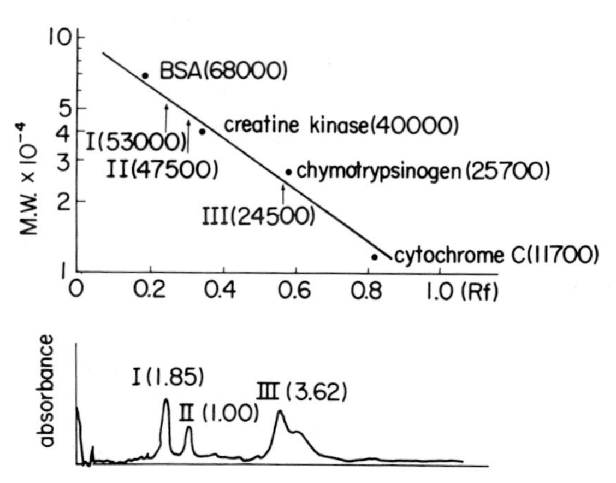

Fig. 4. Scanning of SDS-polyacrylamide gel electrophoretogram of fraction IV_2 and the molecular weight estimation of the stained bands.

Sedimentation profiles and time courses of ^{14}C-L-leucine-incorporation into proteins of various polysomes

The sedimentation profiles of various polysomes obtained from mouse liver and melanoma are shown in Fig. 5. The profiles of total and free polysomes were quite similar in liver and a melanoma, and sufficient amount

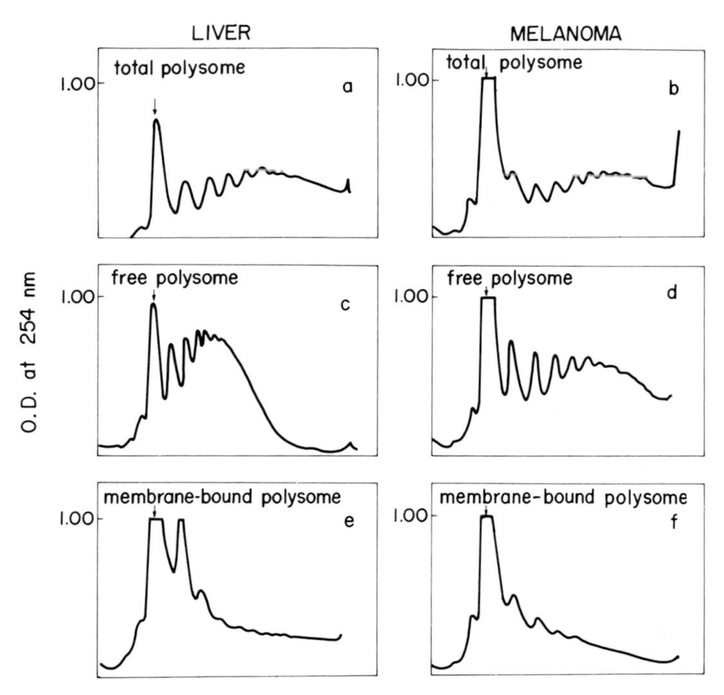

Fig. 5. Sedimentation profiles of various polysomes from mouse liver and melanoma. Total, free, and membrane-bound polysomes were prepared as described in Materials and Methods. About 10 absorbances at 260 nm units of each polysomes were centrifuged in 37 ml of 15–35% sucrose gradient in TKM buffer (50 mM Tris-HCl, pH 7.6, 25 mM KCl and 5 mM $MgCl_2$) in an SW-27 swinging bucket of Beckman Co. at 24,300 rpm for 3 hours at 2°C. UV scanning at 254 nm was performed by ISCO density gradient analyser. a, total polysome of liver; b, total polysome of melanoma; c, free polysome of liver; d, free polysome of melanoma; e, membrane-bound polysome of liver; f, membrane-bound polysome of melanoma. The arrows indicate the position of 80S monosome.

of polysomes larger than pentamer were found to be contained in these preparations. On the other hand, in the case of membrane-bound polysomes, the amount of polysomes larger than pentamer was very small, possibly due to the time consumed for preparation. The time courses of ^{14}C-L-leucine incorporation by various polysomes are shown in Fig. 6. The incorporations of ^{14}C-L-leucine into total and free polysome increased with time, rapidly up to 10 min, thereafter gradually. The amounts of ^{14}C-L-leucine incorporated into total and free polysome of melanomas were much less compared with those of liver. As it was understood from the shapes of profiles of melanoma polysome the contents of monosome in

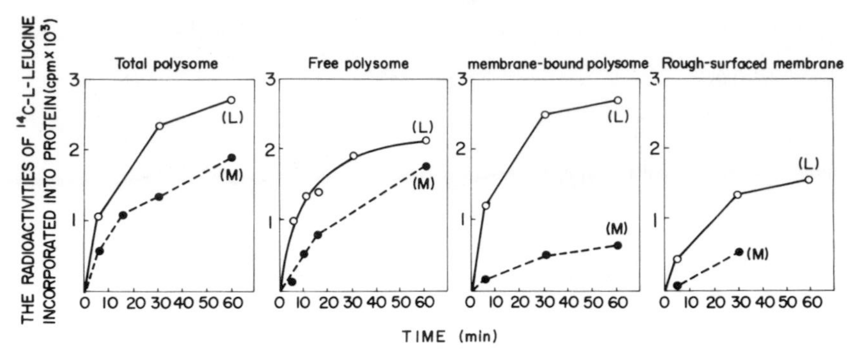

Fig. 6. Time dependent incorporation of ^{14}C-L-leucine into various polysomes and rough-surfaced membrane from melanoma and mouse liver *in vitro*. Various polysomes and rough-surfaced membrane were incubated in cell-free systems at 30°C for the time indicated as described in Materials and Methods. ^{14}C-L-leucine incorporated into protein by each cell-free system was determined as described in Materials and Methods.

melanoma were greater than those in liver, thus the incorporation would be less in melanoma. The incorporations of ^{14}C-L-leucine into melanoma membrane-bound polysome and rough surface membrane were very low as compared with those of liver.

Inhibitory effect of fraction II_1 on the protein biosynthesis in the cell-free system

Figure 7 shows dose response curves of the inhibitory effects of fraction II_1 on various cell particles of melanoma and liver. It was found that the inhibitory effects increased with increasing amounts in all cases, particularly in melanoma. The curves of various polysome preparations of melanoma and liver were very much similar and there appeared to be no differences between melanoma and liver. On the other hand, the curves of microsomes and rough surface membrane were similar and the inhibitory effects were much larger in melanoma than in liver. These results might indicate that the membrane component of the liver disturbed the inhibitory effect of fraction II_1.

Inhibitory effect of fraction IV_2 on the protein biosynthesis in the cell-free systems

Figure 8 shows dose response curves of the inhibitory effects of fraction IV_2 on the various cell particles of melanoma and liver. Although the effectiveness of fraction IV_2 was much higher than that of fraction II_1, and there were slight differences in individual experiments, in general the

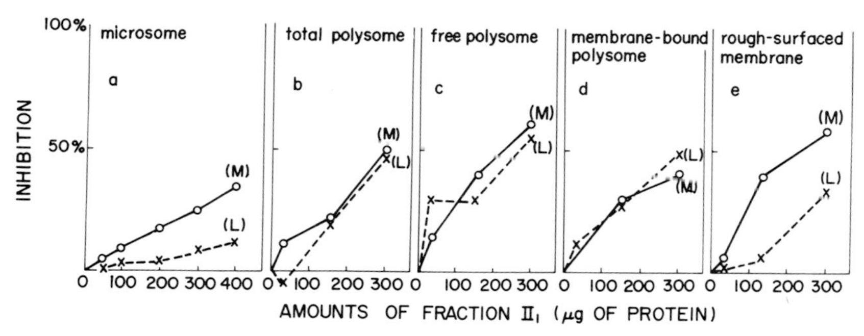

Fig. 7. Inhibitory effects of fraction II_1 on ^{14}C-L-leucine incorporation into proteins by various cell-free systems from melanoma and liver. Inhibitory effects are expressed as % inhibition to the control in which no fraction II_1 was contained. a: microsome, b: total polysome, c: free polysome, d: membrane-bound polysome, e: rough-surfaced membrane. O——O: melanoma, x----x: liver.

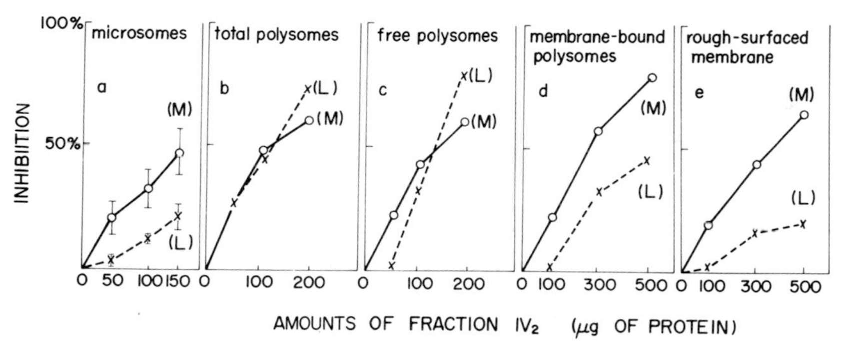

Fig. 8. Inhibitory effects of fraction IV_2 on ^{14}C-L-leucine incorporation into proteins by various cell-free systems from melanoma and liver. Inhibitory effects are expressed as % inhibition to the control in which no fraction IV_2 was contained. a: microsome, the average values of 5 different experiments, b: total polysome, c: free polysome, d: membrane-bound polysome, e: rough-surfaced membrane. O——O: melanoma, x----x: liver.

dose response curves of fraction IV_1 were quite similar to those of fraction II_1. The inhibitory effects on protein biosynthesis by various cell components from melanoma and liver increased with increasing amounts of fraction IV_2. There were no significant differences in total and free polysome of melanoma and liver, and the inhibitory effect was much less in microsome, rough surface membrane, and membrane-bound polysome of liver. Quite similar results were obtained in both fraction II_1 and fraction IV_2, except for membrane-bound polysome, although the inhibitory effect of the latter was much larger than that of the former. In Fig. 8a, the average values of five experiments were recorded in the case of microsomes, since it was noticed, as has been pointed out, that the inhibitory effect of fraction IV_2 varied depending on the preparations.

Nature of inhibitory effects of fraction II_1 and IV_2 on the protein biosynthesis *in vitro*

Time dependent incorporation curves of ^{14}C-L-leucine into total polysome of melanoma with and without addition of fraction II_1 are shown in Fig. 9. The incorporation of ^{14}C-L-leucine into total polysome increased with time but almost ceased at the end of 30 minutes in the presence of fraction II_1, while that of the control up to 60 minutes was determined to continue. In order to examine whether the higher inhibitory effect of fraction IV_2 on melanoma microsome system is dependent on microsome or cell sap, the inhibition by fraction IV_2 was examined in various combinations of microsomes and cell saps from both tissues. As shown in Table 1, the

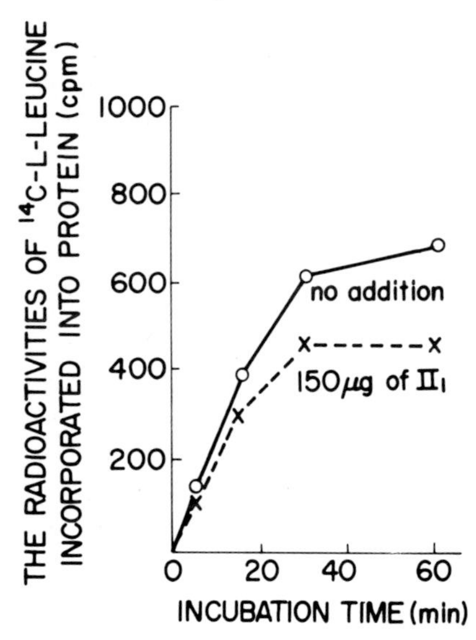

Fig. 9. Time dependent incorporations of ^{14}C-L-leucine into proteins by total polysome from melanoma with or without fraction II_1. Total polysome from melanoma was incubated at 30°C with or without fraction II_1, which contained 150 μg of protein for the time indicated. The radioactivities of ^{14}C-L-leucine incorporated into proteins were determined as described in Materials and Methods. O——O: without fraction II_1, x----x: with 150 μg of fraction II_1.

Table 1. Inhibitory effect of fraction IV_2 on protein biosynthesis in the cell-free system in various combination of microsomes and cell saps isolated from liver and melanoma. The incubation conditions are the same as described for the experiment in Fig. 11 except that a final volume of the incubation mixture was 0.3 ml. Fraction IV_2 used was 150 μg of protein in doses.

Microsome	Cell sap	Inhibitory Effect of Fraction IV
Liver	Liver	11%
Liver	Melanoma	28%
Melanoma	Liver	43%
Melanoma	Melanoma	59%

higher effect of fraction IV_2 on the melanoma system was dependent on tumor microsome. In this particular experiment, 150 μg of the protein of fraction IV_2 was used and numerals shown in the table are expressed as an inhibitory percentage of the control, which did not contain fraction IV_2.

The specificities of fraction II_1 and fraction IV_2 were examined. Since it is now known that fraction II_1 consists mainly of protein, and fraction IV_2 RNA and proteins, the bovine serum albumin and purified yeast RNA were added to the ^{14}C-L-leucine incorporation system of melanoma total polysome. The dose response curves are shown in Fig. 10. As it is shown clearly, no significant inhibitions were observed by albumin nor by yeast RNA.

Comparative dose response studies were carried out on fraction II_1 and fraction IV_2 in the melanoma and liver microsome systems. Figure 11 shows the results obtained. In melanoma, the inhibitory effects of fraction II_1 increased with increasing amounts, as have been observed in previous experiments, but seemed to be limited at 60% even at a high dose, 1200 μg of protein. On the other hand, significant inhibition was clearly observed at a low dose, 60 μg of protein. In liver, the inhibition was only slight at low doses, up to 300 μg. The inhibitory effects of fraction IV_2 increased in both melanoma and liver with increasing amounts used and appeared to be more effective than fraction II_1. In melanoma, the inhibitory effect was definite at as low as 60 μg of protein. On the other hand, in liver only a slight inhibition was observed at the same dose. And the maximum inhibitory effect, 100%, was shown at the concentration of 600 μg, which corresponded to 300 μg in the case of 0.3 ml of the total volume of the reaction mixture. From these experimental results, it was assumed that the inhibitory mechanism of fraction II_1 and fraction IV_2 are quite different from one another.

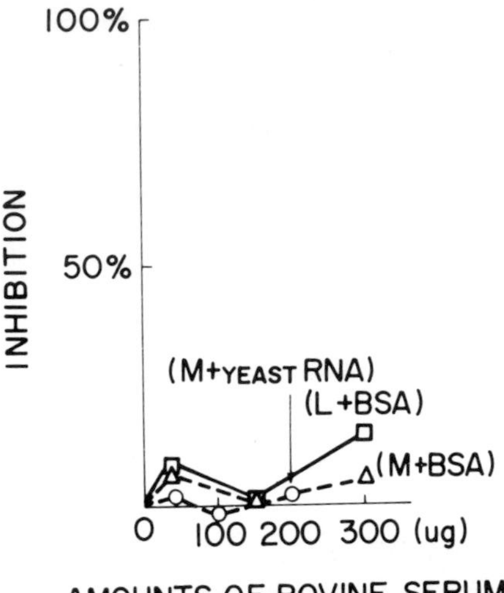

Fig. 10. Effects of addition of bovine serum albumin and yeast RNA on the incorporation of ^{14}C-L-leucine into proteins by total polysome from melanoma and liver. Effects were expressed as % inhibition to the control to which no protein and RNA were added. □——□: liver total polysome with bovine serum albumin, △----△: melanoma total polysome with bovine serum albumin, ○—-—○: melanoma total polysome with yeast RNA.

DISCUSSION

The concept of chalones: The existence of a tissue-specific negative feedback mechanism was conceived by W.S. Bullough and E.B. Laurence after a systemic study of events following a skin lesion produced and observed under strictly controlled experimental conditions.[4,12] And it is now assumed that "mitotic control is maintained by the interaction of a tissue-specific mitosis-inhibiting chalone, which permeates the whole tissue, and a non-tissue-specific mitosis-promoting mesenchymal factor, which originates in the connective tissue and acts only on connective-tissue-adjacent cells." It was shown further that not only in the normal epidermis but also in the epidermal carcinoma, the epidermal cells continued to synthesize the epidermal chalone and to respond by mitotic inhibition to this chalone. The concentration of intracellular chalone was correlated with the growth rate of the tumor; the lower the concentration, the higher the growth rate. The regulatory mechanisms for control of the melanocyte proliferation in the skin have not been revealed as yet. However, there must be present some sorts of controlling mechanisms since it is now well known that melanocytes increase in number when the skin is exposed to

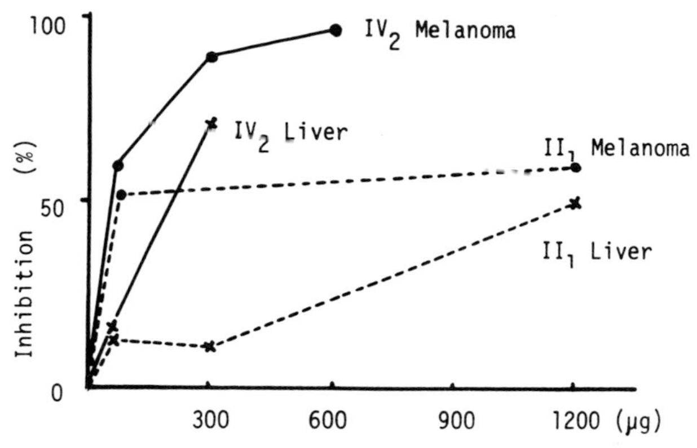

Fig. 11. Dose response curves of the inhibitory effects of fraction II_1 and IV_2 on the *in vitro* incorporation of ^{14}C-L-leucine by liver and melanoma microsome. The inhibitory effects are expressed as a percentage of the control. The complete system consisted of 0.25 M sucrose, 0.4 μCi ^{14}C-L-leucine (331 mCi/m mol), 1.12 mg protein of liver microsome, or 1.44 mg protein of melanoma microsome, 1.48 mg protein of liver cell sap, or 1.06 mg protein of melanoma cell sap, 5 mM creatine phosphate, 3 μg protein of creatine kinase, and 5 mM reduced glutathione in a final volume of 0.6 ml. After 30 minutes incubation at 30°C, the total protein fraction was purified and its radioactivity was measured as described in Materials and Methods.

sunlight, they proliferate during epidermal wound healing, and MSH or c-AMP ceases proliferation and accelerates the melanin formation of melanocytes. Melanocytes distribute in the form of a network between the epidermis and dermis in the skin; therefore, they may be influenced, first, by the non-tissue-specific mitosis-promoting mesenchymal factor which originates in the connective tissue and, second, by the intercellular controlling factor among the melanocytes, which is assumed to permeate the whole melanocyte network. In 1968, Bullough and Laurence first reported the presence of melanocyte chalone in a pig skin crude extract.[1] The water extracts of Harding-Passey mouse melanoma were prepared by Mohr *et al.*[2] and injected subcutaneously to the melanoma-bearing mice and hamsters, resulting in the remarkable suppression of the melanomas. Although they later suggested the possibility that the real antitumor agent was bacterium *Clostridium*, which may have been a contaminant in the original skin extracts,[13] Bullough considered that it was most unlikely since parallel studies of the same skin extracts (which had passed through 80% ethanol and had been freeze-dried) showed that the action on the tumor cells was obvious only after 4 hours; no bacteria could have emerged from sporulation, multiplied, and begun action in that time.[12]

The conclusion that the chalone-like substances were the responsible agent is also supported by the later and more detailed observations of Seiji *et al.*,[3] who used partially purified melanoma extracts, fraction II_1 and fraction IV_2, on the cultured melanocytes. Since the extraction of melanocyte chalone from the melanocyte systems of the normal tissue is almost impossible, they tried to isolate it from the transplantable mouse melanoma. The alcohol precipitate prepared from a water extract of Harding-Passey mouse melanomas contained a melanoma melanocyte cell cycle inhibitor. A single injection of melanoma extract into the peritoneal cavity of mice bearing melanoma caused a 35–65% reduction of ^{14}C-thymidine incorporation into the DNA of melanoma melanocytes. And then an *in vitro* assay method using B-16 mouse melanoma melanocytes in tissue culture was established. Addition of melanocyte extract to the tissue culture medium containing adrenalin and hydrocortisone caused a reduction of the mitotic index, of incorporation of ^{14}C-thymidine into DNA, and of ^{3}H-leucine incorporation into melanocyte protein. It has been warned that when the mitotic activity of a tissue slows down, the specific activity of labeled DNA must not be used as the only criterion of action of the product tested. Furthermore, it was revealed by a team of biochemists working on the hepatic chalone that the inhibiting factor was merely cold thymidine. Considering these mishaps and ignoring them, it is quite obvious that we have to be very cautious in producing an extract with chalone-like activity and to presume that all the properties of this extract are due to the chalone.

In our *in vitro* experiments, the inhibition of DNA and protein biosynthesis due to the melanocyte extract was well correlated. Aside from the inhibitory effect of melanoma extract on DNA biosynthesis in melanocytes which is in controversy, the inhibitory effects of melanoma extract on the protein biosynthesis were subjected to further study. The general protein biosynthesis in the cell is known to occur continuously during the cell cycles except M phase. In an attempt to clarify the inhibitory mechanisms, the effects of fraction II_1 and IV_2 were studied in the *in vitro* incorporation of ^{14}C-leucine by various cell particles of mouse liver and Harding Passey mouse melanoma. The chemical characteristics of fraction II_1 and IV_2 have been studied[3] and additional studies were carried out. As has been indicated previously, fraction II_1 consisted mainly of protein and there were no significant differences in the UV absorption spectrum of fraction II_1 samples prepared at different times. On the other hand, the fraction IV_2 consisted of RNA and proteins and the UV spectrum of the fraction IV_2 varied significantly, depending on the preparations, due to unknown causes. The maximum absorption fluctuated between 253 and 262 nm; it is assumed, therefore, that the inhibitory effect of fraction IV_2 would vary depending on individual samples. And also the yield of fraction IV_2 was very small from a unit of melanoma tissue. The major protein components of the fraction IV_2 were assumed to be small in molecular weight, probably less than 10,000 from the SDS acrylamide gel electrophoretogram and its scanning profiles. And the recent preliminary chemical analysis shows that the nucleic acid component of the fraction IV_2 consisted mainly of a small nucleic acid, the molecular weight of which would be 11,000. The exact nature of this nucleic acid is not clarified yet. Considering these experimental results, in this series of experiments the protein concentration has been used as a parameter of the dose of fraction IV_2, it would be, however, more suitable that the inhibitory effect of the particular fraction IV_2 preparation on the melanoma microsome system is used as a parameter. Polysomes prepared from melanoma and liver and used in this series of experiments were found to be quite similar in their natures. It was clearly shown in the case of melanoma that both fraction II_1 and IV_2 inhibited the incorporation of ^{14}C-L-leucine into proteins of various cell particles *in vitro*. The inhibitory effect of fraction IV_2 was apparently much greater than that of fraction II_1. That 100% inhibition was obtained with fraction IV_2, and 60% with fraction II_1, might indicate that the inhibitory mechanisms of these two fractions would differ from each other. The specificity of these two fractions on the melanoma tissue was shown by the fact that the inhibitory effects of both fractions appeared at the low doses. In the case of liver, the inhibitory effects of both fractions were very low in the microsome and rough surface membrane compared with those of melanoma. This might indicate that the membranes of liver may contain some components which resist the inhibitory effect on the ^{14}C-L-leucine incorporation into proteins of membrane-bound polysomes of both fractions. This phenomenon must not be considered immediately to be due to the tissue specificity of the malanoma extract, since there is no information available at present that the membranes possess the tissue specificity in the *in vitro* system, such as this. It has been also observed that the inhibitory effects of both fractions in the liver system varied depending on mice used.

REFERENCES

1. Bullough, W. S. and Laurence, E. B. Melanocyte chalone and mitotic control in melanomata. *Nature* (London), **220**: 137–138, 1968.
2. Mohr, U., Althoff, J., Kinzel, V., Suss, R., and Volm, M. Melanoma regression induced by "chalone": A new tumor inhibiting principle acting in vivo. *Nature* (London), **220**: 138–139, 1968.
3. Seiji, M., Nakano, H., Akiba, H., and Kato, T. Inhibition of DNA and protein synthesis in melanocytes by a melanoma extract. *J. Invest. Derm.*, **62**: 11–19, 1974.
4. Bullough, W. S., Hewett, C. L., and Laurence E. B. The epidermal chalone. *Exp. Cell Res.*, **36**: 192–200, 1964.

5. Keller, E. B. and Zamecnik, P. C. The effect of guanosine diphosphate and triphosphate on the incorporation of labeled amino acids into proteins *J. Biol. Chem.*, **221**: 45–59, 1956.
6. Wettstein, F. O., Staehelin, T., and Noll, H. Ribosomal aggregate engaged in protein synthesis: characterization of the ergosome. *Nature* (London), **197**: 430–435, 1963.
7. Sugano, H., Watanabe, I., and Ogata, K. Stabilizing effect of ribonuclease inhibitor on structure of polysomes and some properties of four classes of ribosomal particles in rat liver cytoplasm. *J. Biochem*, **61**: 778–786, 1967.
8. Rothschild, J. A. Subfraction of rat liver microsomes. *Fed. Proc.*, **20**: 145, 1961.
9. Lowry, O. H., Rosebrough, N. J., Farr, A. L., and Randall, R. J. Protein measurement with the Folin Phenol reagent. *J. Biol. Chem.*, **193**: 265–275, 1971.
10. Dische, Z. L. Qualitative and quantitative colorimetric determination. *J. Biol. Chem.*, **204**: 983–997, 1953.
11. Weber, K. and Osborn, M. The reliability of molecular weight determinations by dodecyl sulfate-polyacrylamide gel electrophoresis. *J. Biol. Chem.*, **244**: 4406–4412, 1969.
12. Bullough, W. S. Mitotic control in adult mammalian tissues. *Biol. Rev.*, **50**: 99–127, 1975.
13. Mohr, U., Hondius Bondingh, W., and Althoff, J. Identification of contaminating clostridium spores as the oncolytic agent in some chalone preparations. *Cancer Res.*, **32**: 1117–1121, 1972.

Discussion

DR. KARASEK: In your fraction IV, there is a large amount of nucleic acid. Is it certain that it is RNA or could there be some small amounts of DNA in that material as well?

DR. SEIJI: To test this question, nucleic acid was extracted from fraction IV_2 by the Schmidt-Thannhauser-Schneider method. Only the RNA fraction was obtained and it showed a positive reaction with orcinol but a negative reaction with diphenylamine. Therefore, we presume that the nucleic acid in fraction IV_2 is some sort of RNA.

DR. FUKUYAMA: Did you use melanoma growing in mice or melanoma growing in tissue culture?

DR. SEIJI: We have used L-cells and HeLa-cells as well as human melanoma cells in culture for the tissue specificity experiments.

DR. FUKUYAMA: In those experiments, did you also use liver cells in tissue culture?

DR. SEIJI: We did not use liver cells.

DR. FUKUYAMA: Did you see morphological changes in melanoma cells in tissue culture?

DR. SEIJI: The morphological changes in the melanocytes which might occur after the melanoma extracts were added to the culture medium have not yet been studied.

DR. GOLDSMITH: In your inhibitory fraction, you showed that you had many proteins. I am wondering whether the RNA was associated with all the protein components or with just one protein component on those SDS gels which also contained RNA? Was the inhibition removed by RNase?

DR. SEIJI: In order to clarify whether the inhibitory activity of fraction IV_2 depended on the protein moiety or on the RNA moiety, trypsin digestion and RNase digestion of fraction IV_2 were carried out. The inhibitory activity was found to depend on both the protein and the RNA. The nature of the RNA in this fraction has not yet been studied thoroughly but it consists mainly of a small nucleic acid with a molecular weight of about 1,100.

DR. FREEDBERG: It will later become clear that there is much work being done on initiation of protein synthesis in many kinds of systems. The interesting part is that although chalones are quite specific, the translational inhibitors so far have not been very specific and a translational inhibitor in one system usually inhibits in another system. It is a very rapidly changing field at the moment, however. I have a question about the data you presented in which you showed a change in polyribosomal patterns of a) control, b) incubated and c) incubated + inhibitor in which the incubated + inhibitor came between the control and the incubated. Did protein synthesis have to occur during that reaction or was this just a matter of incubation? If you incubated the system without adding one of the essential components, did you change the polyribosomal profile?

DR. SEIJI: The control in which the incomplete reaction system was incubated has not yet been carried out. According to Dr. Ogata, the changes in the polysomal profile which occur during the incubation are most likely specific indicating that translation was proceeding.

Cell Inactivation Effects of Melanoma Extract on Human Melanoma (HMV) Cells *in vitro*

I. Watanabe, T. Kasuga and I. Nojiri

Division of Physiology and Pathology, National Institute of Radiological Sciences, Chiba, Japan

ABSTRACT

The alcohol-precipitated and purified substance (ME II_1) prepared from a water extract of Harding-Passey mouse melanomas has shown significant cell inactivation effects on human melanoma (HMV) cells *in vitro*. Virtually all the cells could survive when the ME II_1-containing medium was removed within the first 12 hours. The cell inactivation effects were beginning to appear after the first 12 hours' noncell inactivation period, and the maximum cell inactivation effects were observed during the next 12-hour period. Further decrease in survival fractions was not significant after the maximum cell inactivation period (plateau phase). The survival fractions at plateau level were dose dependent. The survival curve after 42 hours' ME II_1 administration was characterized by the Dq and Do doses of 31.4 and 15.2 μg/ml, respectively. An experiment on age-dependent response of cells revealed that only the late mitotic and the early half of G_1 phase cells were resistant and the cells in other phases were highly sensitive to ME II_1. Since no bacterial contamination was detected during the experiments, the results may suggest that ME II_1 would be an active agent for cell inactivation.

Tissue-specific growth suppressors (chalone) have been extracted from a number of tissues, such as epidermal,[1] kidney,[2] liver,[3] mature granulocytes,[4] and others. And they have coincidentally indicated that these extracts suppressed mitotic activities in their original tissue.

In 1968 Bullough and Laurence[5] reported that injection of partially purified pig skin extracts into mice bearing Harding-Passey melanomata resulted in an inhibition of the mitotic activity of melanomata. This fact indicates that the pig skin extracts contained melanocyte chalone. Furthermore, Mohr *et al.*[6] have reported that subcutaneous injection of pig skin extracts into NMRI mice bearing Harding-Passey melanomata or into Syrian hamsters with transplanted amelanotic melanomata caused melanoma regression. Later on, however, they proved that the oncolytic activity found in their crude extracts was caused by the contamination of oncolytic bacteria, *Clostridium.*[7] In 1974 Seiji *et al.*[8] extracted chalon-like substances from Harding-Passey melanoma in Swiss mice. Administration of the extracts to tissue-cultured mammalian cells and malignant melanoma of mice resulted in the inhibition of DNA, protein synthesis, and mitotic activity of mouse melanoma cells.[8] After the extracts were purified by Sephadex G-100 gel filtration, two fractions (II and IV) among four were found to be effective for the inhibition of DNA and protein synthesis in mouse melanoma B-16, mouse fibroblast L, human carcinoma HeLa, and human melanoma HMV cells in culture.[8] Recently, we have found cell inactivation effects of the extract on HeLa, B-16, and HMV cells in culture. Similarly, Oikawa (Research Institute of Tuberculosis and Leprosy, Tohoku Univ.) observed a decrease in the cell number of HeLa cells in the presence of the same extract (private communication). Then, we have made an detailed analysis for the effects of the melanoma extract (fraction II_1) on HMV cells in culture.

MATERIALS AND METHODS

Cell line : Human melanotic melanoma, HMV, cells isolated from the vaginal region were used throughout the experiments.[9] The cells were grown in F-10 medium supplemented with 1% heart infusion broth, 5% fetal bovine, and 5% calf sera, and antibiotics (100 unit/ml of penicillin and 100 μg/ml of streptomycin). At 37°C in an atmosphere of 5% CO_2 in air, cells in exponential growth exhibited a generation time of approximately 22 hours with phases of $G_1=7$, $S=10$, $G_2=4$ and $M=1$ hour and the percentages of each phase cells in a population were 34, 50, 12, and 4%, respectively.

Melanoma extract : Purified melanoma extract fraction II_1 (ME II_1) was kindly provided by Dr. M. Seiji (Tohoku Univ. School of Medicine). The ME II_1 was obtained from Harding-Passey mouse melanoma by the water maceration-alcohol precipitation technique and purified through Sephadex G-100 gel column and DEAE Sephadex A-25 column. The detailed description for purification of this fraction has already been reported.[8] Lyophilized ME II_1 was disolved in 1/10 M phosphate buffered saline and sterilized through Millipore HA filter (Millipore Filter Corporation, Bedford, Massachusetts). Protein content of the ME II_1 solution was determined according to Lowry *et al.*,[10] and the administration doses were expressed by the amount of proteins in μg/ml. The solution was stored at $-87°C$. Immediately before the administration, the solution was defrosted and mixed with equal amounts of concentrated F-10 medium (2x) so as to make a normal medium composition. Subsequently, contamination of microorganisms was tested by incubating the ME II_1-containing medium at 37°C.

Autoradiography : The cells grown in Tissue Culture Chamber/Slide (Lab-Tek Products, Westmont, Illinois) were labeled with ^3H-deoxycytidine (2Ci/mM, 2μCi/ml) (Radiochemical Centre Ltd, Amersham) for 20 minutes, washed three times with phosphate-buffered saline and fixed with acetic-alcohol (glacial acetic acid: absolute ethanol, 1:3), and air-dried. The slides were subjected to an autoradiographic process using NR-M2 emulsion (Konishiroku Photo Industrial Co., Tokyo). Developed slides were stained with 1.5% Giemsa solution (E. Merck AG, Darmstadt). The percentage of the labeled cells was determined by counting 1000 cells in each sample.

Mitotic frequency : Mitotic frequency was determined in the same slides used for determination of the labeling index and expressed as the percentage of mitotic cells in 1000 cells.

RESULTS

Time-inactivation curve

Three different doses of ME II_1, namely, 50, 75 and 100 μg/ml, were administered to the cells after they plated and attached to the surface of a plastic dish (usually it takes 4 hours). After various intervals, the cells in the plastic dish were washed once with normal medium and incubated for 10–14 days to allow colony development. Figure 1 shows the time-inactivation curves for these three different doses. Virtually all the cells were able to survive at any dose when the ME II_1-containing medium was replaced with normal medium within the first 12 hours. The cell inactivation effects were beginning to appear after the first 12-hour noncell inactivation period, and the maximum cell inactivation effects were observed during the next 12-hour period. After 24 hours, the inactivation curves gradually reached a plateau. The survival fractions at plateau level were dose-dependent.

Under periodic investigation of the treated cells, it was found that cell division was almost completely diminished during the ME II_1 administration period at high concentration (e.g., 100 μg/ml). These cells became slender and most of them degenerated and floated out from the surface of the dish. At low concentration (e.g., 25 or 50 μg/ml) some cells could grow even in the presence of ME II_1 (proved by low mitotic index) and others began to degenerate. The cells that survived during the ME II_1 administration period could divide after replacement with normal medium; some of these

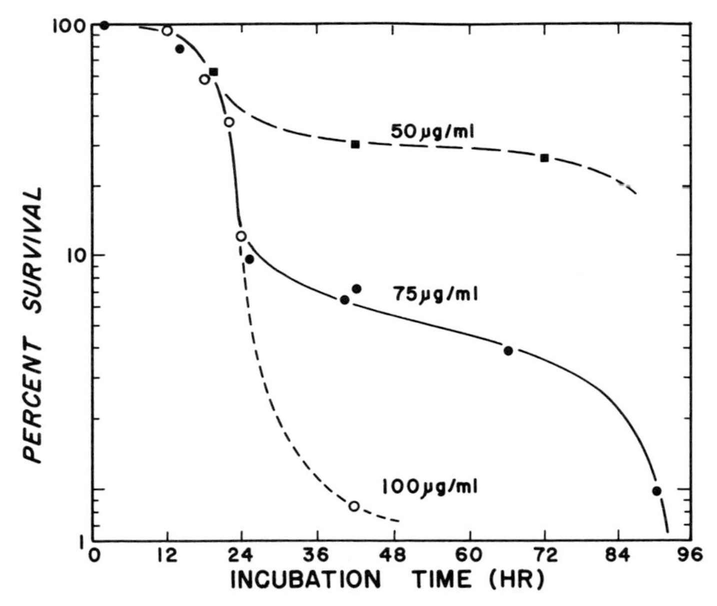

Fig. 1. Time-inactivation curves of HMV cells after various concentrations of ME II_1 administration.

could grow to form colonies, while the rest failed to form a colony by continuous loss of their progeny during the post-treatment incubation (Fig. 2). Since the formation of these abortive colonies implies a manifestation of some transmittable damages given in the cell, it may be suggested that ME II_1 is an active agent for cell inactivation.

Dose-survival curve

Various concentrations of ME II_1 up to 125 μg/ml were administered by replacing normal medium with ME II_1-containing medium at 4 hours after the cells were plated into plastic dishes. The cells were washed once with normal medium after 42 hours continuous administration of ME II_1, and incubated for 10–14 days to allow colony development. As shown in Fig. 3, the survival curve indicated a large shoulder region followed by an exponentially decreasing portion. The Do dose (mean lethal dose, the dose required to reduce survival by a factor e^{-1} along the exponential portion of the survival curve) and the Dq dose (quasi-threshold dose, a measure of the width of the shoulder region of the survival curve) were determined to be 15.2 and 31.4 μg/ml, respectively. In Fig. 3, a similar type of curve was also shown by a particular type of experiment in which ME II_1-containing medium was replaced with normal medium at the 20th hour (dotted line). The Dq dose of the survival curve was much

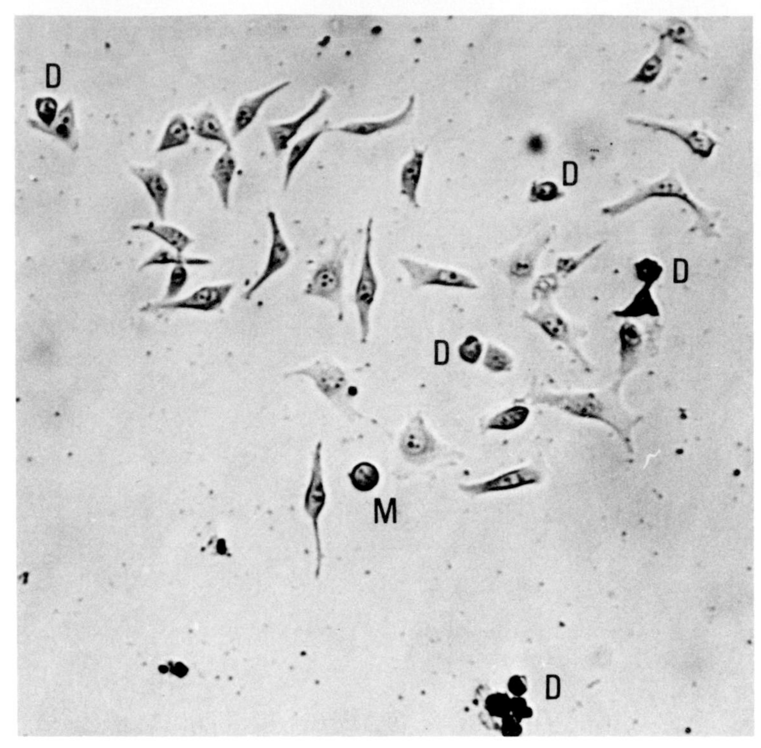

Fig. 2. Microphotograph showing an abortive colony which was developing in normal medium after 42 hours in the ME II_1 administration period (50 μg/ml). Note that some cells are now dividing (M) whereas some others are degenerating (D). The photograph was taken on the 5th day after ME II_1 administration ended.

wider than that obtained from 42 hours treatment (Dq dose 54.7 μg/ml), while Do dose did not change significantly (Do dose 19.8 μg/ml). In other words, mainly Dq dose was decreased during an extended ME II_1 administration period.

Dependence of survival response on cell age

An attempt was made to indicate whether ME II_1 induces inactivation of cells in a random manner or actions correlated with the specific phase of the cell cycle. We have adopted a method in which excess thymidine treatment[11] and selective harvest of mitotic cells[12] were combined.

a. Synchronization of HMV cells Exponentially growing cells were treated with excess thymidine (2.5 mM at final concentration) for 24 hours, washed once with normal medium and incubated for another 10 hours.

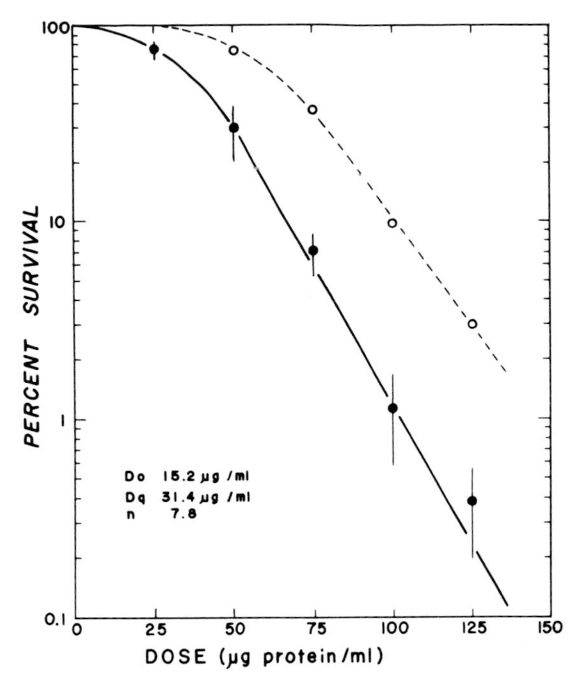

Fig. 3. Dose-survival curves of HMV cells. Dotted line with open circles and solid line with closed circles represent the curves obtained after 20- and 42-hours administration of ME II_1, respectively.

The percentage of S phase cells was periodically measured by labeling cells with ^{3}H-deoxycytidine and scoring labeled cells in autoradiographic samples. As shown in Fig. 4, the percentage of S phase cells increased with increasing incubation time in the presence of excess thymidine and finally reached to 92–95% at 24th hour. This indicates that virtually all the cells were accumulated in the S phase. A rapid decrease in the labeling index was observed after the 28th hour, when the cells were washed at the 24th hour, and reached the minimum level at the 36th hour. The peak of mitotic index was found at the 34th hour (10 hours after release from the excess thymidine blockage), as shown in Fig. 4. A mitotic frequency as high as 20% was consistently obtained in repeated experiments. The mitotic cells at this time were harvested by mechanical agitation and plated into plastic dishes at an appropriate density for the survival assay. The mitotic cells divided into two daughter cells within a few hours. Although double thymidine treatment technique[13] may be applied to synchronize HMV cells, we have adopted the mitotic cell harvest technique after a single thymidine treatment to avoid an unbalanced growth syndrome.

b. Survival response The cells harvested at M phase were plated into

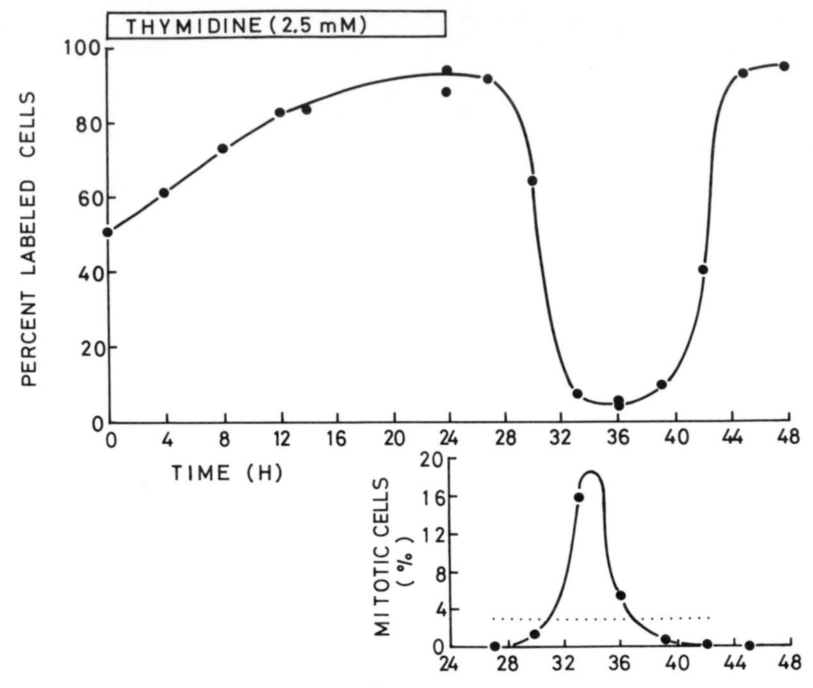

Fig. 4. Change in the percentage of ^{3}H-deoxycytidine labeled cells and the cells in mitosis during a course of partial synchronization.

plastic dishes and allowed to progress along the cell cycle. At various time intervals after plating of M phase cells, the medium was replaced with ME II_1-containing medium and the cells were incubated for 42 hours. ME II_1 administration for M phase cells was done by mixing one part of cell suspension containing M phase cells with equal volume of a two-fold concentration of ME II_1-containing medium and poured into plastic dishes. At the end of administration period, the medium was replaced with normal medium and the cultures were incubated for 10–14 days. The doses used in these experiments were 50, 60, and 90 μg/ml. Under periodic investigation of cells during and after ME II_1 administration, it was found that the cells administered 50 μg/ml of ME II_1 at mitosis completed mitosis at the same rate as that of the control cells. The cells administered ME II_1 at the early G_1 phase continued cell growth and almost all cells divided once or twice during the 42-hour administration period. Most cells in the other phases of the cell cycle (i.e., the late G_1, S, and G_2) ceased cell division; some of them started to degenerate during the latter half of administration period and a greater part of them degenerated after ME II_1 was removed.

Table 1 shows the survivals at each phase of the cell cycle where ME II_1 was administered. The results indicate that only those cells in the late mitotic and the early half of G_1 phases were highly resistant to ME II_1 while those cells in the other phases of the cell cycle were inactivated at an unexpectedly high rate.

Table 1. Survivals assayed at each phase of the cell cycle after 42-hours' continuous administration of ME II_1.

Conc. (μg/ml)	No. of cells per dish	Mitosis		G_1			G_1/S	S			G_2
		meta*	ana** telo	early	middle	late		early	middle	late	
90	1000	—	0.197	—	0.197	0	0	0	0	0	0
60	1000	0	—	3.321	0	0	0	0	0	0	0
50	11500	—	++	+	0·183	0	0	0	0	0	0

Note. The plating efficiency was 50–70%. The percent survivals, represented by numerals, was corrected with the plating efficiency in each experiment.
* Virtually all the cells were spherical in shape.
** Most cells were dumbbell-shaped.
— No experiment at that phase.
0 No colony.
++ Too many cells survived, unable to count the number of colonies.
+ Too many colonies formed, unable to distinguish each colony.

DISCUSSION

Mohr *et al.*[6] have reported that crude extracts of chalone from pig skin showed striking oncolytic activity on melanomata in Syrian hamsters and NMRI mice. They have assumed that melanocytic chalone might be contained in their extracts and the chalone-induced oncolysis in melanomata. Later on, they found oncolytic bacteria, *Clostridium,* in melanomata during and after administration of the extracts by means of histological observation. Also, they found many spore-forming rods in their chalone solution after 3-days incubation at 37°C. After sterile filtration, they could not prove any oncolytic activity in the preparations. In this study, we sterilized ME II_1 solution through Millipore HA filter (0.45 μ). And no bacterial contamination was found during and after ME II_1 administration for 10 to 14 days. Although we have no evidence whether ME II_1 contained microorganisms smaller than 0.45 μ in diameter, the results suggest that ME II_1 played an important role in inducing the inactivation of HMV cells. Another supporting evidence for this view is that the late mitotic and the early half of G_1 phase cells could grow in the presence of ME II_1 and some of them survived without any infectious syndrome. In addition to these, it must be pointed out that melanocytic chalone used by Mohr *et al.*[6,7] was extracted from the normal tissue while ME II_1 was extracted from malignant melanoma cells. The mode of action may be dissimilar to that of the normal melanocytic chalone.

We have shown in this study that the cells were inactivated after 12 hours continuous treatment with ME II_1, the survival curve was characterized by a wide shoulder, and the shoulders decreased during an extended incubation period with ME II_1. These facts imply that either one or two of the following mechanisms induce cell inactivation. That is, (1) ME II_1 primarily induces cumulative type damage in the essential components of the cell and accumulation of the damage causes cell inactivation, or (2), contradictorily, purified ME II_1 itself has no cell inactivation activity on HMV cells but once it resolves into some products during a 12-hour or more incubation period at 37°C, the products become toxic and inactivate the cells. Since epidermal chalone could not survive long in water solution even at 0°C,[14] decomposition of ME II_1 could be the case. An attempt to prove the possibility is now in progress in our laboratory.

Acknowledgement

We would like to express our appreciation to Dr. M. Seiji (Tohoku University School of Medicine) for providing us the melanoma extract together with his valuable suggestions and discussion. The research was supported in part by a grant from the Ministry of Education.

REFERENCES

1. Bullough, W. S. and Laurence, E. B. Mitotic control by internal secretion: The role of the chalone-adrenalin complex. *Exp. Cell Res.*, **33**: 176–194, 1964.
2. Bullough, W. S. The control of mitotic activity in adult mammalian tissues. *Biol. Rev.*, **37**: 307–342, 1962.
3. Saetren, H. A principle of auto-regulation of growth: Production of organ specific mitose-inhibitors in kidney and liver. *Exp. Cell Res.*, **11**: 229–232, 1956.
4. Ryömaa, T. and Kioiniemi, K. Control of granulocyte production. I. Chalone and antichalone, two specific humoral regulators. II. Mode of action of chalone and antichalone. *Cell Tissue Kinet.*, **1**: 329–340, 341–350, 1968.
5. Bullough, W. S. and Laurence, E. B. Melanocyte chalone and mitotic control in melanomata. *Nature* (London), **220**: 137–138, 1968.
6. Mohr, U., Althoff, J., Kinzel, V., Süss, R., and Volm, M. Melanoma regression induced by "chalone": A new tumor inhibiting principle acting *in vivo*. *Nature* (London), **220**: 138–139, 1968.
7. Mohr, U., Boldingh, W. H., and Althoff, J. Identification of contaminating *Clostridium* spores as the oncolytic agent in some chalone preparations. *Cancer Res.*, **32**: 1117–1121, 1972.
8. Seiji, M., Nakano, H., Akiba, H., and Kato, T. Inhibition of DNA and protein synthesis in melanocytes by a melanoma extract. *J. Invest. Derm.*, **62**: 11–19, 1974.
9. Takahashi, I., Kasuga, T., Furuse, T., Kubo, E., Uta, K., Sugano, H., and

Sakamoto, G. Establishment of cell line from human melanotic melanoma and its cellular specificities. *Proc. Jap. Cancer Assoc.* 30th Annual Meeting, Tokyo, 1971, p. 219.
10. Lowry, O. H., Rosebrough, N. J., Farr, A. L., and Randall, R. J. Protein measurement with Folin phenol reagent. *J. Biol. Chem.* **193**: 265–275, 1951.
11. Xeros, N. Deoxyriboside control and synchronization of mitosis. *Nature* (London), **194**: 682–683, 1962.
12. Terasima, T. and Tolmach, L. J. Changes in X-ray sensitivity of HeLa cells during the division cycle. *Nature* (London), **190**: 1210–1211, 1961.
13. Bootsma, D., Budke, L., and Vos, O. Studies on synchronous division of tissue culture cells initiated by excess thymidine. *Exp. Cell Res.*, **33**: 301–309, 1964.
14. Bullough, W. S., Hewett, G. L., and Laurence, E. B. The epidermal chalone: A preliminary attempt at isolation. *Exp. Cell Res.*, **36**: 192–200, 1964.

Discussion

Dr. Karasek: Did you have an opportunity to look at Dr. Seiji's fraction IV as well as his fraction II?

Dr. Watanabe: Since Dr. Seiji is interested in fraction IV right now, we can only get a large amount of fractions I and II. So far we have not worked on his fraction IV.

Dr. Karasek: Did you have an opportunity to look at fraction IV *in vivo*? You alluded to the fact that an earlier extract was inhibitory to the melanoma.

Dr. Watanabe: We would like to try that but we haven't been able to do that as yet because of the limited supply of the material. We also want to know what the specificity of the material is. It would be worthwhile looking at the effect on HeLa cells and mice fibroblasts, or cells from another melanoma.

Dr. Endo: I have one question which must be directed to all the speakers. Dr. Yamaguchi used the name, chalone, for his material, but after that, Dr. Seiji and you, Dr. Watanabe, avoided use of this term for your material. My strong feeling also is, at the moment, that we should not use the word, chalone, when we have no information on the chemical nature of the active principle.

Dr. Watanabe: According to the definition, chalone is a tissue specific and species non-specific suppressor of cell proliferation. It is also known to suppress DNA synthesis which would prolong the G-1 phase. But that type of experiment has not yet been done. I agree with your opinion regarding the use of the term, chalone.

Epidermal Cholesterol as a Barrier for Transepidermal Water Loss

Muneo OHKIDO,* Takashi ABE ** and Itsuro MATSUO*

* *Department of Dermatology, Tokai University School of Medicine, Kanagawa, Japan* and ** *Kanebo Cosmetic Laboratory, Kanagawa, Japan*

ABSTRACT

It is well known that cholesterol is synthesized in the epidermal cells and is a constituent of the cellular or subcellular membranes with phospholipids and proteins. Cholesterol is released free from those membranes during the keratinization process. A question arises as to whether cholesterol in the keratin layer is an end product of cell debris or whether it plays any functional role on the skin surface, from which a large quantity of cholesterol is easily extracted.

Forty healthy Japanese adults were examined for this study. Transepidermal water loss(TWL) was measured by the hygrometric method on the ventral side of a forearm. The volume of casual and replacement lipids on the skin surface was measured by the volumetric method on the other forearm. Free and total cholesterol and squalene were determined by a gas chromatography.

An inverse correlation was seen between TWL and the amount of casual or replacement lipids, and also between TWL and the amount of total cholesterol in casual lipids. In other words, transepidermal water loss is decreased, while the total amount of cholesterol is increased on the skin surface. No correlation was found between TWL and squalene in casual and replacement lipids, total cholesterol in casual lipids, and free cholesterol in casual and replacement lipids.

It is suggested that total cholesterol, mainly esterified cholesterol or epidermal lipids, exerts an inhibitory action on cutaneous insensible perspiration.

The epidermis contains lipids, about 10 to 15% on the dry weight basis, and about 10% of it is sterols, about 10% is phospholipids and 60% is free or combined fatty acids.[1] The quantity of sterols and phospholipids is almost constant in the epidermis and on the skin surface. In our previous measurement,[2] the volume of casual lipids on the skin surface of the back in a normal human adult was 65.45 $\pm$ 26.54 $\mu g/cm^2$. That of total cholesterol in those lipids was 6.40 $\pm$ 2.19 $\mu g/cm^2$. This result confirmed that 10% of the casual lipids on the skin surface consisted of cholesterol.

The volume of replacement lipids on the same skin surface was 56.26 $\pm$ 23.35 $\mu g/cm^2$, in which total cholesterol was 5.02 + 21.2 $\mu g/cm^2$. Again, this result confirmed that 10% of the replacement lipids on the skin surface were cholesterol.

In 1932 Kooyman[3] measured the amount of cholesterol and phospholipids and the ratio of these two lipids in the basal, spinous, and keratin layers of the epidermis. He concluded that these two lipids played an important role during the process of keratinization. Later, Nicolaides[4] reexamined the results of Kooyman and concluded that both cholesterol and phospholipids are the membrane components in the cellular or subcellular units of the epidermis. These two lipids are not contained in sebaceous lipids[5], and therefore his conclusion seemed to be reasonable.

It is well known that both cholesterol[6] and phospholipids[7,8] are biosynthesized in the normal or abnormal epidermis, and their activity of biosynthesis is increased in the hyperkeratotic process, such as in psoriatic epidermis. Recently, Long[9] suggested that phospholipids are decomposed and their decomposed fractions are reutilized in the synthesis of triglycerides in the epidermis. We also observed[10] that glycerol-^{14}C was incorporated into a fraction of triglycerides in adult female rat epidermis. Even if these cholesterol and phospholipids are metabolized during the keratinization process, they were finally released from proteins of the cellular or subcellular membrane units. Then, the complex of lipids and proteins might be very unstable on the skin surface.

A question arises as to whether epidermal lipids in the keratin layer are end products of cell debris or whether they play any functional role on the skin surface from which a large quantity of cholesterol and phospholipids is easily extracted, as mentioned above.

In this paper, we studied how skin surface lipids could contribute to one of the main functions of skin, namely, to protection against transepidermal water loss (TWL).

MATERIALS AND METHODS

Forty healthy Japanese adults aged between 18 and 42 years (20 females and 20 males) were the subjects of this study.

Skin surface lipids were collected on the ventral side of the right forearm of each subject and TWL was measured on the other side. Room temperature was kept at 23 to 24°C and relative humidity was 70 to 80%.

Casual lipids were collected by the cup method (12.56 cm^2 in area) using 3 acetone extractions in a total volume of 15 ml. Replacement lipids were obtained 2 hours afterwards using the same method. Casual and replacement lipids were measured gravimetrically, total and free cholesterol and squalene were estimated by a gas chromatographical method[11] (Table 1).

TWL was measured by the method of electric hygrometry.[12] The result was expressed as $mg/cm^2/hr$ of skin area.

Table 1. The procedure followed in skin surface lipids.

Skin Surface Lipids
(Collected by cup method)
filtrate and concentrate by the evaporation of solvent
Total Lipids
(Gravimetric method)
dissolve in 0.5 ml of *n*-hexane containing 0.01% (W/V) of C_{34} as an internal standard material
Squalene and Free Cholesterol
(Gas liquid chromatography)
hydrolyze with 2 ml of 1N KOH ethanol solution after removal of *n*-hexane
extract of lipids with *n*-hexane
dissolve in 0.5 ml of *n*-hexane containing 0.01% (W/V) of C_{34} after removal of *n*-hexane
Total Cholesterol
(Gas liquid chromatography)

RESULTS

Mean value of casual lipids in 40 healthy volunteers was 60.1 ± 14.6 $\mu g/cm^2$. That in 20 males was 58.9 ± 12.9 $\mu g/cm^2$ and that in 20 females, 61.3 ± 15.9 $\mu g/cm^2$. Mean value of replacement lipids in 40 healthy volunteers was 30.7 ± 12.8 $\mu g/cm^2$. That in 20 males was 28.7 ± 13.9 ug/cm^2 and that in 20 females, 32.7 ± 11.2 $\mu g/cm^2$ (Table 2).

Mean value of TWL in 40 healthy volunteers was 0.238 ± 0.022 $mg/cm^2/hr$. That in 20 males was 0.238 ± 0.023 $mg/cm^2/hr$ and that in 20 females, 0.239 ± 0.021 $mg/cm^2/hr$. This TWL was increased linearly, while skin temperature was elevated. It is shown in Fig. 1. It indicates that a relatively constant value of TWL could be obtained under a certain room temperature.

The relationship between TWL and casual lipids was examined. In a statistical analysis, an inverse correlation was found as $Y = -0.0025X + 0.387$ with a correlation coefficiency $r = 0.6146$ with 1% in significance in

Table 2. Mean values of transepidermal water loss (TWL) and casual and replacement lipids.

	casual lipids			replacement lipids		
	total (40)	male (20)	female (20)	total (40)	male (20)	female (20)
TWL ($mg/cm^2/hr$)	0.238 ±0.022	0.238 ±0.023	0.239 ±0.021			
total lipids ($\mu g/cm^2$)	60.1 ±14.6	58.9 ±12.9	61.3 ±15.9	30.7 ±12.8	28.7 ±13.9	32.7 ±11.2

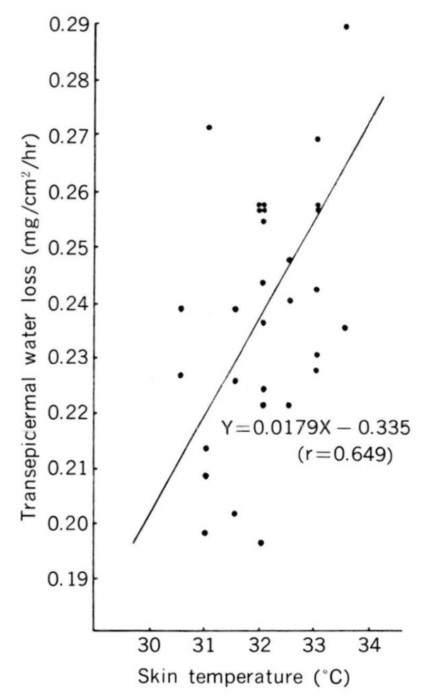

Fig. 1. The relationship between transepidermal water loss and skin temperature.

40 healthy volunteers (Fig. 2). That in 20 males was $Y = -0.0023X + 0.371$ with 1% in significance and that in 20 females, $Y = -0.0025X + 0.393$ with 2% in significance.

An inverse correlation was also found between TWL and replacement lipids as $Y = -0.0028X + 0.322$ with a correlation coefficiency $r = -0.4827$ with 1% in significance in 40 healthy volunteers (Fig. 3). That in 20 males was $Y = -0.0039X + 0.349$ with 10% in significance, but no correlation was found in 20 females. To analyze this relationship further,

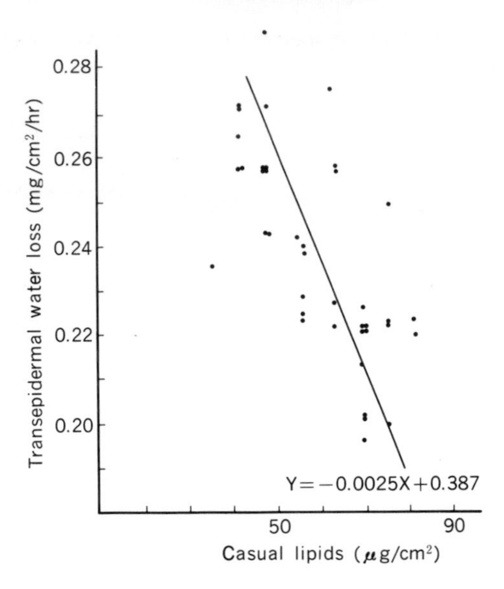

Fig. 2. The relationship between transepidermal water loss and casual lipids.

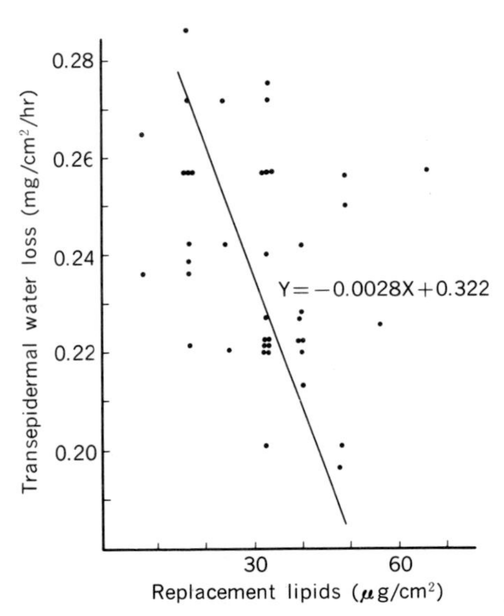

Fig. 3. The relationship between transepidermal water loss and replacement lipids.

squalene and free and total cholesterol in casual and replacement lipids were measured (Table 3). And the correlation between TWL and each lipid was examined.

Table 3. Mean values of squalene and free and total cholesterol in casual and replacement lipids.

	casual lipids		replacement lipids	
TWL ($mg/cm^2/hr$)	0.239 ± 0.021	($n = 23$)		
squalene ($\mu g/cm^2$)	0.74 ± 0.61	($n = 23$)	0.15 ± 0.11	($n = 23$)
free cholesterol ($\mu g/cm^2$)	1.42 ± 0.78	($n = 23$)	0.12 ± 0.13	($n = 23$)
total cholesterol ($\mu g/cm^2$)	2.44 ± 1.44	($n = 18$)	0.32 ± 0.18	($n = 17$)

An inverse correlation was again observed between TWL and total cholesterol in casual lipids as $Y = -0.0075X + 0.260$ with a correlation coefficiency $r = -0.4371$ with 10% in significance (Fig. 4).

No correlation was found between TWL and squalene in casual and replacement lipids, total cholesterol in replacement lipids, and free cholesterol in casual and replacement lipids.

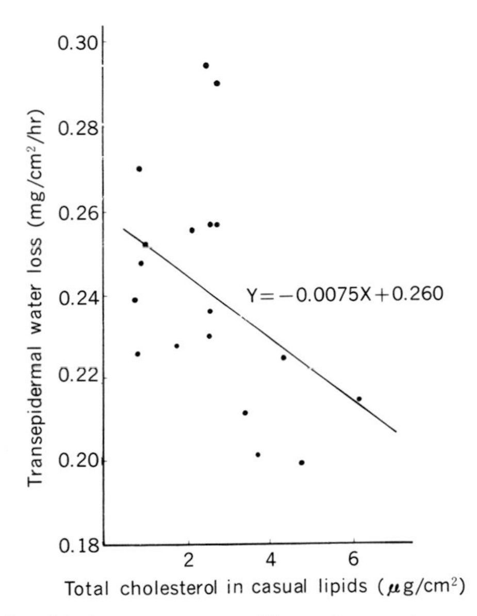

Fig. 4. The relationship between transepidermal water loss and total cholesterol in casual lipids.

DISCUSSION

Hjorth and Fregert[13] proposed that the skin surface film is the first line of defence because the film of sebum was emulsified with sweat and breakdown products from the horny layer. The function of skin against transepidermal water loss was also carried out by this film.

In the results of our present study, an inverse correlation was seen between TWL and the amount of casual or replacement lipids and also between TWL and the amount of total cholesterol in casual lipids. As it is usually defined, replacement lipids were expressed as the reservoir of sebaceous lipids. However, replacement lipids were elucidated to be the remainder of casual lipids in our study. Figure 5 shows a correlation between casual and replacement lipids. This might be the reason why both casual and replacement lipids had a correlation with TWL. In other words, TWL is decreased while the total amount of cholesterol is increased on the skin surface.

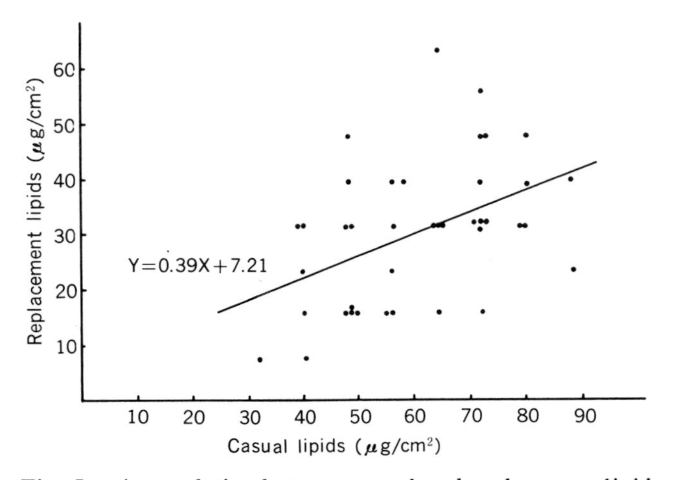

Fig. 5. A correlation between casual and replacement lipids.

On the other hand, squalene, which is synthesized in sebaceous glands, excretes onto the skin surface without decomposition[11,14] and is a marker of sebaceous lipids, had no correlation with TWL.

It is concluded that total cholesterol, mainly esterified cholesterol or epidermal lipids, exert an inhibitory action on cutaneous insensible perspiration.

Acknowledgements

The authors wish to thank Dr. Hidesuke Takahira, Professor of Physiology, Tokai University School of Medicine, for his kind revision of this paper.

REFERENCES

1. Wheatley, V. R. Possible clinical applications of the study of the cutaneous lipids with special reference to a simple method for the rapid quantitative examination of surface-lipid samples. *Bull N.Y. Acad. Med.,* **41**: 353–369, 1965.

2. Ohkido, M., Suzuki, K., Sugihara, I., and Mizuno, N. Effects of ultraviolet irradiation on human skin lipids. *Acta Dermatovener.,* **54**: 223–226, 1974.
3. Kooyman, D. J. Lipids of the skin. Some changes in the lipids of the epidermis during the process of keratinization. *Arch. Derm. Syphil..,* **25**: 444–450, 1932.
4. Nicolaides, N. Lipids, membranes and the human epidermis. In The Epidermis (W. Montagna and W. C. Lobitz, Jr. eds.), Academic Press, New York, 1964, pp. 511–538.
5. Kellum, R. E., Human sebaceous gland lipids, Analysis by thin-layer chromatography. *Arch. Derm.,* **95**: 218–220, 1967.
6. Ohkido, M., Matsuo, I., Usuki, K., and Hatano, H. Lipid biosynthesis in psoriatic epidermis. *Jap. J. Derm.,* ser. B, **79**: 424–427, 1969.
7. Ohkido, M., Matsuo, I., Usuki, K., Hatano, H., and Kanno, T. Phospholipids biosynthesis in psoriatic epidermis. *Jap. J. Derm.,* ser. B, **80**: 80–83, 1970.
8. Ohkido, M., Sugihara, I., and Ohno, M. Lipids biosynthesis from acetate in normal rat epidermis. *Jap. J. Derm.,* ser. B, **82**: 80–83, 1972.
9. Long, V. J. W. Changes in the fatty acid composition of the phospholipids, triglycerides and free fatty acids with depth in the cow snout epidermis. *Brit. J. Derm.,* **87**: 227–234, 1972.
10. Ohkido, M., Sugihara, I., and Ohno, M. Synthesis in epidermal triglycerides (in Japanese). *Jap. J. Derm.,* ser. A., **82**: 734–738, 1972.
11. Hanaoka, H., Ohkido, M., Hattori, Y., Maruta, T., and Arai, T. Reexamination of the sebaceous function (in Japanese). *Jap. J. Derm.,* ser. A., **81**: 259–263, 1971.
12. Baker, H. and Kligman, A. M. Measurement of transepidermal water loss by electrical hygrometry. *Arch. Derm.,* **96**: 441–452, 1967.
13. Hjorth, N. and Fregert, S. Contact dermatitis. In Textbook of Dermatology, vol. 1, 2nd ed. (A. Rook *et al.* eds.), Blackwell Scientific Pub., Oxford, 1972, pp. 305.
14. Ohkido, M., Matsuo, I., and Abe, T. Functional analysis of sebaceous gland activity in normal and pathological human skin. In Biology and Disease of the Hair (T. Kobori and W. Montagna, eds.), University of Tokyo Press, Tokyo, 1975, pp. 575–579.

Discussion

DR. KUROSUMI: Do you think there is any difference in the lipid content——particularly in the cholesterol level——between the epidermis and the hair? I think that if the lipid is derived from lamellar granules, the content in the epidermis must be higher than it is in the hair.

DR. OHKIDO: Generally speaking, the cholesterol level over the whole body surface seems to be about the same, even including the hair. The total lipids, however, are quite different on each part of the body surface since they derive from sebaceous glands. But cholesterol and phospholipids must be the same over the body.

DR. GOLDSMITH: If you could block sweating you might be able to do these studies on the palms or soles where the sebaceous glands would just not be a factor. I wonder if you have done any studies on the palms and the soles?

DR. OHKIDO: Yes, we have done it. But the transepidermal loss of water is so high in this area that it is difficult to be sure that lipids have not been lost thereby.

DR. ASO: Some people believe that humidity plays an important role in proliferation. For instance, psoriatic proliferation is sometimes inhibited in higher humidities, for example, as in a country like Japan where the humidity is very high. Have you studied the lipid composition of psoriatic skin? The squalene appears to be the same but cholesterol and phospholipids are increased on the surface of the psoriatic skin. There appears to be no direct relationship between these effects and humidity.

DR. OHKIDO: We have not studied psoriatic skin in this regard.

DR. ODLAND: I would like to ask you what you consider to be the proportional contribution of the surface lipids to the function of the total lipid in whatever kind of rate-limiting process the horny layer may posses? What part of the total resistance to water diffusion do you consider the surface lipids to provide? Do you think that surface lipids are the major component or just one factor?

DR. OHKIDO: The effect of surface lipids must be very limited.

DR. ODLAND: I wanted to know what you thought about this. I'm in mind of the interesting studies that Prottey has done in England. Patients, who for reasons of parenteral alimentation, had developed essential fatty acid deficiencies and had not only developed abnormal keratinization, a curious scaly skin, but had also exhibited striking changes in TWL (transepidermal water loss) under those circumstances. When they corrected the systemic deficiency of essential fatty acids, the water loss was corrected but, if I recall correctly, it was not corrected when they applied the essential fatty acids through the skin which ultimately corrected the systemic deficiency.

DR. MATOLTSY: I think that when we talk about epidermal water loss, we should distinguish between normal and pathologic skin and if we think in terms of psoriasis or perhaps the wounded skin, we have an altered keratinization process which results in a parakeratotic horny layer. Parakeratotic horny cells are loaded with lipid droplets. As far as I know, nobody knows what kinds of lipid there are. These are inside the cells, so in pathologic conditions we have a complex situation, altered structure, a different horny layer and lipids which are, perhaps, of different nature. I would like to know whether you studied the lipids in psoriatic horny cells?

DR. OHKIDO: No, we do not know how to extract the lipid droplets without contamination from parakeratotic horny layers in psoriasis.

DR. GOLDSMITH: One of the things that I don't understand is that when people analyze this by looking at transport rates and if you find a linear correlation with lipid concentration, I wonder if this might not represent saturation kinetics, i.e. a Michaelis-Menten phenomenon. After you have a pound of cholesterol in your skin, putting 10 pounds more in this spot is not going to change anything. I just wonder if people who really think about transepidermal water loss have ever looked at their data from the point of view of some kind of transport or saturation phenomenon?

DR. OHKIDO: No, we have not carried out such analyses. The concentration of lipids is not linearly related to the water transport in the skin.

III. CONTROL MECHANISMS OF EPIDERMAL DIFFERENTIATION

Regulation in Epidermal Differentiation*

I. A. Bernstein, L. Sachs, R. Ball and G. Walker

Departments of Biological Chemistry and of Environmental and Industrial Health, The University of Michigan, Michigan, U.S.A.

INTRODUCTION

The initial appearance of a protein at a particular stage in differentiation provides a convenient system for investigating the regulatory mechanisms which govern differentiative changes. Formation of a unique, histidine-rich, sulfur-poor protein (HP) in the cutaneous epidermal granular cell of the newborn rat represents such a system. In the biogenesis of this protein, the polypeptide which is the ribosomal gene-product[1], is quickly polymerized into large complexes of about 6×10^4 to greater than 10^6 daltons.[2] Since HP is not made until the keratinizing cell reaches the granular layer, a mechanism must exist to restrict this synthesis to the proper time in the differentiative process.

Protein synthesis could be controlled at the stage where the genetic information in DNA is transcribed into mRNA or where mRNA is translated into protein.[3] In differentiation, regulation could be imposed at either or both of these levels. This question can be directly investigated by determining whether the mRNA for HP is present in epidermal cells at all levels of differentiation (i.e., in basal, spinous, and granular cells) or only in granular cells. Technically, the problem is complicated by the fact that assay of a specific species of epidermal mRNA is presently limited to analysis of the mRNA in polysomes isolated from the cytosol.[4] New technology would have to be developed in order to determine the presence of mRNA for HP in the nucleus. It is possible, for example, that mRNA for HP is synthesized and is present in the nuclei of cells in all layers but is transferred to the cytosol, becomes polysomal, and is translated only in the cells of the granular layer. There is evidence[3] for the nuclear processing of mRNA prior to cytoplasmic translation.

* The research reported in this paper from the authors' laboratory has been supported in part by Grants Number AM 05268, AM 15206, and GM 00187 (Traineeship for L. Sachs) from the United States Public Health Service.

The results of preliminary studies have suggested that in the epidermis of the newborn rat, mRNA for HP is present in polysomes only in the granular cell (cf 4). HP contains a high level of histidine but no leucine.[5] A protein synthesizing system derived from rat ascites cells, incorporated into acid-insoluble form a higher ratio of [^{3}H]histidine to [^{3}H] leucine when mRNA from granular cells was present in the reaction mixture than when mRNA from basal cells was used.[4] Studies are in progress to demonstrate by immunologic techniques that HP-peptides are synthesized only in the presence of mRNA from granular cells.

If mRNA for HP were made only in the granular cell, then the physical or chemical environment of the DNA coding for HP-gene-product would have to be different in the chromatin of the granular nucleus as compared with the chromatin of all the other nuclei. Experiments have been initiated to determine whether chromatin isolated from the different viable cellular layers would show different transcriptional characteristics *in vitro* and whether the RNA transcripts obtained from granular chromatin would direct the synthesis of HP-peptide in a cell-free protein synthesizing system. Initial findings[6] are that RNA transcripts made upon DNA in chromatin isolated from granular cells have a different base composition than those made upon chromatin obtained from basal cells and that the ratio of cytosine to uracil in the RNA transcripts of granular chromatin is consistent with that calculated for mRNA coding for HP.

NATURE AND BIOSYNTHESIS OF HP IN THE EPIDERMIS OF THE NEWBORN RAT

The intraperitoneal administration of [^{3}H]histidine to newborn rats results in an initial concentration of tritiated protein in the epidermal granular cells.[7] This localization of label represents the rapid synthesis of protein in which glutamic acid (and the amide), serine, glycine, arginine, alanine, histidine, threonine, aspartic acid (and the amide), lysine, and proline account for 95% of the amino acid residues; histidine constitutes about 7.6%, and leucine, cysteine, and methionine are absent (Table 1).[2,5] The HP gene-product is a relatively small peptide[1] which is rapidly converted to a very large complex (Fig. 1).[2] Even in short times after the intraperitoneal injection of [^{3}H] histidine, most of the tritiated histidine was found in the large complex and the precursor was not detected. The large complex is broken down to a much smaller unit as the granular cell moves toward the cornified layer.[2] HP (at $\sim 10^5$ daltons) has been isolated from Ugel's[8] "macroaggregates of keratohyalin."[9]

Table 1. Amino acid composition of HP.

Amino acid	% Composition* (a)	Possible codons	$\frac{\Sigma C}{\Sigma C+U+G+A}$ (c/n)**	$\frac{\Sigma U}{\Sigma C+U+G+A}$ (u/n)**
Glutamic Acid	19.5	GAA GAG	0	0
(Glutamine)	—	(CAA CAG)	—	—
Serine	15.8	UCU UCC UCA UCG AGU AGC	0.33	0.33
Arginine	13.0	CGU CGC CGA CGG AGA AGG	0.28	0.06
Glycine	12.0	GGU GGC GGA GGG	0.08	0.08
Alanine	10.3	GCU GCC GCA GCG	0.41	0.08
Histidine	7.6	CAU CAC	0.50	0.17
Threonine	6.0	ACU ACC ACA ACG	0.41	0.08
Aspartic Acid	5.1	GAU GAC	0.17	0.17
(Asparagine)	—	(AAU AAC)	—	—
Proline	4.5	CCU CCC CCA CCG	0.75	0.08
Lysine	1.1	AAA AAG	0	0
Tyrosine	0.5	UAU UAC	0.17	0.50
Isoleucine	0.6	AUU AUC AUA	0.11	0.33
Valine	0.6	GUU GUC GUA GUG	0.08	0.41

In mRNA for HP:
Total molecules of C per 300 nucleotides = $\Sigma 3ac/n$.
Total molecules of U per 300 nucleotides = $\Sigma 3au/n$.
*Residues/100 residues × 100
**n = total C+U+G+A in all possible codons for each amino acid
c and u = sum of all C and U, respectively, in all possible codons for each amino acid

MOLECULAR HISTORY OF HP

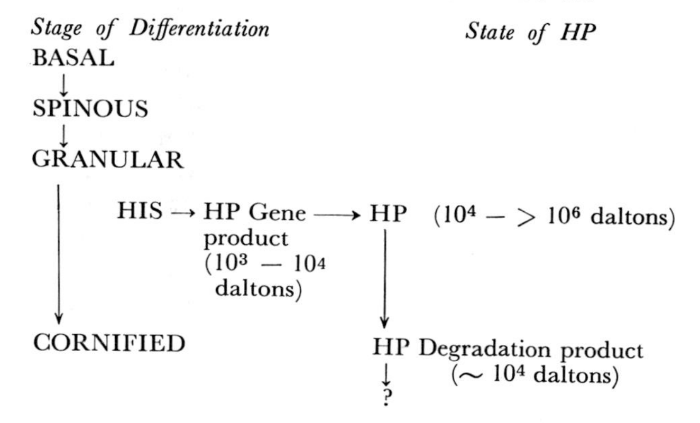

Fig. 1

IN VITRO FORMATION OF RNA TRANSCRIPTS UPON DNA IN CHROMATIN FROM DIFFERENT EPIDERMAL CELL LAYERS

Incubation in 0.24 M NH_4Cl, pH 9.5, at 0° can be used to separate dermis and epidermis obtained from the skin of the newborn rat.[10] Careful treatment of the excised skin at 5, 15, or 30 min under these conditions yielded epidermal preparations (Fig. 2) consisting of a) basal, spinous, and granular cells, b) spinous and granular cells, or c) granular cells, respectively.[10] The cornified layer was included in every case.

Chromatin was isolated from the basal-spinous-granular and from the granular preparations using the method of Bonner *et al*[11]. The DNA in both preparations was transcribed[6] using a purified bacterial DNA-dependent, RNA-polymerase under the following reaction conditions: 310 mM $(NH_4)_2SO_4$, 1.2 mM $MnCl_2$, UTP, CTP, ATP, GTP at 0.2 mM each, chromatin containing 7.2 μg (as DNA), 40 mM Tris-HCl, pH 8.0, 0.4 μCi of [^{3}H]CTP or UTP and 12 μg of enzyme (as protein) in 0.1 ml final volume at 37°. After various times, the acid-precipitable ^{3}H (in 20% trichloroacetic acid) was determined as a measure of the formation of RNA transcripts. Figure 3 presents the results of this experiment. The ratios of the rate of incorporation of ^{3}H from [^{3}H]CTP to that from [^{3}H]UTP was 1.1 for the basal-spinous-granular preparation and 2.4 for the granular preparation, respectively. Since the specific activities of the two labeled precursors were the same, the ratios of the base compositions in the two types of RNA transcript are indicated by the ratios of the incorporation of ^{3}H from the two labeled precursors: that is, the ratio of cytosine to uracil in the transcript from the basal-spinous-granular chromatin would have been 1.1, while that from the granular chromatin would have been 2.4. The values are clearly different. Interestingly, the rate of incorporation of ^{3}H from CTP was not markedly different in the two preparations. The rate of incorporation of ^{3}H from UTP, however, decreased by about 50% in the granular preparation as compared with the whole epidermal preparation.

One can *estimate* the ratio of the concentrations of cytosine to uracil in the mRNA for HP by *assuming* an equal use of all possible codons for the amino acids in HP and calculating the ratio of the number of molecules of cytosine and uracil which would have to be present in a chain of 300 ribonucleotides which would code for a random chain of 100 amino acids in HP.[4] Assuming that all glutamate and aspartate residues exist in HP as the acids and not the amide, the calculated value for cytosine to uracil in mRNA for HP is 2.2 (Table 1). To the extent that glutamine is present, the ratio would increase toward a maximum of 2.8 since the codons for glutamine contain cytosine but do not contain uracil. By similar calcula-

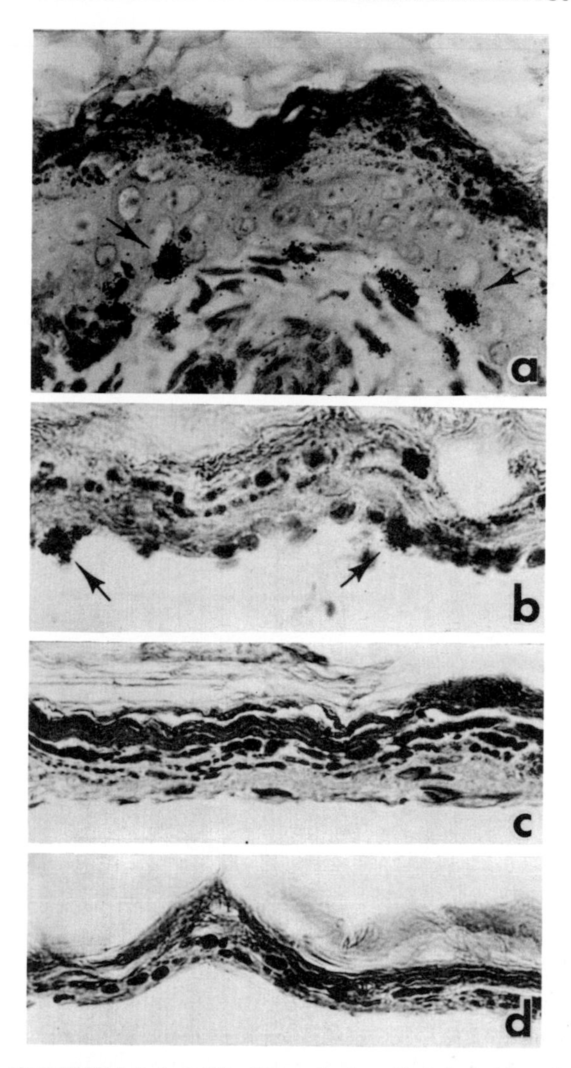

Fig. 2. (a) [^{3}H]-TdR labeled skin. Note the heavily labeled basal cells (arrows) as well as the labeled fibroblasts in the dermis. × 420

(b) Epidermis separated after 5 min of NH_4Cl treatment. This skin was labeled with [^{3}H]-TdR *in vitro* before the epidermal separation. Note the labeled basal cells (arrows). × 570

(c) Epidermis separated after 15 min of NH_4Cl treatment. Basal cells were removed and the epidermis is comprised of spinous, granular, and cornified cell layers. × 570

(d) Epidermis separated after 30 min of NH_4Cl treatment. Basal and most spinous cells were removed by the NH_4Cl treatment and the remaining epidermis is comprised of granular and cornified cell layers. × 570

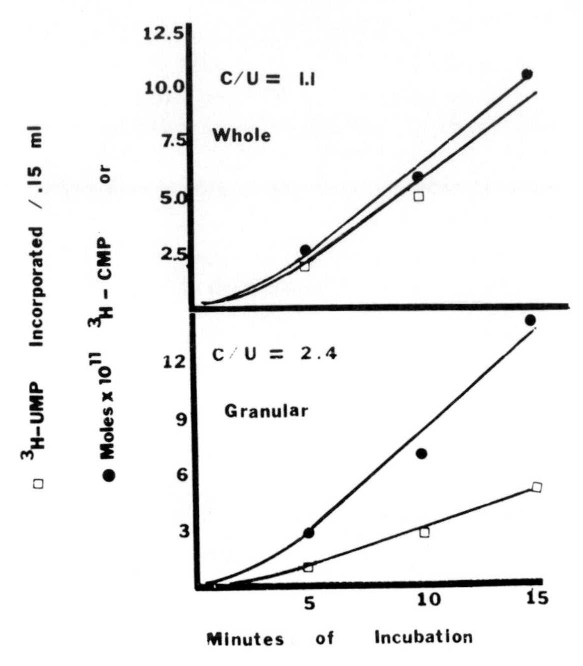

Fig. 3. Rates of incorporation *in vitro* of CMP and UMP into RNA transcripts. For experimental details see the text. Templates were chromatins isolated from the whole epidermis (upper) and a granular preparation (lower).

tions based upon an animo acid composition of urea-extractable protein from the whole epidermis, the cytosine to uracil ratio should be about 1.1 (cf. 4).

The heteropolymerio nature of the RNA transcripts was confirmed by degradation with RNase of suitable specificities and by the demonstration that for full activity, the polymerase required the presence of all four ribonucleotide triphosphates.[6] To confirm that the ratio of the incorporation of ^{3}H from [^{3}H]CTP and [^{3}H]UTP was indicative of the actual ratio of cytosine to uracil in the RNA transcripts formed from granular chromatin, this polynucleotide was isolated and the base composition was determined. The cytosine content was 40% and the uracil level was 18%, giving a ratio of 2.2.[6] Since this high ratio of cytosine to uracil might be indicative of the formation of ribosomal RNA, which is also reported[12] to have a high ratio of cytosine to uracil, the guanine to cytosine ratio was considered. The ratio of guanine to cytosine was about 0.75 or less—a significant difference from the value of about 1.5 for ribosomal RNA as isolated from *Escherichia coli*. Both ribosomal RNA and these RNA transcripts had a high percentage of guanine plus cytosine (54% for ribosomal RNA and 70% for thc RNA transcripts.).

The data now available support the hypothesis that regulation of protein synthesis during epidermal differentiation involves control at the level of transcription, as appears to be the case in erythroid cell differentiation[13–15] and in the formation of α-foetoprotein in the developing mouse liver.[17] Specifically, in regard to the synthesis of HP, the mechanism responsible for activating this biosynthetic process involves transcription of DNA. It seems that either repression is released or induction is imposed. The availability of two species of chromatin which differ in their susceptibility to transcription of a particular sequence—presumably the gene for HP—provides an opportunity to investigate the specific molecular mechanisms which control differentiative events. By comparing the chemical and functional characteristics of these two chromatins, the modulating influence of histones, acidic problems, RNA, and other substances found in chromatin can possibly be elucidated. It should be possible in such a study to determine how a particular differentiative event is triggered. An understanding of the normal regulating mechanisms will also provide a basis for investigating the molecular lesions in epidermal hyper- and neo-plasias.

REFERENCES

1. Sugawara, K. and Bernstein, I. A. Biosynthesis, *in vitro* of "histidine-protein"—a biochemical marker in epidermal differentiation. *Biochim. Biophys. Acta,* **238**: 129–138, 1971.
2. Ball, R. D. Histidine-rich protein II in the stratum corneum—a new marker of epidermal differentiation. Ph. D. Thesis, University of Michigan, 1976.
3. Gross, P. R. Biochemistry of differentiation. *Ann. Rev. Biochem.,* **37**: 631–660, 1968.
4. Bernstein, I. A., Kaman, R. L., Malinoff, H., Sachs, L., and Gray, R. H. Translation of polysomal messenger RNA during epidermal differentiation. *J. Invest. Derm.,* **65**: 102–106, 1975.
5. Hoober, J. K. and Bernstein, I. A. Protein synthesis related to epidermal differentiation. *Proc. Natl. Acad. Sci., USA,* **56**: 594–601, 1966.
6. Sachs, L. Transcription of epidermal chromatins at different stages of differentiation. Ph. D. Thesis, University of Michigan, 1976.
7. Fukuyama, K., Nakamura, T., and Bernstein, I. A. Differentially localized incorporation of amino acids in relation to epidermal keratinization in the newborn rat, *Anat. Rec.,* **152**: 525–536, 1965.
8. Ugel, A. R. and Idler, W. Further characterization of bovine keratohyalin, *J. Cell Biol.,* **52**: 453–464, 1972.
9. Sibrack, L. A., Gray, R. H., and Bernstein, I. A. Localization of the histidine-rich protein in keratohyalin, morphologic and macromolecular marker in epidermal differentiation, *J. Invest. Derm.,* **62**: 394–405, 1974.
10. Walker, G., Sachs, L., Sibrack, L. A., Ball, R. D., and Bernstein, I. A. Separation of epidermal layers of the newborn rat. *J. Invest. Derm., in Press.*

11. Bonner, J., Chalkby, G., Dahmus, M., and Widholm, J. Isolation and characterization of chromosomal nucleoproteins. Methods in Enzymology 12 B, 3, 1968.
12. Aviv, H., Packman, S., Ross, J., Swan, D., Grelin J., and Leder, P. DNA complementary to globin and immunoglobulin mRNA: A probe to study gene expression. *Adv. Exp. Med. Biol.,* **44**: 141–152, 1974.
13. Reeder, R. H. and Brown, D. D. Transcription of the ribosomal RNA genes of an amphibian by the RNA ploymerase of a bacterium. *J. Mol. Biol.,* **51**: 361–377, 1970.
14. Ross, J., Gielen, J., Packman, S., Ikawa, Y., and Leder, P. Globin gene expression in cultured erythroleukemic cells. *J. Mol. Biol.,* **87**: 697–714, 1974.
15. Marks, P. A., Rifkind, R. A., and Bank, A. Control of gene expression during erythroid cell differentiation. *Adv. Exp. Med. Biol.,* **44**: 221–242, 1974.
16. Selander, R. and De La Chapelle, A. The control of differentiation in erythroid cells. *Biochim. Biophys. Acta,* **247**: 141–152, 1971.
17. Koga, K., O'Keefe, D. W., Iio, T., and Tamaoki, T. Transcriptional control of α-foetoprotein synthesis in developing mouse liver. *Nature* **252**: 495–497, 1974.

Discussion

Dr. Goldsmith: My real concern is that in the synthesis of ovalbumin, which is another example of a major protein being produced in a tissue, the amount of chromatin that codes for the messenger for ovalbumin, is a small percentage of the total chromatin being transcribed. Your results would suggest that in the case of the histidine-rich protein a rather major component of the chromatin is being transcribed. I just wonder if there is some kind of resolution of the differences between the ovalbumin system and the histidine-rich protein system.

Dr. Bernstein: We have been concerned about what appears to be the formation, under these circumstances, of RNA transcripts only of histidine-protein on chromatin derived from the granular cell. The only thing I can say is that we are talking about terminal differentiation here. I have the feeling, although I can not document it to your satisfaction, that there really isn't much more being made here. This is a cell programmed, yes, and doing its thing, yes, but not doing everything all the other cells probably are doing. In contrast, the ovalbumin system is a viable system which intends to live a long time. The messenger that we are talking about here ought to be similar to the messenger for fibroin, globin and for other proteins participating in terminal differentiation.

Dr. Freedberg: It's absolutely clear to you, to me and to us all that there are assumptions in here which are unprovable and that the final answer will only come when one gets a hand on the material. I would like to ask two questions. One, is it possible that the results you see are related to some kind of destruction of the chromatin itself? You are looking at terminal differentiation. Is it conceivable that the changes represent destruction of the chromatin in the granular layer? Two, you discussed an interesting part of this problem, namely, that the synthesis of HR protein occurs as a small molecule and later becomes a big molecule and then a small molecule again. Why does the Lord go through that pathway, do you think?

DR. BERNSTEIN: I think that it is most easy to answer the last question. Of course, I don't know. Intuitively, I believe that the protein has a functional role in the formation of keratin.

DR. FREEDBERG: I think that it works as the big thing and not as the little thing.

DR. BERNSTEIN: That is my assumption. Having completed its job, it is degraded as is the DNA in the nucleus. As far as the question of degradation of chromatin is concerned, first of all, it is hard for me to believe that if our results are a function of degradation of chromatin that we are dealing with an artifact of isolation. If so, then the two species of chromatin with which we work, appear to be inherently different. Both chromatins are handled by the same techniques although there is a difference in the time of exposure to NH_4Cl. We have done the controls trying to see what happens with chromatin alone and with polymerase alone. We are now in the process of dissociating and reassociating the chromatins to see if we can turn off—if you will allow me to use that terminology—the histidine-protein gene by mixing the protein of the lower cell chromatin with the chromatin DNA of the upper cell. If we can shut off the histidine-protein gene then I think the argument that you have just presented would not be a viable alternative. Do you agree? The other important question is what is the situation in a disease such as psoriasis where histidine-protein is not seen and where it is distinctly possible that the gene is not turned on? Another possible explanation for the absence of histidine-protein in this disease is that polymerization does not occur. At least we can differentiate between these two possibilities by technology which we now have in hand.

DR. FREEDBERG: Have you looked to see if there is ribonuclease activity in histidine-rich protein?

DR. BERNSTEIN: There is a very small amount of RNase present. Exactly what that means, I don't know.

DR. STEINERT: The thing I would like to stress about this is that it is extremely important that you prove sometime in the future that you are not synthesizing ribosomal RNA in your system because ribosomal RNA does have a C to U ratio of almost 3. And until you can prove that point unequivocally I think your results possibly could be interpreted differently.

DR. BERNSTEIN: I think that is absolutely correct and all that we can say is, as I mentioned before, that the ribosomal composition from *E. coli* does not match this. But we are going to isolate ribosomal RNA from the epidermis.

Quantitative Autoradiographic Studies of Proteins in Keratohyalin Granules*

Kimie Fukuyama and William L. Epstein

Department of Dermatology, University of California School of Medicine, San Francisco, California, U.S.A.

ABSTRACT

Keratohyalin granules (KG) in mammalian epidermis portend the late stages of differentiation in keratinocytes. Normally, KG retain a fairly constant shape and number which is characteristic for each type of epithelium and each species. Histochemical and ultrastructural studies have demonstrated that KG consist of a group of heterogeneous components which largely determine the morphology of KG by the way they combine. The arrangement of KG subcomponents can be altered rapidly and drastically under experimental and pathological conditions; the morphology of KG thus signifies changes in metabolic function of the keratinocytes.

This review will consider the laboratory observations indicating heterogeneity of KG, including comparative studies of certain epithelia in different species and quantitative electron microscopic autoradiographic data. Furthermore, alterations in KG morphology will be described under selected experimental conditions and in some disease states. The findings will be related to current views of how KG participate in the process of epidermal keratinization.

I. MORPHOLOGICAL VARIABILITY OF KERATOHYALIN GRANULES

The ultrastructural appearance of KG in newborn rats differs considerably from that in man and guinea pigs. While the granules of man and

* This study was supported by Grant No. AM 12433 from the National Institutes of Health and in part by a special appropriation from the State of California for the study of Psoriasis, and in part by the Skin Disease Research Foundation of San Francisco, California.

guinea pig demonstrate co-existing fibrillar and interfibrillar substances, the granules of newborn rats exhibit two distinct osmiophilic elements, one of which is very electron dense and often locates at the periphery of large KG as described earlier by Bonneville[1] (Fig. 1a, b). These dense homogeneous deposits (DHD) have also been reported in tongue epithelium of rats by Farbman[2] and Jessen.[3] Because not all investigators have observed DHD during their preparations (e.g. 4), we have investigated the effects of different fixation procedures on the appearance of DHD.[5] The fixatives used were:

1) 2% OsO_4 in Sorensen's phosphate buffer for 2 hours, after fixation for 12 hours in 3.5% glutaraldehyde in Sorensen's phosphate buffer.
2) 3.5% glutaraldehyde in Sorensen's phosphate buffer for 12 hours.
3) 2% OsO_4 in Sorensen's phosphate buffer for 2 hours.
4) 1.33% OsO_4 in S-collidine buffer for 2 hours.
5) 2% OsO_4 in Palade's buffer for 2 hours.

These fixatives were adjusted to pH 7.4 and used at 4°C. The tissues were washed in the same buffer for each fixative, dehydrated in alcohol, and embedded in a mixture of Araldite and Epon. Thin sections were cut and stained with uranyl acetate and lead citrate.

DHD in most specimens appeared very electron dense with glutaraldehyde/OsO_4 double fixation. In some specimens, DHD were not as electron dense as others, but they were distinguishable from other parts of KG by their homogeneous appearance in contrast to the somewhat granular appearance in other areas of KG (Fig. 1a,b). DHD in specimens fixed with S-collidine buffer were less electron dense and showed a somewhat more granular appearance (Fig. 2a) than those prepared by double fixation. The electron density was considerably reduced in DHD after fixation with glutaraldehyde alone (Fig. 2b), and DHD were not readily observable with phosphate buffered OsO_4 fixation alone because the densities of DHD and other parts of KG appeared identical. The same procedures also were used to examine DHD and other constituents of KG in the buccal epithelium, the cheek epidermis and the transitional region in newborn rats.[6] KG in these 3 areas showed a specific morphology for each region which seemed to be determined by differences in the way DHD, other KG components, and tonofilaments related to each other. While DHD appeared at the edge of KG in cheek epidermis, they were associated with the fibrillar components in the transitional region (Fig. 3a,b) and were located in the center portion of KG in buccal mucosa (Fig. 4a,b). The existence of single patches of DHD, without apparent attachment to other KG components, also has been reported.[3] The sizes, shapes, and number of such isolated DHD are variable in newborn rat, and in

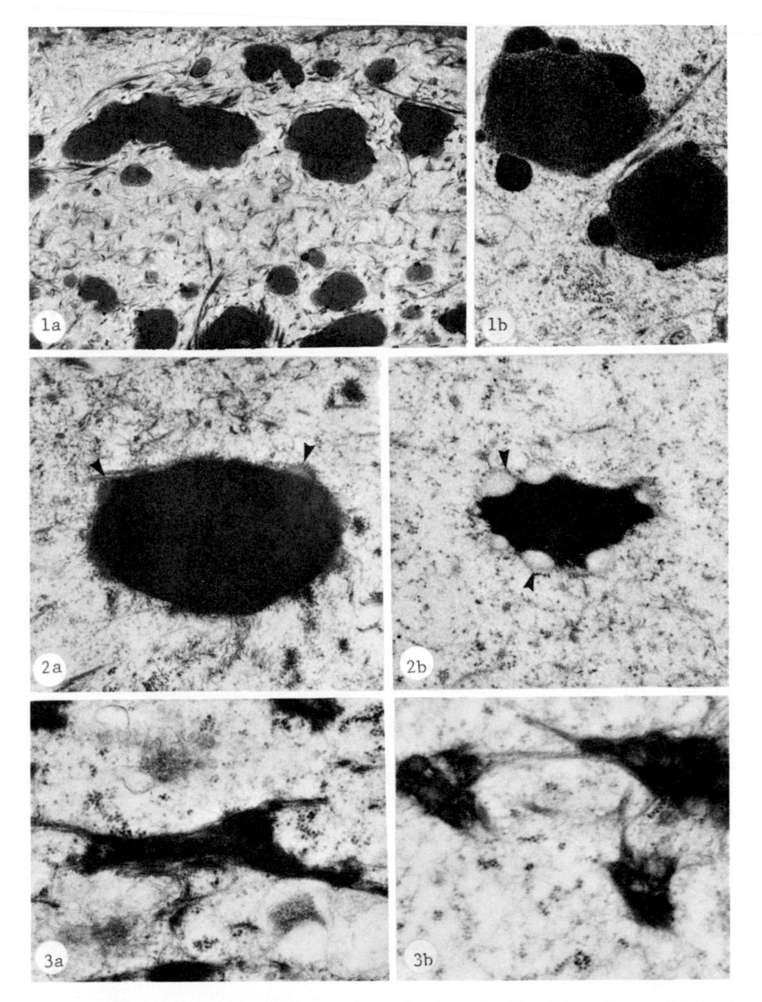

Fig. 1a, b. KG of newborn rat epidermis double fixed in glutaraldehyde and OsO_4: both buffered with Sorensen's phosphate pH 7.4. DHD appear electron dense and locate at the periphery of KG. (a) ×4,000 (b) ×11,400.

Fig. 2a. KG of newborn rat epidermis fixed with S-collidine buffered OsO_4. DHD (↓) are not as electron dense as those in Fig. 1. ×17,100.

Fig. 2b. KG of newborn rat epidermis fixed with phosphate buffered glutaraldehyde alone. DHD (↓) show electron lucent appearance. ×16,600.

Fig. 3a, b. KG of the epidermal-epithelial transitional region of newborn rat cheek. They are composed with filaments and electron dense substances (a) by glutaraldehyde-single fixation; (b) presence of DHD are readily observed. ×21,400.

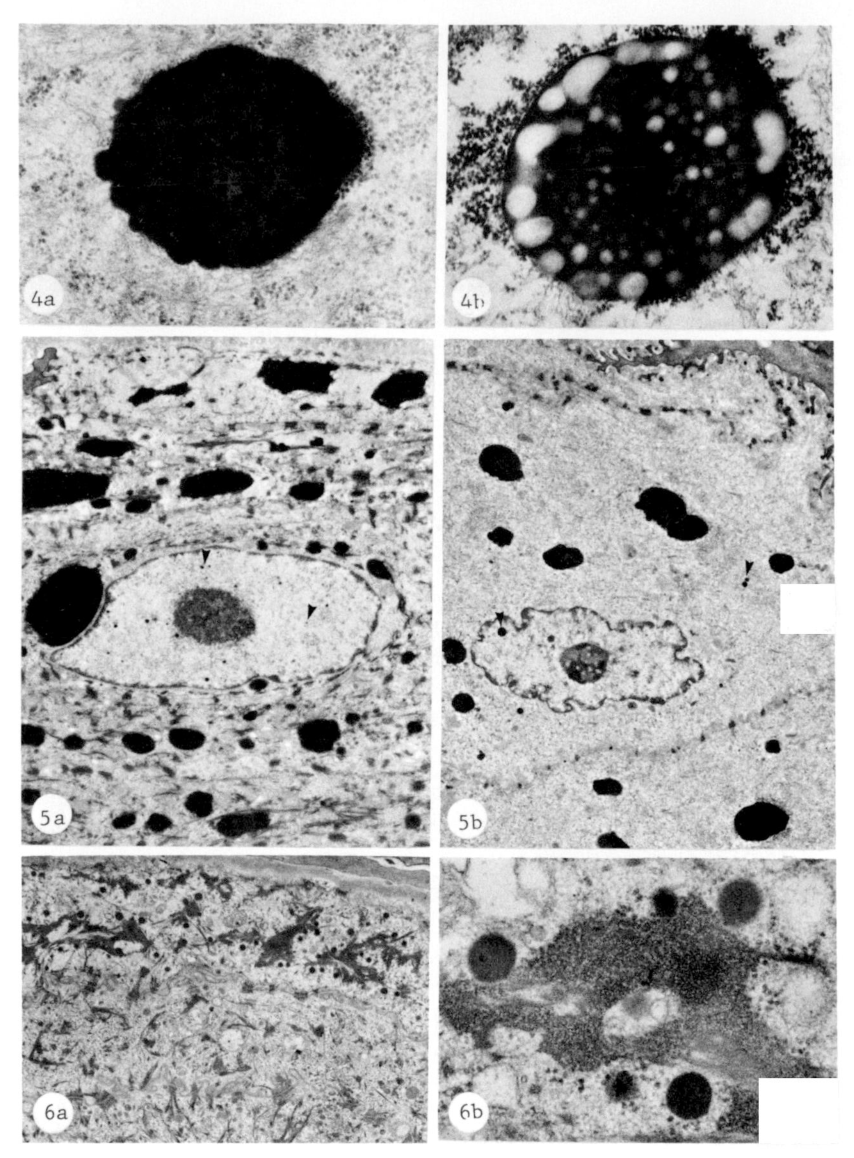

Fig. 4a, b. KG of newborn rat oral mucosum. DHD are present within KG as small patches (a) fixed with glutaraldehyde and OsO_4; and (b) fixed with glutaraldehyde alone. ×25,100.

Fig. 5a, b. The epidermis (a) and oral epithelium (b) of newborn rats. Some DHD (↓) are present singly in both the cytoplasm and nucleus. ×3,800.

Fig. 6a, b. DHD of hairless mouse epidermis do not attach to fibrillar component of KG. (a) ×4,000 (b) ×33,500.

some cases DHD alone attain the size of small KG. Single DHD also are seen in the nuclei (Fig. 5a,b). In hairless mouse epidermis, DHD are mostly of the isolated type, and locate near KG consisting of mixed fibrillar and electron dense material (Fig. 6a,b). Consequently, the total KG of hairless mouse looks grossly different from KG of newborn rats, even in low power electron micrograms.

In one study, the percentage of area occupied by total KG as well as calculations of the individual areas of DHD and other edge and center portions of KG were measured.[7] Electron micrographs of granular cells located in the two outermost layers were enlarged three times. A lattice with regularly spaced points was superimposed over the electron micrograms. The number of lattice points found in each KG compartment and in other cytoplasmic and nuclear areas were computed. The results obtained from more than 180,000 square microns are summarized in Table 1.

The majority of investigators consider that the chemical nature of KG is a mixture of substances such as proteins, RNA, polysaccharide, lipids (e.g., 8–12), although not all the "evidence" presented has been convincing. Histochemical studies demonstrated that KG are composed of proteins with high histidine content,[13] and that DHD contain sulfur containing amino acids.[14] Exact locations of polysaccharide and other moieties of KG are not known, but ribosomes are abundant around KG as a source of RNA. We have used a dilute alkaline (0.02 N NaOH, at 20°C for 48 hours) to extract RNA from glutaraldehyde fixed newborn rat epidermis at 1 and 6 hours after injection of cytidine-^{3}H. 75% of radioactivity was removed from the epidermis in which the ultrastructure of DHD remained unchanged, but other parts of KG lost their electron dense appearance (Fig. 7). In contrast, about 6% radioactivity was extracted, using the same technique, from tissues after injection of various amino acids-^{3}H. Table 2 indicates the percent of radioactivity extracted from tissues after 24 and 48 hours treatment. The concentration of RNA and protein in each extract was also measured: RNA by the Orcinol method,[15] and protein by the Lowry method[16] (Table 3). The results seem to indicate the alkaline soluble components, one possibly being RNA, are not the major components of DHD.

Tonofilaments can be seen embedded in KG in the skin of man and guinea pig. Brody[17] considered that the filaments transform into electron dense KG, and Bonneville[1] believes that the DHD found in newborn rat may represent the transformed filaments. A problem remains to demonstrate the presence of filamentous material in KG of the epidermis and mucous membrane of species such as newborn rat, despite the fact that transitional oral epithelium of the same species tonofilaments in KG are readily demonstrable.[6] Furthermore, Lavker and Matoltsy[4] using "high resolution electron microscopy" reported that fine granular material,

Table 1. Percent of area taken by subcompartment of keratohyalin granules in the cytoplasm and that taken by the DHD-like inclusion in the nucleus.

Cytoplasm				Nucleus	
Keratohyalin granules			Other cytoplasm	Inclusion	Other karyoplasm
DHD	Edge	Center			
0.98 ±0.17	1.7% ±0.3	12.2% ±1.5	85.0% ±1.5	2.2% ±0.9	97.8% ±0.9

Table 2. Percent of soluble radioactivity appearing in each extract from the fixed epidermis.

^{3}H-Precursor	Time After Injection (hours)	Percent of Total Radioactivity Extracted: 1st—24 hrs. 0.02 N NaOH at 20°C	2nd—24 hrs. 0.02 N NaOH at 20°C
Cytidine	1	65.8 ± 3.5	10.3 ± 3.1
	6	61.9 ± 5.0	10.5 ± 3.0
Histidine	1	4.1 ± 2.0	2.1 ± 0.4
	6	3.0 ± 0.6	3.0 ± 0.7
Arginine	1	1.8 ± 0.6	2.9 ± 1.0
	6	2.2 ± 1.3	5.2 ± 1.7
Lysine	1	0.8 ± 0.4	0.4 ± 0.1
	6	0.9 ± 0.1	0.3 ± 0.1
Leucine	1	2.2 ± 0.2	1.3 ± 0.2
	6	2.1 ± 0.2	1.2 ± 0.2
Methionine	1	5.7 ± 0.4	1.9 ± 1.3
	6	4.0 ± 0.6	1.6 ± 1.0
Proline	1	1.6 ± 0.1	1.5 ± 0.9
	6	1.9 ± 1.4	2.2 ± 0.8
Tyrosine	1	4.0 ± 3.2	1.7 ± 0.6
	6	5.4 ± 3.3	1.4 ± 0.5

Table 3. Ratio of RNA and protein measured in each soluble fraction and insoluble residue.

	Percent Soluble: 1st—24 hrs. 0.02 N NaOH at 20°C	2nd—24 hrs. 0.02 N NaOH at 20°C	Insoluble
Protein	11.2 ± 2.7	3.5 ± 0.6	85.3 ± 3.1
RNA	80.1 ± 5.0	12.6 ± 2.8	7.3 ± 3.2

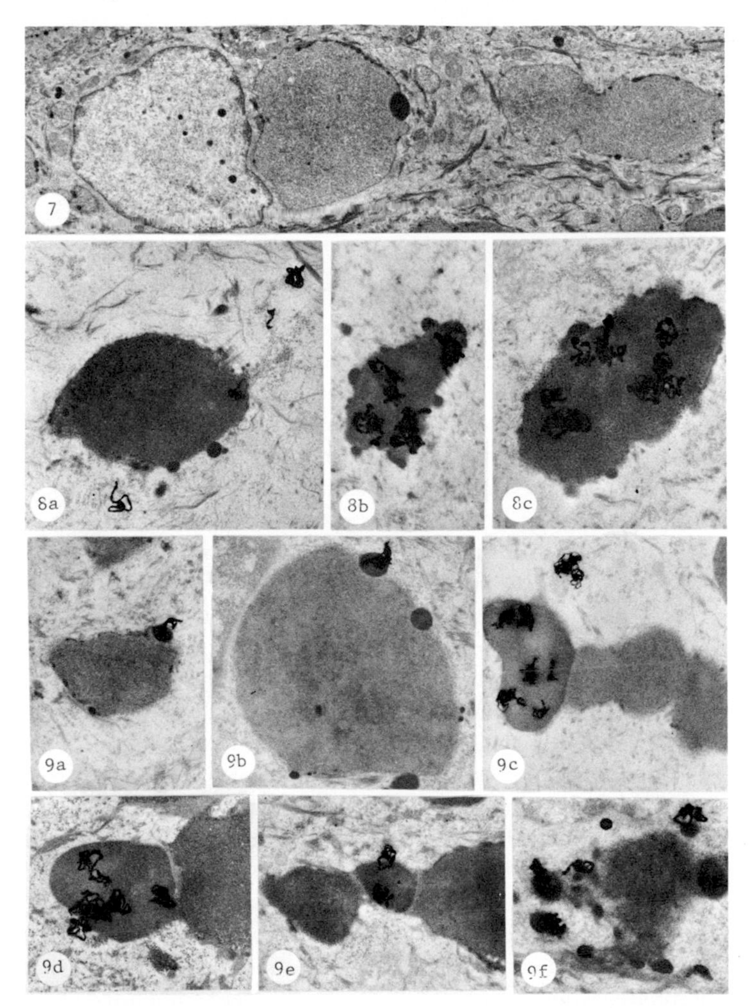

Fig. 7. Newborn rat granular cell: glutaraldehyde fixed, and treated with 0.02 N NaOH at 20°C for 48 hours. Electron density remains in DHD and nuclear inclusions, but not other parts of KG. ×4,500.

Fig. 8. Localization of silver grains in relation to keratohyalin granules of newborn rat epidermis at 1 hour (a), and 6 hours (b, c) after injection of histidine^3. ×7,400.

Fig. 9a-e. Silver grain(s) over DHD of newborn rat epidermis observed at 1 hour (a) and 6 hours (b-e) after injection of cystine-^{3}H. ×7,000.

Fig. 9f. After injection of puromycin, silver grains were also observed over DHD which appeared to be prevented from attacking KG. ×7,000.

consisting of particles with dimensions in the range of 13–35 Å, was the primary constituent of epidermal KG of newborn rats. Further studies are required to define subcomponents of epidermal KG responsible for their unique morphological appearance in some species.

II. ELECTRON MICROSCOPIC AUTORADIOGRAPHY STUDY OF INCORPORATION OF HISTIDINE-^{3}H, ARGININE-^{3}H, CYSTINE-^{3}H, AND PROLINE-^{3}H INTO KG PROTEINS

While electron microscopy and cytochemistry demonstrated KG to be heterogeneous in nature, biochemists have isolated proteins of different chemical composition from "KG" by various techniques. Bernstein and coworkers used 8M urea and 0.1N $HClO_4$ and isolated a "sulfur-poor" and "histidine-rich" protein from human and newborn rats (e.g., 10, 18, 19). Ugel also extracted a protein which contained significant amounts of histidine from bovine hoof with 1M phosphate buffer.[20] However, KG extracted by Tezuka and Freedberg from newborn rats in the presence of deoxycholate did not contain high levels of histidine.[9] Furthermore, an analysis by Matoltsy and Matoltsy indicated that keratohyalin granules isolated in a solution of citric acid containing Brij 35 and solubilized by use of dithiothreitol contained large amounts of half cystine and proline residues and only small amounts of histidine.[21]

In order to investigate the possibility that conflicting chemical findings result from the isolation of different components of KG, and to further elucidate the chemical natures of DHD and other parts of KG, we used electron microscopic autoradiography.[7,22] In this study we injected intradermally 20 uc of histidine-^{3}H, cystine-^{3}H or proline-^{3}H in 0.1 ml saline into a total of 40 newborn rats. Skin specimens obtained from the injected sites at 1 and 6 hours later* were cut into small pieces and fixed in 3% glutaraldehyde and postfixed in OsO_4; both fixatives were buffered in Sorenson's phosphate. After dehydration, the tissues were embedded in a plastic mixture, cut at 600 Å and filmed with Ilford L4 emulsion by a loop method. The specimens were developed and fixed after 9 weeks of exposure at 40°C, and stained with uranyl acetate and lead citrate. Three random blocked tissue pieces were cut, and 4 to 6 sections prepared from each tissue; a total of 12–18 sections for each animal were examined by electron microscopy. Since cornified cells are not sites of protein synthesis, specimens showing more than two grains per field over cornified cells were considered to have a high background and were omitted.

Puromycin, diluted with saline to a concentration of 0.5 mg/0.1 ml,

* In the case of histidine-^{3}H and cystine-^{3}H, we also studied the interval at 1/2 and 3 hours after injection.

was injected intradermally into 3 additional animals at 30 minutes after injection of each amino acid-^{3}H, because puromycin aids in the observation of DHD (ref. Section III). Skin specimens from these animals were taken at 1 hour after injection of puromycin. Photographs ($\times$4000) of all granular cells in the two outermost layers were enlarged 3 times. The number of silver grains over KG, DHD-like inclusions in the nucleus and over other parts of the cytoplasm and karyoplasm were counted. The total area studied and the total number of grains obtained per animal ranged from 12,000–18,000 sq. microns, and between 250–700 grains respectively. The position of tritium source points in the tissue was determined by a technique similar to that used by Ross and Benditt[23] and Petrik and Collet.[24] Grain counts were then analyzed by two different methods at various time intervals after isotope injection. The first analysis was achieved computing the percent of the total grain counts from the number of grains located over KG, KG subcompartments (DHD, other edge and center areas), nuclear inclusions and other parts of cells. The second analysis was performed by computing the specific concentration of grains over each subcompartment, and the specific grain concentration was obtained by dividing the number of silver grains over each subcompartment by the percent area occupied by each compartment.

Incorporation of histidine-^{3}H, arginine-^{3}H, cystine-^{3}H, and proline-^{3}H into proteins of granular cells was observed at all time intervals studied, and the number of grains over KG increased during the first 6 hours. The most striking concentration of silver grains over KG was seen after injection of histidine-^{3}H: The grains first appeared at the edge of the granules and then over the center, but rarely over DHD (Fig. 8a, b,c). Arginine-^{3}H also showed a labeling pattern similar to that of histidine-^{3}H. In contrast, most of the labels found over KG after injection of cystine-^{3}H located at the edge, especially in association with DHD (Fig. 9a,b). A relationship between labels and DHD was further emphasized by the appearance of large DHD as either a part of KG, or isolated from KG (Fig. 9c,d,e,f). Labels observed after injection of proline-^{3}H appeared rather diffusely in the cells, but an association of some labels with DHD was noted at one hour after injection. These findings based on qualitative observations were further confirmed by quantitative analysis. Table 4 summarizes the percent of labeling. Labeling was highest at the edge of KG and lowest over DHD at 1 hr after the injection of histidine-^{3}H and arginine-^{3}H, while it was highest over DHD after the injection of cystine-^{3}H. After proline-^{3}H injection, half the silver grains appeared at the edge and one-quarter were located over DHD and the center of the granules. At 6 hr after the injection of all amino acids, the number of grains over the center of the granules increased, but the degree of increase varied according to the isotopes, being greatest after histidine-^{3}H and least after cystine-^{3}H.

Specific grain counts, on the other hand (Table 5), showed that histidine-^{3}H-containing protein concentrated at the edge of KG and subsequently distributed throughout the granules; at no time, however, did it concentrate in DHD. Labeling patterns after arginine-^{3}H injection somewhat simulated those of histidine-^{3}H, except that the concentration of grains over DHD and the area outside of KG was higher after arginine-^{3}H than after histidine-^{3}H. In contrast, cystine-^{3}H-containing protein

Table 4. Percent label over different subcompartments of granular cells at 1 and 6 hours after injection of histidine-^{3}H, arginine-^{3}H, cystine-^{3}H, and proline-^{3}H.

	Keratohyalin granules			Other areas of cytoplasm
	DHD	Edge	Center	
Histidine-^{3}H				
1 hr	1.17 ± 0.52	24.55 ± 2.9	6.7 ± 2.7	67.4 ± 6.1
6 hr	0.87 ± 0.65	11.0 ± 0.99	67.8 ± 5.8	20.1 ± 4.5
Arginine-^{3}H				
1 hr	0.9 ± 0.3	17.0 ± 9.9	8.0 ± 5.3	74.0 ± 15.5
6 hr	1.8 ± 0.9	7.6 ± 3.4	53.2 ± 7.6	36.9 ± 7.1
Cystine-^{3}H				
1 hr	18.2 ± 4.3	9.4 ± 1.8	7.4 ± 0.89	64.6 ± 3.8
6 hr	13.6 ± 3.6	10.6 ± 1.1	15.4 ± 2.2	60.4 ± 3.6
Proline-^{3}H				
1 hr	4.4 ± 1.1	8.7 ± 1.6	4.4 ± 2.3	82.3 ± 3.7
6 hr	1.1 ± 0.3	14.3 ± 9.4	31.2 ± 6.9	53.2 ± 2.7

Table 5. Percent specific grain counts over different subcompartments of granular cells at 1 and 6 hours after injection of histidine-^{3}H, arginine-^{3}H, cystine-^{3}H, and proline-^{3}H.

	Keratohyalin granules			Other areas of cytoplasm
	DHD	Edge	Center	
Histidine-^{3}H				
1 hr	6.88 ± 2.2	85.22 ± 2.0	3.14 ± 0.88	4.73 ± 1.1
6 hr	6.78 ± 4.6	49.26 ± 3.9	42.16 ± 5.0	1.77 ± 0.34
Arginine-^{3}H				
1 hr	7.89 ± 1.7	78.61 ± 6.6	5.02 ± 0.87	8.48 ± 5.8
6 hr	14.60 ± 5.3	41.05 ± 11.0	40.30 ± 10.2	4.04 ± 1.2
Cystine-^{3}H				
1 hr	73.44 ± 5.9	20.63 ± 5.8	2.64 ± 0.75	3.27 ± 6.5
6 hr	59.94 ± 5.7	30.87 ± 6.0	6.35 ± 1.1	3.38 ± 0.78
Proline-^{3}H				
1 hr	41.16 ± 4.8	46.54 ± 3.5	3.24 ± 1.6	9.04 ± 1.9
6 hr	8.28 ± 1.6	62.56 ± 7.8	22.76 ± 13.7	5.38 ± 2.4

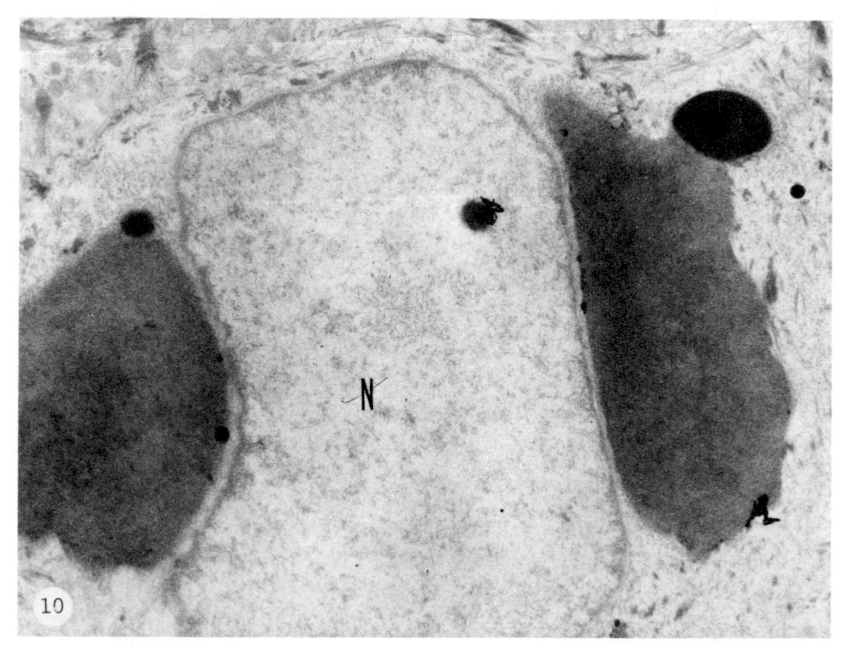

Fig. 10. Labeled DHD-like inclusion in the nucleus (N) and DHD observed in granular cell of newborn rat at 6 hours after injection of cystine-^{3}H. ×9,200.

concentrated heavily in DHD. The distribution of silver grains after the injection of proline-^{3}H was less clear-cut and showed a more even distribution of grains over the subcompartments of KG and outside the granules. Radioactivity from proline-^{3}H initially concentrated over DHD but also appeared at the edge. At 6 hr after injection, the concentration over DHD declined and that over the center increased. These findings coincide with cytochemical and electron microscopical observations that KG are composed of heterogeneous components, and provide further evidence on chemical changes of KG subcomponents. It is possible that different proteins isolated by the biochemists[9,10,20,21] may originate from different subcomponents of KG.

Similarly, grain counts were made of cystine-^{3}H nuclei of granular cells which contained DHD-like inclusions after injection. About 5–10% of the nuclei in the granular layer showed the presence of such inclusions (Fig. 10). Measurement of the area occupied by the inclusions revealed they represent approximately 2% of the total nuclear space (Table 1). Table 6 summarizes the frequency of silver grains associated with the inclusion, and the specific concentration of the silver grains was calculated to be 8.3

Table 6. Summary of grain counts over nuclear inclusions

Time After Injection (hours)	Percent Label		Spec. concen. over inclusion / Spec. concen. over others
	Inclusions	Other Karyoplasm	
½	26	73	8.3 ± 7.4
1	33	66	24.5 ± 16.7
3	35	64	24.0 ± 16.9
6	42	58	58.9 ± 28.0

times that of the rest of the karyoplasm at 30 minutes after injection of cystine-^{3}H. The density of grains over the inclusions increased with time, so that by six hours the inclusions showed 58.9 times greater concentration of grains than elsewhere. The results seem to confirm cytochemical findings that DHD and the nuclear inclusions are chemically similar.

III. MORPHOLOGICAL VARIABILITY UNDER EXPERIMENTAL AND PATHOLOGICAL CONDITIONS

Autoradiographic studies have demonstrated that granular cells of human[25] and animals[26,27] synthesize both RNA and proteins and have provided a laboratory tool for investigation of the metabolic function of differentiated cells in experimental and pathological conditions (Fig. 11). Incorporation of cytidine-^{3}H into RNA and histidine-^{3}H and cystine-^{3}H into protein was greatly reduced in newborn rats after injection of actinomycin D and puromycin, respectively.[28] The ultrastructure of KG in the epidermis in which synthesis of RNA or protein was suppressed by these pharmacological agents, manifested unique changes. Actinomycin D produced "medusa head-like" KG (Figs. 12a,b,17d) and puromycin prevented enlargement of KG by interfering with the insertion of tonofibrils and the addition of DHD. We found KG morphology also varies in conditions where synthesis of both RNA and protein are inhibited. For example, by 3 hours after ultraviolet light (UVB) irradiation, KG of hairless mice no longer exhibited the fibrillar KG and isolated DHD nearby. Instead KG became patchy in appearance and the DHD were found at the periphery as seen in normal newborn rat epidermis (Fig. 13). Similarly, synthesis of RNA and protein were suppressed in granular cells of human palm immediately after friction blister injury[29] and the ultrastructure of KG suddenly changed after the injury (Fig. 14a). We have also reported morphologic changes observed in newborn rat epidermis after injection of concanavalin A and phytohemagglutinin which also inhibit KG formation in granular cells.[30] The cells in which RNA and protein synthesis is inhibited morphologically appeared to stop their differentiation process so that

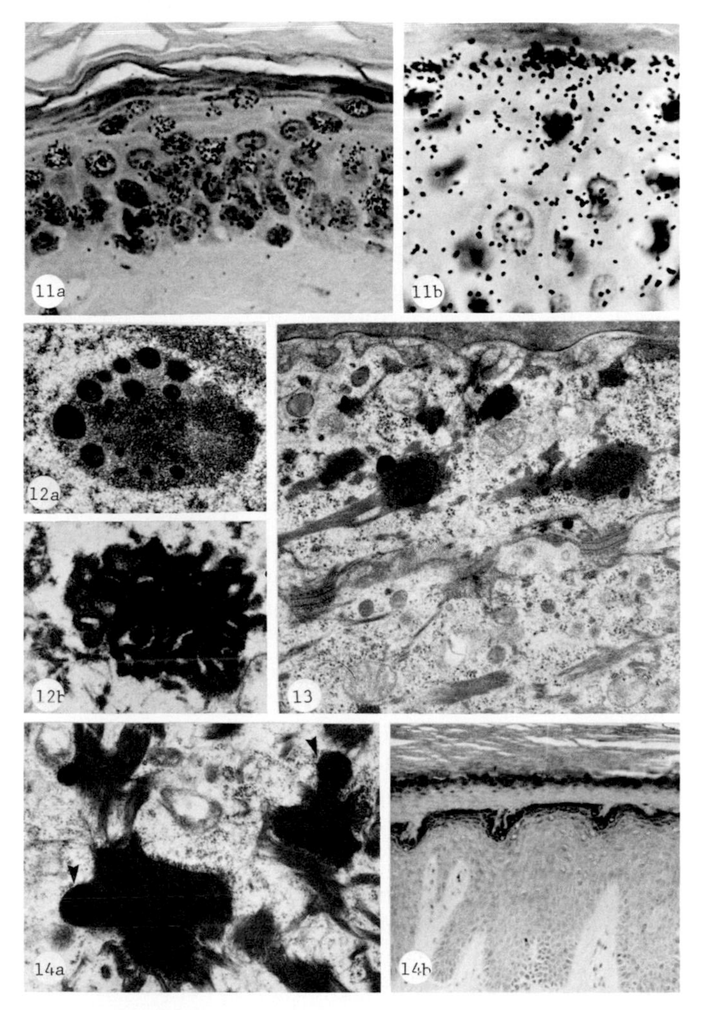

Fig. 11a, b. Light microscopic autoradiography demonstrating incorporation of cytidine-^{3}H and histidine-^{3}H into RNA and protein, respectively, of human skin.

Fig. 12a, b. Nucleolus (a) and KG (b) of newborn rat after injection of actinomycin D. $\times 15,100$.

Fig. 13. Hairless mouse granular cell 3 hours after UVB irradiation. $\times 15,700$.

Fig. 14a. KG of human palm observed at 1 hour after friction blister. DHD (↓) are noted. $\times 12,200$.

Fig. 14b. Light microscopy of human palm at 48 hours after friction blister. The cells in which protein synthesis had been discontinued did not complete their keratinization.

they can often be recognized in the cornified layer, because they retain KG and nuclei (Fig. 14 b). In contrast, upon recovery from these injuries, KG increase in both size and number and the number of granular cells also increase.

Disease states of human epidermis exhibit morphological variability in KG. However, in these situations, granular cell functions are more complicated than under experimental conditions. Rapidly proliferating epidermal cells of the psoriatic lesion present variable types and stages of differentiation. Cytological changes and their sequence also are influenced by therapy.[31,32] Brody[33] has classified psoriatic lesions into three types: 1) parakeratosis without keratohyalin granules; 2) parakeratosis with keratohyalin granules; and 3) hyperkeratosis with keratohyalin granules. This degree of variation requires that investigation of protein synthesis and morphologic study of KG should be carried out with the same individual cells rather than the skin lesion as a whole, if meaningful results in granular cell function are to be obtained in psoriasis. We injected 10 uc of histidine-^{3}H monohydrochloride (Spec. Act. 3.42 C/mM) intradermally into chronic, patchy psoriatic lesions of 3 volunteers. The patients had not received any systemic treatment for at least one year, and the test sites had never been treated topically. Clinically, all lesions resembled each other, presenting typical silvery scales on a reddish, slightly elevated base. Four mm punch biopsies taken 30 minutes after injection of histidine-^{3}H were prepared for light microscopy autoradiography, and the rate of histidine-^{3}H incorporation into protein(s) of cells at the different levels of the epidermis was determined by counting grains by the technique reported previously.[34]

Histopathologically, all three cases represented classical psoriatic lesions. The number of grains appearing over cells in the upper layers of the epidermis and over Malpighian cells were compared with that of basal cells and with the comparative counts made in normal epidermis.[25] Radioactivity was distributed regularly throughout the lower part of the epidermis, whereas in the upper part the concentration of labels differed considerably from area to area (Table 7). There were at least 3 different states of keratinization occurring simultaneously, which are readily seen in a 4 mm diameter biopsy of psoriatic lesions. Area # (1) does not contain KG and shows a reduction in protein synthesis (Fig. 15a,b). Area # (2) shows KG with fairly high concentration of silver grains indicating that protein synthesis continues, probably accompanied by more KG formation (Fig. 15c), and area # (3) exhibits KG, but protein synthesis is no longer taking place (Fig. 15 d). This situation is comparable to that occurring in granular cells after various injuries. It appears the latter cells had discontinued keratinization prematurely.

Ishibashi and Klingmüller (e.g., 35), Anton-Lamprecht and her co-

Table 7. Ratio of silver grains appearing in different histological lesions of psoriasis, as compared with histology of normal epidermis after injection of histidine-^{3}H.

Ratio	Basal cells	Sites without keratohyalin granules		Sites with keratohyalin granules: Low labeling		Sites with keratohyalin granules: High labeling	
		Malpighian cells	Granular cells	Malpighian cells	Granular cells	Malpighian cells	Granular cells
Psoriasis							
Case 1	1.0	1.0	0.4	1.2	0.4	1.7	3.5
Case 2	1.0	0.6	0.3			1.3	2.0
Case 3	1.0	1.0	0.5	1.7	0.2	1.0	1.7
Normal*	1.0					1.3 ± 0.1	3.9 ± 1.1

*Previously reported; see reference[25]

workers (e.g., 36, 37) and others have reported a series of electron microscopic studies on skin diseases classified as the ichthyotic type. A lack of KG, or the appearance of unusual shapes of KG, was demonstrated. Electron dense material accumulates in both a quantitatively and qualitatively abnormal fashion (Fig. 16a,b), but the biochemical changes in epidermal cells responsible for these differences remain unknown. Human viral warts represent another disease state in which KG show abnormalities in size, shape and number. In 1/3 of the cases we studied,[38] KG were composed of ribosome-like granules forming serpentine figures which were identical to KG observed in newborn rat epidermis after injection of actinomycin-D.[28] (Figs. 17a,b,c)

The evidence which overwhelmingly supports the view that KG are composed of heterogeneous components is presented. The manner of arrangement of KG subcompartments helps to explain comparative variations within epithelia and species. Quantitative analysis of electron microscopic autoradiography confirmed the heterogenic nature of KG. Furthermore, the results suggest that different chemical analyses obtained by various biochemical studies may represent protein moiety of KG subcompartments. The experimental studies point out the rapidity with which remarkable morphologic changes can occur in KG. The pathologic findings underscore the unknown and complex facets of KG formation awaiting more sophisticated investigation. But taken together these data emphasize that, while keratinization may be an orchestrated and relatively organized affair, the biochemical and functional events will almost certainly require investigation of changes in single cells if they are to be properly understood. Functional studies of epidermal sheets, especially in disease states, may lead to confusing and contradictory results.

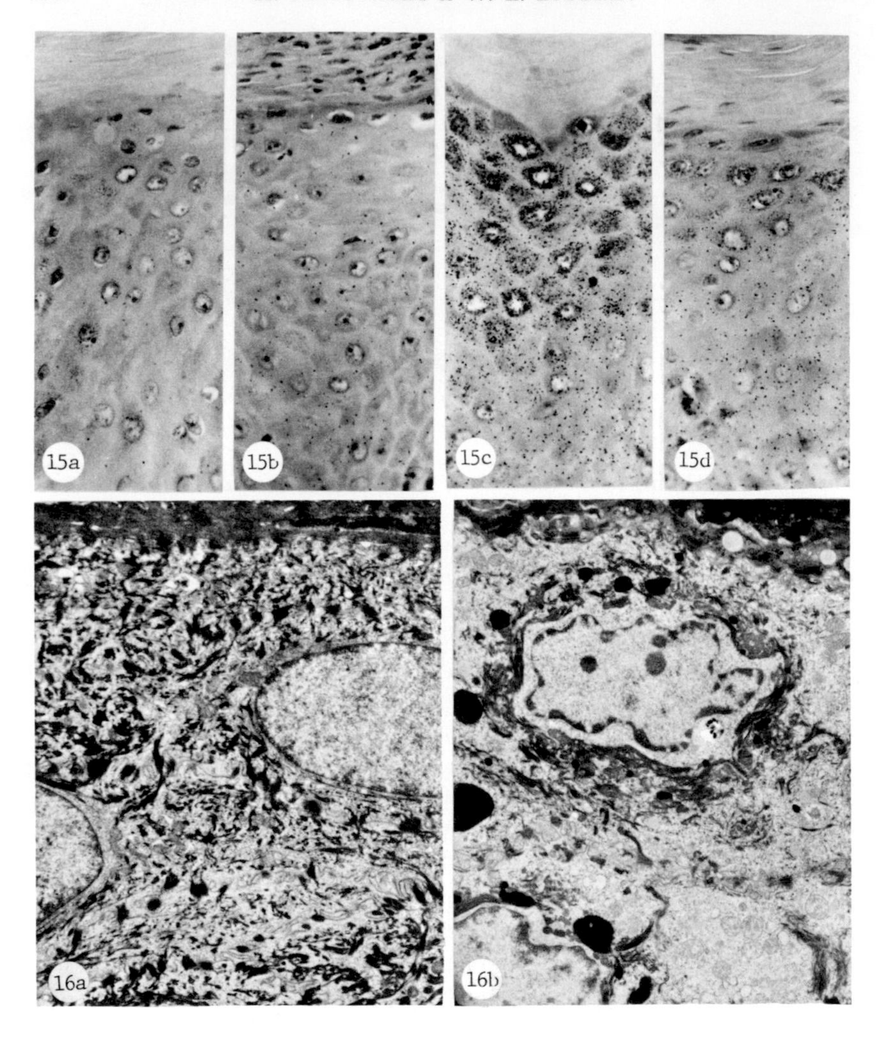

Fig. 15a-d. Light microscopic autoradiography of psoriatic lesion observed in 4 mm punch biopsy at 1 hour after injection of histidine-^{3}H (a and b) shows the areas without KG and (c and d) demonstrate the areas with KG, but incorporation of histidine-^{3}H does not take place in the area (d).

Fig. 16a, b. Skin lesions of patients with ichthyosis vulgaris (a) and (b) epidermolytic hyperkeratosis. $\times$4,000.

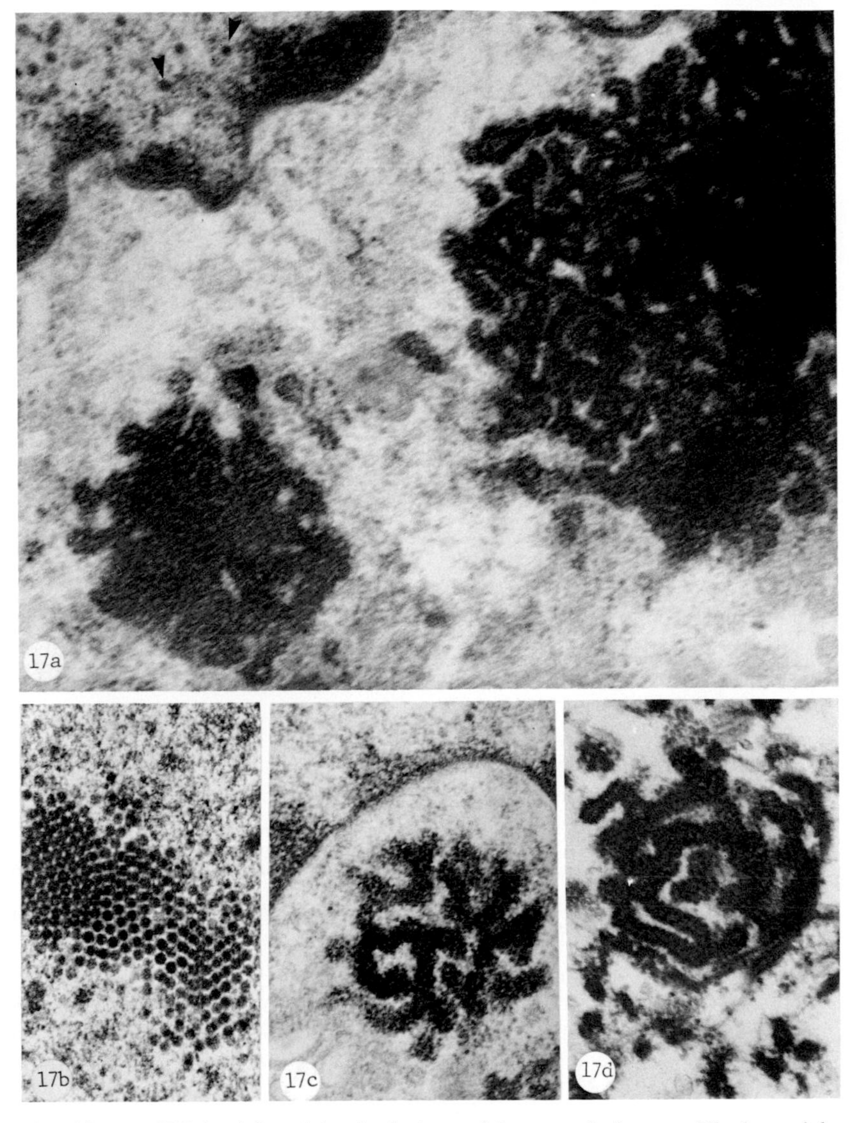

Fig. 17a-c. KG (a-c) found in the lesions of human viral wart. Viral particles (b) and (a ↓) in the nucleus. (a) ×35,700, (b) ×32,600 and (c) ×35,700.

REFERENCES

1. Bonneville, M. A. Observations on epidermal differentiation in the fetal rat. *Am. J. Anat.*, **123**: 147–164, 1968.
2. Farbman, A. I. Morphological variability of keratohyalin. *Anat. Rec.*, **154**: 275–286, 1966.
3. Jessen, H. Two types of keratohyalin granules. *J. Ultrastr. Res.*, **33**: 95–115, 1970.
4. Lavker, R. M. and Matoltsy, A. G. Substructure of keratohyalin granules of the epidermis as revealed by high resolution electron microscopy. *J. Ultrastructure Res.*, **35**: 575–581, 1971.
5. Fukuyama, K., Wier, K. A., and Epstein, W. L. Dense homogenous deposits of keratohyalin granules in newborn rat epidermis. *J. Ultrastr. Res.*, **38**: 16–26, 1972.
6. Fukuyama, K. and Epstein, W. L. Heterogeneous ultrastructure of keratohyalin granules: A comparative study of adjacent skin and mucous membrane. *J. Invest. Dermatol.*, **61**: 94–100, 1973.
7. Fukuyama, K. and Epstein, W. L. A comparative autoradiographic study of keratohyalin granules containing cystine and histidine. *J. Ultrastr. Res.*, **51**: 314–325, 1975.
8. Ugel, A. R. and Idler, W. Further characterization of bovine keratohyalin. *J. Cell Biology*, **52**: 453–464, 1972.
9. Tezuka, T. and Freedberg, I. M. Epidermal structural proteins: I. Isolation and purification of keratohyalin granules of the newborn rat. *Biochimica et Biophysica Acta*, **261**: 402–417, 1972.
10. Sibrack, L. A., Gray, R. H., and Bernstein, I. A. Localization of the histidine-rich protein in keratohyalin: A morphologic and macromolecular marker in epidermal differentiation. *J.Invest. Dermatol.*, **62**: 394–405, 1974.
11. Spearman, R. I. C. and Hardy, J. A. Some ultrastructural observations on keratohyalin granules of guinea pig epidermis. *Arch. Derm. Forsch.*, **250**: 149–158, 1974.
12. Singh, B., McKinney, R. V., and Kolas, S. Histochemistry of the keratohyalin granules in human oral leukoplakia. *J. Oral Pathology*, **4**: 59–66, 1975.
13. Reaven, E. P. and Cox, A. J. Histidine and keratinization. *J. Invest. Dermatol.*, **45**: 422, 1965.
14. Jessen, H. Electron cytochemical demonstration of sulfhydryl groups in keratohyalin granules and in the peripheral envelope of cornified cells. *Histochemie*, **33**: 15–29, 1973.
15. Kamali, M. and Manhouri, H. A modified orcinol reaction for RNA determination. *Clin. Chem.*, **15**: 390–392, 1969.
16. Lowry, O. H., Rosebrough, J. J., Farr, A. L., and Randall, R. J. Protein measurement with the Folin phynol reagent. *J. Biol. Chem.*, **193**: 265–275, 1951.
17. Brody, I. An ultrastructural study on the role of the keratohyalin granules in the keratinization process. *J. Ultrastructure Res.*, **3**: 84–104, 1959.
18. Hoober, J. K. and Bernstein, I. A. Protein synthesis related to epidermal differentiation. *Proc. Natl. Acad. Sci. USA*, **56**: 594–601, 1966.
19. Voorhees, J. J., Chakrabarti, S. G., and Bernstein, I. A. The metabolism of

"histidine-rich" protein in normal and psoriatic keratinization. *J. Invest. Dermatol.,* **51**: 344–354, 1968.

20. Ugel, A. R. Keratohyalin: Extraction and *in vitro* aggregation. *Science,* **166**: 250–251, 1969.
21. Matoltsy, A. G. and Matoltsy, M. N. The amorphous component of keratohyalin granules. *J. Ultrastr. Res.,* **41**: 550–560, 1972.
22. Fukuyama, K. and Epstein, W. L. Heterogenous proteins in keratohyalin granules studied by quantitative autoradiography. *J. Invest. Derm.,* **65**: 113–117, 1975.
23. Ross, R. and Benditt, E. P. Wound healing and collagen formation. V. Quantitative electron microscope radioautographic observations of proline-^{3}H utilization by fibroblasts. *J. Cell Biol.,* **27**: 83–106, 1965.
24. Petrik, P. and Collet, A. J. Quantitative electron microscopic autoradiography of *in vivo* incorporation of ^{3}H-choline, ^{3}H-leucine, ^{3}H-acetate and ^{3}H-galactose in non-ciliated bronchiolar (Clara) cells of mice. *Am. J. Anat.,* **139**: 519–534, 1974.
25. Fukuyama, K. and Epstein, W. L. Synthesis of RNA and protein during epidermal cell differentiation. *Arch. Derm.,* **98**: 75–79, 1968.
26. Fukuyama, K. and Bernstein, I. A. Site of synthesis of ribonucleic acid in mammalian epidermis. *J. Invest. Derm.,* **41**: 47–52, 1963.
27. Fukuyama, K. and Epstein, W. L. Protein synthesis studied by autoradiography in the epidermis of different species. *Am. J. Anat.,* **122**: 269–274, 1968.
28. Fukuyama, K. and Epstein, W. L. Inhibition of RNA and protein synthesis in granular cells by actinomycin-D and puromycin. *J. Invest. Dermatol.,* **56**: 211–222, 1971.
29. Epstein, W. L. and Fukuyama, K. Autoradiographic study of friction blisters. RNA, DNA, and protein synthesis. *Arch. Derm.,* **99**: 94–106, 1969.
30. Stegman, S. J., Bonfilio, N. D., Fukuyama, K., and Epstein, W. L. Effects of phytohemagglutinin and concanavalin- A on mammalian epidermal cells. *Cell Differentiation,* **3**: 71–79, 1974.
31. Suurmond, D., Histologic changes in treated and untreated psoriatic lesions. *Dermatologica,* **131**: 357–366, 1965.
32. Gordon, M., Johnson, W. C., and Burgoon, C. F. Histopathology and histochemistry of psoriasis, II. Dynamics of lesions during treatment. *Arch. Path.,* **84**: 443–450, 1967.
33. Brody, I. Cytoplasmic components in the psoriatic horny layers with special reference to electron-microscopic findings. In The Epidermis (W. Montagna, and W. C. Lobitz, eds.), Academic Press, New York, 1964, pp. 551–572.
34. Fukuyama, K. and Epstein, W. L. Epidermal keratinization: Localization of isotopically labeled amino acids. *J. Invest. Dermatol.,* **47**: 551–560, 1966.
35. Ishibashi, Y. and Klingmüller, G. Erythrodermia ichthyosiformis congenita bullosa Brocq. Über die sogenannte granulöse degeneration. III. Elektronenmikroskopische untersuchung der granularschicht. *Archiv für klinische u. experimentelle Dermatologie,* **233**: 11–32, 1968.
36. Anton-Lamprecht, I. Zur Ultrastruktur hereditärer Verhornungsstörungen. I. Ichthyosis congenita. *Arch. Derm. Forsch.,* **243**: 88–100, 1972.

37. Anton-Lamprecht, I. and Hofbauer, M. Ultrastructural distinction of autosomal dominant ichthyosis vulgaris and X-linked recessive ichthyosis. *Humangenetik,* **15**: 261–264, 1972.
38. Epstein, W. L. and Fukuyama, K. Human wart virus proliferation and its influence on keratinization in epidermal cells. Presented at the Seminar on the Wart Virus, Lyon, France, Dec. 15 and 16, 1975, sponsored by Fondation Mérieux. To be published in the Proceedings of the Seminar on the Wart Virus.

Discussion

Dr. McGuire: How does this compare with the amino acid composition of the keratohyalin oligomers that Dr. Ugel purified 3 or 4 years ago?

Dr. Fukuyama: As far as I can remember, the protein isolated by Dr. Ugel and by Dr. Bernstein's associates was high in histidine and arginine and low in cysteine with the concentration of proline in between, so our data are quite consistent with their findings.

Dr. Baden: I assume that when you show the label moving into the interior this represents molecules originally on the surface of the granules. If that were the case, one might expect that in your early observations, the small keratohyalin granules would show the label in the middle rather than at the edge. When you do this statistical analysis, how do you handle the problem of the small particles as opposed to the large particles or do you only look at the large particles?

Dr. Fukuyama: I look at all the keratohyalin granules, including all sizes present in the granular cells. You are right, the very small ones do have grains in the center. Those are counted as being in the center.

Dr. Baden: It's an important point, otherwise one might think this is not a biosynthetic process but some sort of chemical reaction, particularly with cysteine. You might get some disulfide exchange and end up with something that wasn't really linked into the proteins.

Dr. Fukuyama: That is right. Even at 1 hour, some labeling appears in the center of the small granules.

Dr. Bernstein: I recall that, some years ago, you tested the possibility of indiscriminate addition of cysteine when you treated your tissues with acid.

DR. FUKUYAMA: That is right. We came to the conclusion that we had to do two things. We had to inject puromycin and determine that amino acid incorporation did not occur under circumstances when protein synthesis was inhibited. This was published in the Journal of Investigative Dermatology last year. We also had to ascertain that the amino acid was not linked in a way other than by peptide bond. Exposure to acid did not affect the amount of protein-bound cysteine.

Wound Healing—Some Biochemical Aspects

Akira OHKAWARA* and Takashi AOYAGI**

* *Department of Dermatology, Asahikawa Medical College, Hokkaido, Japan* and** *Department of Dermatology, University of Hokkaido, School of Medicine, Hokkaido, Japan*

ABSTRACT

The following results were obtained through this study:

1. Acanthosis of the epidermis was not prominent throughout the experiments; however, moderate partial acanthosis was noticed around 48, 72, and 96 hours after the wound.
   Increase of mitosis in the germinative cell layer began to be noticed at 24 hours and had reached its peak at 48 hours.
2. Glycogen began to accumulate in the lower layers of epidermis at around 12 hours and its maximum concentration was observed at 48 hours.
3. The imbalance of increased glycogen synthetase, especially in the active form (I-form), and practically no change in phosphorylase activity was responsible for the accumulation of glycogen.
   Decreased percentage of I-form in the total was considered to be due to the increased cyclic AMP level and to increased glycogen itself.
4. Glucose-6-phosphate dehydrogenase and lactate dehydrogenase showed increased activities around 12 hours but isocitrate dehydrogenase did not show any remarkable change during the experiments.
5. Intracellular cyclic AMP level was constant during wound healing except for the sharp rise (40% increase) in concentration at 48 hours after the wound.

INTRODUCTION

In 1962 Lobitz and his associates reported a beautiful piece of work concerning basal cell glycogen and epidermal mitosis after several well-controlled stimuli.[1] According to their report, 1) basal cell glycogen accumulation was most pronounced at 12 hours after the stimuli (regardless of the kinds of stimuli used), 2) highest mitotic counts in the epidermis were observed 48 hours after the alteration of epidermal structure, 3) glycogen accumulation was not related to the mitosis in the epidermis (glycogen was not utilized as an energy source).

At that time, however, glycogen metabolism and the controlling mechanism of cell kinetics in the epidermis were not adequately understood.

Recently, cyclic nucleotides, especially cyclic adenosine monophosphate (cyclic AMP) and cyclic guanosine monophosphate (cyclic GMP), have come to be known to play an important role in cell proliferation and differentiation in the various kinds of tissues[2–5] including epidermis.[6,7]

In addition, glycogen metabolizing enzymes, glycogen synthetase and phosphorylase are known to be regulated through cyclic AMP-sensitive protein kinase in the human epidermis[8,9] as shown in Fig. 1, 2.

In order to study the interrelation among the intracellular cyclic AMP levels, glycogen metabolism, and epidermal proliferation, the epidermal cells which were adjacent to the wound edges were chosen for the experimental materials.

MATERIALS AND METHODS

Materials

Pigs, each weighing approximately 20 kg, were used as experimental animals. The pigs were anesthetized with ketamine and the backs of the pigs were shaved with a motor-driven clipper. Linear wounds of one inch wide and 0.5 mm in depth were made by an electric motor-driven keratome. The specimens, not directly from the advancing repairing edges of the wounded skins, but from approximately 2 mm away from the edges of the wounds, were obtained at various times after the wound with a 7 mm punch (Fig. 3).

Before obtaining the specimens, biopsied areas were frozen with liquid nitrogen and tissues were harvested while frozen. Immediately after their removal, they were cut into halves with a razor blade. Each half was processed for the histochemical and histological studies. The other halves were sectioned 20μ in thickness in a cryostat set at –25°C and vacuum dried at –25°C overnight. The freeze-dried tissues were vacuumed and stored at –25°C until use. The samples of almost pure epidermis were dissected from the freeze-dried tissues at room temperature and they were used for the biochemical studies.

Methods

1) Glycogen Determination: PAS positive material was identified as glycogen by digesting it with diastase. Quantitative assay of glycogen was followed by the method previously described.[10] Glycogen was broken

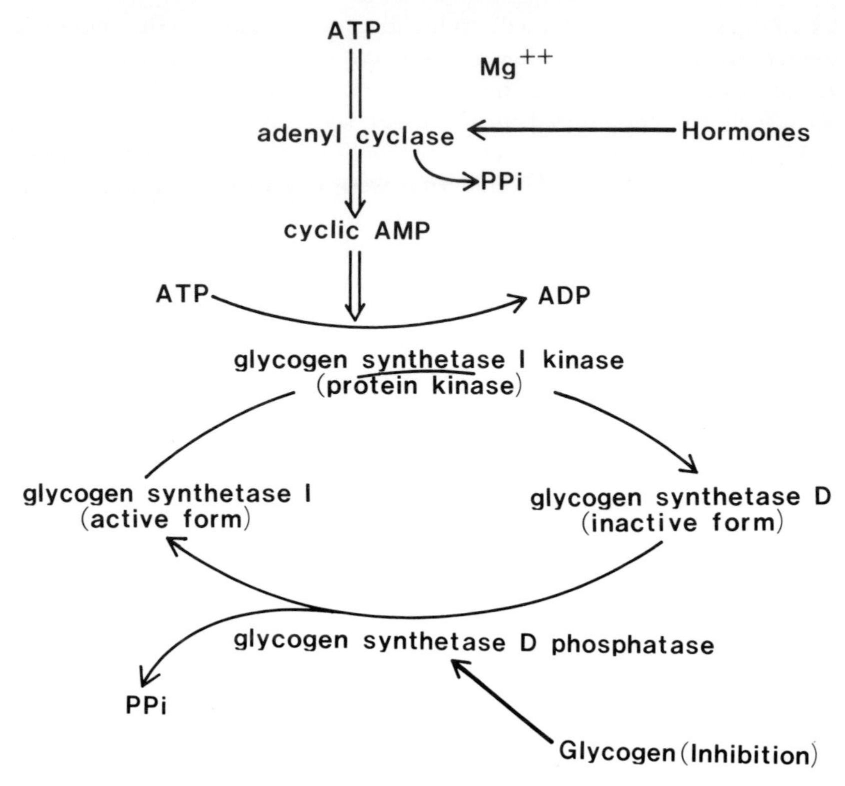

Fig. 1. General metabolic map of glycogen synthetase. Cyclic AMP is involved in the system through protein kinase.

down to glucose by α-1, 6-amyloglucosidase (Boeringer) by incubating the freeze-dried dissected tissues with the enzyme in 0.1M acetate buffer, pH 4.0, overnight at 4°C. Glucose formed was measured by hexokinase (HK) and glucose-6-phosphate dehydrogenase (G6PDH), both obtained from Boeringer by the method described elsewhere.[10]

2) Glycogen Synthetase Activities: Glycogen synthetase activities were assayed by the previousyl described method.[11] The rate of incorporation of ^{14}C uridine diphosphate glucose (^{14}CUDPG) (NEN) into primer glycogen was measured. The total synthetase activities were assayed in the presence of G6P, and the I-form (tissue active form) activities were analysed in the absence of G6P.

3) Phosphorylase Activities and Other Carbohydrate Metabolizing Enzymes: The activities of these enzymes were assayed using the freeze-dried dissected tissues by the method described previously.[12] Phosphorylase-a (tissue active form) was assayed in the absence of exogenous 5′-AMP and

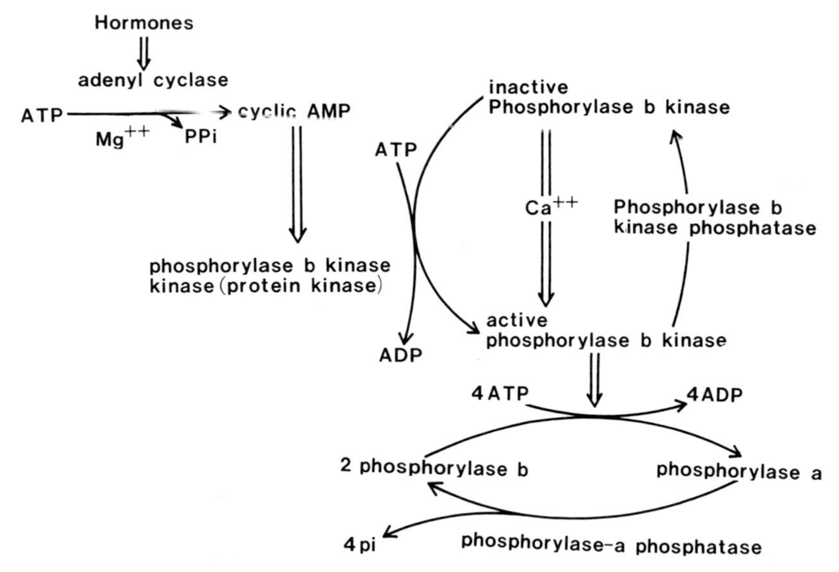

Fig. 2. General metabolic map of phosphorylase. As in glycogen synthetase, cyclic AMP is involved in the system through protein kinase.

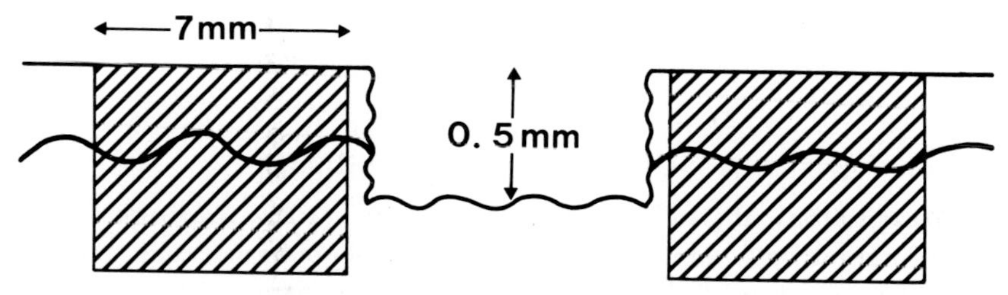

Fig. 3. Schematic illustration of wound formation and biopsy.

the total phosphorylase activities by the addition of 5′-AMP to the reagent mixture. The activities of glucose-6-phosphate dehydrogenase (G6PDH), lactate dehydrogenase (LDH), and NADP-isocitrate dehydrogenase (NADP-ICDH) were assayed essentially by the methods of Lowry *et al.* with a little modification.[13,14]

4) DNA Contents: The analysis of DNA contents followed the fluorometric method described by Santoianni *et al.*[15]

5) Determination of Cyclic AMP Contents: Approximately 100μg of pure epidermal tissues were dissected out of the freeze-dried tissues under the dissection microscope at room temperature. The extraction of cyclic AMP with 0.1N HCl was carried out according to Yoshikawa and Adachi.[16]

30 μl of 0.1N HCl was added to small test tubes which contained the dissected materials. They were boiled for 5 minutes, cooled, and centrifuged in a refrigerated centrifuge at 5,000 rpm for 20 minutes. The supernatants obtained were saved, and to the precipitates 20μl of 0.1N HCl was again added. They were boiled again for 5 minutes, cooled, and spun as before. The supernatants were collected and added to the previous respective supernatants. All the supernatants were lyophylized. The lyophylizates were dissolved in 100μl of distilled H_2O.

The measurement of cyclic AMP in those dissolved samples was carried out with a radioimmunoassay technique by the courtesy of Professor Ui and Dr. Homma, from the Department of Chemical Pharmacology, Faculty of Pharmaceutical Science, Hokkaido University. The sensitivity of the method is quite satisfactory and it is possible to measure as little as 10 fmole of cyclic AMP.

6) Mitotic Index: Mitotic index was determined by counting the mitotic cells in 1,000 germinative cells using H&E slides.

RESULTS AND DISCUSSIONS

1. Morphological (H&E) and histochemical (PAS) findings

Normal pig back skin consists of 4–5 layers of epidermal cells (Fig. 4). Complete absence of the epidermis in the wounded areas was confirmed by biopsying the wound (Fig. 5). Although acanthosis of the epidermis was not prominent throughout the experiments, mild acanthosis began to be noticed at 24 hours, and it became most pronounced at 48 hours after the surgery (Fig. 6). The mild acanthosis remained until 5 days after the wound.

PAS positive diastase digestable material (glycogen) was barely detectable before the wound, but it began to accumulate in the cytoplasma of the basal cells and the lower layers of prickle cells at 12 hours (Fig. 7) and its accumulation had reached its height at 48 hours (Fig. 8). Glycogen

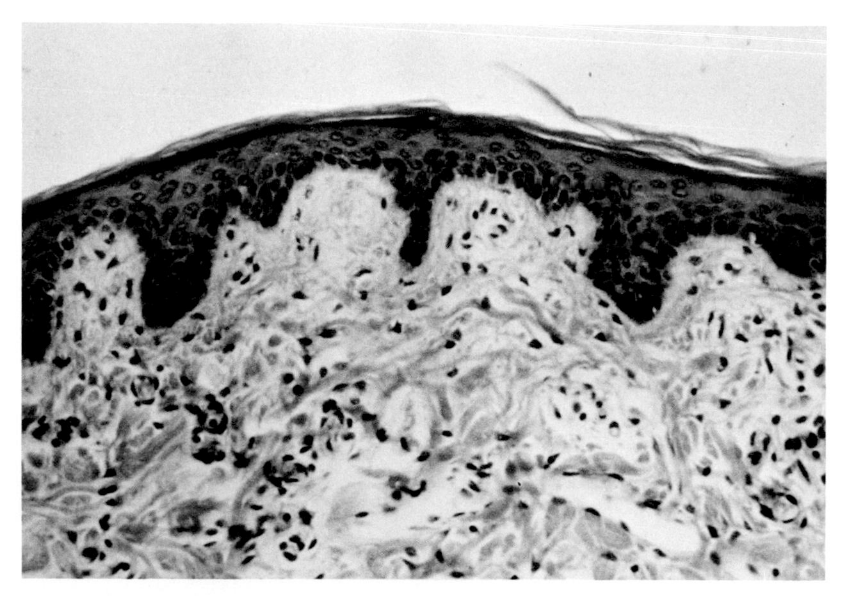

Fig. 4. Normal pig back skin (H&E). Normal pig skin consists of 4–5 layers of epidermis.

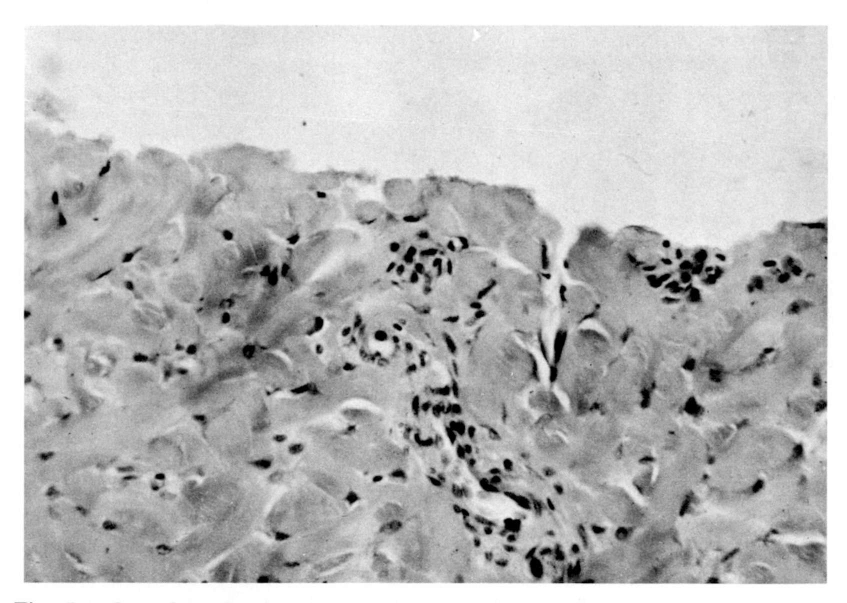

Fig. 5. Complete absence of epidermis was confirmed (H&E) in the wound.

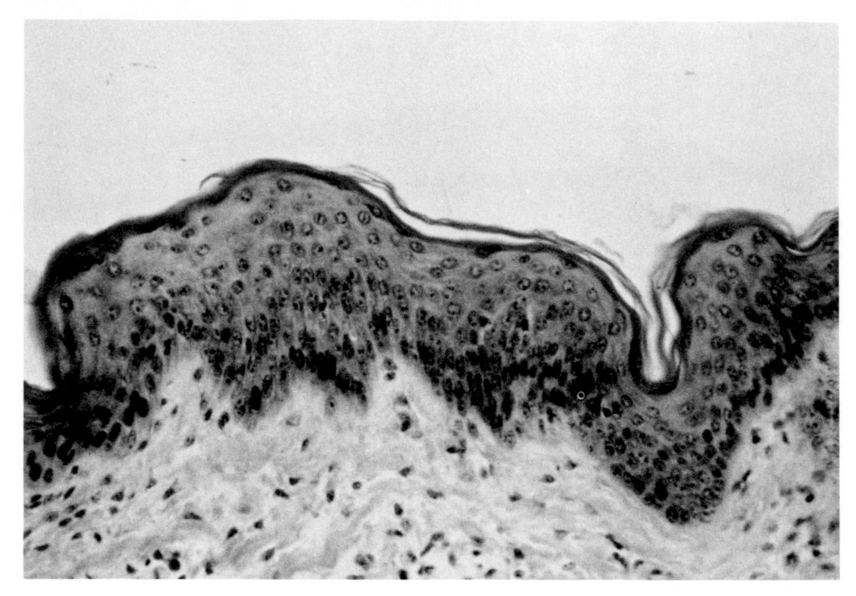

Fig. 6. Mild acanthotic change was observed around 48 hours after the wound (H&E).

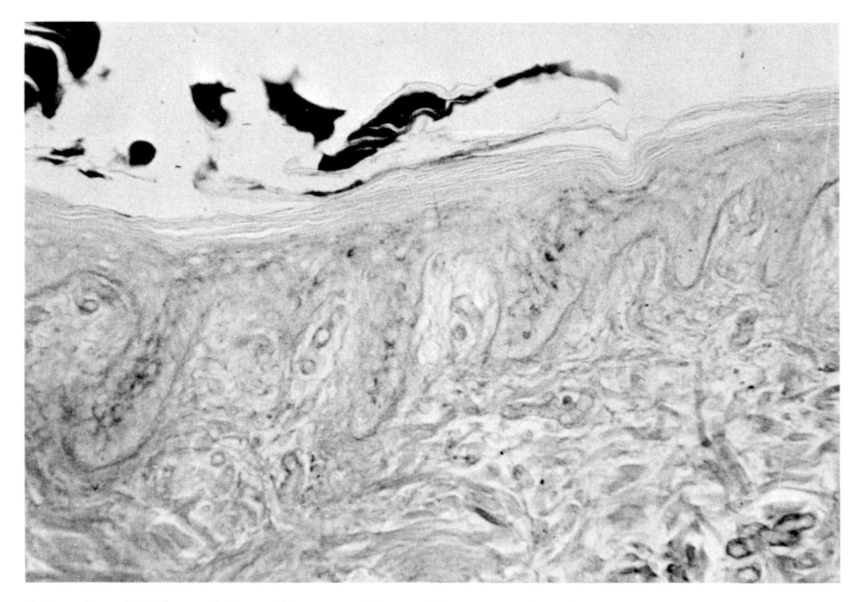

Fig. 7. PAS positive diastase digestable material began to be noticed in the lower portions of epidermis at 12 hours.

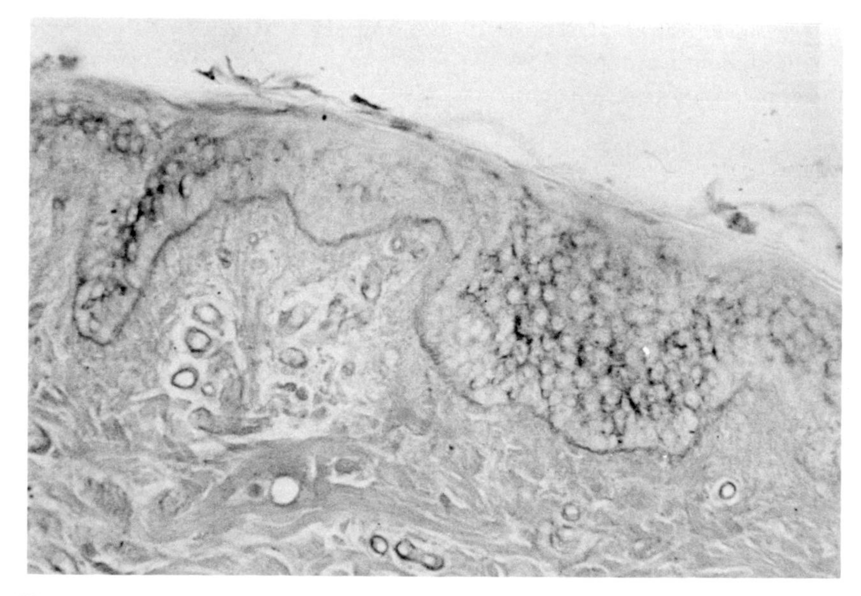

Fig. 8. Most pronounced accumulation of PAS positive material was noticed at 48 hours after the wound.

accumulation had returned to almost normal level by 120 hours. Its accumulation seemed to be rather spotty and it tended to localize in the acanthotic epidermis, but those sites were not related with the epidermal appendages such as hair follicles and eccrine sweat ducts. The results obtained through PAS stainings were well correlated with biochemical glycogen (Table 1).

2. Biochemical glycogen

The results are shown in Table 1. There was a definite peak of gly-

Table 1. Glycogen contents.
Glycogen was measured as described in the text and the results are summarized in the table. Biochemical glycogen exhibited a good correlation with the results obtained by PAS staining.

		PAS
Pre-O	60	(±)
Post-24hr	90	(++)
48hr	105	(+++)
72hr	88	(++)
96hr	65	(±)
120hr	70	(±)

Glycogen concentration is expressed as mg/100g dry weight

cogen concentration at 48 hours. Since, as previously mentioned glycogen, accumulations were spotty, if only those particular areas were compared, the changes in concentration would have been magnified.

3. Glycogen synthetase activities

The changes in glycogen synthetase activities at various times after the wound are summarized in Table 2. The highest activities of both total and active (I-form) forms of the enzyme were noticed at 24 hours, which preceded the maximum glycogen concentration. The most interesting change, however, was observed at 48 hours, in which, though the total activity remained still high as compared to that of the normal epidermis, the tissue active I-form returned to completely normal. This normalized I-form activity indicates that no further abnormal building of glycogen was occurring in the cells. As to what keeps the I-form activity low in those cells is discussed later in this text. The total activities remained slightly higher until 120 hours, but the I-form showed no higher activities after 48 hours.

Table 2. Glycogen synthetase activities
Both total and active forms of the glycogen synthetase activities were assayed as described in the text. Both forms showed the highest activities at 24 hours, which preceded the highest glycogen accumulation. Note diminished active form (I-form) activity at 48 hours.

	Total Activity	I–Form Activity	% I–Form
Pre-O	43.1 ± 14.6(6)	10.1 ± 3.0(7)	23.4
Post-1/2hr.	45.1 ± 3.0(2)	8.4 ± 4.7(2)	18.6
3hr.	46.9	15.0	32.0
4hr.	41.1	18.0	43.8
10hr.	48.7 ± 15.7(5)	12.4 ± 2.6(4)	25.5
12hr.	106.9 ± 18.3(4)	28·1 ± 13.8(5)	26.3
24hr.	225.0 ± 20.6(3)	41.3 ± 12.0(4)	18.4
48hr.	69.4 ± 22.4(2)	9.4 ± 0.2(2)	13.5
72hr.	86.3 ± 35.5(3)	9.8 ± 2.8(3)	11.4
96hr.	72.5 (2)	15.3 (2)	21.1
120hr.	62.8 (2)	10.9 (2)	17.4

Activities: pm of UDPG incorporated into glycogen per min. per mg wet weight at 30°C
( ) denotes number of sample assayed.

4. Phosphorylase activities

Phosphorylase activities were assayed as described in the text and the results are shown in Table 3. Unlike the situations of glycogen-building enzyme (glycogen synthetase), phosphorylase did not exhibit any remarkable change in activities until 120 hours. A delayed increase in phosphorylase activities in the proliferating epidermis of the wound was reported

Table 3. Phosphorylase activities
Phosphorylase activities were assayed as described in the text. Phosphorylase activities did not show remarkable change throughout the experiments.

	Total Activities	a-Form	% a-Form
Pre Wound	0.034	0.018	53
Post 10hrs	0.028	0.022	78
24hrs	0.034	0.015	44
48hrs	0.024	0.018	75
72hrs	0.023	0.014	61
96hrs	0.020	0.013	65
120hrs	0.020	0.013	65

Units: Mole per kg dry weight per hour at 37°C

by Im *et al.*[17] The fact that phosphorylase was not participating in the earlier biological changes needed for repairing the wound indicates that at least epidermal glycogen was not utilized as an energy source for wound healing.

5. G6PDH, LDH, and NADP-ICDH activities

The results are summarized in Table 4. An earlier increase in G6PDH and LDH activities observed up to 24 hours was in good agreement with the previous report.[17] The early increases of those enzymes indicate that they were involved in supplying energy and preparing ribose for nucleic acid synthesis in the epidermal cell proliferation. G6PDH tended to remain high in activities in the longer periods. Even decreased activity of NADP-ICDH during the process of wound healing was reported.[17]

6. Cyclic AMP concentration

The changes in the levels of cyclic AMP are shown in Table 5. Cyclic AMP levels were rather constant throughout the experiments except

Table 4. Dehydrogenase activities
Dehydrogenase activities were measured as described in the text. Those activities showed an earlier increase as compared to those of glycogen synthetase and phosphorylase.

	G6PDH	LDH	NADP-ICDH
Pre Wound	0.81	12.49	0.89
Post 8hrs	1.27	15.23	1.08
10hrs	1.24	20.67	1.01
24hrs	1.18	19.07	0.82
48hrs	1.38	15.53	0.94
72hrs	1.14	14.05	0.99

Units: Mole per kg dry weight per hour at 37°C.

Table 5. Cyclic AMP contents
The extraction of cyclic AMP from the freeze-dried tissues and its measurement were followed as described in the text. There was a definite peak in the concentration at 48 hours after the wound.

Pre Wound		0.48
Post	4hrs	0.43 (−10%)
	8hrs	0.40 (−17%)
	24hrs	0.46 (− 4%)
	48hrs	0.67 (+40%)
	72hrs	0.43 (−10%)
	96hrs	0.50 (+ 4%)
	120hrs	0.47 (− 2%)

Units: pm of cyclic AMP per mg wet weight

for a definite increase at 48 hours and a possible decrease around 8 hours. At 48 hours approximately 40% increase in concentration was observed. How is this increase in the level of cyclic AMP at 48 hours related to glycogen metabolism and epidermal proliferation? As has been shown, cyclic AMP regulates epidermal proliferation possibly through epidermal chalone by inhibiting either G_1 or G_2 phases of the cell cycle.[18–21] Pleiotypic action of cyclic AMP and cyclic GMP has also been suggested.[6,7]

At 48 hours, when the highest cyclic AMP concentration was observed, percent of tissue active form of the synthetase (I-form) was much lower (13%) than the normal state (23%), indicating that cyclic AMP was acting through protein kinase in a direction to lessen the amounts of I-form of the enzyme. High glycogen contents at this stage (Table 1) seemed to be also responsible for the low percentage of I-form since glycogen per se is known to inhibit glycogen synthetase D phosphatase which converts D-form to I-form (Fig. 1).

Mitotic counts were also highest around 48 hours as shown later in the text. Although G_1 inhibition can be measured by the decrease in thymidine uptake and G_2 inhibition by the decreased mitotic counts, it is not possible to conclude through this study in which phase of the cell cycle cyclic AMP is involved until we know the exact length of the each phase of the cell cycle in pig epidermis and participation of other cell growth controlling factors, especially cyclic GMP.

7. DNA contents

The amounts of DNA expressed as dry weight are summarised in Table 6. A slight decrease in the concentration was noticed at 96 hours. The decrease of DNA contents at 96 hours may indicate that cells became slightly larger at this stage.

Table 6. DNA contents
DNA contents were measured as described in the text and are summarized in the table. There was a slight decrease in concentration around 96 hours after the wound.

Pre Wound		11.8
Post	10hrs	11.2 (− 5%)
	24hrs	11.3 (− 4%)
	48hrs	13.4 (+14%)
	72 hrs	13.0 (+10%)
	96 hrs	8.8 (−25%)
	120 hrs	12.5 (+ 6%)

Units: μg of DNA per mg dry weight

8. The highest mitotic counts were obtained at 48 hours after the wound. The mitotic counts showed slightly higher values until 120 hours.

REFERENCES

1. Lobitz, W. C., Jr., Brophy, D., Larner, S. E., and Daniels, F., Jr. Glycogen response in human epidermal basal cell. *Arch. Derm.*, **86**: 207–211, 1962.
2. Zeilig, C. E., Johnson, R. A., Friedman, D. F., Kumar, K., and Sutherland, E. W. Regulatory influences of cyclic AMP in synchronized HeLa cells (abstr). Advances in Cyclic Nucleotide Research, Vol 5, (G. I. Drummond, P. Greengard, and G. A. Robison, eds.) New York, Raven, 1975, p. 832.
3. Coffino, P., Gray J., and Tomkins, G. M. Dibutyryl cAMP synchronizes S49 mouse lymphoma cells in G_1 (abstr). Advances in Cyclic Nucleotide Research, Vol 5, (G. I. Drummond, P. Greengard, and G. A. Robison, eds.) New York, Raven, 1975, p. 831.
4. Teel, R. W. and Hall, R. G. Effect of dibutyryl cyclic AMP on the restoration of contact inhibition in tumor cells and its relationship to cell density and the cell cycle. *Exp. Cell Res.*, **76**: 390–394, 1974.
5. Froehlich, J. E. and Rachmeler, M. Inhibition of cell growth in the G_1 phase by adenosine 3′,5′-cyclic monophosphate. *J. Cell Biol.*, **60**: 249–257, 1974.
6. Voorhees, J. J., Duell, E. A., Stawiski, M., and Harrell, E. R. Cyclic nucleotide metabolism in normal and proliferating epidermis. Advances in Cyclic Nucleotide Research, Vol 4, (P. Greengard and G. A. Robison, eds.) New York, Raven, 1974, pp. 117–162.
7. Voorhees, J. J. and Duell, E. A. Imbalanced cyclic AMP-cyclic GMP levels in psoriasis. Advances in Cyclic Nucleotide Research, Vol 5, (G. I. Drummond, P. Greengard, and G. A. Robison, eds.) New York, Raven, 1975, pp. 735–758.
8. Ohkawara, A., Halprin, K. M., and Levine, V. Glycogen synthetase I and D interconversion in the normal human epidermis. *J. Invest. Dermatol.*, **59**: 182–186, 1972.

9. Ohkawara, A., Halprin, K. M., and Levine, V. Phosphorylase-b kinase in the human epidermis. In preparation.
10. Ohkawara, A., Halprin, K. M., and Levine, V. Glycogen metabolism following ultraviolet irradiation. *J. Invest. Dermatol.*, **59**: 264–268, 1972.
11. Ohkawara, A., Halprin, K. M., and Levine, V. Glycogen synthetase of human epidermis. *J. Invest. Dermatol.*, **59**: 177–181, 1972.
12. Ohkawara, A., Halprin, K. M., and Levine, V. The enzymes of glycogen metabolism in the various cell fractions from the normal human epidermis. *J. Invest. Dermatol.*, **59**: 187–191, 1972.
13. Hershey, F. B., Lewis, C., Murphy, J., and Schiff, T. Quantitative histochemistry of human skin. *J. Histochem. Cytochem.*, **8**: 41–49, 1960.
14. Cruickshank, C. N. D., Hershey, F. B., and Lewis, C. Isocitrate dehydrogenase activity of human epidermis. *J. Invest. Dermatol.*, **30**: 33–37, 1958.
15. Santoianni, P. and Ayala, M. Fluorometric ultramicroanalysis of deoxyribonucleic acid in human skin. *J. Invest. Dermatol.*, **45**: 99–103, 1965.
16. Yoshikawa, K. and Adachi, K. Personal communication.
17. Im, M. J. C. and Hoopes, J. E. Enzyme activities in the repairing epithelium during wound healing. *J. Surgical Res.*, **10**: 173–179, 1970.
18. Voorhees, J. J., Duell, E. A., and Kelsey, W. H. Dibutyryl cyclic AMP inhibition of epidermal cell division. *Arch. Dermatol.*, **105**: 384–386, 1972.
19. Marks, F. and Rebien, W. Cyclic 3′,5′-AMP and theophylline inhibit epidermal mitosis in G_2-phase. *Naturwissenschaften*, **59**: 41–42, 1972.
20. Harper, R. A., Flaxman, B. A., and Chopra, D. P. Mitotic response of normal and psoriatic keratinocytes *in vitro* to compounds known to affect intracellular cyclic AMP. *J. Invest. Dermatol.*, **62**: 384–387, 1974.
21. Delescluse, C., Colburn, N. H., Duell, E. A., and Voorhees, J. J. Cyclic AMP elevating agents inhibit proliferation of keratinizing guinea pig epidermal cells. *Cell Differentiation*, **2**: 343–350, 1974.
22. Elgjo, K. Epidermal chalone and cyclic AMP: an in vivo study. *J. Invest. Dermatol.*, **64**: 14–18, 1975.

Discussion

DR. ODLAND: Why did you do biopsies at the margin distal to the wound?

DR. OHKAWARA: We had no particular reason. As the next step in our work, we are now planning to study the advancing edge.

DR. GOLDSMITH: I have never understood what all the glycogen is doing there. Is it just a secondary phenomenon or is it in some way intrinsic to the process of wound healing? Are there experimental ways of preventing that accumulation of glycogen to see whether there is a resulting affect on the wound healing process?

DR. OHKAWARA: The answer to the second question is. I do not think there is a way. In answer to your first question, I believe it is probably a secondary phenomenon.

DR. ADACHI: Basically, I agree with Dr. Ohkawara, I think as far as glycogen in skin is concerned, there is no direct evidence which suggests that the glycogen is important to biological function. The only exception may be in the case of sweat secretion. Otherwise, I can't see any direct evidence.

DR. KARASEK: Some direct evidence that glycogen is not important comes from use of inhibitors reported in the Journal of Biological Chemistry awhile back. I don't remember the reference exactly. Those workers were studying human organ culture models of wound healing in which they measured the amount of time it took for the epidermis to surround an explant and, at least under their conditions, when inhibitors of glycogen were present and the accumulation of glycogen was prevented, they saw absolutely no effect on the rate at which these experimental wounds re-epithelialized so their data were consistent with glycogen not being involved.

DR. OHKAWARA: I agree that there is no direct evidence that the glycogen which does accumulate under certain circumstances is used as an energy source except maybe in the sweat gland, so it would be a secondary phenomenon.

DR. ADACHI: If I may add one thing, I think generally there is plenty of glucose so why do we need glycogen? We use energy to make glycogen; so as long as glucose is present, glycogen is not needed.

DR. BERNSTEIN: Apropos of that last comment of Dr. Adachi, a number of times here in the seminar, people have raised the issue of an enzyme with no function. This idea doesn't surprise me. I don't see why all enzymes that are present have to be currently in use.

DR. MCGUIRE: We often assume that if the body does something, it had something brilliant in mind. In fact, there are a number of examples where there is no parsimony of nature. Many responses of the body are not necessarily physiologic and I think the accumulation of glycogen in this situation is probably one of those examples. The accumulation is probably in response to a cue which may have some benefit in a different animal or a different cell. I don't think we necessarily have to believe that every kind of response like this is physiologic or directed toward the benefit of the organism.

Epidermal Cyclic AMP System and Its Possible Role in Proliferation and Differentiation

Kenji Adachi

Department of Dermatology, University of Miami School of Medicine, Miami, Florida, U.S.A.

SUMMARY

Many studies during the past few years have suggested that cyclic AMP is specifically involved in the regulation of proliferation and differentiation of various types of mammalian cells *in vitro*. If so, the compounds which regulate the adenylate cyclase system may have important effects on cell function. Despite its potential significance, the basic system which controls intracellular cyclic AMP in skin is not yet fully understood.

Thus the main purpose of our studies was to investigate various stimulators of the epidermal adenylate cyclase system. Among the many biogenic amines, peptide hormones, and other drugs tested, β-adrenergic agonists, histamine, prostaglandin (E series), and adenosine (and adenine nucleotides) were found to stimulate the epidermal adenylate cyclase system to cause a rapid accumulation of intracellular cyclic AMP. This increase in cyclic AMP is transient; further addition of the same stimulator will not increase it (state of the refractoriness), but the addition of other stimulators increases it in additive fashion. These results indicate that the epidermis has four distinctly independent adenylate cyclase systems.

This paper reviews the pertinent literature on cyclic AMP and epidermal cell proliferation and differentiation. One of the *in vitro* studies (Flaxman and Harper, *J. Invest. Derm.*, **65**: 52–59, 1975), using an outgrowth cell culture system of human skin, reported that epinephrine, histamine, and adenine nucleotides inhibited mitosis. Our results, which showed that epidermal adenylate cyclase was markedly stimulated by these drugs and chemicals, concur with these *in vitro* data and support the hypothesis that in an *in vitro* system, the transient rise of cyclic AMP in epidermal cells can be associated with the inhibition of proliferation.

Despite the seemingly important role of cyclic AMP *in vitro*, many studies using *in vivo* systems report contradictory results, that is, that a high cyclic AMP level is associated with cell proliferation. These discrepancies must be resolved before the biological roles of cyclic nucleotides in skin can be fully understood.

INTRODUCTION

The biological and clinical significance of the cyclic nucleotides (cyclic AMP and cyclic GMP) which act as regulating signals in the processes of secretion, differentiation, membrane transport, locomotion, protein and ribonucleic acid syntheses, immunity, and mitosis (and/or cancer) becomes increasingly obvious. This paper will discuss two aspects of this role: (1) What are the basic characteristics of cyclic AMP in the epidermis? (2) What is its biological significance in skin? During the past three years, our laboratory has concentrated on finding answers to the first question. Although this project is still in progress, we are presenting the data we have in the first part of this article. So far, the limited studies of the biological significance of the cyclic AMP system in skin have been unsuccessful. Data from other laboratories, however, strongly suggest that this system regulates both proliferation and differentiation in the epidermis. Some of these data are contradictory, particularly those which result from *in vivo* vs. *in vitro* studies.

Research in the role of cyclic AMP, particularly its role in skin, began only five years ago and hence is still in the trial-and-error stage. The details of how it functions remain to be elucidated. Nevertheless, it appears that cyclic AMP plays a key role in both healthy and pathological skin conditions just as it does in other organs and tissues.

EPIDERMAL CYCLIC AMP SYSTEM

A. β-Adrenergic receptor-adenylate cyclase system

Various peptide hormones and biogenic amines act on specific receptor sites on the cell membrane to activate the adenylate cyclase system (Fig. 1). This causes an intracellular accumulation of cyclic AMP, which, as the second messenger in the cell, regulates its metabolism to produce the specific response required by the original messenger at the cell surface.[1] The degradation of the accumulated cyclic AMP by specific phosphodiesterase which follows is inhibited by theophylline, papaverine, or their related compounds (Fig. 1).

Since 1970, several investigators have tested the effects of catecholamine on adenylate cyclase in the skin. As Table 1 indicates, adenylate cyclase was stimulated by catecholamine in various degrees: the hormone responsiveness was lost after homogenization and the purified membrane fractions responded only poorly to stimulation, whereas in *in vivo* experiments, marked stimulation was observed (the skin was not removed before the catecholamine activation). Our data[13] in an *in vitro* system show a remarkable stimulation by both epinephrine and isoproterenol, which may have been due, at least in part, to our experimental system.[12,13] Since

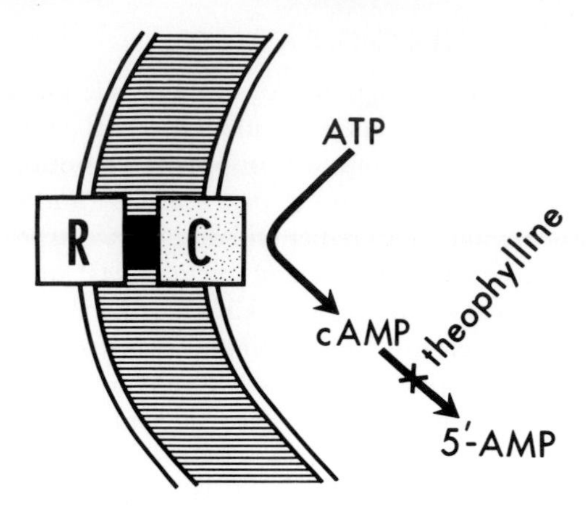

Fig. 1. A schematic model for a receptor-adenylate cyclase system in cell membrane. R = receptor, C = catalytic site of adenylate cyclase.

practically the same technique was used in later experiments to test the effects of prostaglandins,[14] histamine,[15] and adenosine,[16] I shall summarize the procedures here.

Samples of human and pig epidermis were obtained by a keratome without extensive "pre-washing" of the skin surface and without anesthesia. Both pig and human epidermis consists of several layers of epidermal cells, unlike rodent epidermis which consists of less than two layers and can easily be contaminated by dermis of various degrees. The epidermis was cut into 5 × 5 mm and incubated for 20 min at 37°C in Hank's salt solution to standardize the endogenous levels of cyclic AMP and adenosine 5'-triphosphate (ATP). After preincubation, two pieces of epidermis were transferred to Hank's salt solution to which were added experimental drugs and/or hormones. The tissues were then incubated for 5 min at 37°C, frozen, and kept at −60°C until the cyclic AMP assay, which was based on the protein binding method of Gilman[17] with a modified extraction procedure.[12,18]

Our procedures on a semimicro scale have yielded consistently reproducible results. The effects of various adrenergic agonists and antagonists are summarized in Fig. 2. The fact that epinephrine markedly increases the cyclic AMP content in pig epidermis indicates that epinephrine is bound to the receptor site of epidermal adenylate cyclase, which is then activated to produce more cyclic AMP from ATP. Obviously, the addition of theophylline together with epinephrine increases the accumulation of cyclic AMP, whereas the addition of theophylline alone has little effect. Isoproterenol is as effective as epinephrine in inducing cyclic AMP accumulation, while norepinephrine is less effective. Propranolol, the β-

Table 1. Catecholamine stimulation of epidermal adenyl cyclase*.

References	Samples	B-Agonist	Stimulation	Remarks
Mier & Urselman[2] 1970	guinea pig whole skin	epinephrine	0%	*in vitro* homogenate
Bronstad *et al.*[3] 1971	hamster whole skin	epinephrine	1,500%	*in vitro* ^{14}C-ATP pre-labelled
Powell *et al.*[4] 1971	rat epidermis	isoproterenol	300%	*in vitro* slice
Duell *et al.*[5] 1971	newborn rat whole skin	isoproterenol	40%	particulate (17,000 × *g*)
Marks & Rabien[6] 1972	mouse whole skin	epinephrine isoproterenol	20 to 50%	particulate (1,200 × *g*)
Marks & Grimm[7] 1972	mouse whole skin	isoproterenol epinephrine norepinephrine	1,000% 300% 150%	*in vivo* experiment
Hsia *et al.*[8] 1972	human epidermis	epinephrine	40%	particulate (650 × *g*)
Wright *et al.*[9] 1973	human epidermis	epinephrine	300%	*in vitro* slice ^{3}H-ATP pre-labelled
Voorhees *et al.*[10] 1973	human epidermis	isoproterenol	100%	*in vitro* slice no theophylline
Mui *et al.*[11] 1975	human epidermis	epinephrine	1,600%	*in vitro* slice ^{3}H-ATP pre-labelled
Yoshikawa *et al.*[12,13] 1975	human & pig epidermis	epinephrine isoproterenol norepinephrine	2,000% 500%	*in vitro* slice

*From Yoshikawa *et al.*[12]

adrenergic blocker, markedly inhibits the stimulatory effect of epinephrine, but phentolamine, the α-adrenergic blocker, produces no effect. Essentially the same data were obtained with human epidermis.[13] Thus the order of the stimulatory effects by the agonists (isoproterenol $\geqq$ epinephrine $\gg$ norepinephrine), as well as the effect of the blocking agents, indicates that both human and pig epidermis possesses a β-adrenergic receptor-adenylate cyclase system.

B. Prostaglandin (E) receptor-adenylate cyclase system

Despite the early discovery of prostaglandins,[19,20] their relation to the cyclic AMP system has only recently been suggested. A few studies indicate that prostaglandin E increases the cyclic AMP level in epidermis. In 1973, Voorhees *et al.*[10] reported that prostaglandin E_2 caused a transient increase in cyclic AMP whereas prostaglandin E_1 produced no effect. That same

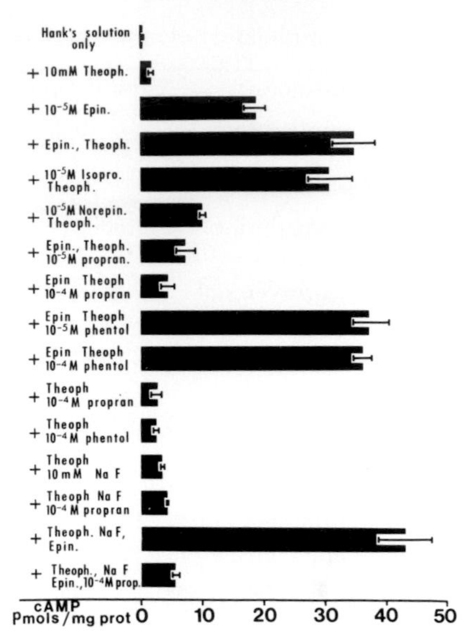

Fig. 2. The effects of adrenergic agents, α and β-adrenergic blocking agents, theophylline and NaF on the cyclic AMP accumulation in pig skin. Data are expressed as the average and individual results in two different series of experiments as indicated by small bars. Two pieces of skin were used for each measurement. (Yoshikawa *et al.*[13])

year, Hsia *et al.*[21] reported that prostaglandin E_2 stimulated epidermal adenylate cyclase. Except for these two preliminary reports, only two other papers have described in some detail the relationship between prostaglandins and cyclic AMP: one by Aso *et al.*,[22] the other by our group.[14] According to these data, which generally agree, not only prostaglandin E_2 but also E_1 activates epidermal adenylate cyclase; prostaglandin $F_{2\alpha}$ is totally ineffective. The main purpose of their study was to determine the difference in prostaglandin response by normal and psoriatic epidermis; ours was to study the basic nature and characteristics of prostaglandin action in skin.

Because homogenized epidermis can no longer respond to prostaglandins,[6] we expected a fragile membrane-bound receptor-adenylate cyclase system to occur. Therefore, we incubated human or pig epidermis with various prostaglandins to test their effectiveness in increasing intracellular levels of cyclic AMP. The prostaglandins were dissolved in 99% ethanol before each experiment, but particular care was taken that the final concentration of ethanol did not exceed 0.1% (Yoshikawa *et al.*[23] had found that short-chain alcohols activate epidermal adenylate cyclase).

Moreover, the incubation period was limited to 5 min since in our system the cyclic AMP level decreases very rapidly after 5 min.

The results of our experiments on stimulatory effects on prostaglandin E_1 and E_2 are shown in Table 2. These data suggest that pig epidermis has a prostaglandin E receptor-adenylate cyclase system. What then is the relation between this receptor and the epinephrine receptor system? Evidence that the two adenylate cyclase systems are different is found in Table 3. If the two sites are different, the sum of the increase by prostaglandin and epinephrine has to equal the cyclic AMP level that is increased by the simultaneous addition of prostaglandin and epinephrine (the concentration of both chemicals should be at their level of saturation). Table 3 demonstrates that the effects of prostaglandin and epine-

Table 2. Effects of prostaglandins on cyclic AMP in pig epidermis.

PG		cAMP picomoles per mg protein	X increase of control
None			
	0 min	1.22	——
	5 min	3.09	X1
PGA_1		3.60	NS
PGA_2		3.88	NS
PGE_1		5.96	X2.5
PGE_2		7.42	X3.2
$PGF_{2\alpha}$		3.89	NS

Final concentration of prostaglandins was 3×10^{-6}M each.

Table 3. The effects of prostaglandin E_2 (PGE_2), epinephrine and β-blocker (propranolol) on cyclic AMP formation.

Additions		cAMP picomoles/mg protein			
		Exp. I	Exp. II	Exp. III	Exp. IV
1. None	0 min	1.1	3.1	2.7	4.3
2. (Control)	5 min	1.6	2.7	3.0	6.5
3. PGE_2		5.4	7.7	5.3	15.2
4. $PGE_2 + \beta$		7.8	7.0	4.9	12.6
5. Epi		9.0	27.6	24.5	33.1
6. Epi + β		0.5	2.7	3.4	10.0
7. PGE_2 + Epi		18.2	30.6	—	54.9
8. PGE_2 + Epi + β		5.7	3.9	7.8	15.6

Concentrations of epinephrine was 2.3×10^{-5} in Exp. I and 4.5×10^{-6}M in others, those of PGE_2 3.3×10^{-5}M in Case 1 and 3×10^{-6}M in others and those of propranolol 10^{-3}M in Case I and 10^{-4}M in others.

phrine are approximately additive; the results suggest an independent receptor-adenylate cyclase for both drugs. Table 3 further demonstrates that adding a β-blocker to the [prostaglandin + epinephrine] blocked the epinephrine effect and left only the prostaglandin effect. Adding an α-blocker, such as priscoline, had no effect on stimulation by either prostaglandin or epinephrine.

C. Histamine receptor-adenylate cyclase system

Although histamine activates adenylate cyclase in guinea pig brain and causes cyclic AMP to accumulate, Voorhees *et al.*[24] reported no such stimulation by histamine in mouse skin. In pig (and human) epidermis, on the contrary, we observed marked activation of the adenylate cyclase system by histamine and counteractivation after the addition of antihistamines.[15] Some of our recent data on this histamine receptor-adenylate cyclase system in the epidermis are summarized here.

The changes in the intracellular content of cyclic AMP in response to histamine are shown in Fig. 3. The concentration increases rapidly, reaches its maximal level by 5 min and gradually decreases over 60 min. This effect was greatly accelerated by the addition of theophylline (5 mM) and by another potent cyclic AMP-phosphodiesterase inhibitor, papaverine (250 μM), but neither papaverine alone nor theophylline alone had much effect. The combination of 5 mM theophylline and 250 μM papaverine with histamine produced no further increase in the cyclic AMP level beyond

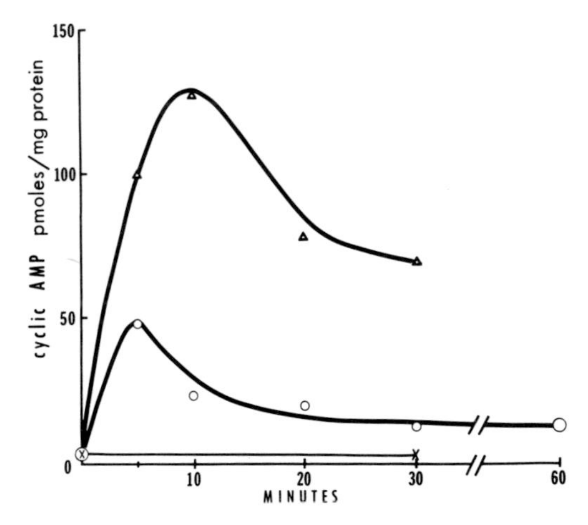

Fig. 3. Time course of the effect of histamine on the intracellular cyclic AMP content. The concentration of histamine added was 100 μM and that of theophylline 5 mM. ($\triangle$ = histamine + theophylline; $\bigcirc$ = histamine only; and $\times$ = control (no addition). (Iizuka *et al.*[15])

that produced by theophylline alone. These data suggest that theophylline and papaverine act at the same site (i.e., the phosphodiesterase enzyme). The increase in cyclic AMP was dose dependent, the maximal effect being obtained at 200 μM. Lineweaver-Burk plots show that the apparent Ka for histamine was about 6×10^{-5}M.

All antihistamines tested prevented histamine from increasing cyclic AMP, but to a different degree. Diphenhydramine, an ethanolamine derivative, was the weakest inhibitor; metiamide, the recently developed H_2 receptor inhibitor, was the strongest, completely blocking histamine stimulation. The phenothiazines (perphenazine, fluphenazine, acetophenazine, promethazine) exercised various degrees of inhibition, promethazine being the strongest inhibitor within this category.

Is the histamine-adenylate cyclase system specific and independent of the epinephrine- and prostaglandin E- adenylate cyclase system? Our data, summarized in Fig. 4, indicate that it is because 1) the cyclic AMP increase caused by the simultaneous addition of histamine and epinephrine was equal to the sum of the increases caused by the single addition of histamine and epinephrine (at their respective saturation concentrations); 2) the effect of histamine was not inhibited by the β-adrenergic blocker propranolol; 3) the cyclic AMP increase caused by the addition of epinephrine was not inhibited by the antihistamine metiamide, which completely

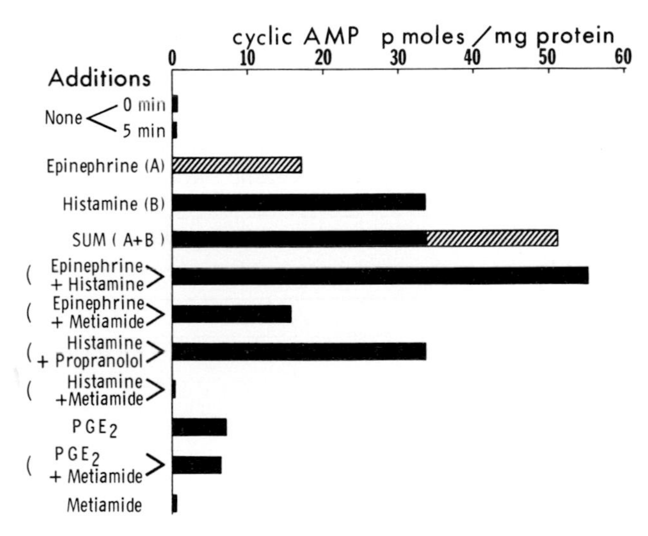

Fig. 4. The effects of epinephrine, propranolol, histamine, metiamide, and PGE_2 on the cyclic AMP level in pig epidermis. Concentrations of the drugs added to the media were epinephrine = 50 μM, propranolol = 50 μM, histamine = 1 mM, metiamide = 500 μM, PGE_2 = 50 μM. Results are the average of two or three experiments. No cyclic AMP-phosphodiesterase inhibitors were added in these series of experiments. (Iizuka *et al.*[15])

inhibited the activation by histamine; 4) the cyclic AMP accumulation caused by prostaglandin E_2 was not inhibited by metiamide.

D. Adenosine receptor-adenylate cyclase system

It has been recently recognized that adenosine and adenine nucleotides are involved in the regulation of cellular processes through the cyclic AMP system. In the fat cells of rats,[25] adenosine inhibits the accumulation of cyclic AMP, but in cerebral slices of guinea pigs,[26] mouse neuroblastoma,[27] or astrocytoma cells,[28] adenosine stimulates adenylate cyclase to cause a marked accumulation of cyclic AMP. In epidermal slices from hairless mice, Voorhees *et al.*[29] reported about a 200% activation of cyclic AMP 5 min after the addition of adenosine. The adenosine concentration varied from 5×10^{-4}M to 10^{-2}M, but dose dependency was not clearly demonstrated. In our most recent work with pig and human epidermis, not only adenosine, but also such adenine nucleotides as AMP, ADP, and ATP activated adenylate cyclase and caused a 2,000–5,000% dose-dependent increase in cyclic AMP. The results of these studies are summarized here.

The time course of adenosine activation is similar to that described in the foregoing sections on epinephrine, prostaglandin E, and histamine activation. The maximum increase was attained after 5 min and then gradually decreased. The addition of papaverine but not of theophylline heightened the effect of adenosine. Table 4 shows the effect of adenosine and the adenine nucleotides (1 mM), all of which markedly increased the intracellular concentration of cyclic AMP. The fact that the addition of these four compounds together had no additive effect on the accumulation of cyclic AMP suggests that they acted on the same site. When ^{3}H-labelled adenine nucleotides were added to the same experimental system, more

Table 4. The effects of adenine, adenosine, AMP, ADP and ATP on the cyclic AMP level.

Additions		Cyclic AMP
		p moles/mg protein
none		
0 min		0.6
5 min		0.2
adenine	1 mM	1.8
adenosine	1 mM	25.7
5′-AMP	1 mM	25.2
5′-ADP	1 mM	26.4
5′-ATP	1 mM	29.5
adenosine + AMP + ADP + ATP 1 mM each		27.3

The cyclic AMP increase during the 5 min incubation period is shown. The results are the average of three experiments.

than 95% of the radioactivities could be recovered in their original forms without hydrolysis. Since adenine nucleotides are not readily permeable through the cell membrane, the receptor site must be on the outer surface of the cell membrane.

An unusual feature of this adenosine-adenine nucleotides-receptor is the inhibition of cyclic AMP accumulation by theophylline, which normally increases accumulation by inhibiting the specific phosphodiesterase. Adenine alone did not have any effect on the adenylate cyclase system, but it inhibited activation by adenosine. The structural similarity of adenine and theophylline to adenosine apparently causes the competition with adenosine at the cell surface receptor site to inhibit cyclic AMP accumulation.

The structural requirement of the activation of this receptor system is shown in Table 5. Apparently the full activation requires a phosphoric

Table 5. The structural requirement for the activation of adenosine receptor-adenylate cyclase system.

Compounds (1 mM)	*	Cyclic AMP
Adenosine skeleton (NH_2, N, N, N, N; * $-CH_2$O; OH OH)		p moles/mg prot.
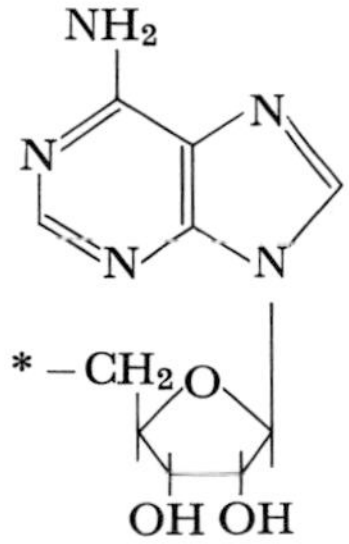	OH	25.7
	HO—P(=O)(—OH)—O—	25.2
	—P—P—	26.4
	—P—P—P—	29.5
	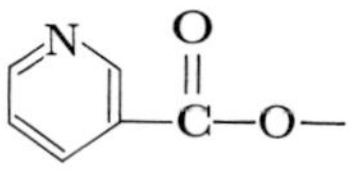 pyridyl(N)—C(=O)—O—	0.6
	HO—S(=O)(—OH)—O—	3.6
	H_2N—P(=O)(—ONa)—O—(with papaverine)	41.5

* indicates the substitution of terminal 5′-phosphoric acid group. The experimental conditions are the same as Table 4. The compounds tested were, from top to bottom, adenosine, AMP, ADP, ATP, adenosine 5′-mononicotinate, adenosine 5′-monosulfate, and adenosine 5′-monophosphoramidate.

acid group bound to adenosine at the 5′-position. Neither 2′- nor 3′-adenylic acid stimulated the receptor system. When the terminal 5′-phosphoric acid group was replaced by sulfuric acid (adenosine 5′-monosulfate), the stimulatory effect was no longer observable. The replacement of the OH terminal of 5′-phosphoric acid by NH_2 (adenosine 5′-monophosphoramidate) greatly reduced the stimulatory effect, but the monophosphoramidate together with papaverine markedly stimulated adenylate cyclase. Other nucleotides and nuclosides which did not contain the adenine group failed to stimulate it. The binding site in this epidermal system is, therefore, specific to adenine derivatives.

The simultaneous addition of epinephrine and AMP, or of histamine and AMP at their saturation concentrations, results in additive effects (Table 6). The addition of a β-adrenergic blocker inhibits epinephrine activation but not adenosine stimulation just as metiamide counteracts

Table 6. The effects of AMP, epinephrine, histamine, and PGE_2 on the cyclic AMP level.

A.

Additions		Cyclic AMP
		p moles/mg protein
none, 0 min		0.2
none, 5 min		0.1
epinephrine	(50 μM)	20.9
AMP	(1 mM)	25.2
histamine	(1 mM)	17.2
epinephrine + AMP		42
AMP + histamine		37.5

B.

Additions		Cyclic AMP
		p moles/mg protein
none, 0 min		0.9
none, 10 min		1.0
AMP	(1 mM)	54.9
PGE_2	(25 μM)	7.8
AMP + PGE_2		67.5

In A series of experiments, no cyclic AMP phosphodiesterase inhibitors were added and incubation time was 5 min.

In B series of experiments, 200 μM papaverine was added and incubation time was 10 min.

Results are the average of two (A) or three (B) experiments.

only histamine activation (Table 7). Furthermore, theophylline inhibits adenosine stimulation but activates epinephrine, histamine, and prostaglandin. Thus, the epidermis possesses four independent specific receptors for adenylate cyclase activation in cell surface membrane (Fig. 5).

Table 7. The effects of propranolol, metiamide, and theophylline on AMP activation.

Additions	Cyclic AMP
	p moles/mg protein
none 0 min	1.1
none 5 min	0.4
AMP (1 mM)	21.1
AMP + metiamide (1 mM)	21.5
AMP + propranolol (100 μM)	19.5
AMP + theophylline (5 mM)	5.2
histamine (1 mM)	33.3
histamine + metiamide (500 μM)	0.8
epinephrine (10 μM)	18
epinephrine + propranolol (100 μM)	3

Incubation period = 5 min at 37°C

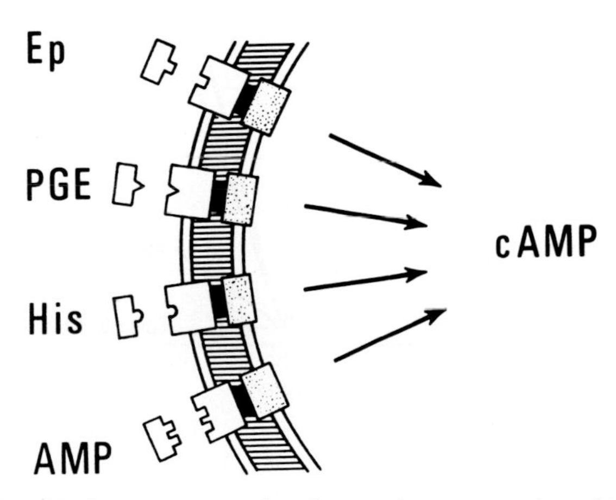

Fig. 5. Various receptor-adenylate cyclase systems in epidermis.

E. Specific refractoriness of the receptor-adenylate cyclase system

In fat tissue, epinephrine, ACTH, and glucagon activate adenylate cyclase in the cell membrane to cause a transient increase in the intracellular concentration of cyclic AMP. Once adenylate cyclase has been activated, however, the addition of any of these hormones will not further elevate the cyclic AMP level. This desensitization or refractoriness is due to the formation and release of a hormone antagonist which inhibits any further activation.[30–32] In more recent studies, we have found that refractoriness to epinephrine also develops in skin, but the underlying mechanism for the refractoriness may be different from that found in fat tissue.

The addition of epinephrine stimulated skin adenylate cyclase and caused a rapid accumulation of cyclic AMP, but within 10 min the cyclic AMP level returned to its original level (Fig. 6). This low level 10 min after epinephrine activation suggests that its rapid decrease is at least partly due to the activation of the specific cyclic AMP-phosphodiesterase. However, the addition of even 10 mM of theophylline (twenty times more than its Ki for the skin enzyme[33]) could not completely stop the decrease in cyclic AMP. Thus, a loss of synthetic capacity and activation of degradation occurred simultaneously. Figure 6 clearly indicates that the further addition of epinephrine (20 mM) at 10 min did not elevate the level of cyclic AMP, whereas the addition of histamine and/or AMP at 10 min did cause the rapid accumulation of intracellular cyclic AMP.

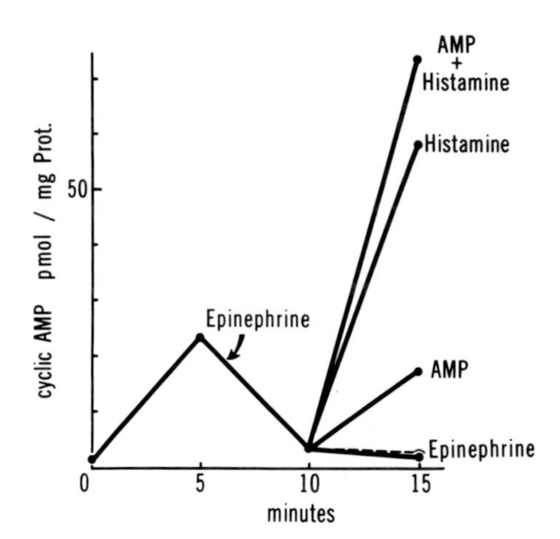

Fig. 6. Cyclic AMP levels after epinephrine activation and subsequent addition of epinephrine, AMP, and histamine. The first activation by epinephrine (20 μM) was initiated at 0 min. After 10 min incubation, the tissues were transferred to fresh Hank's medium containing histamine, AMP, etc. for an additional 5 min incubation. In a separate experiment, the second stimulators were directly added to the Hank's medium containing epinephrine in which skin tissue had been incubated. The results were exactly identical. (······) indicates an additional 20 μM epinephrine (for a total of 40 μM final concentration) compared with (——) the original 20 μM).

In fat tissue, once the refractoriness to epinephrine develops, the fat cells no longer respond to other stimulators such as ACTH or glucagon;[30,31] that is, the desensitization is nonspecific. In skin, however, adenylate cyclase did respond to the addition of other stimulators such as AMP or histamine; in other words, the desensitization of skin adenylate cyclase was specific for epinephrine.

In the fat cell system, the inhibitor responsible for the refractoriness (modulator) leaks out of the membrane and accumulates in the incubation medium. In the skin system, on the other hand, the modulator appears to attach firmly to either the receptor or catalytic site of adenylate cyclase system. Thus, when epinephrine activated skin was washed several times with fresh Hank's medium and incubated with epinephrine, no further increase in cyclic AMP occurred. When skin kept in Hank's medium without a 15-min exposure to epinephrine was incubated in the medium containing epinephrine which had failed to reactivate the tissue previously exposed to epinephrine, the intracellular increase in cyclic AMP was as good as that observed in the "control fresh" tissues incubated in fresh medium with epinephrine. The specificity of refractoriness in the skin system is further shown in Fig. 7. When the tissues were activated by AMP, subsequent stimulation by histamine and/or epinephrine, but not AMP, elevated the cyclic AMP level. The addition of theophylline or papaverine alone or of [AMP + theophylline] did not stimulate the AMP-inhibited adenylate cyclase system.

Of possible physiological relevance is the hypothesis that the refractoriness may act as an important feed-back regulator of hormonal action. If

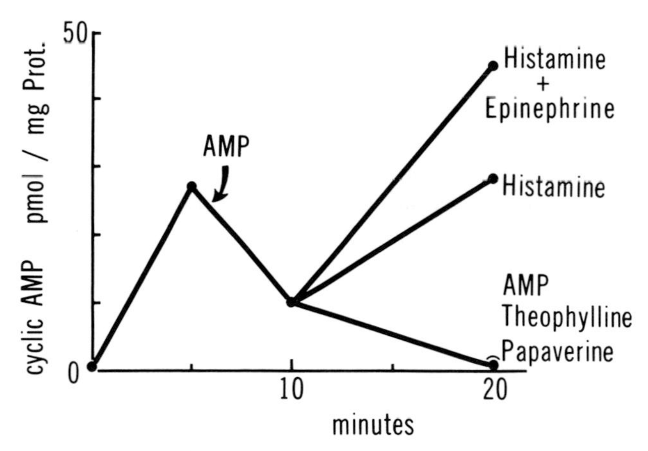

Fig. 7. Initial activation and consequent refractoriness to AMP. The experimental conditions were the same as in fig. 6. Concentrations: theophylline, 5 mM; papaverine, 100 μM; epinephrine 20 μM.

the same data apply to *in vivo* experiments, the general theory that the cyclic AMP level *in vivo* is a good indication of the activity of adenylate cyclase (and of specific phosphodiesterase) should be revised; that is, with the same relatively low cyclic AMP level, the adenylate cyclase can, because of refractoriness, be either active or inactive. Then perhaps the most important molecular response to hormone activation is the transient increase (spiking) in cyclic AMP rather than the static cyclic AMP level itself.

CYCLIC AMP IN EPIDERMAL PROLIFERATION AND DIFFERENTIATION

A. *in vitro* studies

We have studied in detail the epidermal slice system in pigs (cf. Section II). Using this model system, we can now manipulate and predict the intracellular levels of cyclic AMP after stimulation by various hormones. Some of our attempts to test whether cyclic AMP plays a biological role in the pig skin slice model have failed, probably because of an improper organ culture technique. In this section, I will briefly review the work done by other investigators on biological roles of cyclic AMP and correlate their data with ours.

What are the biological roles of cyclic AMP in epidermis? One attractive hypothesis is that cyclic AMP may help to regulate proliferation and differentiation in epidermal cells. The first evidence of a regulatory role in mitosis was introduced by Bürk[34] and by Ryan and Heidrick[35] in 1968. Their independent studies showed that the exogenous addition of cyclic AMP and theophylline inhibited the division of cultured cells. Numerous later works by others appear to indicate that, at least in an *in vitro* system, a low cyclic AMP level is associated with the proliferation and a high level with the maturation of cells. (For a recent review, cf. ref.[36]). In *in vitro* studies of skin, Voorhees *et al.*[37] and Marks and Rebien[38] found that dibutyryl cyclic AMP and/or β-adrenergic agonist inhibited mitosis in the mouse ear. Recently, Flaxman and Harper[39] reviewed their own data on the inhibitory action of various chemicals and hormones on mitosis using the outgrowth culture of human epidermis (Table 8). Their data on adrenergic drugs generally agree with those of Voorhees *et al.*[37] Table 8 also lists our data on adenylate cyclase stimulation by the same compounds. The degree of cyclic AMP accumulation in pig epidermis by β-adrenergic agonist is directly related to the degree of mitotic inhibition.

The data of Flaxman and Harper[39] on the effect of histamine are contrary to those of Voorhees *et al.*[40] who reported slight mitotic stimulation by histamine (and no change in cyclic AMP level in rodent skin),[24] whereas Flaxman and Harper observed a marked mitotic inhibition after hista-

Table 8. Mitotic inhibition *in vitro* and cyclic AMP increase by adrenergic drugs and histamine.

Compound	Mitosis	Cyclic AMP
	(% inhibition)	(p moles/mg prot.)
epinephrine (10^{-9}M)	65	(10^{-5}M) 34
isoproterenol	53	31
norepinephrine	31	10
epinephrine (4.5×10^{-8}M)		
+phentolamine	64	36
+propranolol	22	4
histamine ($10^{-2} \sim 10^{-6}$M)	68	34
histidine (10^{-3}M)	21	0
imidazole acetate (10^{-3}M)	22	0

Data on mitosis are obtained from Flaxman and Harper[39] and those on cyclic AMP from Yoshikawa *et al.*[13] and Iizuka *et al.*[15]

mine activation. Neither the precursor nor the metabolite of histamine activated adenylate cyclase, and neither of them inhibited mitosis (Table 8).

Data on the effect of various nucleotides are summarized in Table 9. Again mitosis and adenylate cyclase activation are fairly well correlated. Adenosine and adenine nucleotides caused a rapid increase in cyclic AMP in our pig epidermal system, and the same compounds caused marked mitotic inhibition in human epidermis. Other nucleotides besides adenine compounds had no effect on either mitosis or adenylate cyclase activity.

As far as differentiation of the epidermis is concerned, the only available

Table 9. Mitotic inhibition *in vitro* and cyclic AMP increase by nucleotides.

Compound	Mitosis	Cyclic AMP
(10^{-3} M)	(% inhibition)	(pmoles/mg prot.)
cyclic AMP	45	—
dibuty. cAMP	55	—
AMP	59	24.4
ADP	79	25.8
ATP	58	28.9
adenosine	50	25.1
adenine	(+6)	1.2
GMP	13	0
cGMP	(+4)	—
CMP	12	—
TMP, IMP, UMP	0	0

Data on mitosis are obtained from Flaxman and Harper[39] and those on cyclic AMP from Iizuka *et al.*[15]

experimental data are those of Delescluse *et al.*[41] Using a primary culture system of epidermal keratinocytes, they reported that cyclic AMP stimulated keratinization which was demonstrated by increased rhodamin B stainability and by increased histidine incorporation into protein fraction. It would be interesting to see also whether the addition of epinephrine, histamine, or adenine nucleotides would also stimulate keratinization.

B. *In vivo* studies

The results of these *in vitro* experiments are consistent with the view that cyclic AMP inhibits proliferation and stimulates differentiation in epidermal cells. But it must be emphasized that all of these positive data are the result of *in vitro* experiments. *In vivo* experiments, on the contrary, indicate that increased cyclic AMP level is associated with increased cell proliferation. For example, Grimm and Marks,[42] who applied tumor-promoting phorbol esters to mouse skin *in vivo,* observed that the increased cyclic AMP level was associated with cell proliferation. Furthermore, the application of retinoic acid to guinea pig ear[43], ultra-violet light exposure on the back of the pig,* and the intradermal injection of prostaglandin E_2 in normal human skin,[44] all stimulated epidermal mitosis and increased, rather than decreased, the level of cyclic AMP. Since the last three examples have already been discussed by Halprin,[45] I will not elaborate further.

Only one *in vivo* work may support the original hypothesis of a low cyclic AMP for active mitosis: that of Marks and Grimm[7] observed diurnal changes in the level of cyclic AMP in mouse epidermis: high during the day, low at night (at dark). They cited the work of Tvermyr[46] on mitotic activity in the dorsal epidermis of mice and concluded that the low cyclic AMP level corresponded to the high rate of mitosis and vice versa.

However, according to Bullough,[47] the mitotic rate is high during rest or sleep when the epinephrine level is low. If we compare Bullough's rather than Tvermyr's data with those of Marks and Grimm[7], the low cyclic AMP level at "dark" (active stage for mice) is associated with high epinephrine levels and a low mitotic rate. Conversely, a high cyclic AMP level corresponds to a high mitotic rate. Thus, if we take Bullough's data instead of Tvermyr's this *in vivo* experiment also contradicts the original hypothesis.

This apparent discrepancy between *in vitro* and *in vivo* data is also found in other disease conditions such as psoriasis and cancer. Voorhees *et al.*[10] proposed that the same hypothesis applies, namely, that in psoriasis a low level of cyclic AMP is responsible for the high mitotic rate and the incomplete differentiation. Later, however, Härkönen *et al.*[48] found that the

* Kechijian, P. Unpublished experiments.

cyclic AMP level in psoriasis is not low but rather high. More recently, Yoshikawa *et al.*[49] using a microtechnique showed that on a dry weight or protein basis, the cyclic AMP level was 20–25% higher in psoriatic lesions than in uninvolved skin, but on a DNA basis was essentially the same. (Our assay technique had two advantages over previous studies: the tissue was frozen before removal and the epidermis was microdissected from the dermis free of contamination.)

The possible roles of cyclic nucleotides in relation to cancer were recently reviewed by Ryan and Heidrick[36] who cited the discrepancies between *in vivo* and *in vitro* results, adding that "most investigators point to elevated levels of cyclic AMP and cyclic GMP levels in Morris hepatomas' different stages of growth and reported that both cyclic nucleotides were much higher in tumors than in normal controls."

Despite these contradictions between *in vitro* and *in vivo* results, a seemingly important biochemical abnormality is common to these pathological conditions. In psoriasis[51] and in phorbol ester-treated skin[42], epinephrine no longer causes a dramatic accumulation of cyclic AMP; that is, the epinephrine receptor sites of adenylate cyclase have a malfunction. Similarly in Morris hepatomas, hormone sensitivity to glucagon is either lost or weakened, that is, glucagon receptor-adenylate cyclase system appears to be malfunctioning.[52]

At any rate, the contradiction between *in vivo* and *in vitro* data should be an important problem in future experiments on proliferation and differentiation. The discrepancy may be explained by changes in cyclic GMP (with or without accompanying changes in cyclic AMP) or other cell surface receptor(s) which may control mitosis without the aid of cyclic nucleotides. Since the data on cyclic GMP and other receptors on skin surface membrane have been extremely limited, the subject can be discussed meaningfully only after further studies.

Acknowledgements

I am grateful to my colleagues K. Yoshikawa, H. Iizuka, K. M. Halprin, and V. Levine for their collaboration in these studies on cyclic nucleotides in skin. These studies were supported by the National Institute of Health grant (AM 17179) and funds from the Dermatology Foundation of Miami.

REFERENCES

1. Robison, G. A., Butcher, R. W., and Sutherland, E. W. Cyclic AMP, Academic Press, New York, 1971.
2. Mier, P. D. and Urselmann, E. The adenyl cyclase of skin. I. Measurements and properties. *Brit. J. Derm.*, **83**: 359–363, 1970.

3. Brønstad, G. O., Elgjo, K., and Oye, I. Adrenaline increases cyclic 3′,5′-AMP formation in hamster epidermis. *Nature, New Biol.*, **223**: 78–79, 1971.
4. Powell, J. A., Duell, E. A., and Voorhees, J. J. Beta adrenergic stimulation of endogeneous epidermal cyclic AMP formation. *Arch. Derm.*, **104**: 359–365, 1971.
5. Duell, E. A., Voorhees, J. J., Kelsey, W. H., and Hayes, E. Isoproterenol-sensitive adenyl cyclase in a particulate fraction of epidermis. *Arch. Derm.*, **104**: 601–610, 1971.
6. Marks, F. and Rebien, W. The second messenger system of mouse epidermis. I. Properties and β-adrenergic activation of adenylate cyclase *in vitro*. *Biochim. Biophys. Acta*, **284**: 556–557, 1972.
7. Marks, F. and Grimm, W. Diurnal fluctuation and β-adrenergic elevation of cyclic AMP in mouse epidermis *in vivo*. *Nature, New Biol.*, **240**: 178–179, 1972.
8. Hsia, S. L., Wright, R., Mandy, S. H., and Halprin, K. M. Adenyl cyclase in normal and psoriatic skin. *J. Invest. Derm.*, **59**: 109–113, 1972.
9. Wright, R. K., Mandy, S. H., Halprin, K. M., and Hsia, S. L. Defects and deficiency of adenyl cyclase in psoriatic skin. *Arch. Derm.*, **107**: 47–53, 1973.
10. Vorhees, J., Kelsey, W., Stawiski, M., Smith, E., Duell, E., Haddox, M., and Goldberg, N. Increased cyclic GMP and decreased cyclic AMP levels in the rapidly proliferating epithelium of psoriasis. In The Role of Cyclic Nucleotides in Carcinogenesis (J. Shultz and H. G. Gratzner, eds.), Academic Press, New York. 1973, vol. 6, pp. 325–373.
11. Mui, M. M., Hsia, S. L., and Halprin, K. M. Further studies on adenyl cyclase in psoriasis. *Brit. J. Derm.*, **92**: 255–262, 1975.
12. Yoshikawa, K., Adachi, K., Halprin, K. M., and Levine, V. Cyclic AMP in skin: Effects of acute ischemia. *Brit. J. Derm.*, **92**: 249–254, 1975.
13. Yoshikawa, K., Adachi, K., Halprin, K. M., and Levine, V. The effects of catecholamine and related compounds on the adenyl cyclase system in the epidermis. *Brit. J. Derm.*, **93**: 29–36, 1975.
14. Adachi, K., Yoshikawa, K., Halprin, K. M., and Levine, V. Prostaglandins and cyclic AMP in epidermis. *Brit. J. Derm.*, **92**: 381–388, 1975.
15. Iizuka, H., Adachi, K., Halprin, K. M., and Levine, V. Histamine (H_2) receptor-adenylate cyclase system in pig skin (epidermis). *Biochim. Biophys. Biophys. Acta*, **437**: 150–157, 1976.
16. Iizuka, H., Adachi, K., Halprin, K. M., and Levine, V. Adenosine and adenine nucleotides stimulation of skin (epidermal) adenylate cyclase. 1976 in press.
17. Gilman, A. G. A protein binding assay for adenosine 3′,5′-cyclic monophosphate. *Proc. Natl. Acad. Sci. USA*, **67**: 305–312, 1970.
18. Yoshikawa, K., Adachi, K., Levine, V., and Halprin, K. M. Micro-determination of cyclic AMP levels in human epidermis, dermis and hair follicles. *Brit. J. Derm.*, **92**: 241–248, 1975.
19. Goldblatt, M. W. A depressor substance in seminal fluid. *J. Soc. Chem. Indust. London*, **52**: 1056–1057, 1933.
20. Von Euler, U.S. Über die spezifische blutdrucksenkende substanz des menschlichen Prostata-und samenblasensekretes. *Klinische Wochenschrift*, **14**: 1182–1183, 1935.

21. Hsia, S. L., Wright, R. K., and Halprin, K. M. Adenyl cyclase of human skin and its abnormalities in psoriasis. In The Role of Cyclic Nucleotides in Carcinogenesis. (J. Shultz and H. G. Gratzner, eds.), Academic Press, New York, 1973, vol. 6, pp. 303–323.
22. Aso, K., Orenberg, E. K., and Farber, E. M. Reduced epidermal cyclic AMP accumulation following prostaglandin stimulation: Its possible role in the pathophysiology of psoriasis. *J. Invest. Derm.*, **65**: 375–378, 1975.
23. Yoshikawa, K., Adachi, K., Halprin, K. M., and Levine, V. Effects of short chain alcohols and hydrocarbon compounds on the adenylate cyclase of the skin. *Brit. J. Derm.*, **94**: 611–614, 1976.
24. Voorhees, J. J., Colburn, N. H., Stawiski, M., Duell, E. A., Haddox, M., and Goldberg, N. D. Imbalanced cyclic AMP and cyclic GMP levels in the rapidly dividing, incompletely differentiated epidermis of psoriasis. In Control of Proliferation in Animal Cells (C. Clarkson and R. Baserga, eds.) Cold Spring Harbor Laboratory, New York, 1974, pp. 635–648.
25. Fain, J. N., Pointer, R. H., and Ward, W. F. Effects of adenosine nucleosides on adenylate cyclase, phosphodiesterase, cyclic adenosine monophosphate accumulation, and lipolysis in fat cells. *J. Biol. Chem.*, **247**: 6866–6872, 1972.
26. Sattin, A. and Rall, T. W. The effect of adenosine and adenine nucleotides on the cyclic adenosine 3′, 5′ phosphate content of guinea pig cerebral cortex slices. *Mol. Pharmacol.*, **6**: 13–23, 1970.
27. Blume, A. J. and Foster, C. J. Mouse neuroblastoma adenylate cyclase. Adenosine and adenine analogues as potent effectors of adenylate. *J. Biol. Chem.*, **250**: 5003–5008, 1975.
28. Clark, R. B. and Gross, R. Regulation of adenosine 3′,5′ -monophosphate content in human astrocytoma cells by adenosine and the adenine nucleotides. *J. Biol. Chem.*, **249**: 5296–5303, 1974.
29. Voorhees, J. J., Marcelo, C. L., and Duell, E. A. Cyclic AMP, cyclic GMP, and glucocorticoids as potential metabolic regulators of epidermal proliferation and differentiation. In The Epidermis, vol. 16, Advances in Biology of Skin (W. Montagna, A. G. Matoltsy, K. D. Wuepper, D. D. Knutson, and J. P. Bentley, eds.), *J. Invest. Derm.*, **65**: 179–190, 1975.
30. Ho, R. J. and Sutherland, E. W. Formation and release of a hormone antagonist by rat adipocytes. *J. Biol. Chem.*, **246**: 6822–6827, 1971.
31. Ho, R. J. and Sutherland, E. W. Action of feedback regulator on adenylate cyclase. *Proc. Nat. Acad. Sci. USA*, **72**: 1773–1777, 1975.
32. Fain, J. N. and Shepherd, R. E. Free fatty acids as feedback regulators of adenylate cyclase and cyclic 3′,5′-AMP accumulation in rat fat cells. *J. Biol. Chem.*, **250**: 6586–6592, 1975.
33. Adachi, K., Levine, V., Halprin, K. M., Iizuka, H., and Yoshikawa, K. Multiple forms of cyclic nucleotide phosphodiesterase in pig epidermis. *Biochim. Biophys. Acta*, **429**: 498–507, 1976.
34. Bürk, R. R. Reduced adenyl cyclase activity in a polyoma virus transformed cell line. *Nature*, **219**: 1272–1275, 1968.
35. Ryan, W. L. and Heidrick, M. L. Inhibition of cell growth by adenosine 3′, 5′ -monophosphate. *Science*, **162**: 1484–1485, 1968.
36. Ryan, W. L. and Heidrick, M. L. Role of cyclic nucleotides in cancer. In:

Advances in Cyclic Nucleotide Research, vol. 4 (P. Greengard and G. A. Robison, eds.) Raven Press, New York, 1974, pp. 81–116.

37. Voorhees, J. J., Duell, E. A., and Kelsey, W. H. Dibutyryl cyclic AMP inhibition of epidermal cell division. *Arch Derm.*, **105**: 384–386, 1972.
38. Marks, F. and Rebien, W. Cyclic 3′, 5′ -AMP and theophylline inhibit epidermal mitosis in G_2 -phase. *Naturwissenschft.*, **59**: 41–42, 1972.
39. Flaxman, B. A. and Harper, R. A. *In vitro* analysis of the control of keratinocyte proliferation in human epidermis by physiologic and pharmacologic agents. In The Epidermis, vol. 16, Advances in Biology of Skin (W. Montagna, A. G. Matoltsy, K. D. Wuepper, D. D. Knutson, and J. P. Bentley, eds.), *J. Invest. Derm.*, **65**: 52–59, 1975.
40. Voorhees, J. J., Duell, E. A., Kelsey, W. H., and Hayes, E. The cyclic AMP system in normal and psoriatic epidermis. *J. Invest. Derm.*, **59**: 114–120, 1972.
41. Delesculse, C., Fukuyama, K., and Epstein, W. L. Dibutryl cyclic AMP-induced differentiation of epidermal cells in tissue culture. *J. Invest. Derm.*, **66**: 8–13, 1976.
42. Grimm, W. and Marks, F. Effect of tumor-promoting phorbol esters on the normal and the isoproterenol-elevated level of adenosine 3′, 5′ -cyclic monophosphate in mouse epidermis *in vivo*. *Cancer Res.*, **34**: 3128–3134, 1974.
43. Aso, K., Rabinowitz, I., Wilkinson, D. I., and Farber, E. M. Prostaglandins and cAMP in retinoic acid-induced epidermal hyperplasia (abst). *J. Invest. Derm.*, **64**: 204, 1975.
44. Eaglstein, W. H. and Weinstein, G. D. Prostaglandin and DNA synthesis in human skin: Possible relationship to ultraviolet light effects. *J. Invest. Derm.*, **64**: 386–389, 1975.
45. Halprin, K. M. Cyclic nucleotides and epidermal cell proliferation. *J. Invest. Derm.* **66**: 339–343, 1976.
46. Tvermyr, E. M. F. Circadian rhythm in epidermal mitotic activity. Diurnal variations of the mitotic index, the mitotic rate and the mitotic duration. *Virchows Arch. Zellpath.*, **2**: 318–325, 1969.
47. Bullough, W. S. Mitotic and functional homeostasis: A speculative review. *Cancer Res.*, **25**: 1683–1727, 1965.
48. Härkönen, M., Hopsu-Havu, V. K., and Raij, K. Cyclic adenosine monophosphate, adenyl cyclase and cyclic nucleotide phosphodiesterase in psoriatic epidermis. *Acta Dermatovener.*, **54**: 13–18, 1974.
49. Yoshikawa, K., Adachi, K., Halprin, K. M., and Levine, V. Is the cyclic AMP in psoriatic epidermis low? *Brit. J. Derm.*, **93**: 253–258, 1975.
50. Thomas, E. W., Murad, F., Looney, W. B., and Morris, H. P. Adenosine 3′,5′ -monophosphate and guanosine 3′,5′ -monophosphate: Concentrations in Morris hepatomas of different growth rates. *Biochim. Biophys. Acta*, **297**: 564–567, 1973.
51. Yoshikawa, K., Adachi, K., Halprin, K.M., and Levine, V. On the lack of of response to catecholamine stimulation by the adenyl cyclase system in psoriatic lesions. *Brit. J. Derm.*, **92**: 619–624, 1975.
52. Allen, D. O., Munshower, J., Morris, H. P., and Weber, G. Regulation of adenyl cyclase in hepatomas of different growth rates. *Cancer Res.*, **31**: 557–560, 1971.

Cyclic AMP and Differentiation of Epidermal Cells in Culture: Biphasic Concentration Change during Growth

Kazuo Aso* and Israel Rabinowitz**

* *Department of Dermatology, School of Medicine, Yamagata University, Yamagata, Japan* and
** *Department of Dermatology, Stanford Medical Center, Stanford, U.S.A.*

ABSTRACT

Concentrations of cAMP, cGMP, along with uptake of ^{3}H-thymidine, leucine, and histidine during the growth period of epidermal cells in culture were assayed. Some of these cells were aggregated and organized in a large system which indicated the differentiation of these cells toward keratinization. There was a biphasic change in the concentration of cAMP during the growth period. There was an initial drop, simultaneous with a peak increase in thymidine as well as leucine and histidine uptake, followed by a steady increase, which was in turn followed by increased protein synthesis and differentiation of cells. There was no significant change in cGMP concentration in the epidermal cells when the concentration of cAMP decreased. The results suggest a biphasic function of cAMP in epidermal cell division and differentiation.

INTRODUCTION

In a previous investigation of rapidly proliferating epidermis, stimulated by the topical application of retinoic acid,[1] it was found that the concentration of adenosine 3′,5′-monophosphate (cAMP) increased significantly after the peak increase of tritiated thymidine uptake. No significant change in the concentration of guanosine 3′,5′-monophosphate (cGMP) was seen in the early phases of proliferation. This suggested a role for cAMP in cutaneous epidermal cell differentiation. In order to confirm the suggestion that cAMP plays a role in differentiation, a second model of epidermal cell growth has been examined, that of guinea pig epidermal cells grown in cell culture. Tritiated thymidine, leucine, and histidine uptake were measured throughout the period of cell growth and the concentrations of cAMP and cGMP were assayed.

MATERIALS AND METHODS

The cGMP assay kit was obtained from Collaborative Res. and tritiated cAMP used in the binding protein assay was purchased from New Eng. Nuclear, Inc., as were tritiated thymidine, tritiated histidine and tritiated leucine. All other chemicals used were reagent grade.

Thin keratotome slices (0.2 mm) taken from the ears of approximately 6-month-old male albino guinea pigs were trypsinized in sterile 0.3% trypsin in phosphate buffered saline (PBS), pH 7.4 for 35 minutes at 37°C. After washing with PBS, the epidermal cells were removed from the slices by gently scraping with a scalpel in PBS. The cell suspension was then filtered through sterile gauze. The filtrate of epidermal cells was seeded in culture medium (Eagle minimal essential containing 9% calf serum) in 6 cm Falcon plastic dishes which were preincubated with the medium in the cell incubator for one day before seeding. The approximate concentration of initial cells per dish was 7–8 $\times$ 10^6. Following the procedure of Constable *et al.*,[2] the medium was not changed for the first three days, and subsequent changes were made twice a week. The cells were grown under an atmosphere of 95% O_2, 5% CO_2 for 12 to 13 days.

For cAMP assay, the contents of three dishes were first washed three times with PBS and the enzymatic action stopped by adding 10% trichloroacetic acid (TCA). The attached cells were scraped from the dishes with a rubber policeman and the cells homogenized in 2 ml of 10% TCA per dish using a glass homogenizer. The supernatant, after centrifugation at 10,000 $\times$ *g* for 20 minutes at 4°C was collected, washed three times with cold water, saturated with ether, and lyophilized. After resuspension of the powder in 100 μl of 0.005 M acetate buffer, pH 4.0, the cAMP assay per contents of three dishes was carried out using a previously described binding protein technique.[1] This procedure was done in triplicate (i.e., 9 dishes) for each assay per day from days 3 to 13. For cGMP assay, the contents of 5 dishes were pooled for assay and this was done with 5 to 8 replications per day (25 to 40 dishes) of incubation from days 3 to 6. The extraction method for cGMP from cells in culture was the same as for cAMP. cGMP determination was performed via radioimmunoassay after chromatographic separation of nucleotides as previously described.[1] For histidine and leucine uptake, 1 μCi of tritiated histidine or leucine in PBS was added to each of three different dishes and incubated for 1 hour. At the end of this incubation, the attached cells were washed three times in three changes of buffer. The cells were then scraped loose and centrifuged 3,000 $\times$ *g* at 4°C for 10 minutes. The precipitate was incubated for 1 hour in 2 ml of 0.1% sodium dodecyl sulfate. A 1 ml aliquot of each sample was added to 1 ml of 20% TCA and kept overnight at 4°C.[3] The suspension was then passed through 0.22 μm Millipore paper, the paper dissolved in 1 ml. of methoxyethanol, and the radioactivity counted after adding 10 ml of 20%

methoxyethanol, 80% toluene, containing 5 gm PPO/liter using a Beckman LS-200 scintillation counter. An aliquot of the cell suspension was taken for protein assay according to the method of Lowry.[4] For days 3 to 13, each triplicate protein assay was repeated twice. For thymidine uptake assay, 1 μCi of tritiated thymidine was added to each of three dishes per day and incubated for 1 hour. Reaction was terminated by adding 1 ml of 1 N perchloric acid (PCA) to each dish after washing three times with 5 ml aliquots of 0.3 N PCA. After cell removal and centrifugation, the final precipitate was dissolved in 2 ml of 0.5 N PCA and heated to 80°C for 20 minutes. After cooling, the suspension was assayed for DNA according to Burton.[5] Tritiated thymidine was counted as previously described.[1] The growth of epidermal cells in culture was observed daily under phase contrast microscopy and photographed. Keratinization of the differentiating cells was visualized by staining cells with acid fuchsin-aniline blue-orange G stain mixture as suggested by Dr. Karasek, Stanford University.

RESULTS

As seen by phase contrast microscopy, the cells are round and discrete immediately after inoculation. They began to adhere to the dish and tended to form aggregates of from 5 to 10 cells each after 2 to 3 days in culture. They continued to aggregate and form large and multiple clumps, and finally began to layer after 9 days in culture (Fig. 1). There was an ap-

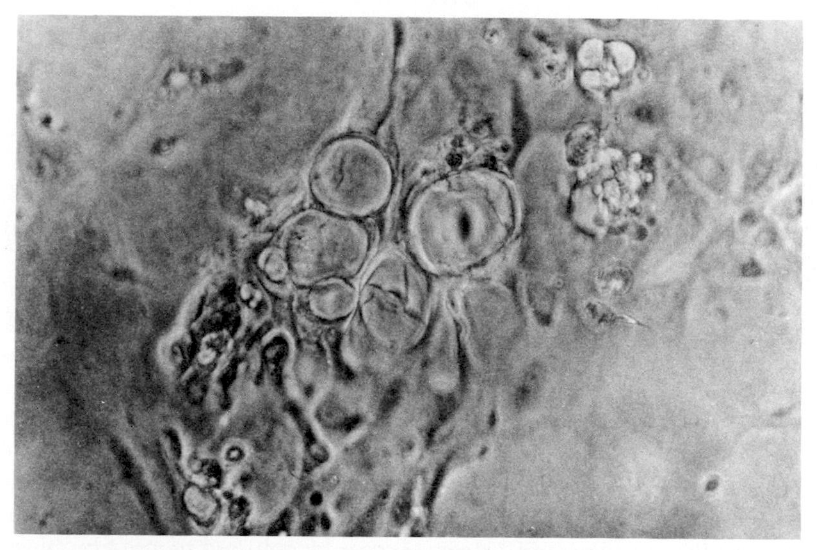

Fig. 1. The epidermal cells aggregated and organized in a layer system, 9 days' culture.

parent increase in cell volume after 10 days and some of the cells were seen to contain fine granules and display signs of keratinization, which was best visualized by the typical pink-red color resulting from the acid stain. At this time, fibroblasts and small numbers of unidentified cells were observed beginning to grow in the dishes. Along with the aggregated and layered cells, some of the cells were seen to form monolayer colonies.

As shown in Fig. 1, the protein content of cells per dish showed a slight increase at days 5 to 6. There was a larger increase starting at day 9. During this period, day 3 through day 9, DNA concentration per dish did not change significantly, and at days 10 to 13, there was a slight decrease, as seen in Fig. 1. Thymidine uptake increased sharply at day 4 and then gradually dropped until another significant increase between days 7 through 9. Leucine and histidine uptake into cellular macromolecules increased significantly at day 4, and a second peak increase was seen at day 9 (Fig. 2). There was no significant difference in histidine uptake compared to that of leucine during the later phases of growth, compared to the earlier days of incubation, as seen in Fig. 2.

There was a statistically significant drop in cAMP concentration on day 4 when thymidine uptake, as well as amino acid incorporation, had peaked (Fig. 3). Not shown in Figs. 2 and 3 is the fact that cAMP concentration was constant from 10 hours after cell seeding up to day 2 of culture.

DISCUSSION

The present study of epidermal cells in culture was similar to the observations of Constable *et al.*,[2] who, by not changing the medium during the first 3 days, showed that cells, aggregated and organized in a layered system, were keratinized. Although the differentiated cells were stained pink-red, typical of keratinization,[6] the suggested criteria of specific histidine incorporation as another index of keratinization,[7] was not satisfied. There was no significant differential increase in histidine uptake over that of leucine.

There was a biphasic change in cAMP concentration in the epidermal cells during the growth period. This suggests a biphasic function of cAMP in cell growth, the first in cell division, followed by its action in differentiation. The initial drop, simultaneous with a peak increase in thymidine uptake as well as amino acid incorporation, can be regarded as one aspect of the pleiotypic response, that is, a coordinated response of a number of essential biochemical functions to a change in the cell's environment.[3] Kram *et al.*[8] have shown that exogenous dibutyryl cAMP inhibits the uptake of uridine, leucine, and 2-deoxyglucose by cultured mouse fibroblasts and suggested a role of cAMP in the negative control of the pleiotypic

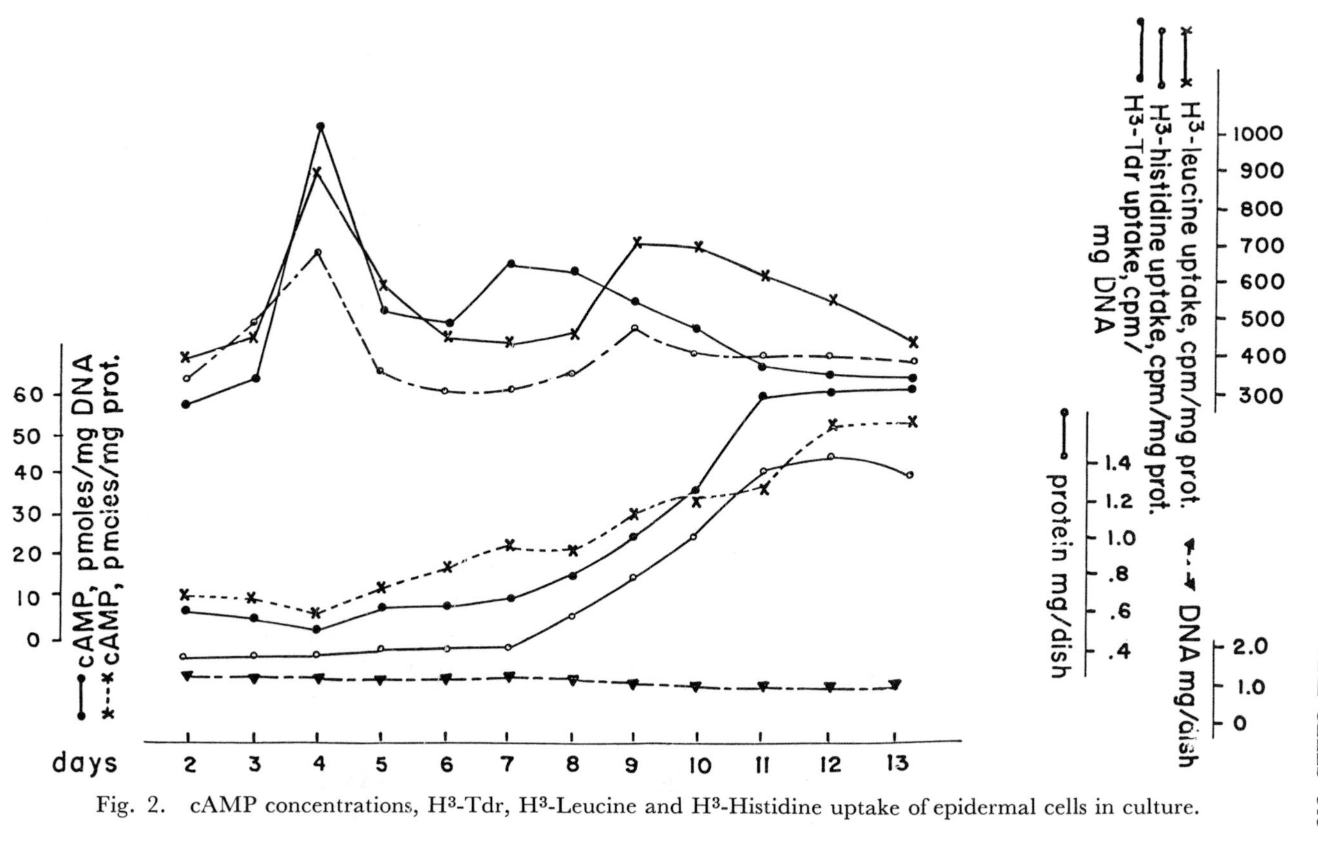

Fig. 2. cAMP concentrations, H^3-Tdr, H^3-Leucine and H^3-Histidine uptake of epidermal cells in culture.

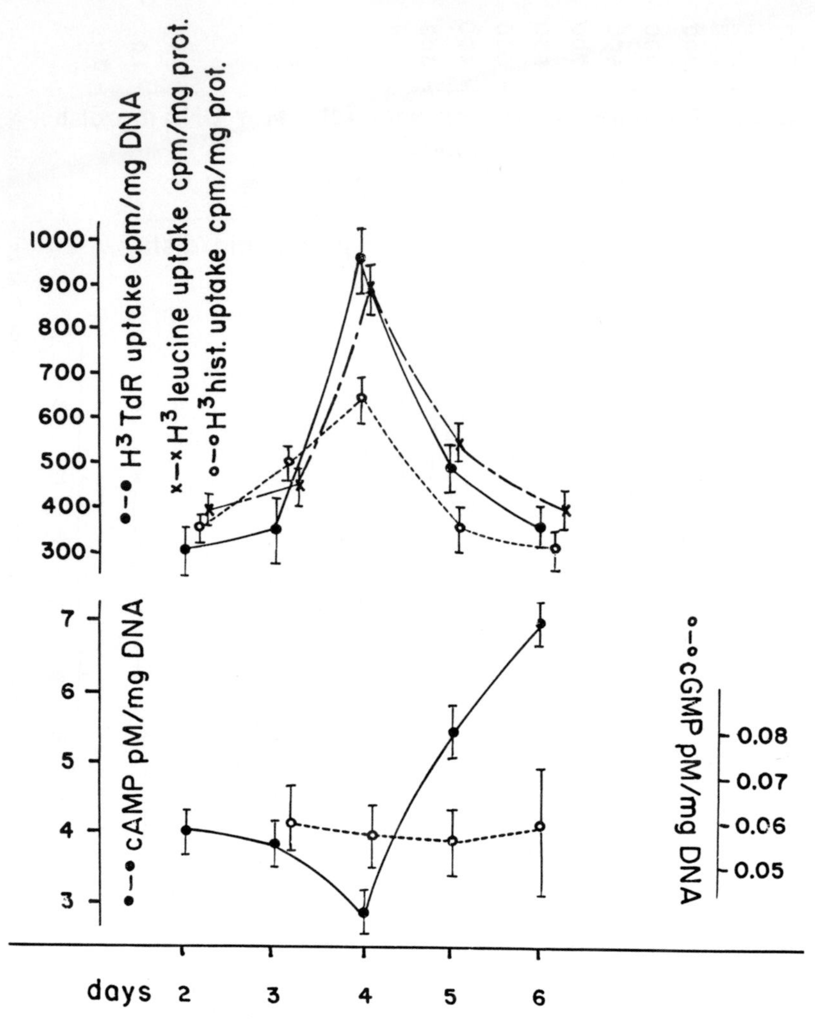

Fig. 3. Concentration of cAMP and cGMP, and uptake of H^3-TdR, H^3-leucine and H^3-histidine by epidermal cells in early period of growth.

response. The decrease in cAMP concentration may be related to changes in cellular permeability and microtubular function.[9,10] The cells, in this view, can utilize substrate for initiating cellular response to environmental change. When this drop in cAMP concentration is prevented by treating the cells with cAMP stimulating agents, there is inhibition of growth[11] or cell division[12] of epidermal cells.

Since significant increase in cGMP concentration with simultaneous decrease of cAMP concentration was not observed during early growth phases, we cannot confirm the suggestion of positive pleiotypic control via cGMP as proposed by Kram and Tomkins[10] or a suggestion that changes in the concentration ratio of cAMP to cGMP can be a dominant factor which triggers epidermal cell division.[13]

Recent studies, however, indicate that cGMP plays a primary role in cell division in other cell types.[14,15] It has been shown that increases in cGMP concentration in fibroblasts were only seen when cultures which were arrested in the G_0 phase were stimulated to grow, or when synchronized growing cells passed through the G_1 phase.[14] The postulated role for increases of cGMP in the cell cycle allows for the possibility that the experimental model used here would not allow detection of this cell cycle-specific change in our nonsynchronized cultures. Immediately after the initial drop of cAMP concentration and the peak uptake of thymidine and amino acids, there was a steady increase of cAMP concentration which was followed by an increase of protein synthesis and differentiation in the cells. Previous work of Aso *et al.*[1] and Grimm and Marks[16] showed that there was an increase of cAMP concentration in the proliferating epidermis of guinea pig ear and mouse back skin, stimulated respectively by topical retinoic acid and phorbol ester. Recently Delescluse *et al.*[17] have shown that exogenous dibutyryl cAMP induces keratinization of guinea pig epidermal cells in culture, suggesting a specific role of cAMP in keratinization. In other cell types, evidence is accumulating that cAMP plays a specific role in differentiation, as in neurite induction of neuroblastoma cells,[18] pigment production of melanoma cells,[17,20] formation of flagella or induction of stalk cells in *Eschericia coli* and *Dictyostelium discoideum*.[21,22] These reports, together with the observations reported here, argue for a dominant role of cAMP in the differentiation of epidermal cells.

Although one can speculate that diverse expressions of differentiation in the cell would necessarily arise from the cofactor role of cAMP in the phosphorylation and activation of protein kinase enzymes, a definite mechanism or involvement of cAMP in epidermal cell differentiation is not clear. Glycogen utilization and synthesis, for which the role of cAMP is well known[23] and which is a distinct feature of proliferating epidermal cells, may, for example, be one of these diverse expressions of epidermal cell differentiation.

REFERENCES

1. Aso, K., Rabinowitz, I. and Farber, E.M. The role of prostaglandin E, cAMP and cGMP on the proliferation of guinea pig ear skin stimulated by topical retinoic acid, *J. Invest. Dermat.*, In press.
2. Constable, H., Cooper, J. R., Cruickshank, C. N. D. and Mann, P. Keratinization in dispersed cell culture of adult guinea pig ear skin, *Br. J. Dermat.*, **91**, 39–48, 1974.
3. Henshko, A., Mamont, P., Shield, R. and Tomkins, G. M. Pleiotypic response, *Nature New. Biol.*, **232**, 206–211, 1971.
4. Lowry, O. H., Rosebrough, N. J., Farr, A. L. and Randall, R. J. Protein measurement with Folin phenol reagent.
5. Burton, K. A study of the conditions and mechanism of the diphenylamine reaction for colorimetric estimation of DNA, *Bioch. J.*, **62**, 315–322, 1956.
6. Personal communication with Dr. M. Karasek.
7. Fukuyama, K. and Epstein, W. L. Epidermal keratinization, localization of isotopically labeled amino acid, *J. Invest. Dermat.*, **49**, 595–604, 1967.
8. Kram, R., Mamont, P. and Tomkins, G. M. Pleiotypic control by cAMP: A model for growth control in animal cells, *Proc. Nat. Acad. Sci. USA*, **70**, 1432–1436, 1973.
9. Sheppard, J. R. and Plagemann, G. cAMP, membrane transport and cell division. 1. Effect of various chemicals on cAMP levels and rate of transport of nucleosides, hypoxanthine and deoxyglucose in several lines of cultured cells, *J. Cell Physiol.*, **85**, 163–172, 1974.
10. Kram, P. and Tomkins, G. M. Interaction with cGMP and possible role of microtubles, *Proc. Nat. Acad. Sci. USA*, **70**, 1659–1663, 1973.
11. Delescluse, C., Colburn, N. H., Duel, E. A. and Voorhees, J. J. cAMP elevating agents inhibit proliferation of keratinizing guinea pig epidermal cells, *Differentiation*, **2**, 343–350, 1974.
12. Voorhees, J. J., Duel, E. A. and Kelsey, W. H. Dibutyryl cAMP inhibition of epidermal cell division, *Arch. Dermat.*, **105**, 387–386, 1972.
13. Voorhees, J. J., Stawaiski, M., Duel, E. A., Haddox, H. K. and Goldberg, N. D. Increased cGMP and decreased cAMP levels in the hypertrophic abnormally differentiating epidermis of psoriasis, *Life Sci.*, **18**, 639–645, 1973.
14. Seifert, W. and Rudland, P. S. Cyclic nucleotides and growth control in cultured mouse cells, correlation of changes in intracellular cGMP concentration with specific phase of the cell cycle, *Proc. Nat. Acad. Sci. USA*, **71**, 4920–4924, 1974.
15. Goldberg, N. D., Haddox, M. K., Nicol, S. E., Glass, O. B., Sanford, S.H., Kuehl, F. A. and Estensen, R. N. Biological regulation through opposing influences of cGMP and cAMP: The Yin Young hypothesis, Advance in Cyclic Nucleotide Res., Vol. 5, edited by Drummond, G. I., Greengard, P., Robinson, G. A.: New York, Raven Press, pp 307–330, 1975.
16. Grimm, W. and Mark, F. Effect of tumor promoting phorbol esters on the normal and the isoproterenol elevated levels of cAMP in mouse epidermis *in vivo*, *Cancer Res.*, **34**, 3128–3134, 1974.
17. Delescluse, C., Fukuyama, K. and Epstein W. L. Dibutyryl cAMP induced

differentiation of epidermal cells in tissue culture, *J. Invest. Dermat.*, **66**, 8–13, 1976.
18. Pasad, K. N. and Kumar, S. Role of cAMP and differentiation of mouse neuroblastoma cells in culture in Role of cyclic nucleotides in carcinogenesis, edited by Gratzner, M. and Shultz, J. pp 207–237, Academic Press, New York, 1973.
19. Moellman, G., McGuire, J. and Lerner, A. B. Intracellular dynamics and the fine structure of melanocytes with special reference to the effect of MSH and cAMP on microtubules and 10mm filament, *Yale J. Biol. Med.*, **46**, 337–360, 1973.
20. Wikswo, M. A. Action of cAMP on pigment donation between melanocytes and keratinocytes, *Yale J. Biol. Med.*, **46**, 592–601, 1973.
21. Yokota, T. and Grots, J. Requirement of cAMP for flagella formation in *Escheria coli* and *Salmonella typhimurium*, *J. Bact.*, **103**, 513–516, 1970.
22. Bonner, J. T. Induction of stalk cell differentiation by cAMP in the cellular slime mold Dictyostelium discoideum, *Proc. Nat. Acad. Sci. USA*, **65**, 110–113, 1970.

Discussion

Dr. Karasek: Can the quantitation be extended to all receptor sites such as those for epinephrine, AMP, PG's, etc.?

Dr. Adachi: Yes, the additive effects can be demonstrated well with epinephrine, AMP and histamine added simultaneously. We have a little difficulty in showing the additive effects by the simultaneous addition of all four. The reason is due to the relatively small increase in cAMP by stimulation with PGE, as compared to the large increase in cAMP by the other three stimulators.

Dr. Freedberg: What is the natural substrate for the cAMP? What is the natural substance which is phosphorylated?

Dr. Adachi: Obviously this is a very important question, but as far as I am aware, there is no clear cut evidence or data concerning the phosphorylation of natural substrates in the epidermis.

Dr. McGuire: The time base of the response of intracellular cAMP in melanoma is similar to the results in the epidermis obtained by Dr. Adachi. There is a peak at about 30 to 45 minutes after MSH or cholera toxin. There are cell types that require cAMP to divide. At higher levels of cAMP there is inhibition of cell division.

Dr. Karasek: Is the receptor site for epinephrine also defective in psoriasis?

Dr. Aso: I have studied the stimulation of the accumulation of cAMP in psoriatic leucocytes by isoproterenol and PGE_1. There was a reduced stimulation of cAMP accumulation by PGE_1, but not isoproterenol. I have not examined the effect of epinephrine in both psoriatic epidermis and psoriatic leucocytes.

Differentiation of Human Keratinocytes in Cell Culture*

Yukio KITANO and Hidehiko ENDO

Department of Dermatology, Osaka University School of Medicine, Osaka, Japan

ABSTRACT

There is no doubt that the dermis plays an important role in the differentiation and maintenance of the epidermis. On the other hand, keratinocytes dispersed from epidermis by trypsinization grow on a glass or a plastic surface in a suitable culture medium, and show some signs of differentiation. The present investigation was undertaken to characterize the differentiation of human keratinocytes in cell culture without dermal components.

The keratinocytes dispersed from the adult human epidermis were grown in a plastic dish and observed for differentiation. About 10 days after the initiation of the culture, fibrillar structures became prominent in the cytoplasm. Intercellular bridges could be clearly seen in the wide intercellular space, and the cells looked spinous. In some areas the uppermost cells became large, took polygonal shape with a clear cellular boundary, and had a pyknotic nucleus. One or two times in subculture did not seem to affect the differentiation of the keratinocytes.

The histological and electron microscopic observations of the sheet of keratinocytes formed in culture revealed that the cells of the lowermost layer were comparable to the basal cells, and the keratinized cells appeared on the surface of the sheet. Thus the keratinocytes differentiated *in vitro* without dermal components, and the process of the differentiation had many similarities to keratinization *in vivo*.

INTRODUCTION

The importance of mesenchyme in the differentiation of epidermis was well established. McLoughlin in 1961 showed that the isolated epidermis

* Supported by research grants from the Ministry of Education, Japan

from embryonic chick skin became completely disorganized, showing very few mitoses, and rapidly keratinized or degenerated, while the explants of the embryonic chick skin followed the normal course of differentiation.[1] It was also shown that the isolated epidermis explanted in contact with cell-free intercellular mucoid material from the mesenchyme kept an organized arrangement. Wessels, by his Millipore filter assembly technique, demonstrated that the dermis could maintain the proliferation and differentiation of the epidermis when Millipore filter was inserted between the tissues.[2] Dodson found that the presence of dermal cells was not essential for survival and keratinization of the epidermis, but some components of the dermis were important for these processes.[3] A collagen gel could provide a suitable surface on which the epidermis was capable of normal differentiation when a macromolecular fraction of chick embryo extract was present as a nutrient.[4,5] These studies were done using the embryonic skin. Briggaman and Wheeler proved that similar epidermal-dermal interactions were working in the maintenance of the adult human epidermis.[6,7] Although the role of dermis is emphasized in the normal relationship between epidermis and dermis, it is clear that the living dermis is not essential for the differentiation of epidermis.

It was established that keratinocytes dispersed from epidermis by trypsinization grew on a glass surface or a plastic surface in a suitable culture medium.[8–12] More recent studies have shown different cell types and the organized arrangement of cells in the culture of dissociated keratinocytes or in the epithelial outgrowth of the explanted skin. Friedman-Kien *et al.* and Prose *et al.* recognized keratinization in the epithelial outgrowth from the adult human skin explanted on a plasma clot.[13,14] Tonofilaments, desmosomes, and lamellar granules were observed. Flaxman *et al.* reported organized structures in epithelial outgrowth on a glass or a plastic surface. In the dispersed cultures of keratinocytes Constable *et al.*[15] described, the keratinocytes tended to aggregate into clumps in which there were spaces. The keratinocytes were organized bordering onto the spaces and showed a progression to keratinization.[16] Fusenig and Worst demonstrated that dispersed keratinocytes proliferated, and formed stratified squamous epithelium on a plastic surface in the absence of dermal substrate or dermal constituents.[17] The present study is concerned with the differentiation of keratinocytes in the absence of the dermal influence.

MATERIALS AND METHODS

Human skin was obtained from adults by a dermatome set at 15/1000 inches, and was soaked in Eagle's minimal essential medium. When the skin of full thickness was used, the dermal side was trimmed with fine iris scissors as far down as possible. The samples were cut into pieces of about

$2 \times 10\ mm^2$, treated with 0.02% EDTA solution for 5 minutes at room temperature, and kept in 0.25% trypsin (Difco) in Ca- and Mg-free phosphate buffered saline for 12 to 24 hours in a refrigerator (4°C). Thereafter, the epidermis was separated completely from the dermis by peeling off with fine forceps.

Epidermal cell suspension was obtained by dissociating the epidermal sheet with forceps and by pipetting. After centrifugation the cells were resuspended in a culture medium, and were distributed into 60 × 15 mm plastic tissue culture dishes (Falcon Plastic). These were incubated in a moisture-controlled chamber with 5% CO_2 in air at 37°C. The medium consisted of Eagle's minimal essential medium supplemented with 30% fetal bovine serum. Subcultures were made by dissociating the sheet of the keratinocytes, formed during cultivation into single cells. In order to harvest the keratinocytes the culture was washed with Ca- and Mg-free phosphate buffered saline, and incubated in a mixture of equivolume of 0.25% trypsin and 0.02% EDTA for a desirable time.

For histological examination the cultures were fixed in 10% formalin, and stained with May-Grünwald Giemsa, or Rhodamin B.[18,19] In some cultures, when the sheet of keratinocytes was well established, the sheet was separated from the dish by pulling with forceps. The separated sheet was fixed in 10% formalin, embedded into paraffine, and histological sections were made. The sections were stained with hematoxylin and eosin (HE) or Rhodamin B.

For electron microscopy the cultures were fixed in 2.6% glutaraldehyde in 0.1 M phosphate buffer (pH 7.4) for 30 minutes at room temperature. After washing in 0.1 M phosphate buffer for 20 minutes, the cultures were postfixed in 1% osmium tetroxide in 0.1 M phosphate buffer (pH 7.4) for 30 minutes. The specimens were then dehydrated in an ethanol series followed by hydroxypropylmethacrylate, and embedded in Epon 812. After polymerization the resin was separated from the culture dish mechanically. Ultrathin sections were cut with LKB ultramicrotome, mounted on copper grid-meshes, and were stained with saturated uranyl acetate solution and lead citrate solution. The sections were observed with Hitachi HU-12 electron microscope.

RESULTS

Within 24 hours of cultivation, the keratinocytes attached to the bottom of the culture dishes singly or in small aggregates. The single keratinocyte migrated around, and upon collision with another keratinocyte they connected with each other. The aggregates grew out, forming cell islands (Fig. 1). After several days, the islands formed a confluent sheet in which the keratinocytes took polygonal shapes, and were arranged in a pave-

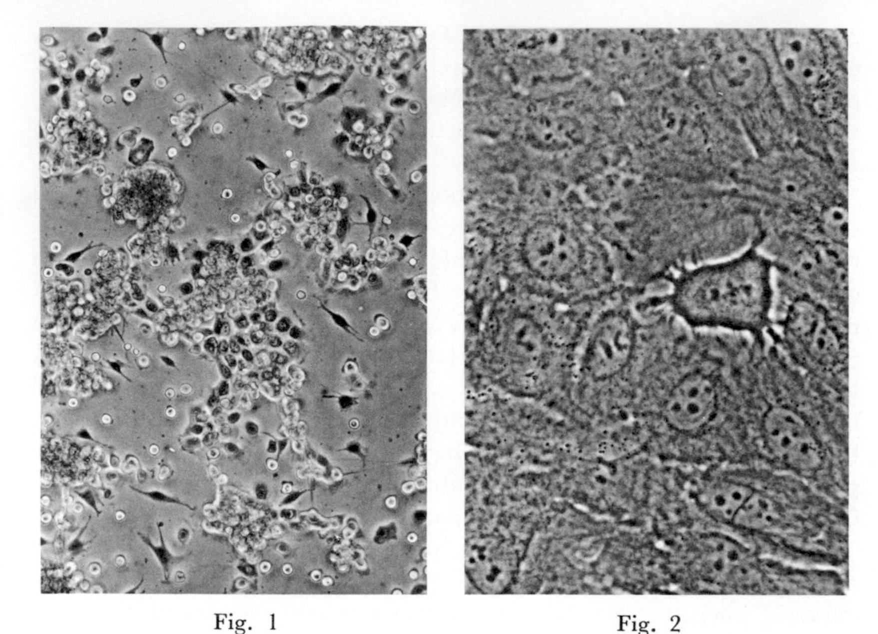

Fig. 1 Fig. 2

Fig. 1. 48 hours after initiation of culture. Keratinocytes attached to the bottom of the culture dish singly or in small aggregates, and began to form cell islands. Melanocytes with long processes were scattered among the cell islands. Phase contrast. ×70

Fig. 2. Confluent sheet of keratinocytes. The keratinocytes had a homogeneous cytoplasm with several granules. One of the keratinocytes was in mitosis. 9 days after the second subculture. Phase contrast. ×300

mentlike mosaic. At first the sheet consisted of monolayer of the keratinocytes which had a large nucleus with prominent nucleoli and a homogeneous cytoplasm with several granules in the perinuclear area (Fig. 2). Numerous mitotic figures were observed, and then the keratinocytes stratified. About 10 days after the initiation of the culture, fibrillar structures became prominent in the cytoplasm (Fig. 3). Intercellular bridges could be clearly observed in the wide intercellular spaces, and the keratinocytes looked spinous (Fig. 4). In some areas the uppermost cells became large, took polygonal shape with a clear cellular boundary, and had a pyknotic nucleus (Fig. 5). No intercellular bridge could be seen among these cells. Desquamated cells were observed occasionally (Fig. 6).

Subcultures were made from the cultures in which the keratinocytes made a confluent sheet and had not stratified yet. After 20 to 30 minutes of the treatment with trypsin and EDTA, the keratinocytes lost the inter-

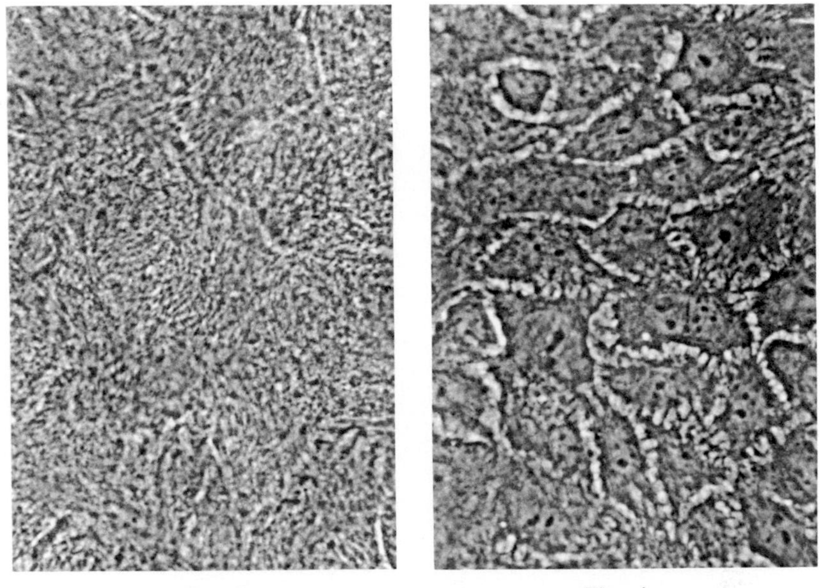

Fig. 3 Fig. 4

Fig. 3. Intricate fibrillar structures in cytoplasm of keratinocytes. 24-day-old primary culture. Phase contrast. ×300

Fig. 4. Prominent intercellular bridges in the wide intercellular space. The cytoplasm looked darker than that of figure 2, and fibrillar structures could be seen. 11 days after the second subculture. Phase contrast. ×300

cellular connection, and became spherical (Fig. 7). There appeared holes in the sheet due to dissociation of the cells. The holes became larger, and finally most keratinocytes came off as single cells. When transferred to a new dish, the cells proliferated in a similar way to the primary culture. The keratinocytes attached to the culture dish singly or in small aggregates, and grew to form a confluent sheet. The keratinocytes were polygonal in shape, and were arranged in a pavementlike mosaic. When the keratinocytes in the sheet stratified, fibrillar structures became prominent in the cytoplasm, and the intercellular bridges were conspicuous. We have so far progressed up to the fourth subculture to find the same pattern of proliferation and differentiation as in the primary culture.

When the keratinocytes in the sheet have stratified, and the features of maturation become distinct, the sheet could be separated from the culture dish by pulling with forceps. In some areas the lowermost layer of the sheet was left on the bottom of the culture dish (Fig. 8). The keratinocytes in the lowermost layer were rather small, and did not have prominent

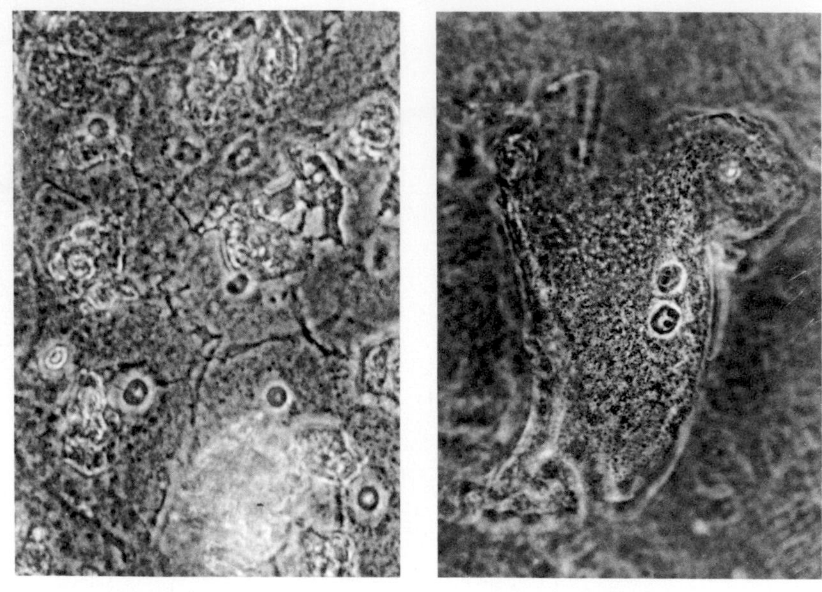

Fig. 5 Fig. 6

Fig. 5. Large polygonal cells with clear cellular margins. The nuclei were pyknotic and intercellular bridges were not observed. Degenerated cells or cellular debris were observed on top. 18-day-old primary culture. Phase contrast. ×300

Fig. 6. Desquamated cell. This cell was distended flatly, and had two pyknotic nuclei. 18 days after the second subculture. Phase contrast. ×300

fibrillar components in the cytoplasm. Many melanocytes were seen scattered among the keratinocytes.

The histological sections of the separated sheet were cut perpendicular to the surface. The sheet consisted of several layers of keratinocytes (Fig. 9). The cells in the lowermost layer were cuboidal, and had somewhat basophilic cytoplasm in HE staining. The upper layer consisted of flatly distended eosinophilic cells with or without a nucleus. The nucleus, where present, was pyknotic or ghostlike with faint eosinophilic nuclear membrane and inner structures. The keratinocytes in the intermediate layer were fusiform or spindle-shaped in cross section and had an elliptical nucleus. The cytoplasm was eosinophilic and fibrillar. A few keratinocytes near the upper layer contained fine basophilic granules which resembled keratohyalin granules (Fig. 10). Rhodamin B stained preparations revealed orange-colored fluorescence, which was intense in the upper layer (Fig. 11). Rhodamin B staining of the unsectioned sheet showed orange-colored fluorescence increasing with the daily lapse of cultivation time.

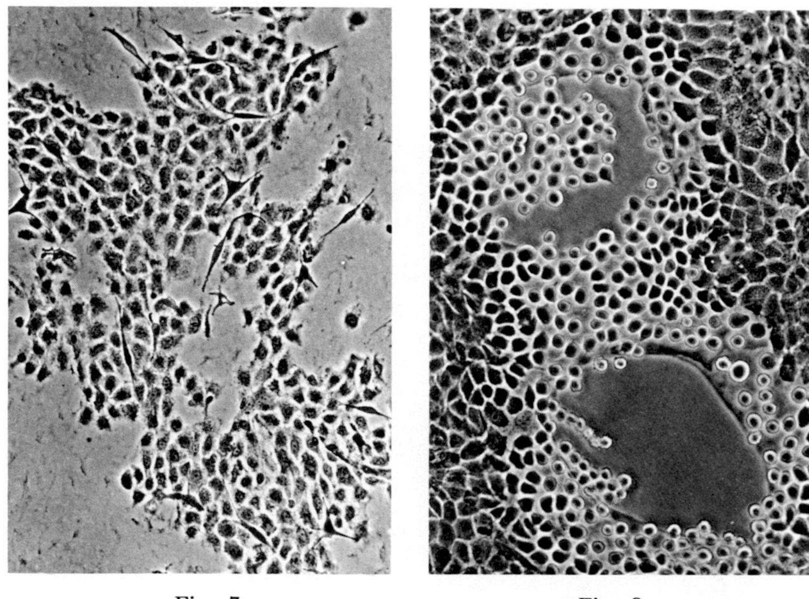

Fig. 7 Fig. 8

Fig. 7. Dissociation of sheet of keratinocytes into single cells. The connection among the keratinocytes was lost by the treatment with trypsin and EDTA, and there appeared holes in the sheet. The polygonal keratinocytes became spherical. Phase contrast. ×70

Fig. 8. Lowermost layer of sheet of keratinocytes after mechanical separation of upper layers. Many bipolar or tripolar melanocytes with cytoplasmic processes were observed among the keratinocytes. Phase contrast. ×70

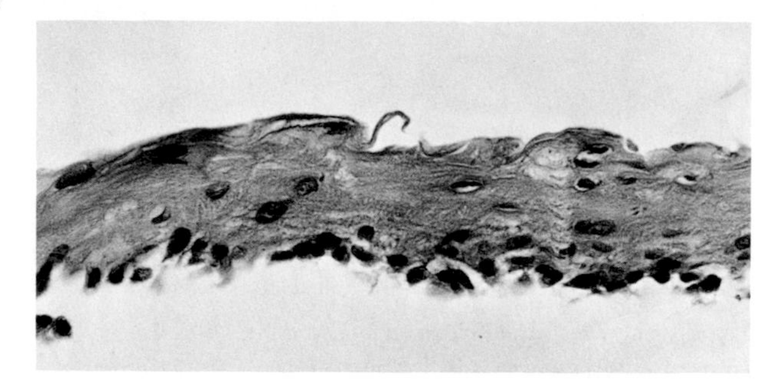

Fig. 9. Cross sectional view of sheet. The cells in the lowermost layer were small and cuboidal. The upper layer consisted of flatly distended cells, and resembled a horny layer. 27-day-old primary culture. HE. ×80

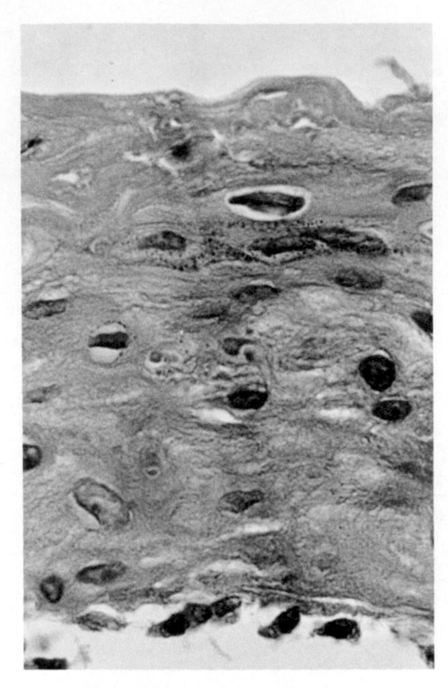

Fig. 10. Cross sectional view of sheet. The fibrillar structures were evident in the cytoplasm of the cells of the intermediate layer. A few cells in the upper intermediate layer contained keratohyalin granules. 27-day-old primary culture. HE. ×150

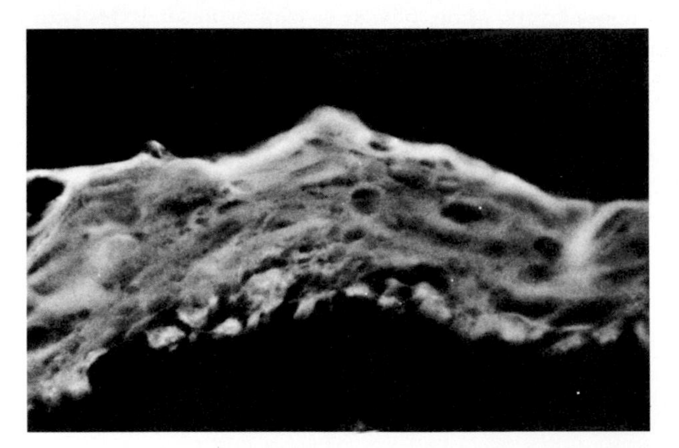

Fig. 11. Rhodamin B stained section viewed in fluorescent light. Orange-colored fluorescence was intense in the upper layer. ×130

The second subcultures grown for 19 days were served for electron microscopic observations. The differences of the keratinocytes in individual layers were well shown in the sections cut perpendicular to the surface of the sheet of keratinocytes. Intercellular spaces in the lower layers slightly widened, and interdigitation existed in some areas. Keratinocytes were connected with each other by desmosomes, which seemed to decrease in number as the cells went up. The keratinocytes in the lowermost layer, however, lacked half-desmosomes, and the basal lamina did not exist at the bottom (Fig. 12). The nuclei of keratinocytes in the lower layers sometimes had deep indentations (Fig. 13). The cytoplasmic appearance of keratinocytes in the lower and intermediate layers was quite different from that in the upper layers. The cytoplasm of keratinocytes in the lower and intermediate layers were rich in organelles such as free ribosomes, mitochondria, microtubules, melanosomes, small vesicles, secondary lysosomes, and glycogen granules (Figs. 14, 15). Mitochondria were usually elongated. Tonofilaments were sparse in the lower layer, and gradually increased as the cells moved upward. Keratohyalin granules were also observed in the cytoplasm of keratinocytes in the intermediate layer (Fig. 16). Membrane-coating granule was not found so far as we investigated.

Ultrastructural features changed abruptly as the cells ascended from the intermediate to the upper layers. The cytoplasm of keratinocytes in the upper layer was less electron dense, and contained only a few organelles, whereas tonofilaments rather increased in number (Fig. 17). No ribosomes, mitochondria, and microtubules were observed. Indentation was seen in the cytoplasm of the uppermost layer. A thickening of the cytoplasmic membrane was noticed The fine structures of the keratinocytes in the upper layers thus resembled that of keratinized cells of the human epidermis *in vivo*.

DISCUSSION

The keratinocytes grown in cell culture formed the well-differentiated epidermis when the cells were grafted back onto appropriate recipients.[20–22] Yuspa *et al.* reported the appearance of epidermal appendages from the donor epidermis.[21] Bauer and de Grood showed that subcultured keratinocytes formed a new stratum corneum under organ culture condition on a Millipore filter.[23] These results clearly indicated that the keratinocytes in cell culture kept the potential to form the epidermis when transferred to a suitable circumstance.

In the present study we could demonstrate that the keratinocytes proliferated, differentiated, and organized to form a stratified squamous epithelium in culture without the dermal elements. Increase of the fibrillar structures in the cytoplasm was observed during the cultivation. In the

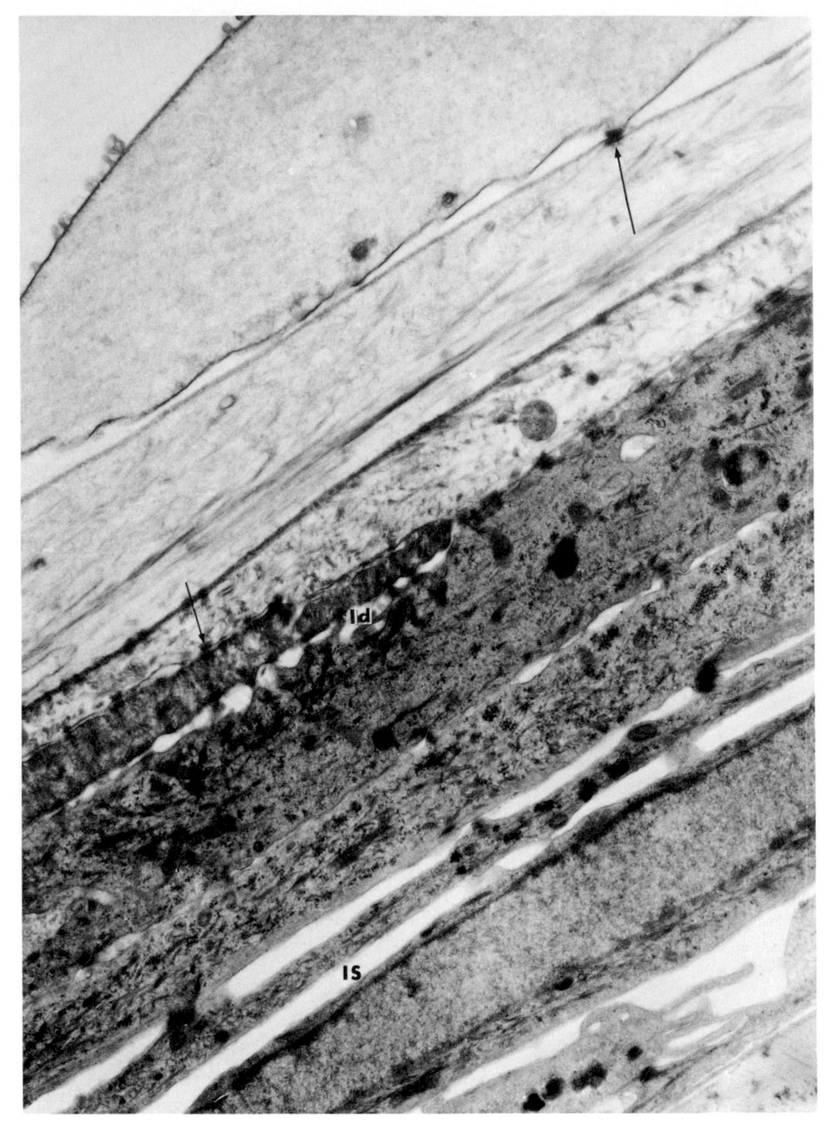

Fig. 12. Electron micrograph of stratified keratinocytes. Keratinocytes were connected with each other by desmosomes (arrows). Intercellular space (Is). Interdigitation (Id). ×6,300

stratified squamous epithelium formed in culture the lowermost layer consisted of cuboidal cells, and the cells in the upper layer were often flattened with a pyknotic nucleus. Electron microscopic observations clearly showed

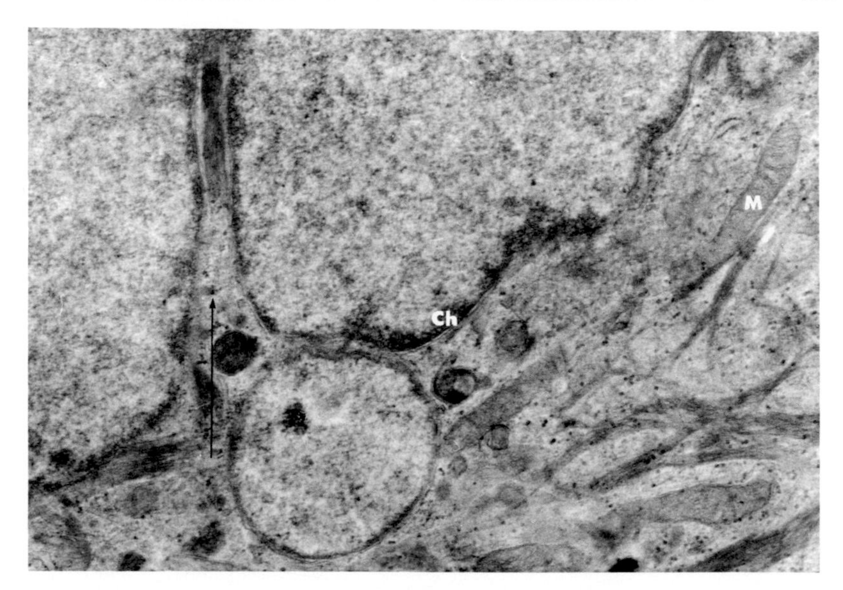

Fig. 13. Electron micrograph of keratinocyte in lower layer. The nucleus was indented (arrow). Perinuclear chromatin (Ch). Elongated mitochondria (M). ×13,200

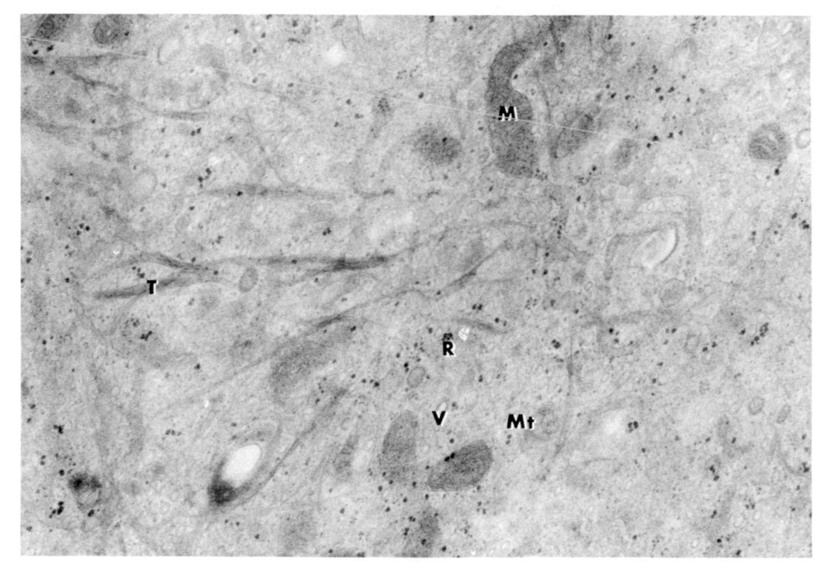

Fig. 14. Electron micrograph of keratinocyte in lower layer. The cytoplasm contained numerous free ribosomes (R), mitochondria (M), microtubules (Mt), tonofilaments (T), and small vesicles (V). ×13,200

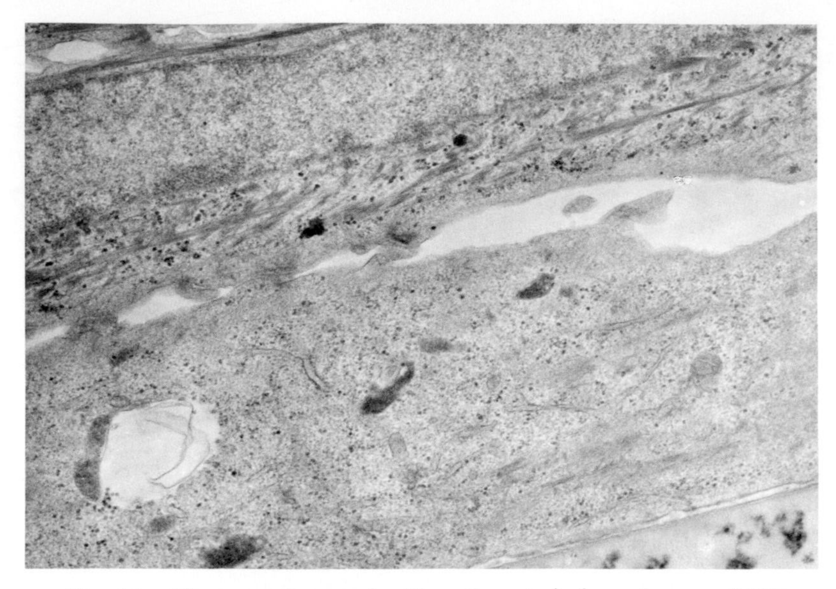

Fig. 15. Electron micrograph of keratinocyte in lower layer. ×8,600

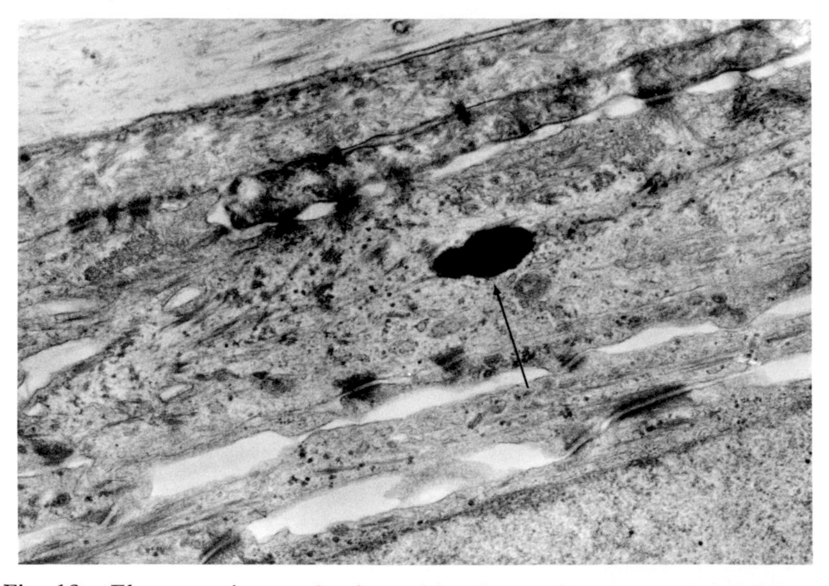

Fig. 16. Electron micrograph of transitional zone between upper and intermediate layers. A keratohyaline granule (arrow) was observed in the keratinocytes of the intermediate layer. Note the abrupt structural change of the cytoplasm of keratinocytes of intermediate and upper layers. ×13,200

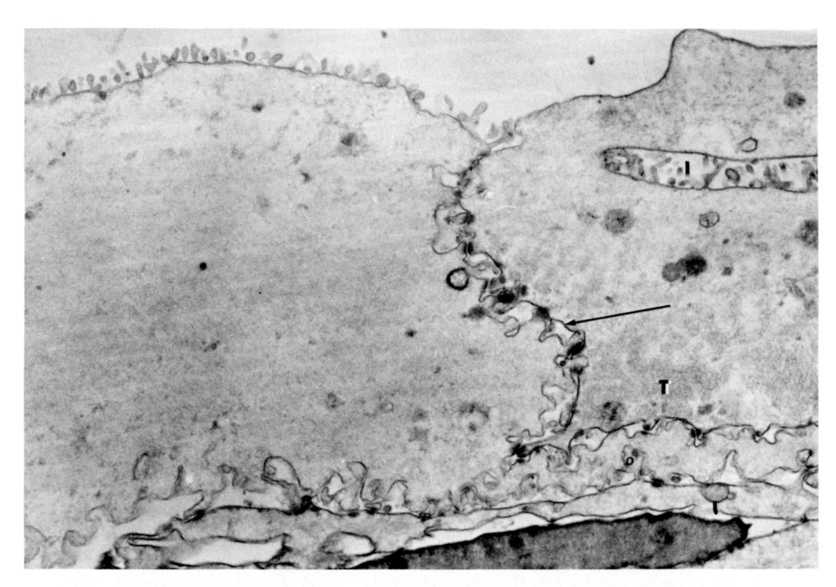

Fig. 17. Electron micrograph of keratinocyte in upper layer. Note a thickening of the cytoplasmic membrane (arrow). Cytoplasmic indentation (I). Tonofilaments (T). ×4,800

maturation of the keratinocytes as these went up from the lower to the upper layers. The cells in the upper layer contained more tonofilaments and less cytoplasmic organelles such as mitochondria and ribosomes than those in the lower layer. Keratohyalin granules were observed. The cells in the uppermost few cell layers thickened cytoplasmic membrane, and appeared keratinized. Thus the differentiation of keratinocytes was achieved *in vitro* without dermal components. Although the differentiation was not so complete, it had many similarities to keratinization *in vivo*. The cells of the lowermost layer on the plastic surface which was supposed to be biologically inactive were comparable to the basal cells, and the keratinized cells appeared on the surface of the sheet.

One or two times of subculture did not seem to affect the differentiation of the keratinocytes *in vitro*. Cruickshank *et al.* described the appearance of fusiform cells which were quite unlike the keratinocytes of the initial cultivation, and concluded that the fusiform cells were dedifferentiated keratinocytes.[10] Bauer and de Grood also showed that the subcultured keratinocytes took fibroblastlike appearance.[23] In our study the keratinocytes kept the original polygonal shape and the pavementlike arrangement after the subculture.

Microtubules, ribosomes, and elongated mitochondria were observed in the keratinocytes of the lower layer, and these were not prominent *in vivo*. The deep indentation of the nucleus was quite similar to the cleft or channel observed by Constable in monolayer culture of adult human epidermal cells, and was not seen in the epidermis.[24]

CONCLUSION

The keratinocytes dispersed from the adult human epidermis were grown in a plastic dish, and observed for the differentiation in cell culture without dermal components. About 10 days after the initiation of the culture, fibrillar structures became prominent in the cytoplasm. Intercellular bridges could be clearly seen in the wide intercellular spaces, and the cells looked spinous. In some areas the uppermost cells became large, took polygonal shape with a clear cellular boundary, and had a pyknotic nucleus. One or two times of subculture did not seem to affect the differentiation of the keratinocytes.

The histological and electron microscopic observations of the sheet of keratinocytes formed in culture revealed that the cells of the lowermost layer were comparable to the basal cells, and the keratinized cells appeared on the surface of the sheet. Thus the keratinocytes differentiated *in vitro* without dermal components, and the processes of the differentiation had many similarities to keratinization *in vivo*.

REFERENCES

1. McLoughlin, C. B. The importance of mesenchymal factors in the differentiation of chick epidermis. I. The differentiation in culture of the isolated epidermis of the embryonic chick and its response to excess vitamin A. *J. Embryol. Exp. Morph.*, **9**: 370–384, 1961.
2. Wessels, N. K. Tissue interactions during skin histodifferentiation. *Develop. Biol.*, **4**: 87–107, 1962.
3. Dodson, J. W. On the nature of tissue interactions in embryonic skin. *Exp. Cell Res.*, **31**: 233–235, 1963.
4. Wessels, N. K. Substrate and nutrient effects upon epidermal basal cell orientation and proliferation. *Proc. Nat. Acad. Sci. U. S. A.*, **52**: 252–259, 1964.
5. Dodson, J. W. The differentiation of epidermis. I. The interrelationship of epidermis and dermis in embryonic chicken skin. *J. Embryol. Exp. Morph.*, **17**: 83–105, 1967.
6. Briggaman, R. A. and Wheeler, C. E. Epidermal-dermal interactions in adult human skin: Role of dermis in epidermal maintenance. *J. Invest. Derm.*, **51**: 454–465, 1968.
7. Briggaman, R. A. and Wheeler, C. E. Epidermal-dermal interactions in adult human skin. II. The nature of the dermal influence. *J. Invest. Derm.*, **56**: 18–26, 1971.
8. Perry, V. P., Evans, V. J., Earle, W. R., Hyatt, G. W., and Bedell, W. C. Long-term tissue culture of human skin. *Am. J. Hyg.*, **63**: 52–58, 1956.

9. Wheeler, C. E., Canby, C. M., and Cawley, E. P. Long-term tissue culture of epithelial-like cells from human skin. *J. Invest. Derm.*, **29**: 383–392, 1957.
10. Cruickshank, C. N. D., Cooper, J. R., and Hooper, C. The cultivation of cells from adult epidermis. *J. Invest. Derm.*, **34**: 339–342, 1960.
11. Briggaman, R. A., Abele, D. C., Harris, S. R., and Wheeler, C. E. Preparation and characterization of a viable suspension of postembryonic human epidermal cells. *J. Invest. Derm.*, **48**: 159–168, 1967.
12. Karasek, M. *In vitro* culture of human skin epithelial cells. *J. Invest. Derm.*, **47**: 533–540, 1966.
13. Friedman-Kien, A. E., Morrill, S., Prose, P. H., and Liebhaber, H. Culture of adult human skin: *in vitro* growth and keratinization of epidermal cells. *Nature*, **212**: 1583–1584, 1966.
14. Prose, P. H., Friedman-Kien, A. E., and Neistein, S. Ultrastructural studies of organ cultures of adult human skin. *Lab. Invest.*, **17**: 693–716, 1967.
15. Flaxman, B. A., Lutzner, M. A., and Van Scott, E. J. Cell maturation and tissue organization in epithelial outgrowths from skin and buccal mucosa *in vitro*. *J. Invest. Derm.* **49**: 322–332, 1967.
16. Constable, H., Cooper, J. R., Cruickshank, C. N. D., and Mann, P. R. Keratinization in dispersed cell cultures of adult guinea-pig ear skin. *Brit. J. Derm.*, **91**: 39–48, 1974.
17. Fusenig, N. E. and Worst, P. K. M. Mouse epidermal cell cultures. II. Isolation, characterization of epidermal cells from perinatal mouse skin. *Exp. Cell Res.*, **93**: 443–457, 1975.
18. Liisberg, M. F. Rhodamin B as an extremely specific stain for cornification. *Acta Anat.*, **69**: 52–57, 1968.
19. Clausen, F. P. and Dabelsteen, E. Increase in sensitivity of the rhodamin B method for keratinization by the use of fluorescent light. *Acta Path. Microbiol. Scand.*, **77**: 169–171, 1969.
20. Karasek, M. A. Growth and differentiation of transplanted epithelial cell cultures. *J. Invest. Derm.*, **51**: 247–252, 1968.
21. Yuspa, S. H., Morgan, D. L., Walker, R. J., and Bates, R. R. The growth of fetal mouse skin in cell culture and transplantation to F_1 mice. *J. Invest. Derm.*, **55**: 379–389, 1970.
22. Worst, P. K. M., Valentine, E. A., and Fusenig, N. E. Formation of epidermis after reimplantation of pure primary epidermal cell cultures from perinatal mouse skin. *J. Natl. Cancer Inst.*, **53**: 1061–1064, 1974.
23. Bauer, F. W. and de Grood, R. M. Alteration of subcultured keratinocytes to keratinizing epithelium. *Brit. J. Derm.*, **89**: 29–32, 1973.
24. Constable, H. Ultrastructure of adult epidermal cells in monolayer culture. *Brit. J. Derm.*, **86**: 27–39, 1972.

Discussion

Dr. Fukuyama: When you subculture the cells, do melanocytes remain in subculture? Do they still make melanin pigment? Do you see the black dendritic cells in the subculture?

DR. KITANO: Yes, we do.

DR. FUKUYAMA: We have a problem subculturing melanocytes which we don't see in subculturing keratinocytes. The melanocytes don't come together. Is there any difference in the rate of growth between keratinocytes and melanocytes in your hands? Is the rate of growth the same in primary and in subcultures?

DR. KITANO: Melanocytes can be separated from the culture dish more easily than keratinocytes and can be transferred to the new dish easily. Melanocytes are observed in the subculture and I think the cells synthesize melanin as in the primary culture. However, the rate of proliferation is faster in the case of keratinocytes than in melanocytes. So in the second or third subculture, the number of melanocytes decreases.

DR. FREEDBERG: Dr. Kitano, I want to clarify something for myself. You seem to have done something which others have found difficult to do and it's not clear to me what you did differently from other workers to permit this to occur. If I understand correctly, you've taken adult human epidermal cells, have trypsinized them, have put them into a culture dish with Eagle's MEM and 30% calf serum and have been able to carry that culture without contamination through a number of subcultures. Is that correct?

DR. KITANO: Yes. We have carried them through three or four subcultures.

DR. FREEDBERG: But there's nothing in there besides MEM and 30% fetal calf serum. These are not growing on collagen gels or some kind of base?

DR. KITANO: No.

DR. KARASEK: Dr. Kitano, can I ask a question about the keratohyalin granules? Approximately what percentage of the cells that you see contain these electron dense particles?

DR. KITANO: Very few. Approximately 15 percent.

DR. STEINERT: I am much impressed with your sections and your microscopy. It really does look as though you have keratinization *in vitro* from your morphological evidence. What I would really like to see is some biochemistry done now to confirm the presence of epidermal keratin proteins. The reason why I would like to see this is that I have observed similar results with mouse epidermal cells in culture. But they do not make the keratin proteins and I am really interested to find out whether, in fact, your system does this.

DR. KITANO: I haven't done any biochemistry yet.

DR. ODLAND: I don't want this to sound like a platitude, but I simply wanted to say that it seems to me that this is a rather remarkable technical achievement. I would welcome hearing from other people who are doing this as to whether they, too, have been able to make cells develop to this apparent extent *in vitro*.

DR. MATOLTSY: Did you observe lamellar bodies or membrane coating granules?

DR. KITANO: So far we have not observed them.

DR. MATOLTSY: I would like to say, too, that I think you have really achieved here a very good technique where parakeratosis is not as advanced as has usually been reported in other studies of this type. If you have membrane coating granules formed, then you have produced all the major differentiation products which are formed normally in the epidermis.

DR. OGAWA: Have you ever done your subculturing with a mitogen such as phytohemagglutinin or pemphigus serum?

DR. KITANO: No, I haven't. Once I added pemphigus serum to the culture just to see whether it had any effect. I did not see any change morphologically.

Keratinization of Epidermal Cells in Cell Culture

Marvin A. Karasek and Su-Chin Liu

Department of Dermatology, Stanford University School of Medicine, Stanford, California, U.S.A.

ABSTRACT

Epithelial cells in outgrowth culture display 3 distinct growth phases, and in monolayer culture 3 growth stages. The physiology and biochemistry of each of these stages have been analyzed by light and electron microscopy, by the isolation of fibrous proteins, and by the incorporation of the appropriate isotopes for measurement of protein and lipid synthesis.

As cells mature *in vitro*, birefringent proteins accumulate. These proteins stain with reagents specific for the completely keratinized cell *in vivo*. Fibrous proteins resistant to solubilization by alkali and proteolytic enzymes are also formed. During this process the incorporation of histidine relative to leucine is increased. In the presence of retinoic acid, histochemical reactions for keratin are inhibited, and the ratio of incorporation of histidine to leucine is decreased.

In both outgrowth cultures and in monolayer cultures, epithelial cells communicate by a system of cytoplasmic attachment bridges that permit epithelial cells to behave as a continuum. These bridges may function to transmit biochemical cues required for the coordination of the synthetic and degradative reactions required for the keratinization process.

INTRODUCTION

In 1926 Strangeways and Fell described the formation of a keratinized epidermis when embryonic tissues were cultivated *in vitro*.[1] Since these initial and pioneering studies, organ culture methodology has been used widely to study keratinization in chick embryo skin,[2–4] fetal rat skin,[5] fetal human skin,[6] adult human skin,[7–9] and in the vaginal epithelium of the mouse and rat.[10–11]

Procedures to isolate and maintain postembryonic skin in cell culture as cell outgrowths[12–14] and as monolayers from dissociated epidermis from the adult human,[15–17] guinea pig,[18] newborn mouse,[19] mouse,[17] and rabbit[17] are now available. Cells in both outgrowth cultures,[20–23] and in monolayer cultures[24–30] have been reported to retain their capacity to keratinize under a variety of experimental conditions.

The keratinization process *in vivo* is a complex process involving sequences of degradative and synthetic reactions carefully coordinated with respect to time, and resulting in the formation of both fibrous and matrix proteins. If maturation is to take place normally, these reactions must be integrated. In this study we will examine the evidence for complete and incomplete keratinization in postembryonic mammalian epithelial cells in both outgrowth and monolayer cultures, present some of the growth characteristics of cells that influence maturation, and discuss the potential applications and limitations of cell culture methodology for studies of the biology, biochemistry, and physiology of the keratinization process.

Light microscopy

The growth of epithelial cells from explant cultures takes place in 3 phases.[12,31] Phase I is characterized by a lag phase of approximately 24–48 hours; Phase II by the appearance of cells and linear growth; and Phase III by a plateau in which lateral expansion of the outgrowth slows (Fig. 1). During the latter half of Phase II and during Phase III, vertical upward movement of cells replaces outward lateral movement. The cells

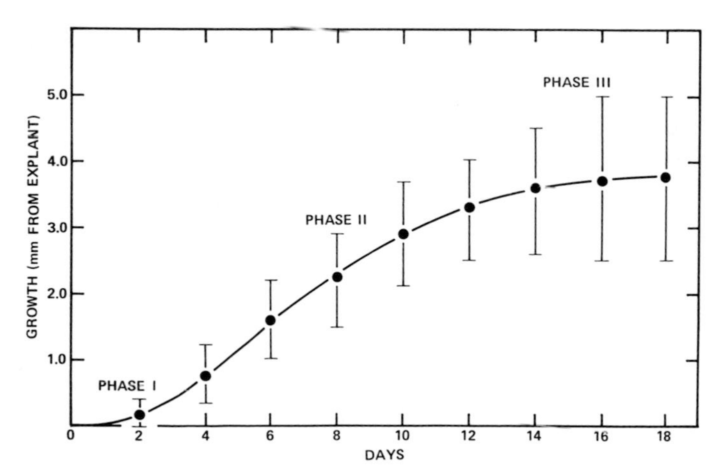

Fig. 1. Growth curve of human skin epithelial cells in outgrowth culture. Three phases in appearance and rate of growth of cells can be measured. (Reproduced from Karasek, M. *J. Invest. Dermatol.*, **65**: 60, 1975.)

in the initial outgrowth are polygonal, show desmosomal attachments, and loosely packed bundles of tonofilaments.[12]

As the cell population increases during Phase II, individual cells increase in size, and cells of up to 4500 u^2 have been measured.[32] The nuclei also enlarge, and each nucleus contains at least two nucleoli. Stratification of cells takes place in Phase II (Fig. 2).

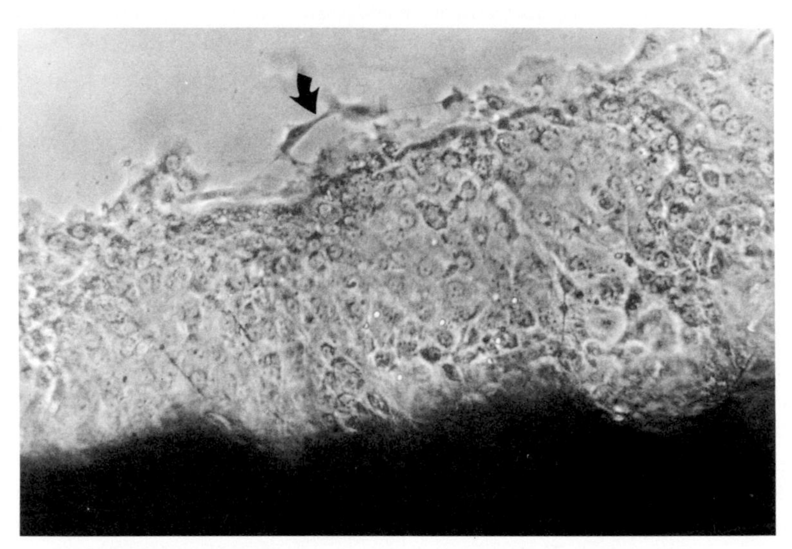

Fig. 2. Phase contrast light micrograph of human skin epithelial cells in Phase II of growth. Cells near the explant are multilayered. Attachment bridges can be seen at the periphery of the growing cell sheet (curved arrow). ×140.

In Phase III, the cultures reach an equilibrium between outward and vertical migration. During Phase III the cells accumulate an amorphous substance in bandlike arrays that is weakly birefringent (Fig. 3).

Electron microscopy

As the cells continue to stratify during Phase III, the basaloid cells show numerous villi which decrease in number and size as the cells move vertically.[33] The tonofilaments reorganize into tonofibrils and cytoplasmic organelles and nuclei are lost.[22] Tonofilaments, although condensed, are of a different density than those observed *in vivo*. Autophagosomes, not seen in normal skin, are frequently observed. The cytoplasmic double membrane is thicker and of approximately 200 Angstroms.[21] The formation of keratohyaline granules is generally not observed during Phase II, and inconsistently during Phase III.

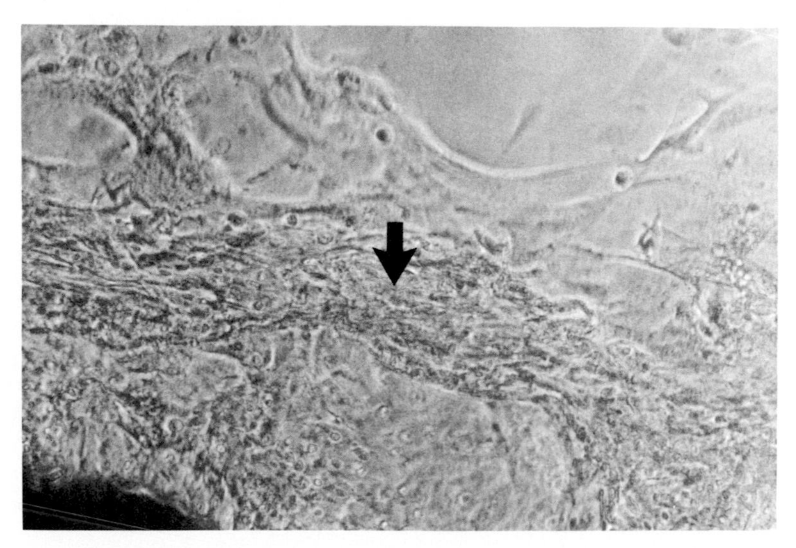

Fig. 3. Phase contrast light micrograph of human skin epithelial cells in Phase III of growth. Horizontal multiplication of cells has slowed, and cells accumulate a band of birefringent amorphous proteins (arrow). ×140.

Biochemistry

Proteins from cells in Phase I of growth are soluble in 0.1 N NaOH. In Phase II and Phase III, cells become increasingly resistant to lysis by dilute alkali and detergents.[34]

The physical and chemical properties of proteins synthesized in Phase III of growth are typical of fibrous proteins. When cells in Phase III are lysed with 2% sodium dodecyl sulfate for 30 minutes at 37°C, a residue of insoluble proteins is recovered. These proteins are insoluble in dilute acid (0.1 N HCl) and base (0.1 N NaOH). When suspended in the appropriate buffers and exposed to proteolytic enzymes (pepsin, trypsin, and chymotrypsin) soluble peptides are not released.[20] DNA and mucopolysaccharides are not present in the residue. These findings are consistent with the synthesis of fibrous proteins during Phase III of growth.

Histochemistry

Histochemical methods to differentiate keratinized from nonkeratinized cells have been described by Kreyberg,[34] Liisberg,[35] and Ayoub and Shaklar.[36] When applied to cells in Phase I of growth, histochemical reactions characteristic of stratum corneum cells are negative. In Phase II and Phase III, reactions identical to those observed in stratum corneum *in vivo* are observed (Fig. 4A).

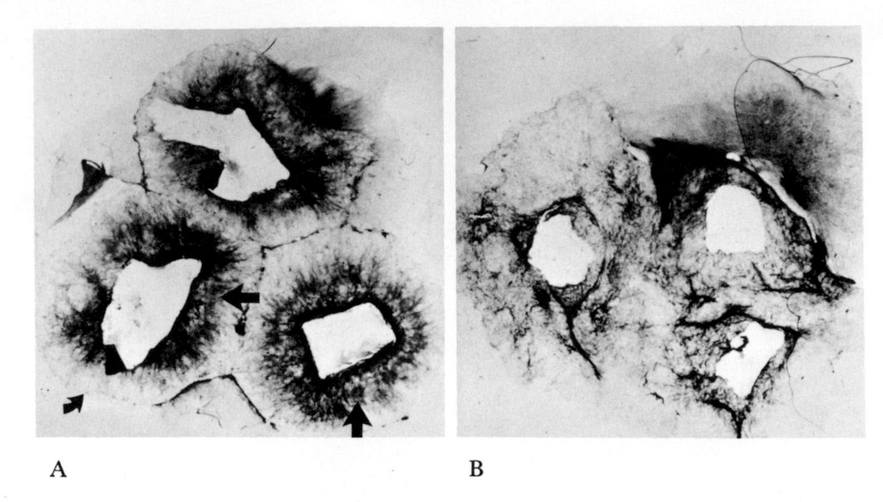

Fig. 4A and 4B.
Comparison of outgrowth in presence (4B) and absence (4A) of retinoic acid. In the absence of retinoic acid, keratinized cells stain a bright red with aniline blue-orange G (arrows). Nonkeratinized cells at edge of explant stain a dark blue (curved arrow). In the presence of retinoic acid (4B), histochemical reactions for keratin are inhibited.

Physiology

As cells appear in Phase II, communication between cells of the outgrowth is maintained by a system of cell contacts referred to as ''attachment bridges'' (Fig. 5).[12] We have suggested that these bridges may act to communicate information between epithelial cells undergoing maturation.[12] That a barrier to surface exchange is present in adult epidermis but not within the epidermis is indicated by the electrical resistance studies of Loewenstein and Kanno.[37]

To study the communication of skin epithelial cells, we have artificially altered the equilibrium established between cells in Phase III of growth, and have recorded the activity of the uninjured and injured cell populations by time-lapse motion picture photography and by direct measurement of growth. When the cell population is perturbed by wounding, signals to initiate movement are transmitted throughout the cell population, and the entire epithelial cell sheet is stimulated to contract. The population reverts immediately to Phase I of growth, and keratinization is not observed until cells reestablish contact and initiate Phase II of growth. As is shown later in this study, similar attachment bridges are also established in cells released by trypsin.

Keratinization *in vivo* is influenced by a number of systemic drugs.[38]

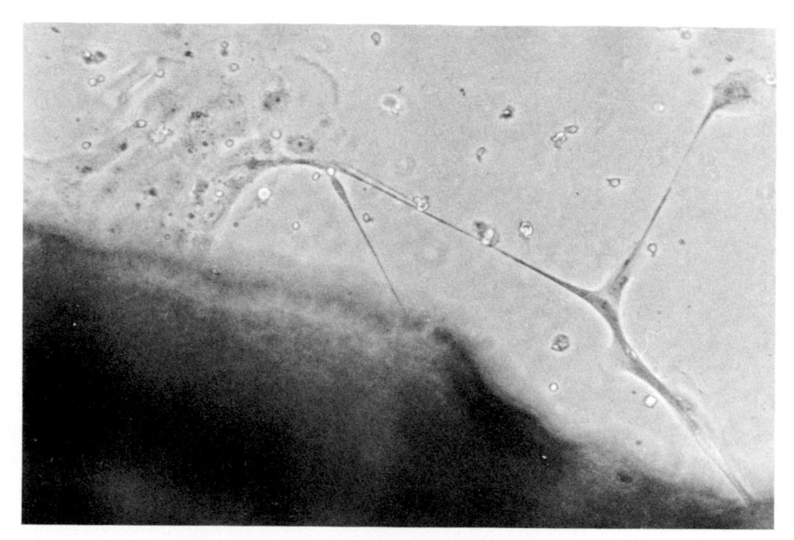

Fig. 5. Phase contrast light micrograph of elaborate network of attachment bridges formed during Phase I and II of outgrowth. ×140.

To determine if agents that alter keratinization *in vivo* produce similar changes *in vitro*, the effect of continuous exposure to all-trans-retinoic acid on three phases of epithelial cell outgrowth was measured. As shown in Fig. 6, the effect of retinoic acid is most evident when cells enter Phase III of growth. When stained with anilin blue–orange G,[36] a marked reduction in keratinized cells is observed (Fig. 4B). At higher concentrations of retinoic acid (50 μg/ml), both epithelial cell growth and keratinization is inhibited. Thus, at lower concentrations of the drug, growth is apparently enhanced due to an inhibition of the keratinization process. At higher concentrations, retinoic acid is toxic to cells. Results with retinoic acid *in vitro* are consistent with *in vivo* studies which demonstrate that moderate doses of the vitamin enhance growth, and high doses depress mitotic activity.[39]

Cell modulation

In outgrowth cultures, rabbit skin epithelial cells do not synthesize keratohyaline granules during Phase I and inconsistently during Phase II and Phase III. To determine if the inability to synthesize keratohyaline granules was a function of the *in vitro* system, rabbit epithelial cells were regrafted to a prepared site on an autologous host during Phase II and Phase III, and the fate of the regrafted cells was observed over a 6 week time period.[40] These studies clearly demonstrated that, although keratohyaline granule formation does not take place consistently in Phase II or Phase III *in vitro*, this function is rapidly reexpressed *in vivo* after grafting.

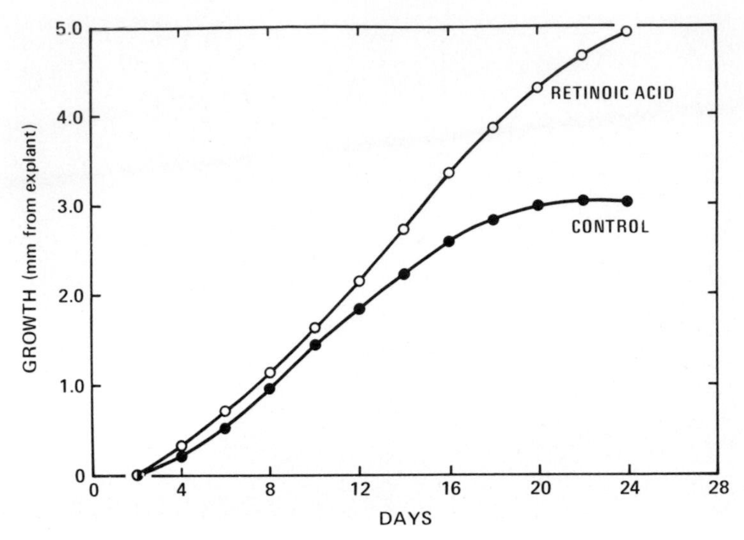

Fig. 6. Growth curves of normal adult human skin epithelial cells in the presence and absence of retinoic acid (10 μg/ml).

KERATINIZATION IN MONOLAYER CULTURES

Isolation of cells

In 1941 Medawar described a method of isolating sheets of pure epidermal epithelium from human skin by mild proteolytic digestion of split-thickness sections.[41] If exposed to a solution of trypsin for longer intervals of time with gentle agitation, further proteolysis takes place, and single epithelial cells are released. Using the trypsin procedure, dispersed populations of viable skin epithelial cells have been prepared from a number of experimental animals[17–19] and from man.[15–17]

Physiology

When plated on a collagen gel, dispersed epithelial cells from mammalian species display similar stages of growth and development characterized by attachment (State I), reassociation (State 2), and vertical movement (State 3). The morphology of the cell populations at each of these growth stages is illustrated in Figs. 7–9 for human skin epithelial cells. When stained with anilin blue–orange G, keratinized cells are observed in Stage 3.

Attachment bridges are formed during State 2 of growth and are shown in Fig. 8.

The composition of the charged surface in direct contact with prolifer-

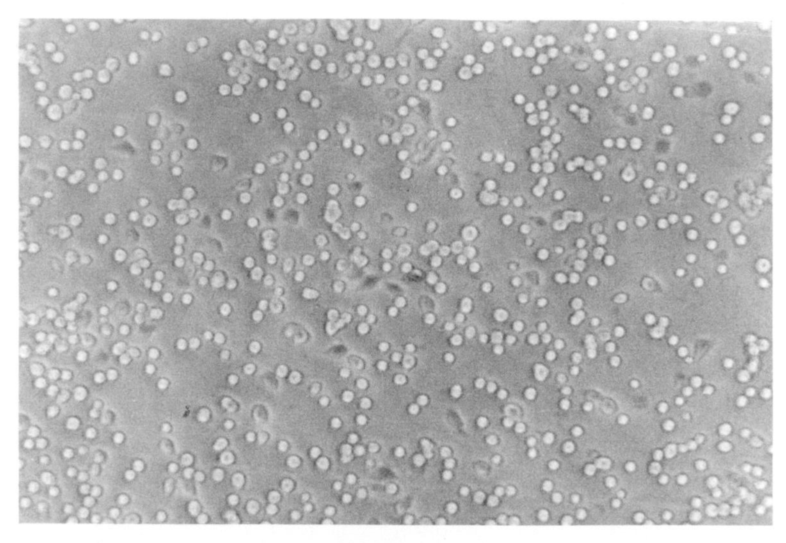

Fig. 7. Phase contrast light micrograph of human skin epithelial cells in Stage I (attachment) of growth. Cells begin to attach and spread within 3 hours and show no tendency to aggregate. ×140.

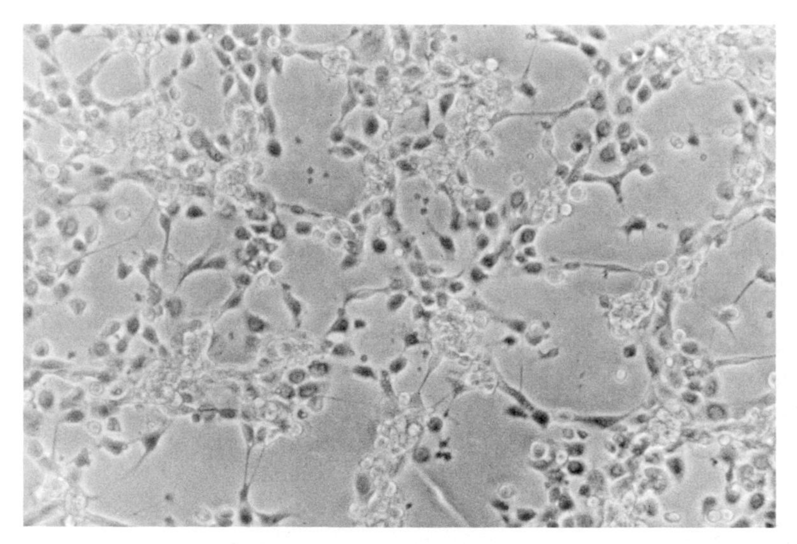

Fig. 8. Phase contrast light micrograph of human skin epithelial cells in Stage 2 (reassociation) of growth. Cells have formed a series of microcolonies joined by an array of attachment bridges. ×140.

ative cells is of major importance in the attachment and growth of keratinocytes. When plated on a glass surface, primary cultures of human skin epithelial cells do not enter Stage 1 of growth, and reaggregation is

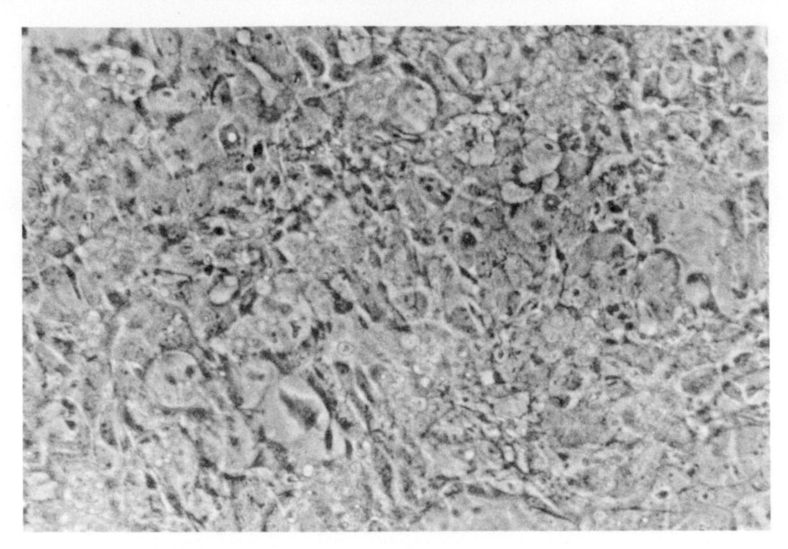

Fig. 9. Phase contrast light micrograph of human skin epithelial cells in Stage 3 (vertical movement) of growth. Cells are multilayered and are irregular in configuration, and cell diameters are increased. ×140.

the principle cell interaction (Fig. 10). On a collagen gel or a collagen-coated surface, cells immediately enter Stage 1 of development. The primary event leading to the initiation of maturation may be the detachment of the basal cell from the basement membrane. This detachment can occur in the absence of cell division.[42]

Ultrastructure

Changes in the ultrastructure of monolayer cultures of human skin and guinea pig epithelial cells have been described in detail by Constable.[24] In intact skin, mitochondria are nearly spherical; in culture, they are elongated and branched. Polyribosomes are increased, and lysosomes are evident. Desmosomes and associated tonofilaments appear to be normal, but the remainder of the network is disorganized and the tonofilaments, as in outgrowth culture, are found to be distributed unevenly. Keratohyaline granules are not seen.

Biochemistry

Lipid metabolism The role of lipids in keratinization is implicated from a number of clinical observations. Lack of essential fatty acids in young animals leads to abnormal keratinization;[43,44] compounds which inhibit cholesterol synthesis can lead to a type of keratinization seen in congenital ichthyosis;[45] and x-ray diffraction techniques show that callus contains lipid in an amorphous form.[46] Shifts in lipid metabolism also

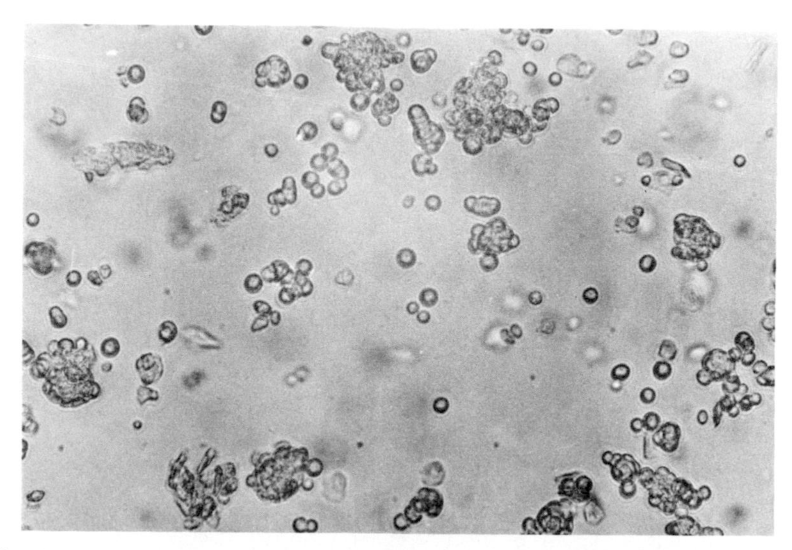

Fig. 10. Phase contrast light micrograph of human skin epithelial cells plated on a glass surface. Cells fail to attach, and reaggregation is the main cellular interaction. ×140.

accompany keratinization in mammalian skin. Sterol esters increase in concentration,[47] phospholipids decrease, and free fatty acids increase.[48] Using developing chick embryos, Freinkel found an enhanced synthesis of sterol esters and wax esters, and a decrease in phospholipids.[49]

To determine if similar changes in lipid profiles take place during growth and maturation of mammalian cells in monolayer culture, lipogenesis was studied at various stages of growth of rabbit skin epithelial cells.[50] Cells were pulsed for 4 hours with ^{14}C-acetate, lipids extracted, separated into neutral lipids and phospholipids, and fractionated chromatographically into classes. The patterns of lipogenesis observed were those generally associated with proliferative cell populations and not keratinizing populations. Cholesterol remained almost totally unesterified, and enhanced triglyceride production was not observed.

Protein metabolism Specific changes in protein metabolism have been reported to take place during keratinization. Histidine and other amino acids are preferentially incorporated into the keratinizing zone,[51,52] and proteins of specific composition have been isolated.[53]

To determine if changes in protein metabolism similar to those observed *in vivo* also take place *in vitro*, rabbit skin epithelial cells were pulsed with ^{14}C-leucine and ^{3}H-histidine, and the ratio of incorporation of each isotope was measured at 3 stages in the growth and maturation of skin cells.

As shown in Table 1, the ratio of histidine to leucine slowly increases as

Table 1. Effect of retinoic acid on incorporation of ^{3}H-thymidine, ^{14}C-uridine, ^{3}H-histidine, and ^{14}C-leucine into rabbit skin epithelial cells.

stage of growth (days)	medium	^{3}H-thymidine total incorp. (cpm)	^{3}H-thymidine % of control	^{14}C-uridine total incorp. (cpm)	^{14}C-uridine % of control	$^{3}H/^{14}C$	^{3}H-histidine total incorp. (cpm)	^{3}H-histidine % of control	^{14}C-leucine total incorp. (cpm)	^{14}C-leucine % of control	$^{3}H/^{14}C$
3	control*	16,005		34,013		0.47	15,651		58,125		0.27
			230		200			160		145	
	R.A.**	36,899		68,051		0.54	25,319		84,432		0.30
7	control	14,654		23,365		0.63	14,444		38,701		0.37
			130		196			156		170	
	R. A.	23,365		45,982		0.42	22,544		65,911		0.34
10	control	21,916		33,812		0.64	15,570		47,752		0.32
			77		105			100		124	
	R. A.	16,863		35,727		0.47	15,434		59,505		0.25

*Control: MEM with 10% calf serum, 1.25% DMSO and serine, 4×10^{-4}M

**R. A. : 0.4 ug/ml in above medium

the cells mature. This ratio is, however, less than would be expected for an actively keratinizing cell population *in vivo*.[54]

Nucleic acid metabolism. DNA synthesis in monolayer cultures is not random. As shown in Fig. 11, a small population of cells is in synchrony, as indicated by the increased uptake of ^{3}H-thymidine at discrete time points. An increase in cell number proportional to the uptake of thymidine is not observed.

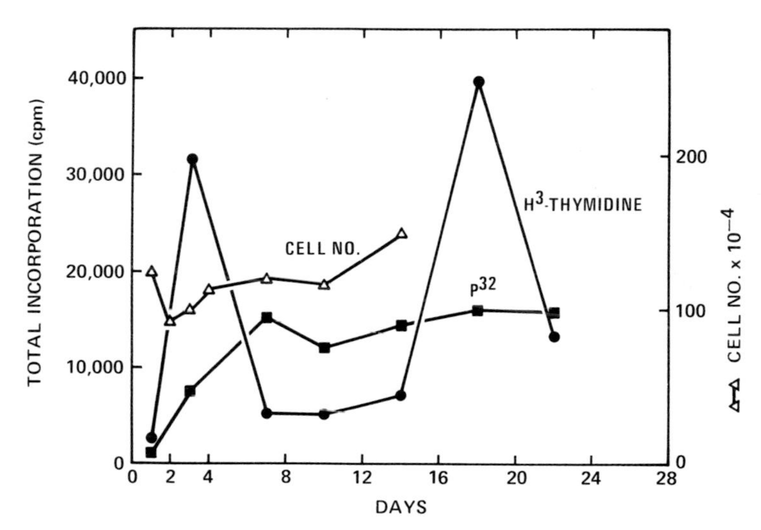

Fig. 11. Relationships between incorporation of ^{3}H-thymidine, ^{32}P, and cell numbers of rabbit skin epithelial cells at 3 stages of growth.

Effect of retinoic acid on keratinization in cell culture

Both human and rabbit monolayer cultures show marked inhibition of keratinization when grown in the presence of retinoic acid. In Table 1, the effect of retinoic acid on the incorporation of histidine and leucine is shown. This table demonstrates the ability of retinoic acid to increase nucleic acid and protein metabolism. Keratinization is inhibited and the ratio of the incorporation of histidine/leucine is decreased.

Reformation of a keratinizing epidermis by transplanted rabbit skin epithelial cells

The formation of an epidermis by serially cultivated rabbit skin epithelial cells (second passage) is shown in Fig. 12. Rabbit epithelial cells were cultivated on collagen gels, and the epithelial cells and gel transplanted to a prepared site on the back of a New Zealand rabbit. An epidermis characteristic of one formed with primary cultures from explant cultures was observed. These findings indicate that serially cultivated rabbit skin epithelial cells do not lose their capacity to differentiate and are able to reform an epidermis with normal histochemical and morphologic characteristics following grafting.

DISCUSSION

Epithelial cells in cell culture display many of the characteristics of cells undergoing the process of keratinization *in vivo*. These characteristics include an increase in size, a change in configuration, loss of nuclei, stratification, and the synthesis of products with histochemical reactions found only in keratinized cells.

Although some morphologic and histochemical parameters are consistent with keratinization in explant and isolated cell cultures, there are important biochemical reactions in these cells that do not take place either in cell or organ culture. These include changes in both protein and lipid metabolism.

In organ culture, keratinization of the metatarsal epidermis is histologically similar to keratinization of the tissue *in vivo*. *In vivo*, 11 protein bands (determined by gel electrophoresis) could be identified; *in vitro*

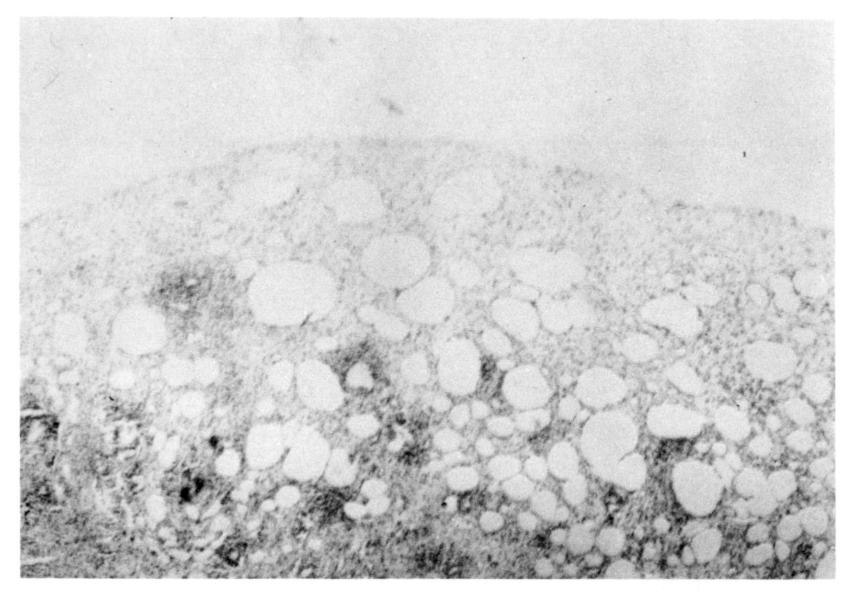

Fig. 12A

only 2 protein bands were observed.[2] Although these biochemical changes were marked, the histology of the epidermis was not altered.

During keratinization lipid profiles are changed in proliferative and nonproliferative cells. Freinkel demonstrated that in embryonic chick skin, enhanced incorporation of ^{14}C-acetate into nonpolar sterol esters correlated with the differentiation of keratinizing cells.[49] In agreement with results found for mammalian tissues,[47,48] a decline in phospholipid production and an increased fatty acid synthesis were observed. In cell culture, the patterns of lipogenesis of rabbit epidermal cells do not correlate with a normally keratinizing epidermis. Cholesterol remains

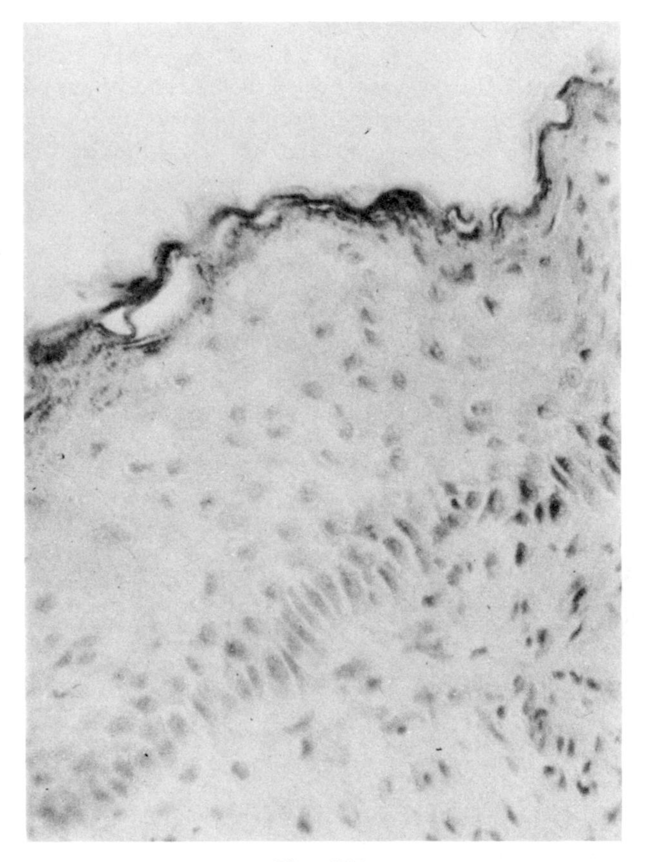

Fig. 12B

Figs. 12A and 12B.
Reformation of an epidermis by transplanted, serially cultivated rabbit skin epithelial cells. (A) Control (B) Second passage rabbit epithelial cells.

almost totally unesterified, and enhanced triglyceride production is not seen. That these biochemical alterations are problems that may be solved with better adjustment of the *in vitro* conditions is suggested by the finding that when epithelial cell populations are regrafted to an autologous host, normal function of cells is restored.

Of particular interest in epithelial cell physiology is the function of cell membranes. In both outgrowth cultures and in isolated epithelial cell cultures, a series of unusual cellular attachment bridges is formed that permits epithelial cells to remain in close contact with each other. Although the exact function of these bridges is not known, it is likely that they may permit the transmission of signals that coordinate movement and metabolism of epithelial cells.

In monolayer cultures, the density and confluence of the cell population plays a role in the maturation of the epithelial cell. When plated at low population densities, epithelial cells tend to form amorphous proteins, and cells do not assume a typical squamous cell configuration. When allowed to grow to confluence, cells survive for extended periods of time, an equilibrium appears to be established, and sheets of keratinized cells are continuously released. When this equilibrium is perturbed by wounding, increased migration and mitotic activity rapidly restore the equilibrium.[31] The increase in population density may cause the accumulation of intracellular cues that permit a more ordered form of maturation to take place. Specific biochemical cues from the dermis to initiate this process do not appear to be required. Of particular importance is the ability of cells to respond to agents that alter keratinization *in vivo*. Retinoic acid induces profound changes in the metabolism of epithelial cells in cell culture resulting in an inhibition of keratinization. Further study of the mechanism of action of retinoic acid should provide valuable insights into the biochemical regulation of the synthesis of fibrous proteins of the keratinocyte.

Acknowledgements

This work was supported by Grant AM 14121 of the National Institute of Health, United States Public Health Service.

REFERENCES

1. Strangeways, T. S. T. and Fell, H.B. Experimental studies on the differentiation of embryonic tissues growing *in vivo* and *in vitro*. I. The development of the undifferentiated limb-bud (a) when subcutaneously grafted into the post-embryonic chick and (b) when cultivated *in vitro*. *Proc. Roy. Soc. London*s B., **99**: 340–366, 1926.
2. Smith, K. The proteins of the embryonic chick epidermis. *Developmental Biol.*, **30**: 263–278, 1973.

3. Wong, Y.C. Ultrastructural features of chick embryo skin cultured in a chemically defined medium. *Anatomical Record.*, **180**: 629–644, 1974.
4. Matoltsy, A.G. Keratinization of embryonic skin. *J. Invest. Dermatol.*, **31**: 343–346, 1958.
5. Trowell, O.A. The culture of mature organs in a synthetic medium. *Exptl. Cell. Res.*, **16**: 118–147, 1959.
6. Bang, F. B. and Niven, J.S.F. A study of infection in organized tissue cultures. *Brit. J. Exptl. Path*, **39**: 317–322, 1958.
7. Matoltsy, A. G. and Sinesi, S. J. A study of the mechanism of keratinization of human epidermal cells. *Anat. Rec.*, **128**: 55–67, 1957.
8. Hambrick, G. W., Lamberg, S., and Bloomberg, R. Observations on keratinization of human skin *in vitro*. *J. Invest. Derm.*, **47**: 541–550, 1966.
9. Reaven, E. P. and Cox, A. J. Organ culture of human skin. *J. Invest. Derm.*, **44**: 151–156, 1965.
10. Hardy, M. H., Biggers, J. D., and Claringbold, P. J. Vaginal cornification of the mouse produced by estrogens in vitro. *Nature*, **172**: 1196–1197, 1953.
11. Kahn, R. H. Effect of estrogen and Vitamin A on vaginal cornification in tissue culture. *Nature*, **774**: 317, 1954.
12. Karasek, M. A. *In vitro* culture of human skin epithelial cells. *J. Invest. Dermatol.*, **47**: 533–540, 1966.
13. Friedman-Kien, A. E., Morrill, S., Prose, P. H., and Liebhaber, H. Culture of adult human skin: *in vitro* growth and keratinization of epidermal cells. *Nature* (London), **212**: 1583–1584, 1966.
14. Flaxman, B. A., Lutzner, M. A., and Van Scott, E. J. Cell maturation and tissue organization in epithelial outgrowths from skin and buccal mucosa *in vitro*. *J. Invest. Dermatol.*, **49**: 322–332, 1967.
15. Wheeler, C. E., Jr., Canby, C. M., and Cawley, E. P. Long-term tissue culture of epithelial-like cells from human skin. *J. Invest. Derm.*, **29**: 383–392, 1956.
16. Briggaman, R. A., Abele, D. C., Harris, B. A., and Wheeler, C. E., Jr. Preparation and characterization of a viable suspension of postembryonic human epidermal cells. *J. Invest. Dermatol.*, **48**: 159–168, 1967.
17. Karasek, M. A. and Charlton, M. E. Growth of postembryonic skin epithelial cells on collagen gels. *J. Invest. Dermatol.*, **56**: 205–210, 1971.
18. Regnier, M., Delescluse, C., and Prunieras, M. Studies on guinea-pig skin cultures. 1. Separate cultures of keratinocytes and dermal fibroblasts. *Acta Dermato-venereologica*, **53**: 241–247, 1973.
19. Fusenig, N. E. and Worst, P. K. M. Mouse epidermal cell cultures. I. Isolation and cultivation of epidermal cells from adult mouse skin. *J. Invest. Dermatol.*, **63**: 187–193, 1974.
20. Karasek, M. A. Effect of all-trans-retinoic acid on human skin epithelial cells *in vitro*. *J. Soc. Cosmet. Chem.*, **21**: 925–931, 1971.
21. Flaxman, B. A., Lutzner, M. A., and Van Scott, E. J. Cell maturation and tissue organization in epithelial outgrowths from skin and buccal mucosa *in vitro*. *J. Invest. Dermatol.*, **49**: 322–332, 1967.
22. Prose, P. H., Friedman-Kien, A. E., and Neistein, S. Ultrastructural studies

of organ cultures of adult human skin. *In vitro* growth and keratinization of epidermal cells. *Laboratory Invest.*, **17**: 693–716, 1967.

23. Barnett, M. L. and Szabo, G. Effect of Vitamin A on epithelial morphogenesis *in vitro. Expt. Cell Res*, **76**: 118–126, 1973.
24. Constable, H. Ultrastructure of adult epidermal cells in monolayer culture. *Brit. J. Derm.*, **86**: 27–39, 1972.
25. Bauer, F. W. and De Grood, R. M. Alteration of subcultured keratinocytes to a keratinizing epithelium. *Brit. J. Dermatol.*, **89**: 29–32, 1973.
26. Constable, H. C. Jr., Cruickshank, C. N. D., and Mann, P. R. Keratinization in dispersed cell cultures of adult guinea-pig ear skin. *Brit. J. Dermatol.*, **91**: 39–48, 1974.
27. Yuspa, S. H. and Harris, C. C. Altered differentiation of mouse epidermal cells treated with retinyl acetate *in vitro. Exptl. Cell Res.*, **86**: 95–105, 1974.
28. Rheinwald, J. G. and Green, H. Serial cultivation of strains of human epidermal keratinocytes: The formation of keratinizing colonies from single cells. *Cell* **6**: 331–343, 1975.
29. Fusenig, N. E., Thon, W., and Amer, S. M. Growth and differentiation in epidermal cell cultures from embryonic mouse skin. *FEBS Symposium.*, **24**: 159–163, 1972.
30. Delescluse, C., Fukuyama, K., and Epstein, W. L. Dibutryl cyclic AMP-induced differentiation of epidermal cells in tissue culture. *J. Invest. Dermatol.*, **66**: 8–13, 1976.
31. Karasek, M. *In vitro* growth and maturation of epithelial cells from postembryonic skin. *J. Invest. Dermatol.*, **65**: 60–66, 1975.
32. Hsu, T. Tissue culture studies on human skin. *Texas Reports Biol. Med.*, **10**: 336–352, 1952.
33. Hashimoto, K. and Kanzaki, T. Surface ultrastructure of tissue cultured keratinocytes. *J. Ultrastruct. Res.*, **49**: 252–269, 1974.
34. Kreyberg, L. Main histological types of primary epithelial lung tumors. *Brit. J. Cancer*, **15**: 206–210, 1961.
35. Liisberg, M. F. Rhodamine B as an extremely specific stain for cornification. *Acta anat.*, **69**: 52–57, 1968.
36. Ayoub, P. and Shklar, G. A modification of the Mallory connective tissue stain as a stain for keratin. *Oral Surg., Oral Med. Oral Pathol.*, **16**: 580–581, 1951.
37. Loewenstein, W. R. and Kanno, Y. Studies on epithelial (gland) cell junction. I. Modifications of surface membrane permeability. *J. Cell Biol.*, **22**: 565–586, 1963.
38. Flesch, P. Inhibition of keratinizing structures by systemic drugs. *Pharmacological Reviews*, **15**: 653–671, 1963.
39. Sherman, B. The effect of Vitamin A on epithelial mitosis *in vitro* and *in vivo*. *J. Invest. Dermatol.*, **37**: 469–480, 1961.
40. Karasek, M. A. Growth and differentiation of transplanted epithelial cell cultures. *J. Invest. Dermatol.*, **51**: 247–252, 1968.
41. Medawar, P. B. Sheets of pure epidermal epithelium from human skin. *Nature* (London), **148**: 783–784, 1941.
42. Etoh, H., Yasuko, T., and Tabachnick, J. Movement of beta-irradiated

epidermal basal cells to the spinous-granular layers in the absence of cell division. *J. Invest. Dermatol.*, **64**: 431–435, 1975.
43. Sinclair, H. M. Essential fatty acids and their relation to pyridoxine. *Biochem. Soc. Symp.*, **9**: 80–99, 1952.
44. Hansen, A. E., Adam, D., Wiese, H., Boelsche, A., and Haggard, M. Essential fatty acid deficiency in infants. *In* Essential Fatty Acids (H. M. Sinclair, ed.), Academic Press, New York, 1958, p. 216.
45. Anchor, R., Winkelmann, R., and Perry, H. Cutaneous side effects from use of triparanol (MER-29): Preliminary data on ichthyosis and loss of hair. *Proc. Mayo Clin.*, **36**: 217, 1951.
46. Swanbeck, G. Macromolecular organization of epidermal keratin. An x-ray diffraction study of the horny layer from normal, ichthyotic, and psoriatic skin. *Acta Derm-vener.* Stock **39**, (suppl.), 43: 1–37, 1959.
47. Kooyman, D. Y. Lipids of the skin: Some changes in the lipids of the epidermis during the process of keratinization. *Arch. Derm.*, **25**: 444–450, 1932.
48. Long, V. J. W. Variations in lipid composition at different depths in the cow snout epidermis. *J. Invest. Derm.*, **55**: 269–273, 1970.
49. Freinkel, R. K. Lipogenesis in epidermal differentiation of embryonic chicken skin. *J. Invest. Derm.*, **59**: 332–338, 1972.
50. Wilkinson, D. I., Nichols, T., and Karasek, M. Lipogenesis during *in vitro* growth of epidermal cells. *Clinical Res.*, **22**: 161, 1974.
51. Fukuyama, K. and Epstein, W. Ultrastructural autoradiographic studies of keratohyalin granule formation. *J. Invest. Dermatol.*, **49**: 595–604, 1967.
52. Fukuyama, K. and Epstein, W. Inhibition of RNA and protein synthesis in granular cells by actinomycin-D and puromycin. *J. Invest. Dermatol.*, **56**: 211–222, 1971.
53. Hoober, J. K. and Bernstein, I. Protein synthesis related to epidermal differentiation. *Proc. Nat. Acad. Sci.*, **56**: 594–601, 1966.
54. Moore, J. T. and Karasek, M. A. Isolation and properties of a germinative and a nongerminative cell population from postembryonic mouse, rabbit, and human epidermis. *J. Invest. Dermatol.*, **56**: 318–324, 1971.

Discussion

Dr. Watanabe: Did you use an inhibitor of a biochemical process during the cell differentiation? I am asking because I would like to know whether during the exposure to the inhibitor, differentiation stopped and then continued after the inhibitor was removed?

Dr. Karasek: Well, the only one we looked at so far is retinoic acid. This material binds strongly to the cells and even when one washes out the chemical, one continues to have the effect. Thyroxin is the compound we would like to use. In subculture, thyroxin produces an almost immediate keratinization of cells. Unfortunately, in explant culture we see no effect of thyroxin, but we think that using agents which are selective in initiating or stopping the system, is one way of beginning to dissect the biochemical events in differentiation.

DR. GOLDSMITH: It seems that until we have more chemistry on these cells, and more immuno-specific reagents, that histochemistry is going to be an important technique and I think it will be important for us to understand why these stains are doing what they do. Now, Dr. McGuire aside, I think some of us think that there is at least some keratin being synthesized in the germinative layers. What is the basis for this keratin not taking those stains, yet once it is in the fully keratinized layers, the staining of keratin is very dramatic? I just wonder what kind of studies have been done in this regard.

DR. KARASEK: Unfortunately, there have not been any. I personally think that it has something to do with the way that lipids are deposited on membranes. The morphologic appearance of the cells do not always coincide with the expected presence of proteins as defined biochemically. For example, Dr. Margaret Smith, using the developing chick embryo in culture, finds that the cells look absolutely beautiful morphologically, but a comparison of the proteins isolated actually from the *in vitro* and *in vivo* systems showed that the proteins were different.

DR. FREEDBERG: People have had trouble growing adult human epidermis and subculturing it. Now, what you've implied is that you are able to grow it and it might be able to be subcultured a little bit. How many generations?

DR. KARASEK: For a maximum of three generations in our hands. The secret appears to be the addition of a small amount of EDTA to the culture. Without the addition of EDTA one gets clots. One can not get dissociated cells and one has difficulty getting a subculture. In the presence of trypsin and EDTA, as Dr. Kitano has pointed out, cells separate very nicely and if you plate them out on a collagen gel, there is a plating efficiency of 80 percent. You can do it a third time. After the third time something happens and the cells do not grow. We simply do not know what the problem is.

DR. FREEDBERG: Dr. Kitano, do you use EDTA?

DR. KITANO: Yes. I would like to ask those who are working on the structure of keratin protein, what band or fraction is specific for keratin and, if that band appears, can you be sure keratinization has occurred?

DR. STEINERT: I cannot. What I would like to ask Dr. Karasek is, have you tried a number of different cell types? Have you tried myoblasts? Do muscle cells give a positive reaction with rhodamine B?

DR. KARASEK: We have not looked at myoblasts.

DR. STEINERT: The reason why I asked is, that in my hands, the cells do not produce keratin but produce loads of actin. Twenty percent of the total protein in the cell is actin.

DR. KARASEK: There is a fair amount of actin in a variety of cells. For example, we know that fibroblasts have levels——

DR. STEINERT: But fibroblasts have less than 1 percent. We're talking about one or two orders of magnitude more.

DR. KARASEK: I certainly don't know, but we have endothelial cells and a variety of other cell types and one cannot screen all the cells. My own guess is that it is not actin.

DR. FUKUYAMA: What is the histochemical reactivity of dyskeratotic cells with Orange G?

DR. KARASEK: I really don't know. We have a number of stains now which operate by different biochemical processes and this gives us some assurance that we are probably looking at keratin.

DR. ODLAND: I really object to the implication that these stains are specific for keratin.

DR. KARASEK: What I meant to say is keratinized cells.

DR. BADEN: I would like to come back to this point again. A number of years ago, we were interested in looking at the problem of staining in a variety of vertebrates. We did a combination study in which we did histochemical staining, electron microscopy and X-ray diffraction. We found that tissues which had entirely different fibrous protein content showed exactly the same staining whereas other tissues which had the same fibrous protein showed different staining characteristics. So, I think that when you say it is a marker for keratinization, I think you have to specify what you are talking about. I do not think that it is a marker for filament synthesis or a specific filament. I think that when you talk about keratinization, people think "keratin" and they think "fibrous protein." I believe that this is the pitfall into which you could fall. We are simply talking about a cell that goes through some morphologic change.

DR. KARASEK: That is correct. We know that if we isolate fibrous protein from these cells, the fibers do not stain.

DR. BADEN: And the second point I would like to make is that we, and several other laboratories, have developed antibodies which are very specific for purified fractions of fibrous protein. These antibodies can be used to look at filaments. So, one can be quite specific without dealing with stains if one chooses to be. The third thing is, how do you know that your gel is not inhibiting keratinization since he (Dr. Kitano) gets what seems to be better keratinization. Maybe your cells grow better or faster, but have you looked at the question of whether the gels are actually inhibiting keratinization?

DR. KARASEK: Yes, in fact, there's something in the gel—a mucopolysaccharide—which doesn't inhibit but which initiates keratinization. If we use, not a gel but a collagen-coated surface, we see an entirely different pattern of incorporation of amino acids.

DR. SEIJI: With respect to the specificity of the Orange G-aniline blue stain, are there any differences between the ordinary cornified cell and the dyskeratotic cell such as the individually keratinized cell in Bowen's disease?

DR. KARASEK: I don't know.

DR. SEIJI: Does the keratinized cell with a positive stain show the ordinary keratin pattern under the electron microscope?

DR. KARASEK: We haven't looked yet.

DR. MCGUIRE: I would like to make several comments and ask a question. I have never said that basal cells do not synthesize keratin, I haven't seen it demonstrated but I wouldn't be surprised if basal cells do synthesize keratin. Okay, that's the comment. I do not want anyone to misunderstand my position on the "actin bandwagon." We have demonstrated clearly that actin is present throughout the epidermis. Where the keratin shows up is for someone else to demonstrate. What surprises me a little bit is Dr. Steinert's remark. I think that his position is clear that chemical demonstration of keratin is essential at this stage of the game in cell culture. However, when pressed on this point, he withdrew. I don't understand why because he has a very good chemical demonstration of keratin and in his analytic polyacrylamide gel system, has been able to identify the keratin peptides and demonstrate their capacity to polymerize and form α-helices. He has also done amino acid analyses. So I think that this is the way to go to get yourself out of this predicament of staining something and saying that it is keratin.

DR. KARASEK: Our biochemistry has not been quite as elegant as doing an amino acids analysis but we were able some years ago to look at the changes in solubility of the cells between stage 1 and 21. In the early stages nearly the entire cell was solubilized. As the cells aged it became increasingly difficult to do so and we could isolate a protein which was extraordinarily resistant to alkaline- or acid-solubilization and almost totally resistant to enzymatic hydrolysis by trypsin, pepsin or chymotrypsin. So there clearly are some very resistant proteins.

DR. MCGUIRE: Let me ask you and Dr. Kitano one brief question and that is, are all of your cells parakeratotic, that is do all of your cells retain nuclei in the outermost level?

DR. KARASEK: In our hands, most of them do.

DR. KITANO: I noted that in the upper layers, the very top layers, there are cells which have lost the nucleus.

DR. KARASEK: Yes, we, too, have been able to pick out cells at the very top which have lost the nuclei.

DR. PECK: I have noted from the scanning picture that your monolayer was flat and smooth. I would like to bring up the work of my collaborator, Dr. Bruce Wetzel, who has shown that the scanning electron microscope picture of monolayer cells varies with the serum concentration. With no serum, they tend to be more spherical and with more surface structure of villi and microridges. With 1% they tend to flatten and, certainly at 10% and higher, they are flattened and smooth so I think that if you are going to look for surface changes with the scanning electron microscope, you have to consider serum concentration. Another point. Did you look at the ultrastructure of the retinoic acid-treated cells? In particular, I presume you did not find mucous granules. In my talk, I showed a long series of events in the metaplastic process and I wondered if you saw any of these events. Did you see increased golgi, desmosomal cleavage or increased rough endoplasmic reticulum? As a correlate to that question, I have not come across any study of retinoic acid-treatment of epidermis cells where mucous granules have formed. In my mind, this as well as

Hardy's work, means that the dermis may do more than just provide a framework. I'd like your comment on that. The last point was, I am aware of two normal human keratinocyte lines, one developed by Dr. Bean in Seattle, and one by Dr. Howard Green at MIT. There is also a new line developed from a squamous cell carcinoma in Colorado. I wonder if you have worked with these or have any comments on them.

Dr. Karasek: I'll have to respond to these privately because of the lack of time.

Adrenocortical Control of Epidermal Keratinization Demonstrated by Combined Morphological and Biochemical Studies of the *in ovo* and *in vitro* Development of Chick Embryonic Shank Skin

Hiroyoshi Endo,* Masanobu Sugimoto,** Kimiko Tajima,*** Asato Kojima,** Yachiyo Atsumi-Hirato * and Akiko Obinata*

* *Department of Physiological Chemistry, Faculty of Pharmaceutical Sciences, University of Tokyo, Tokyo, Japan,* ** *Department of Pathology, National Institute of Health, Tokyo, Japan* and *** *Department of Dermatology, Saitama Medical School, Saitama, Japan*

ABSTRACT

In an attempt to clarify the possible hormonal control of epidermal keratinization, we devised a novel replicate skin culture method to give enough skin explants for a variety of chemical and biochemical determinations of the epidermis.

First, employing 4 days' cultivation in a chemically defined medium of the undifferentiated shank skin of a 13-day embryonic chick, we have morphologically demonstrated the epidermal keratinization dramatically induced by minimal concentrations of hydrocortisone. Second, chemical and biochemical analyses of the cultured epidermis indicated the prompt synthesis after hydrocortisone addition of a group of metabolically stable tonofilamentous proteins through the induction of rather long-lived mRNA. Lastly, these *in vitro* findings were examined for physiological significance. In terms of glucocorticoid "receptor" as a prerequisite to the steroid action, the skin from 10–11-day chick embryos gave virtually no cytoplasmic binding of ^{3}H-glucocorticoids. The receptor activity of the embryonic skin showed a gradual increase at 12–13 days of incubation and finally gave an abrupt rise around 15 days to reach a maximum.

Considered together with the onset of chick embryonic adrenocortical function at 13–15 days and the start of keratinization of the shank skin at 18 days of incubation, our findings would lead to the conclusion that terminal differentiation of epidermal cells (keratinization) is a culmination of their own preceding differentiation (development of glucocorticoid receptor) exactly co-ordinated with adrenocortical cell differentiation (secretion of corticosteroids).

One of the most important functions of the skin is the formation of keratinized layers in the uppermost part of the epidermis for protecting the living organism from the external environment. The epidermal keratinization consists of a series of finely organized processes of cell differentiation, in which synthesis of specific proteins within the cells and formation of specific intracellular structures such as tonofilaments and tonofibrils are accompanied by degradation of common cellular organelles such as mitochondria and nuclei. The process has long been one of the most attractive subjects in tissue culture studies ever since Strangeways and Fell[1] successfully cultured chick embryonic skin by the watch-glass method using the natural medium of a plasma clot. Later, on the natural medium, Fell[2] and Weissmann and Fell[3] observed the accelerating effect of hydrocortisone, though at rather a high concentration of 7.5 μg/ml, upon the keratinization of chick and rat embryonic skin. Although these histological studies suggested that glucocorticoids may play an important role in keratinization of the skin in embryonic development, further phenomenological confirmation of the possible hormonal control of epidermal differentiation and substantial analysis of the control mechanism remained to be carried out.

DEVICE FOR ORGAN CULTURE OF CHICK EMBRYONIC SKIN

The main aims of this work were first to devise a simple replicate culture method for cultivating rather large pieces of the skin in order to obtain a large enough amount of the explants for chemical and biochemical analyses, and secondly to make the cultures successful in a chemically defined medium in order to exclude the participation of any unknown factors contained in biological materials, such as serum or embryo extract used in the above pilot studies.

These objects were fully achieved on chick embryonic shank skin by using the 'Millipore' filter–roller-tube method[4] (Fig. 1) and a chemically defined medium, BGJb[5] supplemented with 50 μg/ml of sodium ascorbate. The shank of a 13-day-old chick embryo from which the distal and proximal portions had been removed (ca 7 $\times$ 7 mm^2) was spread out in a drop of physiological buffered saline solution on a Petri dish by carefully removing the bones, muscle, and loose connective tissue under a dissecting microscope. The 'Millipore' filter (type HA or PH) was placed on a sheet of the skin. The skin was then ready for culture, the dermis being tightly in contact with the filter. The filter with a stretched sheet of the skin was put into a roller-tube containing 1 or 1.5 ml of the medium. The tubes were incubated at 37°C while rotating 10 times per hour in a roller-drum which held them at an angle of approximately 5°. The medium was renewed every other day.

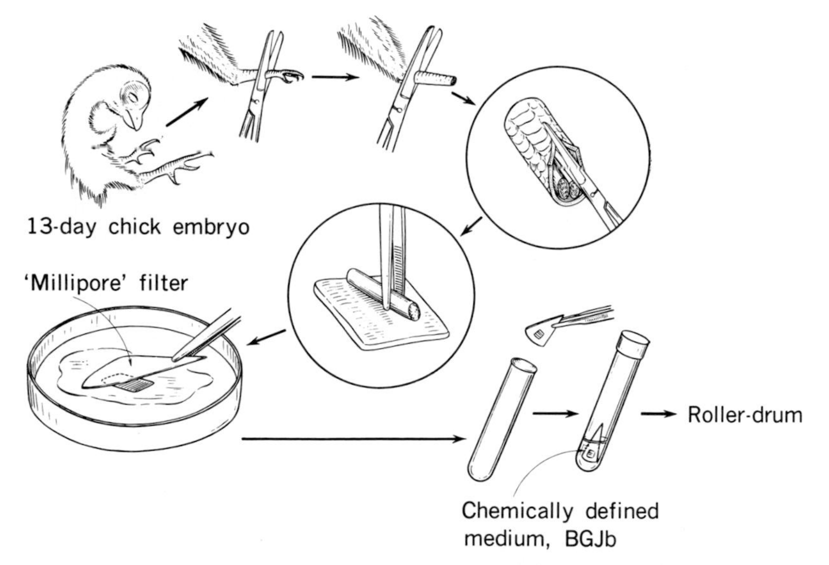

Fig. 1. Diagramatic illustration of procedure of the 'Millipore' filter–roller-tube culture method for chick embryonic shank skin.

DEMONSTRATION OF KERATINIZATION-INDUCING EFFECT OF HYDROCORTISONE

The epidermis of the 13-day chick embryonic shank skin at the start of culture consisted of a single layer of basal columnar cells and on this two layers of intermediate and flattened cells.[4,6] When the skin was cultured in the chemically defined medium without hydrocortisone added, the explant grew well as a whole and the epidermis grew as thick as six to eight cell layers after 4 days' cultivation (Fig. 2a).[4,6] During growth *in vitro*, however, no sign of keratinization could be observed over the epidermis, though eosinophilic granules probably indicating corpusculum cribriformium or peridermal granules appeared in subpericytes as well as in pericytes, and sometimes the swollen cells just below the subpericytes became diffusely eosinophilic. Even after far longer cultivation the keratinized layers could not be discerned effectively all over the explants, except in a few limited areas at the edges, despite the fact that the eosinophilic swollen cells in the epidermis increased in number.[4]

When hydrocortisone was added to the culture medium at concentrations from 0.001 μg/ml upwards, however, a heavily eosinophilic, compact layer devoid of cellular structures was deposited in the uppermost stratum of the epidermis after 4 days' cultivation (Fig. 2b).[4,6] Moreover,

a

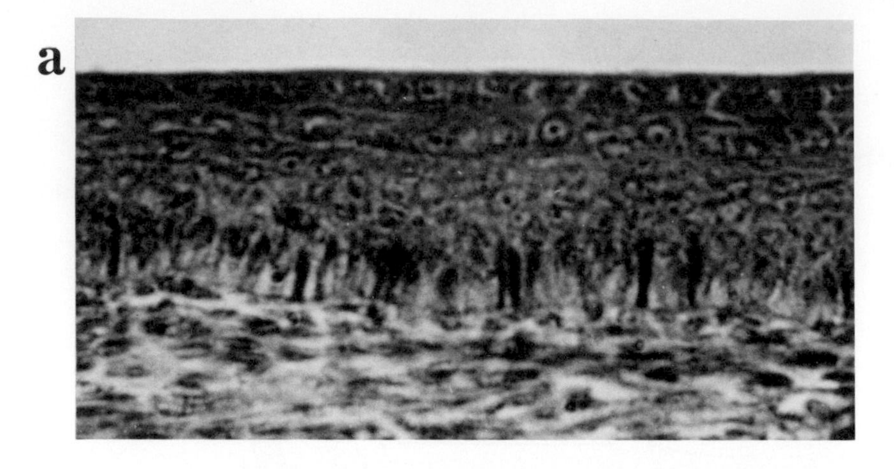

b

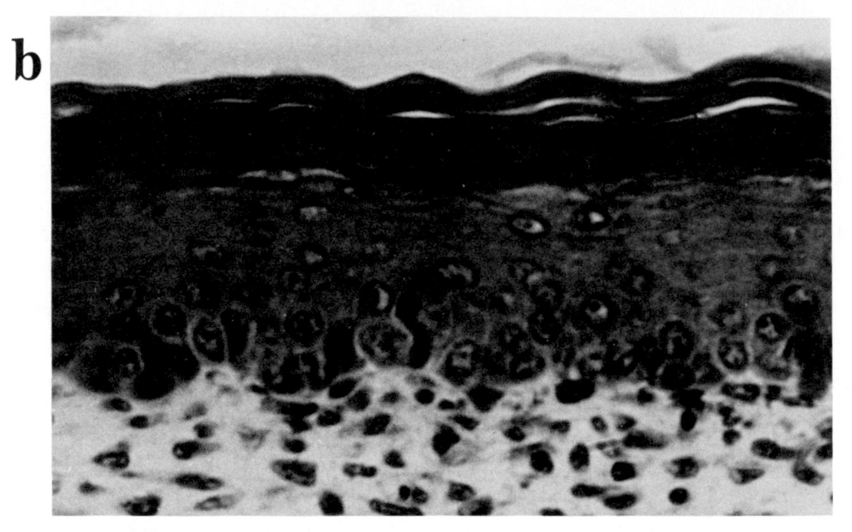

Fig. 2. Induction of epidermal keratinization by hydrocortisone of 13-day chick embryonic shank skin growing *in vitro*.
(a) The skin cultured for 4 days in a chemically defined medium, BGJb supplemented with ascorbate.
(b) The skin cultured for 4 days in the chemically defined medium containing hydrocortisone at a concentration of 0.01 μg/ml.

the basal cell layer of the epidermis was fairly undulating in treated explants, whereas it was rather even in control explants.[4] Furthermore, when examined by electron microscopy, an intense keratinized layer was also observed in the hydrocortisone-treated explants alone (Fig. 3b) as observed by light microscopy, and a remarkable difference other than this was that the treated epidermal cells were rich in tonofibrils (Fig. 3b) compared to the control ones (Fig. 3a).[7] These morphological features observed in the epidermis of the hydrocortisone-treated skin were all very similar to physiologic keratinization in the course of normal development of the chick.

As to the minimal effective concentration and the steroid specificity for such a keratinization-inducing effect of hydrocortisone, even the lowest concentration tested of 0.0001 μg/ml showed a very slight but discernible effect, and at a concentration of 1.0 μg/ml the precursors of hydrocortisone, pregnenolone and progesterone showed no effect at all while a mineral-corticoid, deoxycorticosterone, produced only a slight effect comparable with that of the lowest concentration of hydrocortisone.[4] Therefore, the keratinization-inducing effect of hydrocortisone can be attributed to its glucocorticoid activity and considered as reflecting a possible hormonal control of the epidermal differentiation *in ovo*.

CHEMICAL EXAMINATION OF THE KERATINIZATION-INDUCING EFFECT OF HYDROCORTISONE

One of the most important points to be solved in this problem is whether or not this *in vitro* keratinization induced by hydrocortisone accompanies the chemical changes according with normal embryonic keratinization. With respect to this point, amino acid analyses of the whole epidermal protein from the cultured explants and the normal embryonic and chicken skin showed that changes produced by hydrocortisone in the content of each amino acid were generally in good accord with those found through the normal development of chick embryo. The most striking change was a remarkable increase of glycine content, suggesting that hydrocortisone enhanced the synthesis of some glycine-rich proteins probably involved in the processes of keratinization.[4] In the following experiments, therefore, the hydrocortisone effect on *in vitro* keratinization was further examined chemically.

The epidermis separated from cultured skin, the epidermis obtained from chick embryos of varying developmental stages, and the horny layer of neonatal chicks were solubilized by homogenizing in 8 M urea solution and thereafter by reducing with mercaptoethanol and carboxymethylating with monoiodoacetic acid. After centrifugation, the chemically modified epidermal protein obtained as the supernatant was designated as S-carboxymethylated epidermal proteins (SCMEp).[7,8]

When subjected to polyacrylamide gel disc electrophoresis,[7,8] SCMEp from the keratinized layer of neonatal chicks revealed the existence of more than seven bands (Fig. 4 Ia). These bands were roughly divided into two groups, SCMEp*A* and SCMEp*B*. Amidoschwarz heavily stained the

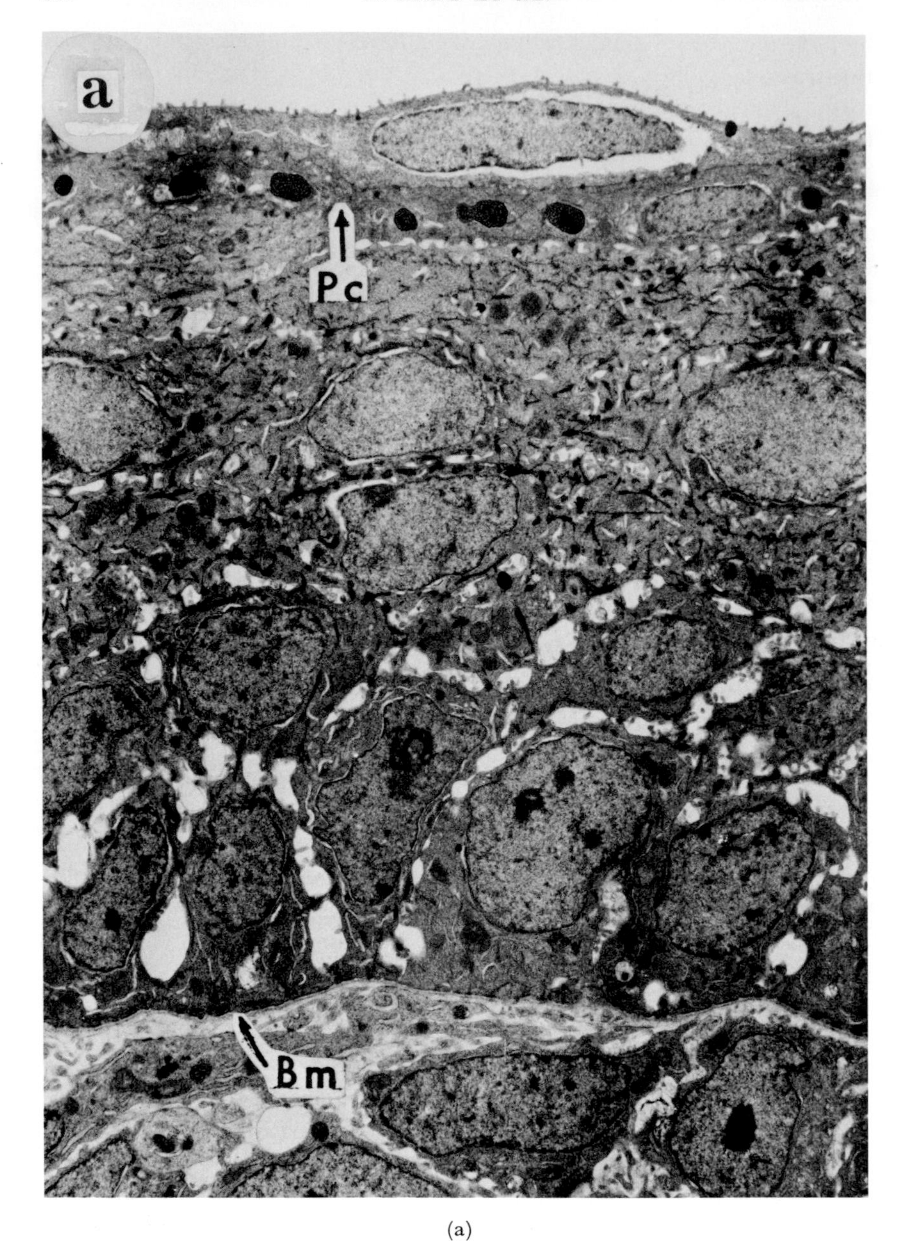

(a)

Fig. 3. Ultrastructure of 13-day chick embryonic shank skin cultured for 4 days in the chemically defined medium with or without hydrocortisone (0.01 μg/ml).

(a) Control explant. Pericyte (*Pc*) remained, and under these cells no

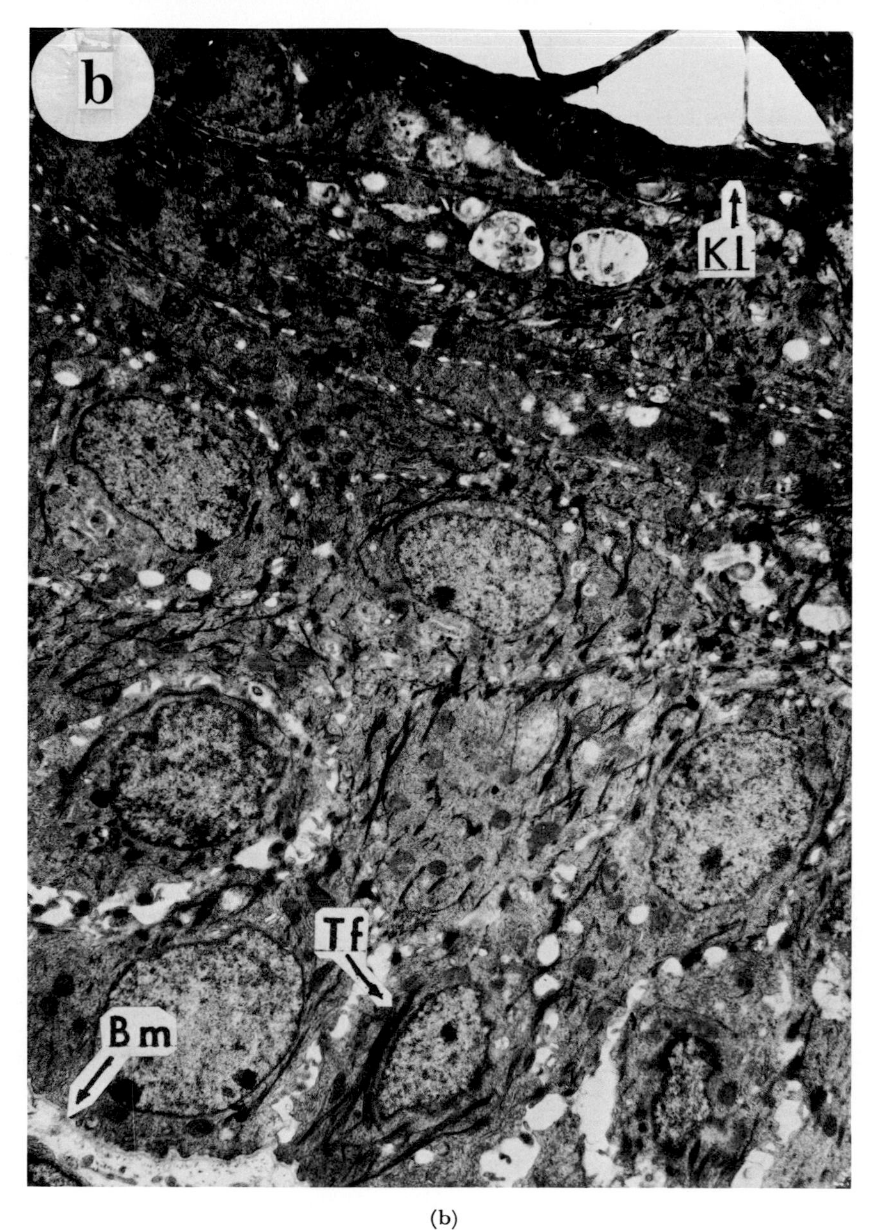

(b)

keratinized layer was found. *Bm*, basement membrane.

(b) Hydrocortisone-treated explant. The keratinized layer (*Kl*) was obvious, and tonofilaments (*Tf*) were common in the basal and intermediate cells.

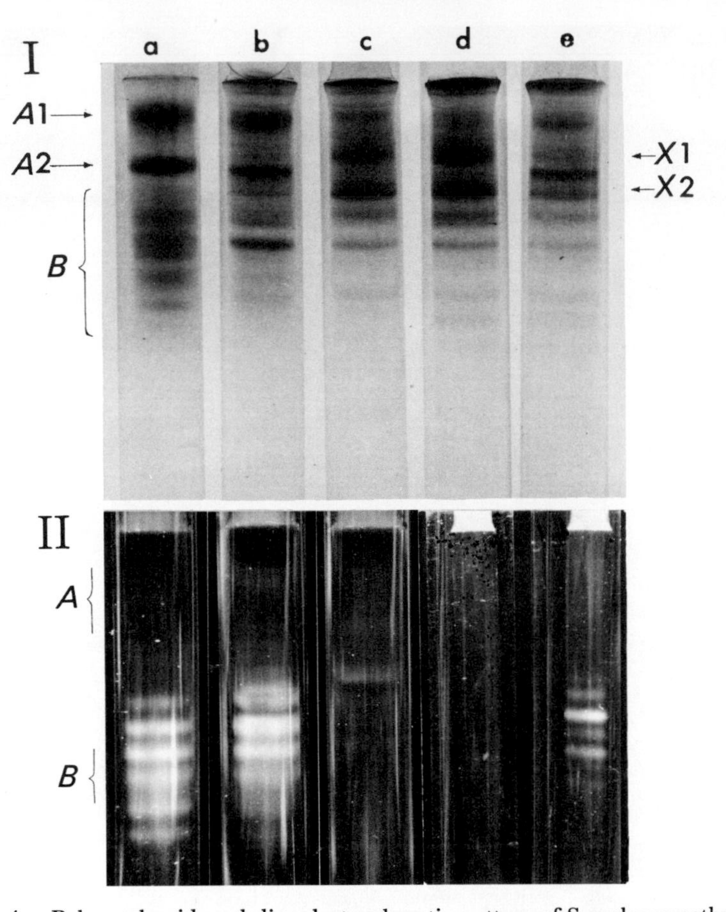

Fig. 4. Polyacrylamide gel disc electrophoretic pattern of S-carboxymethylated epidermal proteins (SCMEp) from chick shank skin.
(I) Amidoschwarz staining, (II) Fixation in 5% perchloric acid. a, 3-day chick's keratinized layer; b, 17-day embryonic epidermis; c, 15-day embryonic epidermis; d, epidermis of 13-day embryonic skin cultured for 4 days in the chemically defined medium alone; e, epidermis of 13-day embryonic skin cultured for 4 days in the chemically defined medium containing hydrocortisone (0.01 μg/ml).

former but weakly the latter, while fixation with perchloric acid revealed only the latter by exhibiting characteristic opalescence (Fig. 4 Ia and 4 IIa). When SCMEp from embryos in various developmental stages was investigated, however, the epidermis of 15-day embryos displaying no sign of the formation of a keratinized layer gave an entirely different electrophoretic pattern; *A*1 and *A*2 could scarcely be seen and instead two other bands, *X*1 and *X*2, were clearly discernible just before and behind the band of *A*2, respectively (Fig. 4 Ic). On the other hand, 17-day embryonic epidermis, which is just about to form a keratinized layer, showed a pattern similar to that of the keratinized layer with regard to *A* bands; *A*1 and *A*2, particularly the latter, appeared abundantly while *X*1 and *X*2 decreased (Fig. 4 Ib). With respect to SCMEp*B*, it was nearly absent in 14- to 15-day embryos, but appeared appreciably in 16- to 17-day embryos and was found in a great amount in the keratinized layer of neonatal chicks (Fig. 4 Ia—Ic and 4 IIa—IIc). From these findings it could be concluded that SCMEp*A* and SCMEp*B* are the derivatives of the epidermal structural proteins relating to embryonic keratinization, because they were numerous in the differentiated epidermis, especially in the keratinized layer, but infrequent in the undifferentiated epidermis.

After 13-day chick embryonic shank skin was cultured for 4 days in the chemically defined medium with or without hydrocortisone, SCMEp of these cultured skin explants was subjected to the same examinations by disc electrophoresis.[7,8] In parallel with the histological and ultrastructural findings mentioned before, only the epidermis of the hydrocortisone-treated skin electrophoretically revealed a differentiated pattern similar to that of the keratinized layer of neonatal chicks with respect to *A* bands (Fig. 4 Ia and 4 Ie), while the epidermis of the control skin gave an undifferentiated pattern like that of 15-day embryos (Fig. 4 Ic and 4 Id); *A* bands were found only in the hydrocortisone-treated epidermis, while *X* bands instead of *A* ones were clearly found in the control epidermis. The effect of hydrocortisone on SCMEp*B* formation was, however, somewhat complicated. When examined by Amidoschwarz staining, all the epidermis of the cultured explants gave undifferentiated patterns irrespective of the hormone treatment (Fig. 4 Id and Ie),[7,8] but when the gel was fixed in 5% perchloric acid, opalescent bands characteristic of *B* bands were detected only in the hydrocortisone-treated explants (Fig. 4 IId and IIe).[10] These findings lead to the conclusion that hydrocortisone dramatically stimulated the accumulation of an epidermal structural protein, SCMEp*A*, but to a lesser degree another epidermal structural protein, SCMEp*B*, in the chick embryonic skin growing *in vitro*.

From amino acid composition of whole epidermal protein of cultured explants, as mentioned before, it was suggested that hydrocortisone accelerated the synthesis of some glycine-rich epidermal proteins involved in the processes of keratinization.[4] SCMEp*A* and SCMEp*B* seem likely to be the supposed glycine-rich proteins, since these two kinds of protein increase with the progress of embryonic keratinization and also accumulate in the hydrocortisone-treated explants. To make sure of this, SCMEp from the keratinized layer of neonatal chick shank skin was submitted to column chromatography of Sephadex G-100 to fractionate into SCMEp*A* and SCMEp*B* fractions and determine their amino acid composition. Thus, it was shown that, confirming the above supposition, SCMEp*A* and SCMEp*B* were both rich in glycine (Table 1).[7] It should be noted here that the amino acid composition of SCMEp*A* and that of the tonofilaments isolated from rat epidermis[9] are very similar to each other, suggesting that SCMEp*A* is a derivative of the tonofilaments of chick epidermis. All the chemical data, considered together with the ultrastructural finding that tonofibrils strikingly increased in the basal cells of the hydrocortisone-treated explants (Fig. 3),[7] strongly suggest a mechanism that minimal level of hydrocortisone induces the basal cell activity for synthesizing epidermal structural proteins, especially SCMEp*A* of tonofilamentous nature, and eventually culminates in keratinization of the epidermal cells after 4 days' cultivation.

Table 1. Amino acid composition of epidermal structural proteins of 3-day chick shank skin.[a)]

Amino acid residue	SCMEp	SCMEp*A*	SCMEp*B*	Tonofilaments of newborn rat epidermis[b)]
Lys	1.6	4.2	0.8	5.5
His	0.9	1.0	1.0	1.4
Arg	5.2	7.0	4.7	5.5
Asp	5.2	7.2	5.3	10.1
Thr	3.5	3.2	3.8	3.6
Ser	11.2	9.9	11.8	11.1
Glu	6.7	12.5	4.8	14.2
Pro	7.2	1.9	9.5	0.7
Gly	25.7	22.6	25.7	20.2
Ala	5.3	4.9	5.5	5.9
Val	5.2	4.3	5.1	4.1
Met	0.2	0.8	0.0	0.8
Ile	3.3	5.0	2.8	3.3
Leu	7.0	8.1	7.5	7.8
Tyr	8.0	4.4	9.9	3.4
Phe	3.7	3.0	3.9	2.7

a) Each figure represents the number of amino acid residues per 100 amino acid residues of the protein.

b) Values of the amino acid composition of tonofilaments of newborn rat were quoted from the report by Tezuka and Freedberg.[9]

BIOCHEMICAL ANALYSES OF THE KERATINIZATION-INDUCING EFFECT OF HYDROCORTISONE

In an attempt to elucidate the above possible mechanism of hydrocortisone action upon keratinization, metabolic features of epidermal protein synthesis in hydrocortisone induced keratinization were studied with the use of radioactive amino acids. The shank skin from developing chick embryos or the skin cultured in the chemically defined medium in the presence or absence of hydrocortisone was pulse-labeled with ^{3}H-glycine and/or ^{14}C-leucine, and sometimes the labeled skin was chase-cultured. The epidermis was separated from the labeled skin and SCMEp was prepared. The labeled SCMEp thus obtained was electrophoresed and the gels were sectioned into 1-mm coronal slices to determine the incorporated or residual radioactivity.[10] The glycine content of the whole epidermal protein increased owing to the accumulation of SCMEp*A* and SCMEp*B* during both normal embryonic keratinization and hydrocortisone-induced *in vitro* keratinization, while the contents of several other amino acids including leucine were maintained rather constant in these processes.[4] Furthermore, the leucine contents of SCMEp*A* and SCMEp*B* were nearly identical to each other (Table 1).[7] In the following metabolic studies, therefore, the incorporation ratio of ^{3}H-glycine to ^{14}C-leucine was used as a major indicator of the synthesis of the epidermal structural proteins.

First, the double-isotope incorporation pattern in gel slices of ^{3}H-glycine- and ^{14}C-leucine-labeled epidermal proteins were compared for the 13- and 16.5-day embryonic epidermis (Fig. 5).[10] For the 13-day epidermis, ^{3}H and ^{14}C incorporation were nearly parallel throughout the gel slices with appreciable peaks of both isotopes in the SCMEp*X* bands (Fig. 5b), which were characteristic of undifferentiated epidermis.[7] Therefore, the incorporation ratio of ^{3}H to ^{14}C was nearly constant throughout the gel slices (Fig. 5a). For the 16.5-day epidermis, on the other hand, increased amounts of ^{3}H were found in the SCMEp*A* and SCMEp*B* bands, whereas ^{14}C was distributed rather evenly in these bands and in SCMEp*X* bands (Fig. 5d). Thus the incorporation ratio of ^{3}H to ^{14}C gave peaks in the gel slices containing SCMEp*A* and SCMEp*B* (Fig. 5c), coupled with the appearance of these new stained bands.[7,8] These results indicate that the glycine-rich epidermal structural proteins, SCMEp*A* and SCMEp*B*, are newly synthesized in large amounts in the developmentally advanced epidermis of chick embryos.

Second, the hydrocortisone-treated and nontreated explants were each double-labeled with ^{3}H-glycine and ^{14}C-leucine (Fig. 6).[10] The incorporation ratio of ^{3}H to ^{14}C was nearly constant throughout the gel slices in the nontreated explants (Fig. 6a) as in the immature 13-day embryonic epidermis (Fig. 5a). This suggests that the pattern of protein synthesis in the

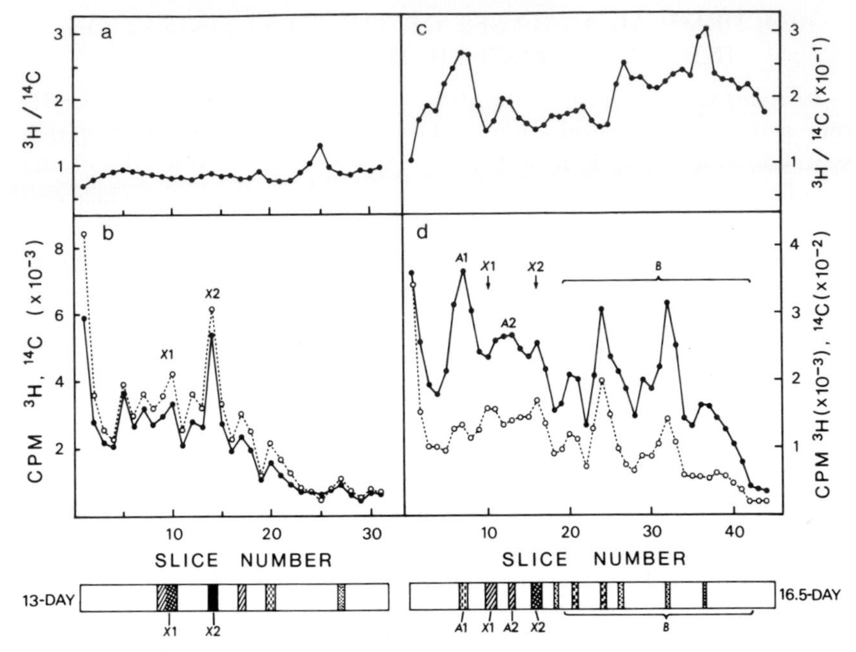

Fig. 5. Electrophoretic patterns of radioactivity of SCMEp from 13- or 16.5-day chick embryonic shank skin double-labeled for 24 hr with ^{3}H-glycine and ^{14}C-leucine. The ratio of ^{3}H-glycine to ^{14}C-leucine: (a) 13-day embryonic epidermis, (c) 16.5-day embryonic epidermis. Amounts of radioactivity of ^{3}H-glycine (●——●), ^{14}C-leucine (○······○): (b) 13-day embryonic epidermis, (d) 16.5-day embryonic epidermis. The diagram at the bottom illustrates the protein band patterns stained with Amidoschwarz.

control explants cultured for 4 days remained at an undifferentiated stage comparable to 13-day embryonic epidermis. On the contrary, in the hydrocortisone-treated explants the pattern of the ratio of ^{3}H to ^{14}C showed three remarkable peaks in the gel slices in bands SCMEp*A*1 and SCMEp*A*2 having large amounts of radioactivities and in the faster-moving, SCMEp*B*-containing region with only small amounts of radioactivities (Fig 6c). These results show that hydrocortisone dramatically accelerates the synthesis of a group of glycine-rich epidermal structural proteins, SCMEp*A*, while to a lesser degree enhancing the synthesis of another group of glycine-rich epidermal structural proteins, SCMEp*B*.

Furthermore, when time-dependent changes in the protein-synthesizing activity of the epidermis were investigated by pulse-labeling the explants

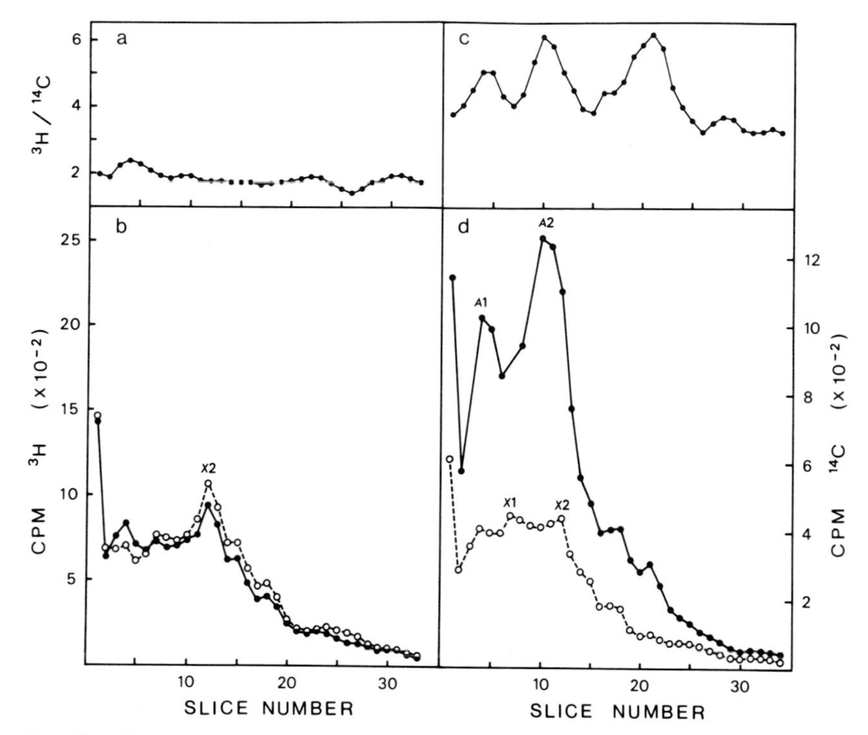

Fig. 6. Electrophoretic patterns of radioactivity of SCMEp from the hydrocortisone-treated and nontreated explants cultured for 3 days and then double-labeled with ^{3}H-glycine and ^{14}C-leucine for 24 hr. The ratio of ^{3}H-glycine to ^{14}C-leucine: (a) nontreated epidermis, (c) hydrocortisone-treated epidermis. Amounts of radioactivity of ^{3}H-glycine (●——●) and ^{14}C-leucine (○- - - -○): (b) nontreated epidermis, (d) hydrocortisone-treated epidermis.

with radioactive amino acids at varying times after the addition of hydrocortisone to the medium, it was shown that stimulation of SCMEp*A* synthesis by hydrocortisone begins about 12–16 hours after addition of the steroid.[10]

In the following experiments, therefore, the hydrocortisone-treated explants after 1 day in culture were pulse-labeled with ^{3}H-glycine. Afterwards, they were chase-cultured in hydrocortisone-containing medium supplemented with cold glycine and cycloheximide. When residual radioactivity in SCMEp*A* bands was compared with that in the whole epidermal protein, radioactivity in SCMEp*A* bands was almost completely maintained with reduction of only 2.7% of the initial amount after a 12 hours' chase-culture (Table 2). On the contrary, radioactivity in the

Table 2. Metabolic stability of SCMEp*A* induced by hydrocortisone.[a)]

Chase-culture time (hr)	Residual radioactivity in total protein (cpm/μg prot.)	%	Residual radioactivity in SCMEp*A* bands cpm	%
0	313.9	100	10613	100
3	329.2	105	11331	107
6	258.7	82.4	10512	99.0
12	187.7	59.8	10324	97.3

a) The hydrocortisone-treated explants after 1 day in culture were labeled with ^{3}H-glycine (10 μCi/ml, 2 Ci/mmole) for 12 hr. Afterwards, they were chase-cultured in hydrocortisone-containing medium supplemented with cold glycine (2 mM) and cycloheximide (4 μg/ml). The amounts of residual ^{3}H in TCA-insoluble precipitates and in the SCMEp*A* bands separated by polyacrylamide gel disc electrophoresis were determined at various times after the chase.

whole epidermal protein including SCMEp*A* reduced progressively throughout the chase-culture period, and 12 hours later approximately 40% of the initial total activity was lost (Table 2). Therefore, it follows that SCMEp*A* is metabolically far more stable as compared with other epidermal proteins.

In the last two experiments of the metabolic studies, behavior of mRNA in hydrocortisone-induced synthesis of SCMEp*A* was studied. Explants cultured for 4 days were incubated with and without added actinomycin D for 9 hours and then pulse-labeled with ^{3}H-glycine for 120 minutes. For the hydrocortisone-treated explants, relatively large amounts of ^{3}H were found in the SCMEp*A*1 and SCMEp*A*2 bands in actinomycin-treated explants (Fig. 7a). On the other hand, for the control explants the incorporation of ^{3}H-glycine into SCMEp*X*1 and SCMEp*X*2 bands was almost completely inhibited by actinomycin D (Fig. 7b). The effect of actinomycin D on SCMEp*A* synthesis was further examined in the hydrocortisone-treated explants. Actinomycin D was added to the hydrocortisone-treated explants after 4 days in culture and at various times after its addition the protein-synthesizing activity of the epidermis was followed by pulse-labeling with ^{3}H-glycine for 60 minutes. The rate of general protein synthesis decreased to approximately 20% of the initial rate within 12 hours (Fig. 8). On the other hand, the rate of SCMEp*A* synthesis in the same explants slowly decreased and, even 12 hours after the addition of actinomycin D, the synthetic activity of SCMEp*A* remained at more than 50% of the initial level (Fig. 8). These findings strongly suggest that the mRNAs for SCMEp*A* induced by hydrocortisone are long-lived compared to those for other general proteins including SCMEp*X*.

It should be stressed that the long life of SCMEp*A* proteins themselves and their mRNAs, which were induced by hydrocortisone, can be considered to be pertinent to the chemical and ultrastructural findings already

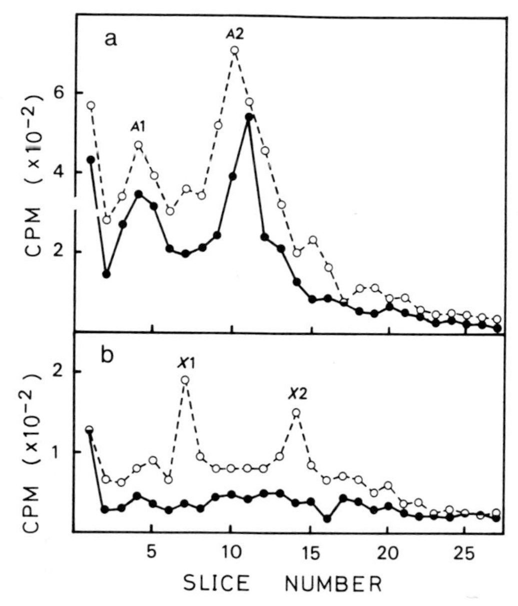

Fig. 7. Electrophoretic patterns of radioactivity of SCMEp from actinomycin D-treated and nontreated explants labeled with ^{3}H-glycine. One of the pair-mate explants of 4-day culture was incubated with (●——●) 4.0 μg/ml actinomycin D or without (○----○) for 9 hr and then labeled with ^{3}H-glycine for 120 min. (a) Hydrocortisone-treated explants; (b) nontreated ones.

mentioned; large quantities of the epidermal structural proteins are accumulated and at the same time a great amount of tonofibrils are formed during the hydrocortisone-induced *in vitro* keratinization as in the physiologic *in ovo* keratinization.

DEVELOPMENTAL SIGNIFICANCE OF THE HYDROCORTISONE-INDUCED IN VITRO KERATINIZATION[11]

Keratinization of the epidermis of chick embryonic shank skin can first be observed morphologically on day 18 of incubation,[4] with the synthesis of the epidermal structural proteins first being detected on day 16.5 of incubation (Fig. 5).[10] On the other hand, a few days before this developmental stage or on days 13 to 15 of incubation, the adrenal cortex of chick embryos begins to secrete corticosteroids.[12] Considering these *in ovo* findings together with the before-mentioned *in vitro* ones, it seems highly probable that glucocorticoids participate in epidermal keratinization in the normal

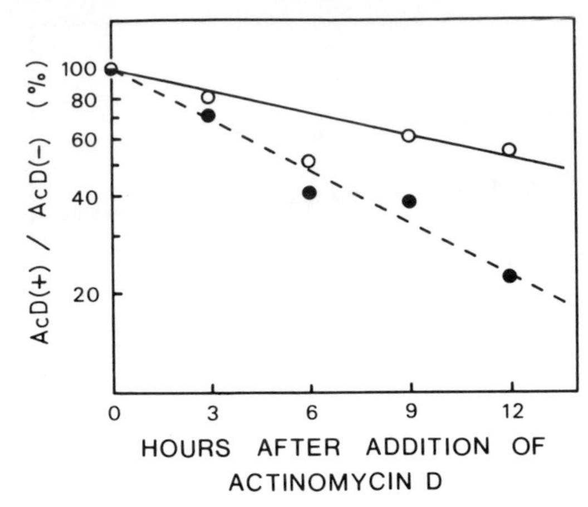

Fig. 8. Effect of actinomycin D on the synthesis of SCMEp*A* and other proteins in hydrocortisone-treated explants. One of the pair-mate explants after 4 days in culture was incubated with actinomycin D (AcD(+)) and the other without the drug (AcD(—)). At the indicated time the explants were pulse-labeled with ^{3}H-glycine for 60 min. Labeled SCMEp was fractionated into SCMEp*A* and other proteins by polyacrylamide gel disc electrophoresis. In each fraction, the ratio of radioactivity of AcD(+) explants to that of AcD(—) ones was expressed as percentages. SCMEp*A* fraction (○——○); the other fraction (●——●).

embryonic development of the chick. Therefore, physiological significance of the *in vitro* findings was further examined in terms of the cytoplasmic glucocorticoid receptor which would probably be prerequisite to the hydrocortisone action.

The shank skin from 10- to 19-day chick embryos developing *in ovo* and 1-day-old chicks after hatching were homogenized. After cytosol fraction of the homogenate was incubated in the cold with ^{3}H-labeled hydrocortisone or dexamethasone, the extent of specific binding of tritiated steroids was examined quantitatively as well as qualitatively. With the use of ordinary sucrose density gradient centrifugation technique, typical profile of the cytoplasmic binding of ^{3}H-hydrocortisone was obtained for 15-day chick embryonic shank skin (Fig. 9). The same pattern was also found confirmed for ^{3}H-dexamethasone binding. Specificity of the glucocorticoid binding was demonstrated by competition with various steroids; cytoplasmic ^{3}H-hydrocortisone binding of 15-day chick embryonic shank skin was successfully competed only by natural and synthetic glucocorticoids and as an exception by progesterone but not by other sex steroids (Table 3).

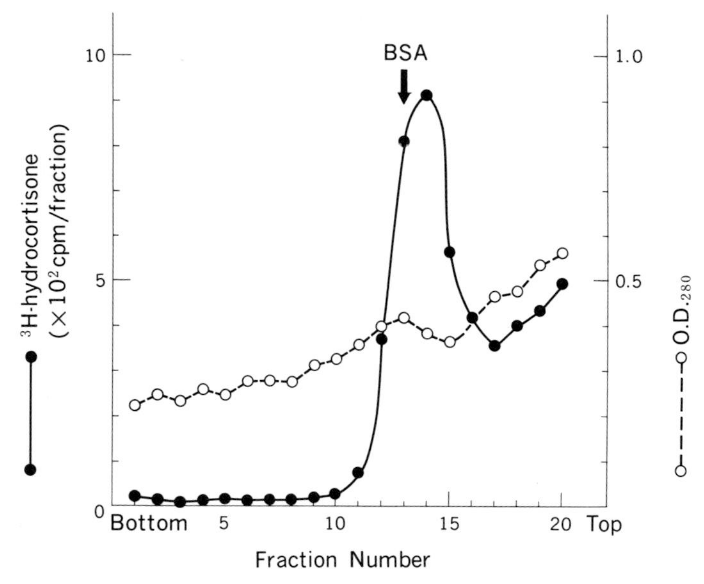

Fig. 9. Sedimentation profile in sucrose density gradient centrifugation of cytoplasmic ^{3}H-hydrocortisone binding of 15-day chick embryonic shank skin.

Table 3. Specificity of cytoplasmic ^{3}H-hydrocortisone binding of 15-day chick embryonic skin as revealed by competition with various steroids.

Nonradioactive steroid added	Concentration (M)	Binding of ^{3}H-cortisol (%)
None*	———	100
Hydrocortisone	1.1×10^{-6}	68.0
	1.0×10^{-5}	3.0
Dexamethasone	1.1×10^{-6}	4.4
Progesterone	1.0×10^{-5}	7.6
Estradiol	0.5×10^{-5}	46.3
	1.0×10^{-5}	59.7
Testosterone	0.5×10^{-5}	64.1
	1.0×10^{-5}	29.7

* The cytosol from 15-day chick embryonic shank skin was incubated with ^{3}H-hydrocortisone (0.3×10^{-9}M) at 0°C for 90 min.

On the basis of these findings, change of the dexamethasone-binding activity of the skin was followed with advancing development *in ovo* and after hatching (Fig. 10). The skin from 10- to 11-day embryos gave virtually negative binding of ^{3}H-dexamethasone. After that however, the binding activity of the embryonic skin showed a marked increase at 12 to 13 days of incubation and gave an abrupt rise around 15 days of incubation to reach a maximum. Thereafter, a rather gradual fall of the binding activity was observed. This fall after 15 days of incubation might be due largely to occupation of the receptor with endogenous glucocorticoids, the level of which increases abruptly just before the developmental stage.[12] It should

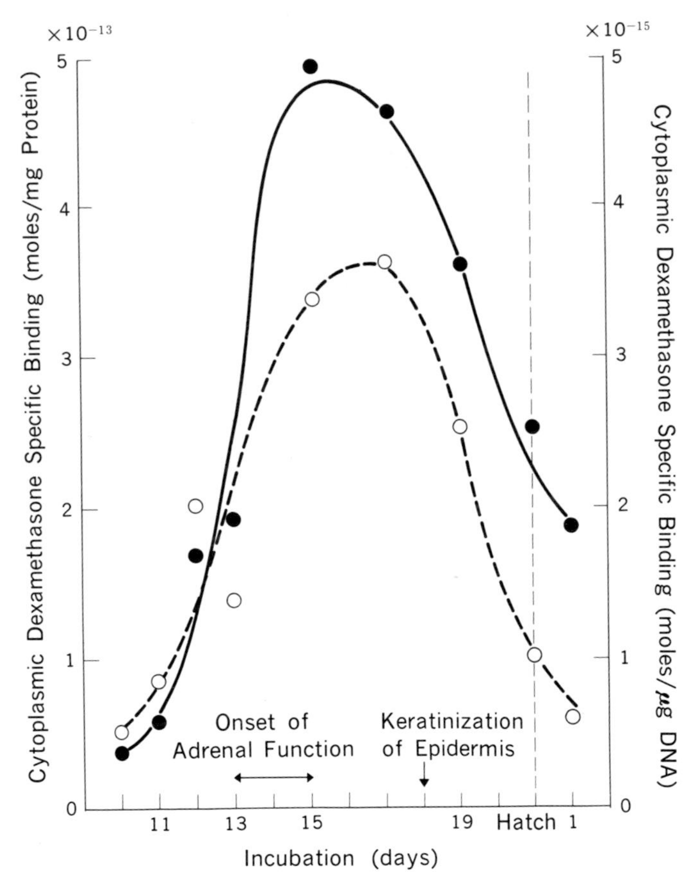

Fig. 10. Change of the cytoplasmic dexamethasone binding activity of chick shank skin with development *in ovo* and after hatching. ●——● specific activity of cytoplasmic dexamethasone binding expressed as moles per mg protein; ○----○ specific activity of cytoplasmic dexamethasone binding expressed as moles per μg DNA.

be noticeable here that in the normal course of chick embryonic development, differentiation of the shank skin in terms of glucocorticoid receptor exactly coincides with onset of adrenocortical function at 13–15 days of incubation and that terminal differentiation of the skin with respect to epidermal keratinization starts just after the developmental stage.

When considered together with the above-mentioned *in ovo* development, the *in vitro* findings from a series of morphological, chemical, and biochemical studies using 13-day chick embryonic undifferentiated shank skin can be taken to indicate that epidermal cells is under adrenocortical control *in ovo* in going through with their differentiation or keratinization.

Acknowledgements

The authors wish to express their sincere thanks to Professor Dame Honor B. Fell of the University of Cambridge, Cambridge, England, for her continued encouragement. This investigation was supported in part by a Grant-in-Aid for Scientific Research from the Ministry of Education of Japan.

REFERENCES

1. Strangeways, T. S. P. and Fell, H. B. Experimental studies on the differentiation of embryonic tissues growing *in vivo* and *in vitro*. I. The development of the undifferentiated limb-bud (a) when subcutaneously grafted into the post-embryonic chick and (b) when cultivated *in vitro*. *Proc. Roy. Soc.,* B **99**: 340–360, 1926.
2. Fell, H. B. The influence of hydrocortisone on the metaplastic action of vitamin A on the epidermis of embryonic chicken skin in organ culture. *J. Embryol. Exp. Morph.*, **10**: 389–409, 1962.
3. Weissmann, G. and Fell, H. B. The effect of hydrocortisone on the response of fetal rat skin in culture to ultraviolet irradiation. *J. Exp. Med.,* **116**: 365–380, 1962.
4. Sugimoto, M. and Endo, H. Accelerating effect of hydrocortisone on the keratinization of chick embryonic skin growing in a chemically defined medium. *J. Embryol. Exp. Morph.*, **25**: 365–376, 1971.
5. Biggers, J. D., Gwatkin, R. B. L., and Heyner, S. Growth of embryonic avian and mammalian tibiae on a relatively simple chemically defined medium. *Exp. Cell Res.*, **25**: 41–58, 1961.
6. Sugimoto, M. and Endo, H. Effect of hydrocortisone on the keratinization of chick embryonic skin cultured in a chemically defined medium. *Nature,* **222**: 1270–1272, 1969.
7. Sugimoto, M., Tajima, K., Kojima, A., and Endo, H. Differential acceleration by hydrocortisone of the accumulation of epidermal structural proteins in the chick embryonic skin growing in a chemically defined medium. *Develop. Biol.*, **39**: 295–307, 1974.
8. Sugimoto, M. and Endo, H. Synthesis of epidermal specific protein selective-

ly stimulated by hydrocortisone in the chick embryonic skin growing in a chemically defined medium. *Endocrinol. Japon.*, **18**: 457–461, 1971.
9. Tezuka, T. and Freedberg, I. M. Epidermal structural proteins. II. Isolation and purification of tonofilaments of the newborn rat. *Biochim. Biophys. Acta*, **263**: 382–396, 1972.
10. Kojima, A., Sugimoto, M., and Endo, H. Epidermal protein metabolism directed toward keratinization by hydrocortisone in the chick embryonic skin growing in a chemically defined medium. *Develop. Biol.*, **48**: 173–183, 1976.
11. Atsumi-Hirato, Y. and Endo, H. Change of glucocorticoid receptor of the skin with development of chick embryos. In preparation.
12. Kalliecharan, R. and Hall, B. K. A developmental study of the level of progesterone, corticosterone, cortisol, and cortisone circulating in plasma of chick embryos. *Gen. Comp. Endocrinol.*, **24**: 364–372, 1974.

Discussion

Dr. Adachi: I wonder if you found nuclear binding in the same way? Have you followed the same change with nuclear binding?

Dr. Endo: Yes, very recently we found the so-called nuclear binding of ^{3}H-corticoids in the chick embryonic skin.

Dr. Adachi: I think the demonstration of nuclear binding is important since you are dealing with new macromolecular synthesis.

Dr. Suzuki: Did the nuclei in the keratinocytes completely disappear in the horny layer after treatment with hydrocortisone?

Dr. Endo: Of course in the upper part of the horny layer, we could not find any sign of nuclei. But in some limited areas we could see pyknotic nuclei in the horny layer of the treated skin.

Dr. Baden: It perhaps would be more precise to describe the shank of the chick as scale from the electrophoretic patterns you were showing.

Dr. Endo: I agree with you. The material was scale. Precisely speaking, our work should be described as dealing with the development of scale.

Dr. Baden: To get more of an idea how the hydrocortisone works in terms of what it is inducing, it might be of some interest to study the scaleless mutant chick which does not make scale keratin but makes true epidermis. It makes an α-protein. Since you would be dealing with only the α-protein you might get a handle on exactly what the hydrocortisone is doing, whether it is inducing the dermal papilla which is then inducing the scale, or is actually having a direct effect on the epidermis.

High Phenylalanine Feed Inhibits Mammalian Hair Keratinization

K. Toda, T. Yoshii and S. Shono

Department of Dermatology, Tokyo Teishin Hospital, Tokyo, Japan

ABSTRACT

In PKU patients, other than liver, peripheral lymphocytes and fibroblasts show deficiency of phenylalanine hydroxylase. Mice fed with high concentrations of phenylalanine showed inhibition of hair growth.

In the culture of the hair from a 6-day-old C-57 black mouse, ^{3}H-glycine uptake was determined by the concentration of tyrosine in the modified MacCoy's 5A medium.

These suggest that not only mental and pigment disorders but also keratinization disorders can be observed in PKU patients and that phenylalanine and its direct metabolic product tyrosine are two important free amino acids in the synthesis of the mammalian hair.

INTRODUCTION

Phenylketonuria (PKU) is an inherited disease. 80% of phenylalanine is oxidized to tyrosine by phenylalanine hydroxylase in normal subjects. In PKU patients, phenylalanine is not properly oxidized to tyrosine because of lack of phenylalanine hydroxylase; blood phenylalanine concentration, therefore, elevates to 10 to 30 times as much as normal, and accumulated phenylalanine is transformed to phenylpyruvic acid excreted in urine.[1,2,3]

The serum concentration of phenylalanine increases because phenylalanine not transformed to tyrosine. This accumulated phenylalanine also inhibits mammalian tyrosinase activity.[4]

The widely accepted pathogenesis of this condition is a marked deficiency of phenylalanine hydroxylase in liver. Deficiency of phenylalanine hydroxylase in organs other than liver, such as peripheral lymphocytes and fibroblasts, however, has been observed by Yoshii *et al.*[5] Actually this condition results mainly from defects in liver, and also every cell in this condition must have a deficiency of phenylalanine hydroxylase because this condition is inherited as a single autosomal recessive gene.

No animal model of this condition has been reported. The purpose of this paper is, however, to reveal abnormality of phenylalanine metabolism in a mouse fed with saturated phenylalanine water *in vivo* and *in vitro* by using the hair tissue culture technique.

MATERIALS AND METHODS

(1) C-57 black mice were depilated and fed with phenylalanine-saturated water for 3 weeks and control animals were fed with regular plain water. Regrown hair was taken, washed, and cut to 5 mm lengths. These hairs were immersed in 8M urea 5 mM Tris-HCl buffer (pH 8.6) for 24 hours and centrifuged at 10,000 $\times$ *g* for 10 minutes. The fractions were dialysed against distilled water for 2 days. Samples were applied to acrylamide gels and run at 5 mA per tube for 2 hours.

(2) Hairs from 6-day-old newborn mice were incubated in the modified MacCoy's 5A medium which did not contain phenylalanine and tyrosine, with 2μC ^{3}H-glycine and tyrosine at various concentrations for one hour at 37°C. After the incubation, hairs were rinsed and solubilized. Uptake of ^{3}H-glycine was counted by a liquid scintillation counter.

RESULTS

Six-week-old C-57 mice fed with phenylalanine-saturated water for 3 weeks after depilation of hair showed marked inhibition of hair growth (Fig. 1). General condition of the mouse appeared to be normal except for abnormality in hair growth.

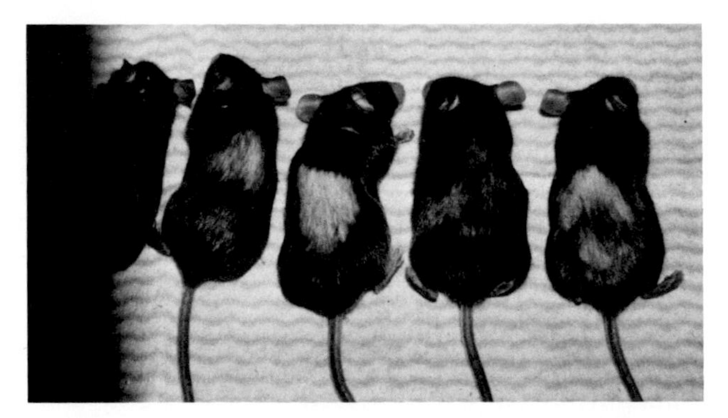

Fig. 1. Mice fed with saturated phenylalanine water for 3 weeks after depilation: marked inhibition of the hair growth can be seen. 4 mice from the right are treated and the mouse at the left is control.

Acrylamide gel electrophoretic pattern of 8 M urea-soluble material from hair of a treated mouse showed almost no fast-moving bands and missed bands in fibrous protein bands (Fig. 2).

Six-day-old newborn mouse hair was incubated in the modified MacCoy's 5A medium with $2\mu C$ ^{3}H-glycine and tyrosine at various concentrations. ^{3}H-glycine uptake increased with concentration of tyrosine in the medium (Fig. 3). This suggests that tyrosine and glycine are closely related in the synthesis of hair protein and also that tyrosine must be one of the aminoacids regulating hair protein synthesis.

Fig. 2. Acrylamide gel electrophoresis; the gel at left is control, almost no fastmoving bands can be seen in the gel at right. Gels are run from the bottom to the top.

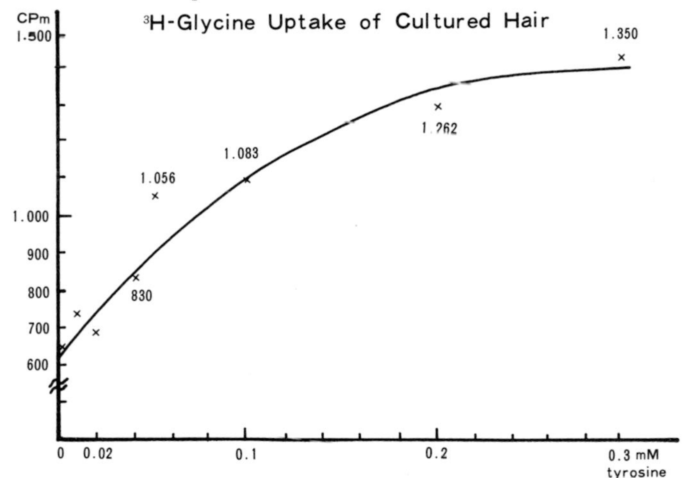

Fig. 3. ^{3}H-glycine uptake in the cultured hair from a 6-day-old newborn C-57 black mouse in the modified MacCoy's 5A medium with tyrosine at various concentrations. H-glycine uptake increases with the concentration of tyrosine.

DISCUSSION

PKU was considered to be a deficiency of phenylalanine hydroxylase only in liver. Yoshii *et al.*, however, have reported deficiency of phenylalanine hydroxylase in the cells other than liver, such as peripheral lymphocytes and fibroblasts from PKU patients, because cultured peripheral lymphocytes and fibroblasts from PKU patients do not incorporate ^{3}H-phenylalanine but incorporate more ^{3}H-tyrosine than those from control.[5]

Deficiency of phenylalanine hidroxylase in animals has not been reported. Our results, however, reveal that animals with high phenylalanine feed and cultured hair from a newborn mouse in the modified MacCoy's 5A medium can be an experimental model of PKU.

Melanogenesis in PKU has been reported by several authors. Miyamoto *et al.* revealed that high concentration of phenylalanine inhibited mammalian tyrosinase *in vitro*.[4] We tried to make hair color of animal dilute by feeding with high concentrations of phenylalanine. No dilution of hair color could be made, but marked inhibition of hair growth could be observed.

The solubilized proteins from the hairs of various kinds of animals consist of three major components. They are low sulfur proteins, high sulfur proteins, and glycine-tyrosine rich proteins. Low sulfur proteins are considered to be fibrous proteins and high sulfur proteins and glycine-tyrosine rich proteins are considered to be matrix proteins.

High concentration of phenylalanine in blood results in deficiency of tyrosine in tissues because substrate inhibition of phenylalanine hidroxylase, which is an enzyme catalyzing reaction from phenylalanine to tyrosine, takes place in the blood. Deficiency of tyrosine in tissues then results in inhibition of protein synthesis. A certain amount of tyrosine is essential to the synthesis of the glycine-tyrosine rich matrix proteins. It can be said from the results shown in Fig. 3 that tyrosine concentration regulates the synthesis of glycine-tyrosine rich proteins. Tyrosine deficiency not directly but indirectly inhibits hair growth.

These data also suggest that in PKU patients not only abnormality in melanogenesis but also abnormality in keratinization can be observed. Direct evidence of abnormality in keratinization of PKU patients is left for future study.

REFERENCES

1. Følling, A. *Ztschr. Physiolo. Chem.*, **227**: 169, 1934.
2. Jervis, G. A. *Proc. Soc. Exp. Biol. Med.*, **82**: 514, 1953.
3. O'Flym, M. F., Tillman, D., and Hsia, D. Y. *Am. J. Dis. Child.*, **113**: 22, 1967.
4. Miyamoto, M. and Fitzpatrick, T. B. *Nature*, **179**: 199, 1957.
5. Yoshii, T. and Toda, K. *J. Dermat*, 1976. In press.

Discussion

Dr. Goldsmith: It is extremely interesting that in the lymphocytes and fibroblasts, you are able to find differences between normal and abnormal. When you try to do very sensitive enzyme assays for this enzyme, or for the tyrosine amino transferase, it looks like this whole enzyme pathway is not present in fibroblasts and lymphocytes. Yet the transport function reflected the disease in your data. Were your patients treated or untreated and did your results make any difference with respect to whether these lymphocytes were washed or treated several times with normal plasma?

Dr. Toda: Yes, these patients were given a diet low in phenylalanine. I think it doesn't make any difference whether they are treated or not since the cell has such low activity. I do want to say that at this time I do not have any positive data as to whether there is a keratin disorder in this type of patient.

Dr. Tezuka: Did you check the hair with the electron microscope?

Dr. Toda: Yes, I did. We checked the hair by both light and electron microscopy but couldn't see any difference between the normal and the patient's hair.

Dr. Tezuka: It seemed to me that the fibrous protein did not enter the gel. Did you use an SDS-gel?

Dr. Toda: No, it was a regular acrylamide gel.

IV. BIOCHEMICAL CHARACTERIZATION OF EPIDERMAL CELL CONSTITUENTS

Intermolecular Cross-Links in Epidermal Differentiation

Kiyoshi SUGAWARA

Laboratory of Biochemistry, Department of Agricultural Chemistry, Faculty of Agriculture, Ibaraki University, Ibaraki, Japan

ABSTRACT

The purpose of this study has been to obtain information on the biochemical processes of late stages in epidermal differentiation. Low cystine content and the presence of epidermal thiol-urea insoluble protein suggest that covalent cross-links other than disulfide bonds would play an important role in epidermal differentiation. This paper reports the identification of γ-glutamyl-ε-lysine cross-links in epidermal protein (0.5N KOH insoluble) and demonstrates the role of the transglutaminase in the formation of the cross-links *in vitro*. The epidermis of newborn rats was fractionated into 8M urea soluble (I), thiol-8M urea soluble (II), and thiol-8M urea insoluble (III) protein fractions by successive extraction. III was further fractionated into 0.5N KOH soluble (IV) at 0°C and the residual (V) protein fractions. After proteolytic digestion and gel chromatography of these fractions, the peptide fractions obtained were analyzed by amino acid analyzer. γ-Glutamyl-ε-lysine (about 5.2 moles/10^5 g of protein) was identified in the digest of V (chemically corresponds to membrane protein), followed by IV (0.4 moles/10^5 g of protein) which would presumably be cross-linked by esterlike bonds, but could not be detected in I (prekeratin) and II (high cystine protein). On the other hand, when minced epidermis was incubated with ^{14}C-lysine, and the labeled V was prepared, digested, and analyzed, the radioactivity was recovered in γ-glutamyl-ε-lysine and free lysine in the ratio of approx. 1:1.2. Furthermore, on incubation of ^{14}C-labeled soluble epidermal proteins with unlabeled residual tissue in the presence of epidermal transglutaminase, the radioactivity was recovered in III. These results suggest that γ-glutamyl-ε-lysine cross-links play an important role in the late stage of epidermal differentiation, especially in the process of structural and chemical modification of the membrane.

INTRODUCTION

As the precursor of "Histidine-Rich Protein"[1] in the epidermis, a protein which is soluble at pH 4.5 (MW, 12,000) has been postulated. This protein, after conversion to "Histidine-Rich Protein,[2] would be keratinized.[3] In general, epidermal protein contains low cystine, especially the Histidine-Rich Protein which has no cystine content.[1,4,5] These findings suggest that, in addition to disulfide bonds, other covalent cross-links would play an important role in the epidermal differentiation. Serveral kinds of cross-links other than disulfide bonds have been reported in wool keratin, collagen, elastin,[6] fibrin, and clotted semen. Especially, γ-glutamyl-ε-lysine peptides were isolated from native wool keratin,[7] medulla of hair and quill,[8] and the inner root sheath cells of hair follicles.[9]

A transglutaminase catalyzes the formation of γ-glutamyl-ε-lysine bond in fibrin[10] and coagulated semenal protein.[11] Similar transglutaminases were detected in the epidermis of several vertebrate species[12,13] and other tissues.[14,15] These data suggest that an epidermal transglutaminase may cross-link epidermal proteins during the keratinization. However, there was no direct evidence for the role of transglutaminase in epidermal keratinization.[13,14] This paper describes the identification of γ-glutamyl-ε-lysine cross-links in the epidermal protein (thiol-urea-insoluble proteins) and demonstrates the formation of the cross-links during the keratinization *in vitro*. Preliminary reports of this work have appeared.[16,17]

MATERIALS AND METHODS

Epidermis and fractionation Three-day-old newborn rats (wistar strain) were used. The epidermis was separated from the dermis by incubating the skin in 0.24 M NH_4Cl (pH 9.0) for 15 min at 0°C. The epidermis was homogenized with 8M urea-0.2M Tris-acetate (pH 8.5) and fractionated according to the diagram (Fig. 1). Fraction I was dialyzed against 0.01N NH_4OH and lyophilized. Fraction II was S-carboxymethylated with monoiodoacetamide at 25°C for 2 hr according to the method of Konigsberg.[18] After the dialysis against water, the fraction was lyophilized. Fraction IV was neutralized with perchloric acid and $KClO_4$ formed at 0°C was filtered off.

Proteolytic digestion of proteins Each fraction was digested at 37°C by the sequential addition of three proteolytic enzymes, according to William-Ashman *et al.*[11] At first 6 mg of pronase P (Kaken Chemicals, Tokyo) was added in 10 ml of 0.1M ammonium bicarbonate (pH 8.0)-1mM calcium acetate, containing about 100 mg of the protein sample. The mixture was incubated for 5 days with four more additions of 3 mg of pronase P at 24 hr intervals. After inactivation of the pronase by heat, the digestion was continued for 48 hr with two additions of 10 units of carboxy-

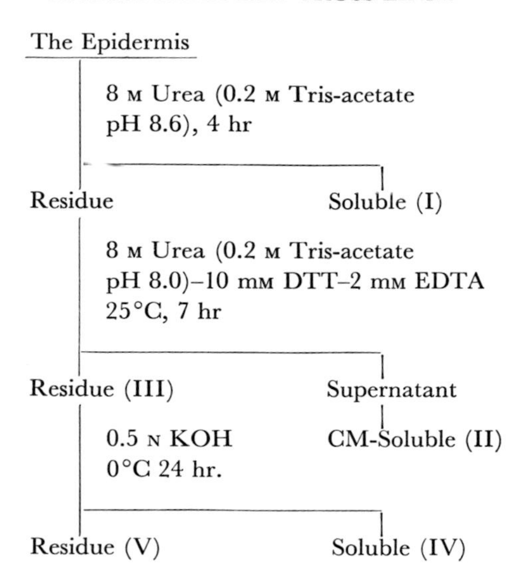

Fig. 1. The fractionation of epidermal protein.

peptidase A (Sigma Chemicals), applied at 24 hr intervals. Finally 0.5 units of aminopeptidase M (Lörm and Haas) were added for a period of 24 hr. The digested mixture was adjusted to pH 4.5 with 1N acetic acid and centrifuged. The supernatant was subjected to gel chromatography. A similarly incubated mixture of the proteolytic enzymes used for the digestion was used as a control.

Gel chromatography The supernatant of the digest was applied to a column of Bio-Gel P-2 (2.1 × 91 cm) equilibrated with 0.1N acetic acid and eluted with the same solution.

Identification and determination of γ-glutamyl-ε-lysine This was carried out essentially according to the method of Lorand *et al.*[19] on a Hitachi KLA-3B amino acid analyzer with a Custom 2612 resin cloumn (0.9 × 52 cm: equilibrated in 0.2N citrate buffer, pH 3.31) at a flow rate of 60 ml/hr at 55°C.

Preparation of [^{14}C]-lysine labeled soluble epidermal protein The epidermis was separated from the dermis by incubating the tissue in 0.01% trypsin in Earle's isotonic solution as previously described.[1] The minced epidermis was suspended in Earle's isotonic solution containing 300 units/ml of penicillin. The incubations were carried out aerobically at 37°C in a shaking incubator for 2 hr after the addition of [^{14}C]-lysine. After the incubation, the tissue was homogenized with 10mM Tris-acetate (pH 7.5)-1 mM EDTA. This and all further operations were carried out at 0 ~ 5°C. The supernatant of the homogenate at 77,000 × *g* for 60 min was applied to a column of Sephadex G-25. The excluded protein fractions were collected and incubated with epidermal transglutaminase.

[^{14}C]-lysine-labeled fraction V The sedimented fraction of [^{14}C]-lysinelabeled tissue was extracted successively with 8M urea, 8M urea-DTT, and 0.5N KOH at 0°C as indicated in Fig. 1.

High-voltage paper electrophoresis The electrophoresis was carried out in pyridine-acetic acid-water (1:10:289, V/V) buffer, pH 3.7, at a potential gradient of 50V/cm for 90 min.

Isolation of epidermal transglutaminase The epidermis was separated from the dermis with 0.24M NH_4Cl and homogenized in 10 mM Tris-acetate (pH 7.5)-lmM EDTA buffer at 0°C. The clear supernatant at 77,000 × *g* for 60 min was submitted to the enzyme purification essentially according to the method of Chung and Folk[20] except that Sephadex G-100 was used for gel filtration instead of Bio-Gel A-5m. Transglutaminase activity was measured by incorporation of [^{14}C]-putrescin into casein.[16]

RESULTS

Identification and estimation of γ-glutamyl-ε-lysine in epidermal proteins

After exhaustive proteolytic digest with pronase P, carboxypeptidase A, and aminopeptidase M, the digests of each fraction of epidermal proteins were applied to a column of Bio-Gel P-2 to separate the peptides from the free amino acids (Fig. 2). The column had previously been calibrated with authentic γ-glutamyl-ε-lysine (Cyclo Chemicals) and a standard mixture of amino acid. The peptide fractions, free from neutral and acidic amino acids, were collected and analyzed by amino acid autoanalyzer. The results of a representative experiment are shown in Fig. 3. The proteolytic digests of the 8M urea-insoluble epidermal protein exhibited a peak that eluted at about 270 min. The peak was located between those of leucine (243 min) and tyrosine (314 min) when the standard amino acid mixture was added. No free dipeptide was detectable when the proteolytic enzymes used for the digestion were incubated in the absence of 8M urea-insoluble epidermal protein. When the authentic γ-glutamyl-ε-lysine dipeptide was added to the proteolytic digest of the urea-insoluble protein, the elution pattern showed the expected augmentation only of the peak eluting at 270 min.

Further confirmation of the results shown in Fig. 3 was obtained by isolation of relcvant γ-glutamyl-ε-lysine dipeptide from a large-scale run. About 30 ml of the effluent between 255 and 280 min was collected and dried. After hydrolysis in redistilled 6N HCI for 22 hr at 108°C and then vacuum distillation, suitable amounts of the hydrolyzate were applied to

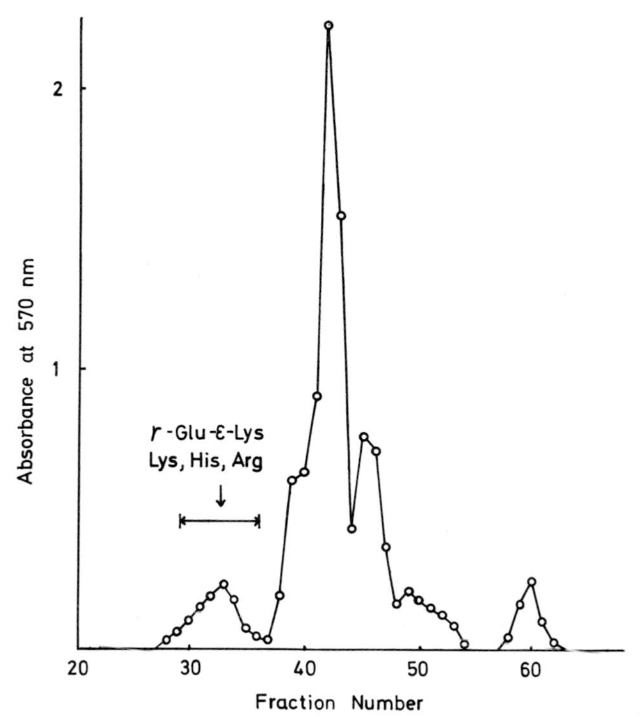

Fig. 2. Gel-filtration of proteolytic enzyme digest of 8M urea-10mM DTT insoluble epidermal protein on a column of Bio-Gel P-2.

short and long columns of the amino acid autoanalyzer. Only the peaks of glutamic acid and lysine were prominent, with mere traces of a few other amino acids visible. The Glu to Lys ratio measured gave a value of 0.87, which supports the original assignment of the dipeptide as γ-glutamyl-ε-lysine. No γ-glutamyl-ε-lysine was identified from the digests of 8M urea-soluble epidermal protein. The urea-insoluble epidermal protein contained 1.55 moles of γ-glutamyl-ε-lysine per 10^5g of the protein (Table 1).

The urea-insoluble protein was fractionated to several fractions successively as indicated in Fig. 1 in order to know the distribution of the γ-glutamyl-ε-lysine cross-links in the urea-insoluble proteins. Table 2 shows a typical example of the data obtained. No γ-glutamyl-ε-lysine was detected in the digests of urea-DTT soluble protein fraction. Whole urea-DTT-insoluble fraction contains 2.77 moles of the isopeptide cross-links per 10^5 g of protein. About two-thirds of this protein was solubilized when treated with 0.5N KOH at 0°C for 24 hr and the 0.5N KOH-soluble protein contains 0.4 moles of the cross-links per 10^5 g protein. Presumably, this protein

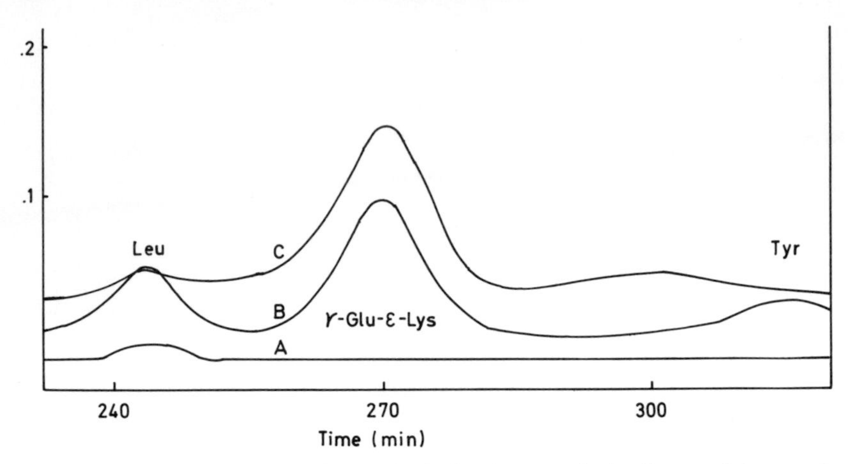

Fig. 3. Identification of γ-glutamyl-ε-lysine in proteolytic digests of 8M urea insoluble epidermal protein. (A) Incubated proteolytic enzyme control. (Proteolytic digest of 8M urea soluble epidermal protein showed the same pattern.). (B) Standard amino acid mixture (100 nmol of each amino acid), added to 250 nmol of authentic γ-glutamyl-ε-lysine peptide. (C) Proteolytic enzyme digest of 12.37 mg of 8M urea-10mM DTT insoluble epidermal protein.

Table 1. Estimation of γ-glutamyl-ε-lysine in the epidermis of newborn rats.

Fraction	% of T Protein	moles per 10^5 g
Urea-soluble (I)	60	0
Urea-insoluble	40	1.55

Table 2. The distribution of γ-glutamyl-ε-lysine cross-links in the urea-insoluble epidermal protein

Fraction	% of T Protein	moles per 10^5 g
Urea-DTT soluble (II)	51	0
Urea-DTT insol. (III)	45	2.77
0.5N KOH sol. (IV)	30	0.4
0.5N KOH insol. (V)	15	5.15

fraction is stabilized by esterlike bonds. On the other hand, 5.15 moles of γ-glutamyl-ε-lysine cross-links per 10^5 g protein were obtained in the 0.5N KOH-insoluble protein fraction. In this fraction, about 67% of the protein was solubilized after the enzymic digestion. Half-cystine content of the urea-DTT soluble, urea-DTT insoluble, and urea-DTT-0.5N KOH insoluble protein fractions were 3.8, 1.5, and 1.4 mole percent of each protein fraction, respectively.

The incorporation of [^{14}C]-lysine into the γ-glutamyl-ε-lysine cross-links in the 0.5N KOH insoluble protein in vitro

[^{14}C]-lysine was incubated with minced epidermis, which consists mainly of stratum and granular layers as reported previously.[2] After being shaken for 2 hr at 37°C, the 0.5N KOH-insoluble protein fraction was prepared and submitted to sequential proteolytic digestion. After gel chromatography on a Bio-Gel P-2 column, the peptide fractions, containing free lysine, were then concentrated and separated by H.V. paper electrophoresis. The radioactivity was detected only in the spots of γ-glutamyl-ε-lysine and lysine. The ratio of radioactivity of γ-glutamyl-ε-lysine to free lysine was 1:1.2 (Table 3). Lysine content of the 0.5N KOH insoluble protein fraction was 5.9 moles per 100 residue of the protein.

Table 3. Two forms of [^{14}C]-lysine in proteolytic digest of fraction V

	CPM	Ratio
Lysine	689	1.2
γ-Glutamyl-ε-lysine	585	1.0

The incorporation of [^{14}C]-lysine-labeled soluble epidermal protein into the keratin fraction in the presence of epidermal transglutaminase

[^{14}C]-lysine-labeled soluble epidermal protein (51,400 dpm/mg protein) and 0.1M Tris-acetate pH 7.5-1mM EDTA-insoluble residual epidermis were dissolved or suspended in 0.1M Tris-acetate, pH 8.6, and heated at 80°C for 20 min to inactivate the transglutaminase. Then the [^{14}C]-lysine-labeled protein(1.03×10^4dpm) was incubated with the residual tissue (6 mg protein) in the presence of isolated epidermal transglutaminase(45 μg protein), 4mM $CaCl_2$ and 1mM DTT in 0.7 ml of 0.1M Tris-acetate, pH 8.6. After the incubation at 43°C for 5 hr, the fraction sedimented at 12,000 x *g* was extracted with 8M urea (0.2M Tris acetate, pH 8.6). The radioactivity of the 8M urea-insoluble fraction was counted in a scintillation spectrometer. Table 4 shows a typical result. About 2.8% of the radioactive soluble protein was converted to 8M urea-insoluble protein fraction in the presence of epidermal transglutaminase.

DISCUSSION

In addition to disulfide bonds, γ-glutamyl-ε-lysine cross-links were identified in the 8M urea-10mM DTT-insoluble epidermal protein of new-

Table 4. Epidermal TGase dependent incorporation of [^{14}C]-labeled soluble protein into the fraction III.

	DPM	%
Complete	285	100
−TGase	27	9.5
$-CaCl_2$	9	3

born rats. The γ-glutamyle-ε-lysine cross-links were localized mainly in 0.5N KOH-insoluble protein fraction (5.15 moles per 10^5 g protein used). This protein fraction corresponded chemically to the membrane protein which was prepared from 0.1N NaOH insoluble fraction of the stratum corneum cell by Matoltsy and Matoltsy.[21] On the other hand, it was postulated that the characteristic insolubility of the stratum corneum is mainly due to the cell membrane rather than their intracellular contents, since physical disruption of the cells after swelling in urea permitted release of polypeptides.[22] In the present experiment about 45% of the [^{14}C]-lysine incorporated into the alkali-insoluble protein was found to be in the cross-linked form when [^{14}C]-lysine was incubated with minced epidermis. Because the epidermis used consisted mainly of stratum corneum and adhering layers of granular cells,[2] the cross-links would be formed in the upper granular layer where the final process of the cytodifferentiation would be taking place. The localization of epidermal transglutaminase in upper malpighian and granular layers also has been reported.[13] These suggest that γ-glutamyl-ε-lysine was formed in the later stage of the epidermal differentiation. Purified epidermal transglutaminase was shown to catalyze the conversion of [^{14}C]-lysine-labeled soluble epidermal protein into 8M urea-insoluble protein. This reaction required the presense of the cell debris fraction. These facts also suggest that the cross-links play an important role in the process of the structural and chemical modifications of the membrane, involving the binding of membrane with granular protein,[23] during the late stage of differentiation; thus the horny cells become highly resistant to chemical attack.

REFERENCES

1. Hoober, J. K. and Bernstein, I. A. Protein synthesis related to epidermal differentiation. *Proc. Natl. Acad. Sci. U.S.A.*, **56**: 594–601, 1966.
2. Sugawara, K. and Bernstein, I. A. Biosynthesis, *in vitro*, of "histidine-protein"-a biochemical marker in epidermal differentiation. *Biochim. Biophys. Acta*, **238**: 129–138, 1971.
3. Katagata, Y. Studies on biosynthesis of epidermal keratin-incorporation of pH 4.5 soluble protein into the keratin by non-ribosomal system. Thesis, Ibaraki University, Ami, Ibaraki-ken, Japan, 1975.

4. Sibrack, L. A., Gray, R. H., and Bernstein, I. A. Localization of the histidine-rich protein in keratohyalin: A morphologic and macromolecular marker in epidermal differentiation. *J. Invest. Dermatol.*, **62**: 394–405, 1974.
5. Tezuka, T. and Freedberg, I. M. Epidermal structural proteins. III. Isolation and purification of histidine-rich protein of the newborn rat. *J. Invest. Dermatol*, **63**: 402–406, 1974.
6. Gallop, P. M., Blumenfeld, O. O., and Seifter, S. Structure and metabolism of connective tissue proteins. *Ann. Rev. Biochem*, **41**: 617–672, 1972.
7. Asquith, R. S., Otterburn, M. S., Buchanan, J. H., Cole, M., Fletcher, J. C., and Gardner, K. L. The identification of ε-N-(γ-L-glutamyl)-L-lysine cross-links in native wool keratins. *Biochim. Biophys. Acta*, **221**: 342–348, 1970.
8. Harding, H. W. J. and Rogers, G. E. The occurrence of the ε-(γ-glutamyl) Lysine cross-link in the medulla of hair and quill. *Biochim. Biophys. Acta*, **257**: 37–39, 1972.
9. Harding, H. W. and Rogers, G. E. ε-(γ-Glutamyl)lysine cross-linkage in citrulline-containing protein fractions from hair. *Biochemistry*, **10**: 624–630, 1971.
10. Lorand, L. Fibrinoligase: The fibrin-stabilizing factor system of blood plasma. *Ann. N. Y. Acad. Sci.*, **202**: 6–30, 1972.
11. Williams-Ashman, H. G., Notides, A. C., Pabalan, S. S., and Lorand, L. Transamidase reactions involved in the enzymic coagulation of semen: Isolation of γ-glutamyl-ε-lysine dipeptide from clotted secretion protein of guinea pig seminal vesicle. *Proc. Nat. Acad. Sci*, **69**: 2322–2325, 1972.
12. Goldsmith, L. A., Baden, H. P., Roth, S. I., Colman, R., Lee, L., and Fleming, B. Vertebral epidermal transamidases. *Biochim. Biophys. Acta*, **351**: 113–125, 1974.
13. Buxman, M. M. and Wuepper, K. D. Keratin cross-linking and epidermal transglutaminase. *J. Invest. Dermatol.*, **65**: 107–112, 1975.
14. Harding, H. W. J. and Rogers, G. E. Formation of the ε-(γ-glutamyl)lysine cross-link in hair proteins. Investigation of transamidases in hair follicles. *Biochemistry*, **11**: 2858–2863, 1972.
15. Bohn, H. Comparative studies on the fibrin-stabilizing factor from human plasma, platelets and placentas. *Ann. N. Y. Acad. Sci.*, **202**: 256–272, 1972.
16. Sugawara, K. and Mogaki, K. Formation of epidermal keratin—Identification of γ-glutamyl-ε-lysine cross-links and role of transglutaminase (abstr). Seikagaku, **46**: 641, 1974.
17. Sugawara, K. Formation of epidermal keratin—Fractionation of the keratin and the distribution of γ-glutamyl-ε-lysine (Abstr). *Seikagaku*, **47**: 539, 1975.
18. Konigsberg, W. Reduction of disulfide bonds in proteins with dithiothreitol. *Methods in Enzymol.*, **25**: 185–188, 1972.
19. Lorand, L., Downey, J., Gotoh, T., Jacobsen, A., and Tokura, S. The transpeptidase system which cross-links fibrin by γ-glutamyl-ε-lysine bonds. *Biochem. Biophys. Res. Comm.*, **31**: 222–230, 1968.
20. Chung, S. I. and Folk, J. E. Transglutaminase from hair follicle of guinea pig. *J. Biol. Chem.*, **69**: 303–307, 1972.
21. Matoltsy, A. G. and Matoltsy, M. N. The membrane protein of horny cells. *J. Invest. Dermatol.*, **46**: 127–129, 1966.

22. Steinert, P. M. The extraction and characterization of bovine epidermal α-keratin. *Biochem. J.*, **149**: 39–48, 1975.
23. Matoltsy, A. G. and Parakkal, P. E. Membrane-coating granules of keratinizing epithelia. *J. cell Biol.*, **24**: 297–307, 1965.

Discussion

Dr. Ogawa: You achieved a high purification of the transglutaminase from the epidermis. What was the specific activity and how did you calculate the enzyme activity in your system?

Dr. Sugawara: The activity was determined in the usual way. The specific activity was shown on one of the slides I presented.

Dr. Baden: I missed one point and that is, when you did the incubations with a purified enzyme, did you use a labeled fraction that was a urea-soluble labeled protein?

Dr. Sugawara: No. It was soluble in 10mm buffer at pH 7.5.

Dr. Baden: So it is truly neutral-soluble proteins which were incubated with the enzyme and cell debris?

Dr. Sugawara: Yes. Cell debris is essential for the incorporation. Without cell debris we cannot get the formation of urea-insoluble protein.

Dr. Goldsmith: Is the protein which was labeled and in which you demonstrated glutamyl-lysine cross-links——in that last incorporation experiment——the same protein that has the label by the initial analysis? Many proteins may be cross-linked by transglutaminase, and my question is, is that the protein which is cross-linked *in vivo*?

Dr. Sugawara: I think so since the ^{14}C was recovered mainly in the urea-thiol-insoluble fraction.

Dr. Tekake: How active is the transglutaminase at normal skin temperature?

Dr. Sugawara: I can't say exactly but it appears to have about half the activity it has at the optimum temperature of 55°.

Dr. Matoltsy: To me this is quite interesting because the alkali-insoluble Fraction V, may correspond to the thickened membrane of the horny cells. You have shown that it contains about 15 percent of the whole material. This is a little bit higher than what we generally find for membranes which is about 5 percent. It has a relatively high proline content which again suggests that it may be the highly resistant membrane. But there is one problem that I don't quite understand. If I understood you correctly, you were labeling Fraction V and then you identified the γ-glutamyl-lysine link in this. If this is a membrane preparation that appears in the dead horny cells, I don't see how it would be possible to label it. How would the amino acid, lysine, incorporate into this protein which is actually appearing in an inactive cell?

DR. SUGAWARA: In order to label the protein, the minced tissue was incubated with ^{14}C-lysine at 37° for 2 hours, then the residual tissue was refractionated to obtain Fraction V. With regard to your comment, the 15 percent was based on only the urea-insoluble fraction whereas you calculated your values based on whole epidermis. If you calculated my values on the basis of the whole epidermis, it would probably be closer to 5 percent. We could not use ^{14}C-glutamine as the label because of the possible metabolism of that compound to other labeled substances. Therefore, we only use the labeled amino donor.

ε-(γ-Glutamyl)Lysine Cross-links in Proteins*

Lowell A. Goldsmith

Duke Medical Center, Durham, North Carolina, U.S.A.

ABSTRACT

The ε-(γ-glutamyl) cross-link is a posttranslational modification of several structural proteins. Those proteins include the α and γ chains of fibrinogen, the basic protein of semen, hair medulla protein, and an as yet unidentified epidermal protein. The methods for identifying and measuring this cross-link are discussed. The structural characteristics of the peptides with this cross-link are discussed. Recent studies suggesting a role for this cross-link in the formation of the cell envelope protein of the epidermis are presented, including the results of incorporation of putrescine into various soluble and insoluble epidermal proteins of newborn mouse epidermis and human stratum corneum.

INTRODUCTION

Most peptide bonds involve the carboxyl group attached to the α-carbon of an amino acid. Peptide bonds involving the carboxyl group attached to the γ-carbon of glutamic acid are less common, but serve unique and important biological roles. The role of ε-(γ-glutamyl)lysine dipeptides or ε-(γ-glutamyl)lysine cross-links is the subject of this review. The formation, measurement, and characteristics of this dipeptide and the methods for identifying the molecules which may contain or be induced to form this bond will be discussed. Some recent studies of this dipeptide in epidermis will be emphasized.

* Research supported by grant AM-17253 from the National Institutes of Health, United States Public Health Service. The author is the recipient of Research Career and Development Award AM00008 from the United States Public Health Service.
Publication number 8 of the Dermatological Research Laboratories of Duke Medical Center.

A. γ-Glutamyl bonds

Gamma-glutamyl peptide bonds were first discovered in the tripeptide glutathione and their presence in a variety of natural peptides has been reviewed by Harrington.[1] This bond occurs as a postribosomal (posttranslational) modification of several structural proteins. Both extracellular proteins (e.g., fibrin and the basic protein of semen), and intracellular proteins (e.g., cell membrane proteins, hair medullary protein, epidermal cell proteins) contain this bond. This bond in some cases (γ chain of fibrinogen) has been shown to be an intermolecular cross-link between two peptide chains, in other cases it may be intramolecular, but no definite natural example of that latter circumstance has been described.

B. Transglutaminases

The characteristics and enzymology of the transglutaminases which form these bonds has been reviewed elsewhere.[2] Transglutaminases are calcium requiring enzymes which deaminate certain protein (substrate 1) bound glutamines forming a substrate 1-acyl enzyme. Certain epsilon amino groups of another (or the same) protein (substrate 2) may react with the substrate 1-acyl enzyme forming a covalent complex of substrate 1 and substrate 2 through an ε-(γ-glutamyl) lysine peptide bond. Some of the transglutaminases are present as zymogens (plasma factor XIII, platelet factor XIII) and require proteolytic activation; others (liver, hair follicle, and epidermal transglutaminase) are not present as zymogens.

C. Site and timing of dipeptide synthesis

The ε-(γ-glutamyl) lysine bond is definitely formed after the completion of protein synthesis in the case of the cross-linking of fibrin and the basic protein of semen since the cross-linking of these proteins occurs extracellularly. In epidermis the fraction which contains the cross-link is keratin (α-fibrous protein, urea-sulfhydryl soluble proteins). Keratin contains the dipeptide in the stratum corneum while the same protein fraction (prekeratin) in the living cell layers (stratum germinativum) of the epidermis does not contain the dipeptide.[3] This is strongly suggestive of the dipeptide being formed after the completion of protein synthesis of the proteins in prekeratin. In mouse fibroblasts (L-cell) in culture, however, fractions containing the dipeptide were found in the cell membrane, microsomal and endoplasmic reticulum fractions.[4] It is, therefore, possible that under some circumstances the dipeptide might be formed while nascent protein chains are still attached to ribosomes.

D. Biological role of cross-link

The ε-(γ-glutamyl) lysine bond is resistant to cleavage by a variety of proteolytic enzymes of mammalian and bacterial origin.[5]* This biological feature may uniquely allow proteins containing this cross-link to be at several crucial biological interfaces between an organism and his environment: sites such as the cell membrane, the epidermis, blood clots, and the coital plug.

ISOLATION AND QUANTITATION OF CROSS-LINK

A. Enzyme digestion

The definite demonstration of ε-(γ-glutamyl)lysine cross-links requires a complete enzymatic digestion of the material to be studied and quantitation of the cross-link by ion exchange chromatography. Various amino acid analyzer programs for detecting ε-(γ-glutamyl)lysine are published.[5,7] Isotope dilution techniques can be used to complement the quantification by ion exchange chromatography.[5] The isolated peptides should yield equimolar amounts of glutamic acid and lysine after acid hydrolysis, and other derivatives of the peptide can be hydrolyzed to prove that the γ-glutamyl and ε-lysine groups of glutamic acid and lysine, respectively, are the ones in peptide linkage.

Various schemes of enzymatic digestion have been used by different investigators. The enzymatic digestion of fibrin included the sequential use of trypsin, pronase twice, leucine aminopeptidase, and prolidase.[5] The digestion of wool[8,9] included the sequential use of trypsin, pepsin, chymotrypsin, pronase, leucine aminopeptidase, and carboxypeptidase A & B; the same authors in later studies omitted the carboxypeptidases and substituted aminopeptidase M for leucineaminopeptidase.[10] For the digestion of wool medullary protein the digestion used included pepsin, pronase, aminopeptidase M and prolidase.[9,10] For the coagulated basic protein of semen, pronase was used twice, followed by Mn^{+2} activated leucine aminopeptidase and carboxypeptidase A[7]; for cell membranes the digestion used pronase twice and leucineaminopeptidase.[4] Mixtures containing the enzymes, but not the protein being studied, are always included as controls. Incomplete digestion, if it leaves some insoluble highly cross-linked material, can lead to a conclusion that the dipeptide was not present, or can produce falsely low values for the dipeptide.

B. Cyanoethylation

Cyanoethylation, a procedure in which free amino groups react with acrylonitrile and form acid stable products, has been used as an indirect measure of blocked amino groups.[5] If all the ε-amino groups in a protein

* There are enzymes which can digest the isolated ε-(γ-glutamyl) lysine peptide in the kidney (highest levels), liver, and small intestine.[6]

react with acrylonitrile it is unlikely that any ε-(γ-glutamyl) lysine cross-links are present. The sensitivity of this obviously depends on the sensitivity of measuring lysine after acid hydrolysis. However, the presence of unreactive amino groups certainly does not indicate the presence of ε-(γ-glutamyl) lysine cross-links: other cross-links, other groups may be attached to the γ-amino group, and stearic blocking or other structural factors may be preventing cyanoethylation. The usefulness of cyanoethylation as a screening procedure, or when working with well-characterized proteins or their peptides is unquestioned, however.

PROTEINS CONTAINING ε-(γ-GLUTAMYL)LYSINE CROSS-LINKS

A. Fibrin

Completely cross-linked fibrin contains 6 moles of cross-link per mole of fibrin[11] as determined by complete enzymatic digestion and ε-(γ-glutamyl)lysine quantitation. The α chains and γ chains of fibrin participate in cross-linking while the β chains does not.[12] The γ chain forms γ dimers and the α-chain multimers.[12]

These are two cross-links per pair of γ chains. The chains are in an antiparallel arrangement when cross-linked.[13] The glutamine residue of the cross-link is near the C terminal end of the molecule, 14 residues in from the end. The lysine residue is 6 residues in from the C terminal end. The cross-linked tryptic peptides of bovine and human γ chains are published (Fig. 1).[13] Structural model shows that the two residues in the cross-link would be closely opposed and that both the residues project from the same side of an alpha-helix in an ideal arrangement for bonding. Two residues of the hydrophobic amino acid leucine, 4 residues from the cross-linking site, may play a role in the orientation of the cross-linking site.[14] This is the most completely known sequence of a natural cross-linking site.

In the α chain there are 4 cross-links per two alpha chains. The α-chain forms polymers of various molecular weights when treated with its normal cross-linking enzyme, plasma factor XIII.[12] Recently, both cross-linking acceptor sites have been identified in a 23,000 molecular weight peptide.[15] Further data on the structure of this peptide should soon be available.

···Leu-Thr-Ile-Gly-Glu-Gly-Gln-Gln-His-His-Leu-Gly-Gly-Ala-Lys-Gln-Ala-Gly-Asp-Val-COOH
HOOC-Val-Asp-Gly-Ala-Gln-Lys-Ala-Gly-Gly-Leu-His-His-Gln-Gln-Gly-Glu-Gly-Ile-Thr-Leu···

Fig. 1. Cross-linked peptide of the γ dimer of fibrinogen.

The antiparallel γ chains are cross-linked at[1] as indicated with ε-(γ-glutamyl) lysine dipeptides. The carboxy-terminal amino acid valine is designated 1. Adopted from.[13]

B. Wool

The presence of blocked amino groups in wool was considered because of ε-amino groups in wool which did not react with dimitrofluorobenzene. Enzymatic digestion of Merino 64 wool revealed 15 umoles of ε-(γ-glutamyl)lysine per gram of wool. The α-keratose (low sulfur) proteins and the α-keratose (high sulfur) proteins had 9 and 8 ugms of cross-link per gm of wool respectively; the very insoluble β-keratose fraction had 30 umoles/gm.[8] The ε-(γ-glutamyl)lysine cross-links in hair (wool) are predominantly in the medulla of hair.[9,10] The amount of these cross-links varied even within the medulla when the medullae of many animals were studied[10] (Table 1). There was no direct relationship between the citrulline content of the medulla and ε-(γ-glutamyl)lysine cross-links.[10] Acid soluble and acid insoluble peptides had the same cross-link content.[9]

Table 1. ε-(γ-glutamyl)lysine cross-links in various hair and medullary proteins. The medullary peptides were isolated by sequential enzymatic digestion with trypsin and then the ε-(γ-glutamyl)lysine released by complete enzymatic digestion and quantitated by ion exchange chromatography.[10]

Animal	Tissue	ε-(γ-glutamyl)lysine Content Residues/1000 Residues
Guinea Pig	Hair Medulla	28.1
Guinea pig	Internal Root Sheath	2.5
Rat	Hair Medulla	29.5
Rabbit	Hair Medulla	25.9
Camel	Hair Medulla	24.5
Corgi	Hair Medulla	9.2
Seal	Hair Medulla	6.6
Possum	Hair Medulla	5.6
Kangaroo	Hair Medulla	5.5
Echidna	Quill	10.1
Porcupine	Quill	2.3

The cross-linked peptide in guinea pig medulla was isolated after thermolysin and trypsin digestion, and one of these peptides was partially purified by gel filtration and ion exchange chromatography. The apparent molecular weight of the cross-linked peptide by gel filtration was 5,000–8,000. Its amino acid composition and cross-link content were determined[16] (Table 2). The peptide after acid hydrolysis has a high content of glutamic acid and citrulline. Essentially 1 in 13 residues is cross-linked in this peptide. Whether these cross-links are intra- or inter-molecular is not known. Since there are 2 N-terminal groups, probably two chains are present; whether the chains are identical or not cannot be assessed at this time.

Table 2. Amino acid composition of peptides and proteins naturally containing ε-(γ-glutamyl)lysine dipeptides[a]

	Guinea Pig Hair Medulla[b]	Guinea Pig Prostatic Vesicular Fluid[c]
Aspartic	3.0 (3)	11
Threonine	1.0 (1)	6
Serine	1.5 (2)	23
Glutamic & Glutamine	33.3 (33)	28
Citrulline	11.1 (11)	—
Glycine	2.4 (2)	19
Alanine	2.2 (2)	6
Valine	0.8 (1)	13
Methionine	0.3	6
Isoleucine	0.4	2
Leucine	4.0 (4)	14
Tyrosine	0.3	0
Phenylalanine	0.5	6
Lysine	6.2 (6)	16
Histidine	0.2	2
Arginine	0.7 (1)	10
Tryptophan	—	1.0
Number of Residues	66	163
Cross-link per Peptide	5.4 (5)	6–7
Amide N	—	21.0

[a]Values are given as residues/molecule

[b]From Harding and Rogers.[16] The figure in ( ) is the original author's figure for residues/peptide.

[c]From Williams-Ashman.[7]

C. Basic protein of semen (guinea pigs)

The postcoital vaginal plug in rodents is formed by polymerization of the basic protein in rodent semen by the action of prostatic transglutaminase.[7] The protein from guinea pigs has been extensively studied. The protein is synthesized in the seminal vesicle. The molecular weight of the unpolymerized protein is 17,900, and its amino acid composition is in Table 2. In the vaginal canal, it is highly cross-linked by a prostatic transglutaminase so that 38–44% of the lysines are present in ε-(γ-glutamyl)lysine cross-links. Further details of the exact sequence around the cross-linking sites are not available.

D. Cell membrane

ε-(γ-glutamyl)lysine cross-links have been demonstrated in the plasma membranes, endoplasmic reticulum, and the microsomal fraction of cul-

Table 3. ε-(γ-glutamyl)lysine dipeptides in mouse fibroblast L-cell components.
The cells were grown in culture in the presence of [^{14}C]lysine or [^{14}C]glutamine plus [^{3}H]lysine and the cell fractions isolated by differential centrifugation, the fraction enzymatically digested, and the dipeptide determined on an automatic amino acid analyzer. Some of the data presented are the means calculated from the author's more extensive data.[4] Cyanoethylation was performed using acrylonitrile according to the method of Pisano.[5]

Tissue Component	Method of Analysis	ε-(γ-glutamyl)lysine peptides per mg protein
		$\times 10^4$
Plasma Membrane	Cyanoethylation	39.5
	Enzyme digestion	17.0
Endoplasmic Reticulum	Cyanoethylation	34.5
	Enzyme Digestion	7.0
Microsomes	Enzyme Digest	3.6

tured mouse L-fibroblasts[4] (Table 3). The exact proteins containing the cross-links were not identified. Interestingly, proteins still on the microsomes contained cross-links. As one might suspect, cyanoethylation gave a higher result for the number of cross-links because it will detect ε-amino groups not available for cyanoethylation for any reason.

E. Collagen

Collagen was the first structural protein in which γ-glutamyl cross-links was considered. This was on the basis of chemical studies.[1] However, careful and complete enzymatic digestions failed to detect any such linkages within the sensitivity of the techniques used[17,18] and possible artifacts during the chemical methods of analysis make those results difficult to interpret.

F. Epidermis

Cyanoethylation studies suggested that the S-carboxymethylated keratin fraction of cow snout epidermis and human epidermis contained 7–9 nmoles of blocked lysine per mg of protein.[3] Complete enzymatic digestion of the S-carboxymethylated keratin (urea-sulfhydryl soluble) fraction of human epidermis and isolation and quantitation of the ε-(γ-glutamyl) lysine cross-links present demonstrated 7–8 nmoles of ε-(γ-glutamyl) lysine per mg protein.[19] Quantitation was by ninhydrin analysis and used isotope dilution techniques. Since there was more than one polypeptide chain in the protein fraction studied the exact protein containing this cross-link is not yet identified. Epidermis is thought not to contain a citrulline-rich protein similar to the citrulline-rich medullary protein of hair which contains this cross-link. Identification of the cross-linked proteins will require further study.

G. The acyl acceptor site as determined by studies of model compounds and liver transglutaminase

An absolute requirement of the acceptor site amino acid in proteins is that it must be glutamine and not asparagine.[2] Extensive studies which characterized the active site of liver transglutaminase showed this requirement was related to the binding of the α-and β-methylene carbons of glutamine in a particular hydrophobic niche in the enzyme and that the change of glutamine to asparagine prevented productive binding.[20] Using the ε-amino donor analogue [^{14}C] glycine ethyl ester and guinea pig liver as a source of enzyme, the incorporation of this ester was equivalent to the glutamine content of insulin and the α and β chains of hemoglobin.[2] In lysozyme and ribonuclease, however, the ester incorporation was only 7% and 15% respectively of the theoretical maximum in both proteins. However, after cleavage of the disulfide bonds in these proteins by oxidation, incorporation of ester equal to the glutamine content was possible. These studies show that disulfide cross-links and probably other features of secondary, tertiary, and quaternary structure can impose constraints on the ability of a peptide bound glutamine to be an acyl acceptor in this reaction. It is also reasonable to suppose that there could be variation in relative abilities of different transglutaminases to deaminate a glutamine residue, incorporate a low molecular substrate via transglutamination, and incorporate a high molecular substrate via transglutamination into the same glutamine residue on an acceptor molecule.

Studies with model substrates and guinea pig liver transglutaminase showed that in order for a glutamine residue to be an effective substrate, glutamine must be at least the third amino acid from the NH_2 terminal end of a peptide. Using a series of model peptides, it was possible to study the effect of amino acid substitution of residues surrounding the glutamine residue on the rate of transglutamination.[21] The data from these studies is summarized in table 4, the native peptide is peptide number 1. If leucine was substituted for glycine (number 2) at the peptide-N of glutamine the reaction rate was significantly reduced. If leucine was substituted for glycine at the peptide-C of glutamine (number 4) the reaction rate was increased. A leucine for glycine substitution (number 3) two amino acids from the peptide-N of glutamine, or an phenylalanine at the peptide-N of glutamine (number 5) did not significantly affect the rate of amine incorporation. The substitutions had only small effects on the Kms of the substrates. These model studies may be relevant to the *in vivo* reactiveness of peptide bound glutamines since in glucagon one of the glutamines in a sequence near the C-terminal end of the molecule-Asp-Phe-Val-Gln-Try-Leu-Met-COOH, is preceded by valine and this glutamine was not transglutaminated with guinea pig liver transglutaminase.[22] The presence of hydrophobic amino acids such as valine, leucine, or isoleucine *in vivo* and *in vitro* may

Table 4. Kinetic constants for amine incorporation into model substrates by guinea pig liver transglutaminase.[a]

The incorporation of methylamine and hydroxylamine into the substrates was measured by the colorimetric $FeCl_3$ procedure for hydroxylamine and using [^{14}C] methylamine. Kinetic parameters were derived by computer analysis of the data.[b]

Number	Peptide Substrate	Amine Substrate Methylamine	Hydroxylamine	
		Vmax	Vmax	Km
1	Formyl-Gly_3-Gln-Gly_3	42 ± 8	135 ± 6	20.2 ± 1.7
2	Formyl-Gly_2-Leu-Gln-Gly_3	2.3 ± 0.4	25 ± 1	21.0 ± 1.1
3	Formyl-Leu-Gly-Gly-Gln-Gly_3	52 ± 5	248 ± 42	25.4 ± 4.4
4	Formyl-Gly_3-Gln-Leu-Gly_2	166 ± 25	383 ± 21	9.8 ± 1.0
5	Formyl-Gly_2-Phe-Gln-Gly_3	34.1 ± 7.2	178 ± 23	35.5 ± 7.7

[a]Data from Gross *et al.*[21]

[b]V_{max} is expressed in umoles/min/umol enzyme

therefore protect certain peptide bound glutamines from transglutamination. Further studies with model compounds and especially more studies with the natural protein substrates for transglutaminases will be necessary to completely determine the structural characteristics of the acyl acceptor in determining specificity.

H. Characterization of the amine donor as determined by model donor substances

A number of primary amines are substrates for transglutaminases and their roles as substrates has been studied. With guinea pig liver transglutaminase the best substrates were (listed in order of the amount of amine incorporated) putrescine, phenylethylamine, glycinamide, histamine, spermine cadaverine, methylamine, ethanolamine, alaninamide, noradrenaline, serotonin, lysine, and aniline.[23]

Similar studies with factor XIII have been performed and used to identify potential inhibitors of cross-linking.[24]

Using peptides or proteins as donors to define the exact characteristics of the donors has not yet been done. The only natural cross-linked protein in which there is information about the structural characteristics of the ε-amino group donor is the cross-linked γ chain of fibrinogen which is discussed above (A) and the structure around the donor lysine is described in Fig. 1.

I. Recent studies of structural proteins using transglutaminase to cross-link proteins or incorporate primary amines

Transglutaminases have been used to incorporate ^{14}C-putrescine and other radioactive amines or the fluorescent amine dansylcadaverine into the hyalin layer of sea urchin embryos,[25] mouse erythrocytes,[26] human erythrocytes,[25] and rabbit sarcoplasmic reticulum.[26] In the studies with rabbit sarcoplasmic reticulum most of the labelled amine appeared in the Ca^{+2}-ATPase and some very highly cross-linked high molecular weight protein which was not identified.[26] In erythrocytes, spectrin and the proteins in bands 1 and 2 were heavily labelled with these markers and spectrin was polymerized into high molecular weight cross-linked aggregates.[25,26] It was suggested that similar cross-linking may occur *in vivo* and have physiological significance.[26]

Recent studies of human clotted plasma have shown that only three proteins incorporated dansylcadaverine: those were α_2-macroglobulin, an unidentified protein with a molecular weight of 110,000, and cold-insoluble globulin (CIG).[27] The first two proteins were not cross-linked by factor XIII (plasma transglutaminase), but CIG was cross-linked to itself to form dimers, trimers, and tetramers and also was cross-linked to the α-chain of fibrin.[28]

Cold insoluble globulin is a protein of unknown function present at levels of 330 ug/ml in human plasma. It has a molecular weight of 390,000 (200,000 after disulfide reduction), sediments at 12–14S, is in Cohn Fraction I, and migrates as a β-globulin.[28] A very similar, immunologically identical protein, is present in the fibroblast plasma membrane.[29] The protein is greatly reduced on the cell surfaces of transformed cells.

J. Epidermis

Several methods have been used to define the location of the substrates for transglutaminase: these include incorporation of radioactive putrescine into the epidermis and determining its location autoradiographically[3] and chemically after tissue extraction[3] and the incorporation of dansylcadaverine into tissue and determining its location with fluorescent microscopy[30] and after tissue fractionation.[31]

Autoradiographic studies[3] of mouse back epidermis *in vivo* showed that after intradermal injection the majority of the isotope was in the lower epidermis at 1 hour and in the stratum corneum at 4 hours. Similar incorporation was found autoradiographically with putrescine incorporation into cow snout epidermis *in vitro*.[3] In cow snout epidermis *in vitro* with dansylcadaverine as a substrate, calcium-dependent, free-sulfhydryl dependent incorporation of fluorescent label was seen over the malpighian and granular layers, and membrane fluorescence was seen in the lower and mid-malpighian layers.[30] Some fluorescence was seen in the lower stratum corneum. Since dansylcadaverine can be noncovalently incorporated into membranes,[32] controls to rule out this possibility are necessary.

The metabolism of putrescine does not involve recycling to amino acids so putrescine incorporation of putrescine is free of that technical difficulty.

Putrescine is synthesized into polyamines which are very basic and may form very strong noncovalent interactions with a variety of acidic substances. Gel filtration alone may not remove noncovalently bound putrescine and/or its metabolites from some proteins.

In vitro incorporation studies with cow snout epidermis were performed incubating cow snout epithelium in Hank's solution with ^{14}C-putrescine for one hour.[3] The tissue was fractionated into prekeratin, keratin, and cell envelope fractions by sequential urea, and urea plus mercaptoethanol extractions followed by a 0.2M NaOH digestion.[3] The specific activity of the proteins in the various fractions were prekeratin, 39 cpm/mg; keratin, 27 cpm/mg; and cell envelope, 135 cpm/mg. Putrescine incorporation was inhibited 84% when the epidermis was preincubated with 50 mM iodoacetamide before the addition of putrescine.

Rat epidermis *in vivo* has been studied with similar results (see Table 5) in that the highest specific activity of labelling was seen in the cell envelope fraction. This fraction had an amino acid composition consistent with a cell envelope fraction with 11.2 residues of proline per 100 residues of amino acids, and contained no hydroxyproline as hydroxylysine. To study the problem further, and to identify possible precursors of the protein in the cell envelope which incorporated putrescine, studies were done with newborn rats.

Newborn rat cells were injected with [^{3}H]-putrescine and the epidermis extracted as indicated in the legend to Table 6. The cell envelope fraction was rapidly labelled with significant radioactivity by 2 hours. The decline

Table 5. *In vivo* [^{3}H-1,4] putrescine incorporation into rat epidermis.

Three days after wax epilation a rat was injected intradermally with 100 uc of [^{3}H-1,4] putrescine (specific activity 182.8 Ci/mole). Twenty-four hours later the animal was sacrificed, the epidermis removed, homogenized in 0.25M sucrose twice and then extracted with 24 hours in 100 ml of the solutions as noted. The sucrose and urea extracts were done in the presence of 200 mg nonradioactive putrescine. After exhaustive dialysis and lyophilization the fractions were weighed and counted in protosol. The protosol blank was 19 cpm. The cpm are after subtraction of background.

Fraction	Extraction Conditions	Amount Protein mg	Specific Activity cpm/mg
1	8M Urea, 0.1M Tris, 1 mM EDTA, pH 8.6	34	34
2	8M Urea, 0.1M Tris, 1 mM EDTA, pH 8.6 with 10 mM Dithiothreitol	20	97
3	Final Pellet after 0.1N NaOH extraction	2.2	525

Table 6. Putrescine incorporation into the cell envelope fraction of newborn rat epidermis.

2.5 uC of [^{3}H] 1,4-putrescine (Specific Activity 182.8 Ci/mole) was injected intraperitoneally into 12-hour-old newborn Charles River CD rats and animals were sacrificed after 2, 6, and 24 hours. The skin was heat separated at 56° for 30 seconds and the epidermis of several animals combined. The wet weight of the 2, 6, and 24 hour samples were 1.3, 1.73 and 1.74 gms, respectively. The epidermis was ground in a glass-glass Duall homogenizer in 0.25M sucrose containing 100 mg of nonradioactive putrescine. The mixture was centrifuged at 15,000 rpm for 15 minutes at 4°C and the pellet extracted with 20 ml of 10% sodium dodecyl sulfate, with 50 mM dithiothreitol, and 0.1 M Tris-Cl (pH 9.0) overnight, in a nitrogen atmosphere at 4°C. The suspension was then centrifuged as above and the pellet extracted with 0.1N NaOH at 4° overnight; recentrifuged, and treated with 0.1N NaOH at 4° for seven days. The mixture was centrifuged as above, the pellet dialyzed, and lyophilized. An aliquot of the lyophilized pellet was counted in Protosol-Aquasol; weight was determined gravimetrically.

Time After Injection	cpm/mg Envelope Protein
2 hr.	831
6 hr.	1030
24 hr.	291

in radioactively by 24 hours probably reflects new envelope synthesis. The amino acid analysis of this fraction showed a high proline content and no hydroxyproline or hydroxylysine. The majority of the epidermal proteins were solubilized with sodium dodecyl sulfate (SDS) and dithiothreitol (methods in legend to Table 6) and then fractionated on Agarose 1.5 m columns. The fractionation of the epidermal proteins 2, 6, and 24 hours after putrescine injection was performed.

At 2 and 6 hours there was radioactivity only in the lower molecular weight protein peaks, and by 24 hours there was more heterogeneity in molecular size of the radioactive peaks (Fig. 2). There was free putrescine in the end volume peak of the column (which is not represented on these figures). In another experiment, performed in a similar fashion, 24 hours after injection of [^{35}S]-cysteine and [^{3}H]-putrescine, the material was fractionated as before on an Agarose 1.5 m column. The radioactivity was counted using settings for double labelled counting and the cpm for [^{3}H] and [^{35}S] calculated. The putrescine labels and cysteine labels do not coincide (Fig. 3). Both cysteine and putrescine were incorporated into the cell envelope fraction: there was 1241 cpm [^{35}S]-cysteine incorporated and 1,000 cpm [^{3}H]-putrescine incorporated per mg protein; the cell envelope fraction contained 7.5 residues of proline per 100 residues of amino acids and no hydroxyproline or hydroxylysine.

Sodium dodecyl sulfate (SDS) electrophoresis was performed on tubes

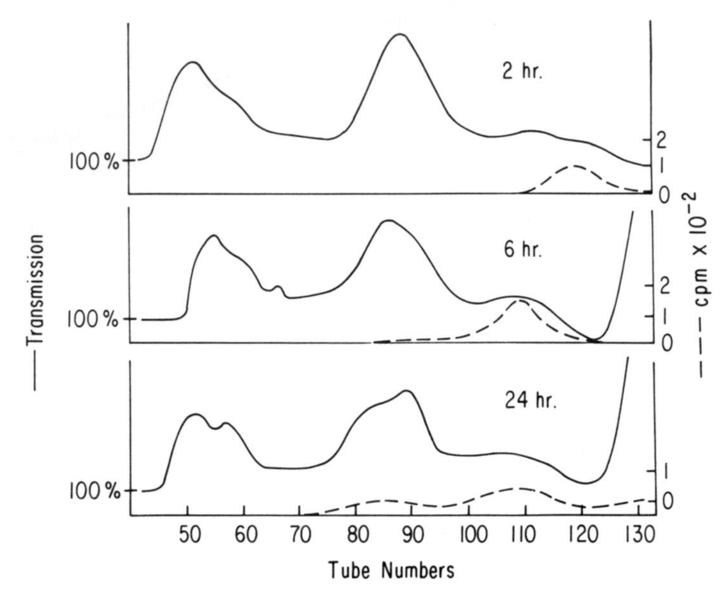

Fig. 2. Agarose 1.5m gel chromatography of rat epidermal protein.

After incorporation of ^{3}H-putrescine for various periods of time the epidermis was separated from the skin and treated as outlined in the text. The SDS extracts were separated on an Agarose 1.5 m column (2.5 $\times$ 85 cm) as indicated. The transmission was continously monitored in a Uvicord II monitor (LKB), and a direct tracing of the strip chart recording is presented. 2.1 ml fractions were collected. 0.5 ml of the fractions were added to 10 ml Aquasol, and the radio-activity measured.

from the peaks of [^{3}H]-putrescine and [^{35}S]-cysteine incorporation. The samples were separated by vertical SDS-acrylamide gel electrophoresis, and then stained with commassie blue. Aliquots of tube 36 (arrow A in Fig. 3) and 63 (arrow B in Fig. 3) were placed on the gel in multiple sample slots, and after staining portions of the gel were cut, and treated with H_2O_2 to solubilize the radioactive counts. The radioactivity was then measured on a Packard 3375 Scintillation spectrometer with double label settings to measure both [^{3}H]-putrescine and [^{35}S]-cysteine. Radioactive cysteine and putrescine were incorporated into different proteins (Fig. 4). Very high molecular weight proteins which did not enter the gel and proteins with a 45–60,000 molecular weight range have incorporated [^{35}S]-cysteine. The putrescine containing proteins from these studies contained low molecular weight proteins with weights of 12–16,000. No cystine incorporation was seen in these proteins in areas of high putrescine incorporation. The protein which incorporated putrescine may represent a small percentage of

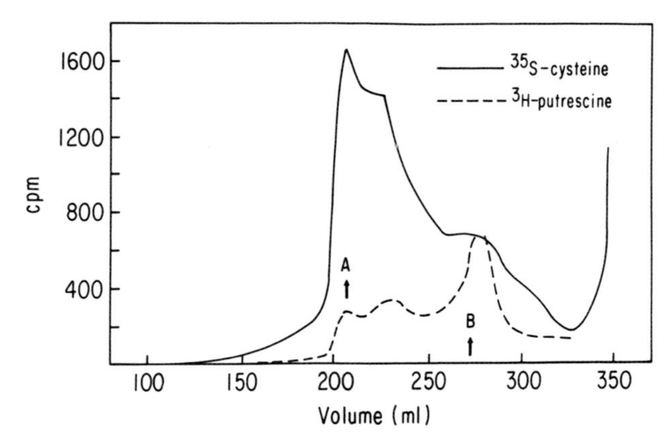

Fig. 3. Agarose 1.5m chromatography of [^{35}S]-cysteine and [^{3}H]-putrescine labelled rat epidermis proteins.

Newborn rats were injected intraperitoneally with 2 uC[^{35}S]-cysteine and 1 uC [^{3}H]-putrescine and sacrificed after 24 hours. The epidermal proteins were extracted as described in the text and applied to a column in 10 mM Tris, pH 9.5, 0.15 M NaCl, 1 mM EDTA. 0.5 ml of the fractions were counted for [^{35}S] and [^{3}H]-putrescine by a double isotope technique. The [^{3}H] counts presented are corrected for any [^{35}S] which was recorded in the [^{3}H] channel; under the counting conditions used, no [^{3}H] entered the [^{35}S] channel. Tube 36 indicated at A, and Tube 63 indicated at B, were applied to slab gels for further analysis as discussed in the text and the legend for Fig. 4.

the protein in this fraction since one of the heaviest staining protein bands was not labelled with putrescine. Enough labelled material for definitive identification of the putrescine labelled protein in this fraction of rat epidermis has not been isolated, although studies are in progress. Such studies with the *in vivo* corporation into living epidermis have the potential of identifying the actual, low molecular weight precursor(s) of cross-linked proteins.

Preliminary work with human epidermis has demonstrated that there are proteins which can incorporate putrescine. These studies were done *in vitro* using proteins extracted from the stratum corneum and, therefore, may not be absolutely comparable to the *in vivo* studies previously described.

To label the intrinsic acceptor proteins in human epidermis extractions of the scales from a patient with lamellar ichthyosis was performed. The scales were extracted with 0.25 M sucrose for 2 hours. The mixture was then centrifuged at 15,000 × *g* for 30 minutes at 4°C. The resulting pellet was extracted with 0.1M Tris-Cl (pH 7.5) for 18 hours and the mixture centrifuged at 15,000 × *g* for 30 minutes at 4°C. The supernatant was concentrated on an Amicon PM-10 filter. This supernatant (2.0 ml) was incubated with 2.0 ml glycine-NaOH (pH 9.5) and 300 ul of [^{14}C]-putrescine

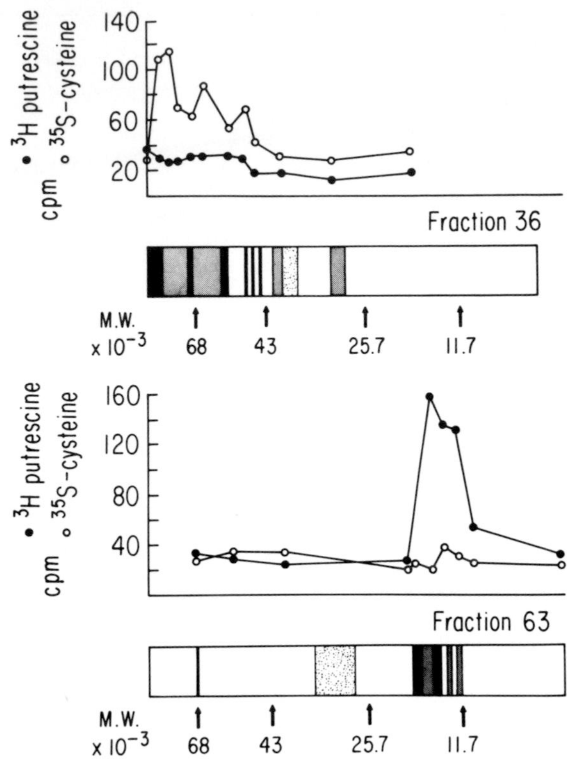

Fig. 4. Epidermal proteins incorporating [^{35}S]-cysteine and [^{3}H]-putrescine. Tube 36 (top graph and gel) and tube 63 (bottom graph and gel) from an SDS 1.5 m Agarose column (in Fig. 3) were electrophoresed in 10% acrylamide gels in Tris-HCl (upper buffer was Tris-Boric acid, spacer gel buffer was Tris-H_2SO_4) as described by Neville.[33] After staining, areas of the gels as indicated were cut, pooled, dried and treated with 50 ul of 30% H_2O_2 at 37° for 18 hours. These materials were then placed in 20 ul Protosol and 10 mls Aquasol, and the radioactivity measured. Presented are diagrams of the SDS gels with the position of the molecular weight markers for serum albumin (68,000), ovalbumin (43,000), chymotrypsinogen (23,700), and cytochrome C (11,700) indicated.

(9.3 uc) at 37°C for 18 hours. 3.4% of the total radioactivity was incorporated into protein which was insoluble in 5% cold trichloroacetic acid. To determine which proteins incorporated putrescine, to 4 mls of the incubation mixture, dithiothreitol to make a final concentration of 50 mM, and sodium dodecylsulfate to make a final concentration of 2% were added. The sample was boiled for seven minutes and applied to the 1.5 m Agarose column which was developed as the previous Agarose columns (Fig. 5). The peaks of radioactivity were dialyzed exhaustively and lyophilized.

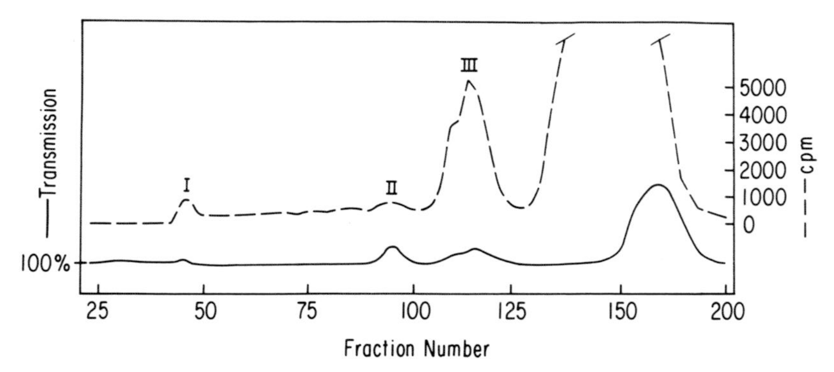

Fig. 5. Gel filtration chromatography of ^{14}C-putrescine incorporation into epidermal proteins.

Human transglutaminase and an acceptor protein(s) were prepared as indicated in the text, treated with sodium dodecylsulfate and putrescine and chromatographed on Agarose 1.5 m under the same conditions as previously used (Fig. 3) and 0.5 ml of each fraction was counted for radioactivity; the transmission was constantly monitored. The Roman numerals refer to peaks pooled which were dialyzed and lyophilized.

Autoradiographs of the SDS gels of this fraction showed heavy labelling of only the proteins in peak I (Fig. 6). Predominantly very high molecular weight material was labelled which did not enter the spacer gel (4% acrylamide); much smaller amounts of protein entered the spacer gel, but did not enter the running gel (10% acrylamide). Studies of the lyophilized fractions in a similar manner confirmed these findings that only very high molecular weight protein was labelled.

Under these incubation conditions the major protein peak (by commassie blue staining) in this fraction did not incorporate putrescine. Similar *in vitro* studies with the germinative layers may help further define the acceptor protein(s) of human epidermis.

Our current studies with putrescine as an donor suggest that the cell envelope fraction is rapidly labelled, a very high molecular protein neutral buffer protein in human stratum corneum is labelled, and a low molecular weight protein(s) in living epidermis (newborn rat) is labelled. The relationship of these proteins to one another, and the possibility that these proteins are cross-linked by ε-(γ-glutamyl)lysine bonds *in vivo* still remains to be definitively determined.

Acknowledgments

The technical assistance of Ms. Marie Hanigan and the secretarial assistance of Ms. Susan Boos are gratefully acknowledged.

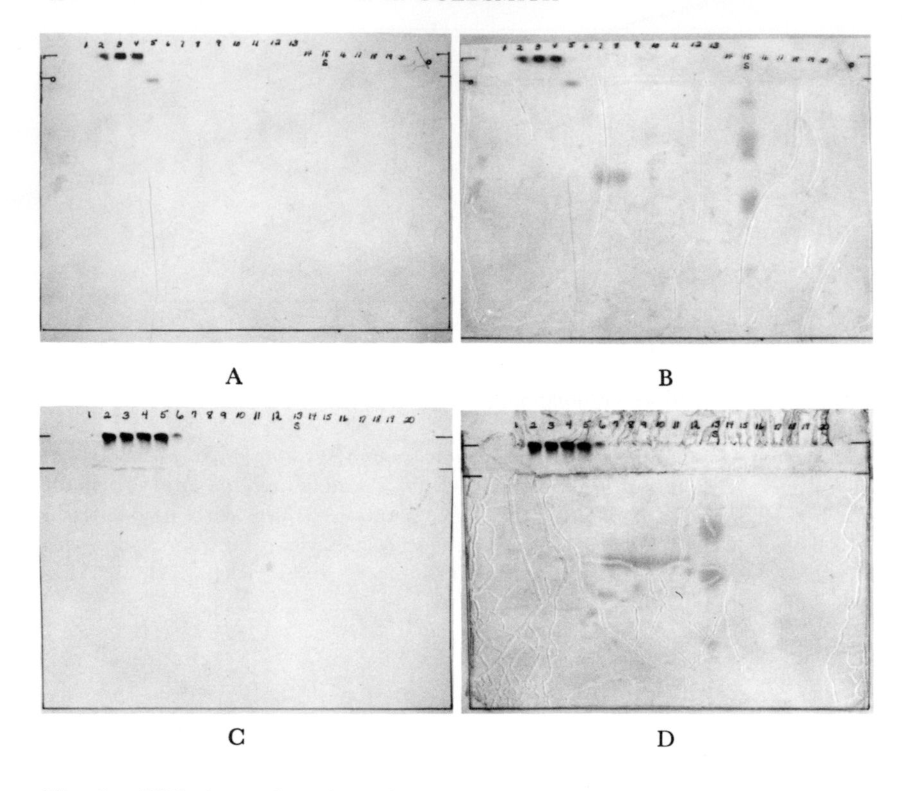

Fig. 6. SDS electrophoresis and autoradiography of [^{14}C]-putrescine labelled epidermal proteins.

Fractions from an SDS-Agarose column (Fig. 5) were studied directly on SDS-gels (6A,B) or studied after pooling, dialysis, and concentration (6C,D). SDS gel electrophoresis was performed according to the techniques of Neville[33] and autoradiography according to the techniques of Bonner and Laskey.[34] Autoradiography results were exposed for 12 days at −70 °C before developing. The autoradiographs themselves are in A and C. In B and D the autoradiograph is placed over a dried, stained, acrylamide gel. Cracks are an artifact of drying the acrylamide gels. The edges of the gel, and position of the spacer gel are as indicated. In A and B, 40 ul fractions of the Agarose column (Fig. 5) were studied. Slots 1, 14, 16, 19, and 20 were empty. Slot 15 had bovine serum albumin, ovalbumin, chymotrypsinogen and cytochrome C standards. Slot 2 is an aliquot of the 44; slot 3 tube 45; slot 4 tube 46; slot 5 tube 70; slot 6 tube 80; slot 7 tube 98; slot 8 tube 98; slot 9 tube 110; slot 10 tube 112; slots 11–13 tube 114; slot 17 tube 116; slot 18 tube 124.

In C and D fractions of the lyophilized peaks (I, II, and III in Fig. 5) were applied to the gels. Slots 1, 12, and 14 were blank. Slot 13 contained ovalbumin, chymotrypsinogen, and cytochrome C. Slot 2 aliquots of peak I (slot 6 with one fifth the material of tubes 2–5); slots 7–11 Peak II; and slots 15–19 Peak III. The material in slots 2–5 had 700 cpm by direct measurement. The results are described in the text.

REFERENCES

1. Harding, J. J. The unusual links and cross-links of collagen. *Adv. in Protein Chem.*, **20**: 109–190, 1965.
2. Folk, J. E. and Chung, S. I. Molecular and catalytic properties of transglutaminases. *Adv. Enzymol.*, **38**: 109–91, 1973.
3. Goldsmith, L. A., Baden, H. P., Roth, S. I., Colman, R., Lee L., and Fleming, B. Vertebral epidermal transamidases. *Biochim. Biophy. Acta*, **351**: 113–125, 1974.
4. Birckbichler, P. J., Dowben, R. M., Matacic, S., and Loewy, A. G. Isopeptide bonds in membrane proteins from eukaryotic cells. *Biochim. Biophys. Acta*, **291**: 149–155, 1973.
5. Pisano, J. J., Finlayson, J. S., and Peyton, M. P. Chemical and enzymatic detection of protein cross-links. Measurement of ε-(γ-glutamyl)lysine in fibrin polymerized by factor XIII. *Biochemistry*, **8**: 871–876, 1969.
6. Raczyński, G., Snochowski, M., and Buraczewski, S. Metabolism of ε-(γ-L-glutamyl)-L-lysine in the rat. *Br. J. Nut.*, **34**: 291–296, 1975.
7. Williams-Ashman, H. G., Notides, A. C., Pabalan, S. S., and Lorand, L. Transamidase reactions involved in the enzymatic coagulation of semen: Isolation of γ-glutamyl-ε-lysine dipeptide from clotted secretion protein of guinea pig seminal vesicle. *Proc. Nat. Acad. Sci. USA*, **69**: 2322–2325, 1972.
8. Asquith, R. S., Otterburn, M. S., Buchanan, J. H., Cole, M., Fletcher, J. C., and Gardner, K. L. The identification of εN-(γ-L-glutamyl)-L-lysine cross-links in native wool keratins. *Biochim. Biophys. Acta*, **207**: 342–348, 1970.
9. Harding, H. W. J. and Rogers, G. E. ε-(γ-glutamyl)lysine cross-linkage in citrulline containing fractions from hair. *Biochemistry*. **10**: 624–630, 1971.
10. Harding, H. W. J. and Rogers, G. E. The occurrence of the ε-(γ-glutamyl)lysine cross-link in the medulla of hair and quill. *Biochim. Biophys. Acta*, **257**: 37–39, 1972.
11. Pisano, J. J., Finalyson, J. S., Peyton, M. P., and Nagai, Y. ε-(γ-glutamyl) lysine in fibrin: Lack of cross-link formation in factor XIII deficiecny. *Proc. Nat. Acad. Sci. USA*. **68**: 770–772, 1971.
12. Schwartz, M. L., Pizzo, S. V., Hill, R. L., and McKee, P. A. Human factor XIII from plasma and platelets. Molecular weights, subunit structures, proteolytic activation and cross-linking of fibrinogen and fibrin. *J. Biol. Chem.*, **248**: 1395–1407, 1973.
13. Chen, R. and Doolittle, R. F. γ-γ cross-linking sites in human and bovine fibrin. *Biochemistry*, **10**: 4486–4491, 1971.
14. Doolittle, R. F. Structural aspects of the fibrinogen to fibrin conversion. *Adv. in Protein Chem.*, **27**: 1–109, 1973.
15. Ferguson, E. W., Fretto, L. J., and McKee, P. A re-examination of the cleavage of fibrinogen and fibrin by plasmin. *J. Biol. Chem.*, **250**: 7210–7218, 1975.
16. Harding, H. W. J. and Rogers, G. E. Isolation of peptides containing citrulline and the cross-link, ε-(γ-glutamyl)lysine, from hair medulla protein. *Biochim. Biophys. Acta*, **427**: 315–324, 1976.
17. Bensusan, H. B. An investigation of the products of an enzymatic hydrolysis of collagens. *Biochemistry*, **8**: 4716–4723, 1969.

18. Bensusan, H. B. The glutamyl linkages in collagen. *Biochemistry*, 8: 4723–4727, 1969.
19. Abernathy, J. and Goldsmith, L. ε-(γ-glutamyl)lysine cross-linking of human epidermal fibrous proteins (abstr). *Clin. Res.*, **23**: 226A, 1975.
20. Gross, M. and Folk, J. E. Mapping of the active sites of transglutaminases 1. Activity of the guinea pig liver enzyme toward aliphatic amides. *J. Biol. Chem.*, **248**: 1301–1306, 1973.
21. Gross, M., Whetzel, N. K., and Folk, J. E. The extended active site of guinea pig liver transglutaminase. *J. Biol. Chem.*, **250**: 4648–4655, 1975.
22. Folk, J. E., and Cole, P. W. Structural requirements of specific substrates for guinea pig liver transglutaminase. *J. Biol. Chem.*, 240: 2951–2960, 1965.
23. Clarke, D. D., Mycek, M. J., Neidle, A., and Waelsch, H. The incorporation of amines into protein. *Arch. Biochem. Biophy.*, **79**: 338–354, 1959.
24. Lorand, L., Rule, N. G., Ong, H. H., Furlanetto, R., Jacobsen, A., Downey, J., Öner N., and Bruner-Lorand, J. Amine specificity in transpeptidation. Inhibition of fibrin cross-linking. *Biochemistry*, **7**: 1214–1223, 1968.
25. Lorand, L., Shishido, R., Parameswaran, K. N., and Steck, T. L. Modification of human erythrocyte ghosts with transglutaminase. *Biochem. Biophys. Res. Commun.*, **67**: 1158–1166, 1975.
26. Dutton, A. and Singer, S. J. Cross-linking and labeling of membrane proteins by transglutaminase-catalyzed reactions. *Proc. Nat. Acad, Sci. USA*, **72**: 2568–2571, 1975.
27. Mosher, D. F. Action of fibrin-stabilizing factor on cold-insoluble globulin and α_2-macroglobin in clotting plasma. *J. Biol. Chem.*, **251**: 1639–1645, 1976.
28. Mosher, D. F. Cross-linking of cold-insoluble globulin by fibrin-stabilizing factor. *J. Biol. Chem.*, **250**: 6614–6621, 1975.
29. Ruoslahti, E. and Vaheri, A. Novel human serum protein from fibroblast plasma membrane. *Nature*, **248**: 789–791, 1974.
30. Buxman, M. M. and Wuepper, K. D. Keratin cross-linking and epidermal transglutaminase. *J. Invest. Derm.*, **65**: 107–112, 1975.
31. Buxman, M. M., Buehner, G. E., and Wuepper, K. D. Isolation of substrates for epidermal transglutaminase from bovine epidermis (abstr). *Clin. Res.*, **24**: 262A, 1976.
32. Pincus, J. H., Chung, S. I., Chace, N. M., and Gross, M. Dansylcadaverine: A fluorescent probe and marker in cell membrane studies. *Arch. Biochem. Biophys.*, **169**: 724–730, 1975.
33. Neville, D. M. Molecular weight determination of protein-dodecyl sulfate complexes by gel electrophoresis in a discontinuous buffer system. *J. Biol. Chem.*, **246**: 6328–6334, 1971.
34. Bonner, W. M. and Laskey, R. A. A film detection method for tritiumlabelled protein and nucleic acids in polyacrylamide gels. *Eur. J. Biochem.*, **46**: 83–88, 1974.

Discussion

Dr. Baden: Is the last soluble protein you discussed, neutral-soluble or does it require SDS and mercaptoethanol?

DR. GOLDSMITH: It was a 0.05M Tris, pH 7.5 extract separated on an Agarose-SDS column. The protein was from the stratum corneum. There is a chance that the protein may incorporate one of these precursors. To identify it as the true natural product I think that one will eventually have to label it and follow its progress through epidermal maturation. The way we work with this protein assay in the laboratory, we essentially cross-link casein with this enzyme system, although this enzyme system has no natural role in casein metabolism.

DR. BADEN: Is the electrophoresis we're talking about SDS-electrophoresis?

DR. GOLDSMITH: Yes. That is SDS-electrophoresis with mercaptoethanol present. The material which has incorporated the label in this experiment is extremely high molecular weight material which is initially extractable in neutral buffer.

DR. BERNSTEIN: Since I didn't have an opportunity to comment after Dr. Sugawara's presentation, I would like to compliment both him and you on producing interesting data in a very difficult area. In trying to correlate the two, I wonder if you would respond to my suggestion that maybe in using the putrescine test you are utilizing an enzyme that happens to be there, linking that with an existent γ-glutamyl group on the protein and that his study might well deal with the natural substrate. One thing I would like to know is the specificity of the γ-glutamyl-linking enzyme. Will it work only with the free amino acid reagent or do you know whether it will work with the whole protein?

DR. GOLDSMITH: I think it's tricky. In terms of the diamine substrates, putrescine and dansylcadaverine are really two of the very best substrates. Lysine by itself will be incorporated. It is incorporated at a much lower rate. It is incorporated at a rate essentially 20 times less than one of these substrates, and I think, that one of the very, very critical points is whether the lysine in Dr. Sugawara's paper was actually incorporated in a peptide chain by synthesis or whether the lysine was incorporated in the same way that putrescine was from the point of view of interpreting how many of the lysine molecules were actually cross-linked. I would just like to expand on that. Under the proper conditions, with this enzyme, you could attach putrescine or you could attach lysine, or you could attach other primary amines to the glutamines in the protein. From interpreting the stoichiometry of something like essentially one out of every two molecules of lysine incorporating this cross-link, it would be very important to know whether that radioactive lysine was actually in a peptide chain, and then was cross-linked or whether the lysine was acting as another primary amine.

DR. OGURA: I'd like to ask a question about incorrect lysyl-glutamyl linkages in the protein. I have studied SH- and SS-binding in proteins. Sometimes incorrect SS-bonds are formed in proteins. In that situation the primary amino acid sequence is right but the protein that is formed is quite different. It may be a non-functional protein. Therefore, are there any incorrect linkages formed between lysine and glutamyl residues in your protein?

DR. GOLDSMITH: That is an extremely important question. Many years ago this cross-link was first described in collagen but it was an artifact because collagen was treated with organic solvents in order to demonstrate this cross-link. It is possible to get various rearrangements from a regular kind of an α-peptide bond going through a cyclic amide to get this bond. This will happen under acid conditions. I think one has to be extremely careful with the enzyme reactions especially when very crude enzyme preparations are used, to assure that appropriate controls are done to ascertain that one is not forming this cross-link as a function of one's methodology. Now, I'd like to ask Dr. Sugawara a question. In your soluble protein where you incorporated the lysine, was this lysine peptide-bound in a chain or was it added on to a glutamine?

DR. SUGAWARA: Transglutaminase activity is, of course, very active at the higher pH, for example, at pH 10. Perhaps you tried it. But at physiological pH, it is not very active for lysine because the pK value for the ε-amino group is 10.2. At physiological pH where the amino group is in the ionized form, the ε-group does not work in the cross-linking reaction. It has to be in the non-ionized form. After the incubation, I have isolated the ^{14}C-protein containing the lysine and not the lysine *per se*.

Epidermal and Other Tissue Transglutaminases*

Hideoki OGAWA** and Lowell A. GOLDSMITH†

***Department of Dermatology, Juntendo University, Tokyo, Japan* and †*Division of Dermatology, Department of Medicine, Duke Medical Center, Durham, North Carolina, U.S.A.*

ABSTRACT

Transglutaminases which are present in many tissues form ε-(γ-glutamyl)lysine bonds in many structural proteins. The characteristics of liver, plasma, and other transglutaminases are reviewed. Recent studies on the isolation and physicochemical characterization of human epidermal transglutaminases are presented in detail. The enzyme is calcium dependent, has a molecular weight of 50–51,000, and cross-links fibrin. The activity of the isolated enzyme is increased 10–15 fold by heating to 56°C or pretreatment with Me_2SO. The activation is temperature dependent but not very dependent on pH. Human hair transglutaminase is chemically and immunologically different from human epidermal transglutaminase.

INTRODUCTION

Transglutaminases can form ε-(γ-glutamyl)lysine bonds in proteins. These bonds cross-link and stabilize protein structures. The enzymes capable of performing transglutamination reactions are of intrinsic interest since their natural substrates are two large proteins which must be approximated and then modified. The characteristics of these enzymes in several tissues, and several recent studies of these enzymes in keratinizing tissues including human epidermis, are the subjects of this article.

* Supported in part by grant #AM-17253 from the United States Public Health Service. Lowell A. Goldsmith is the recipient of Research Career and Development Award #AM-00008 from the United States Public Health Service.
Publication number 9 of the Dermatological Research Laboratories of Duke Medical Center.

TRANSGLUTAMINASES

1. General characteristics and assay techniques

All transglutaminases are calcium-dependent enzymes which catalyze the deamination of a protein-bound glutamine of one substrate and the incorporation of that protein into an acyl-enzyme complex.[1] The epsilon-amino group of a protein-bound lysine is then incorporated into a peptide bond with the glutamyl residue of the first protein substrate. Figure 1 is a diagrammatic representation of the reaction. If a suitable amine donor is not available hydrolysis alone may occur, with the net effect being the

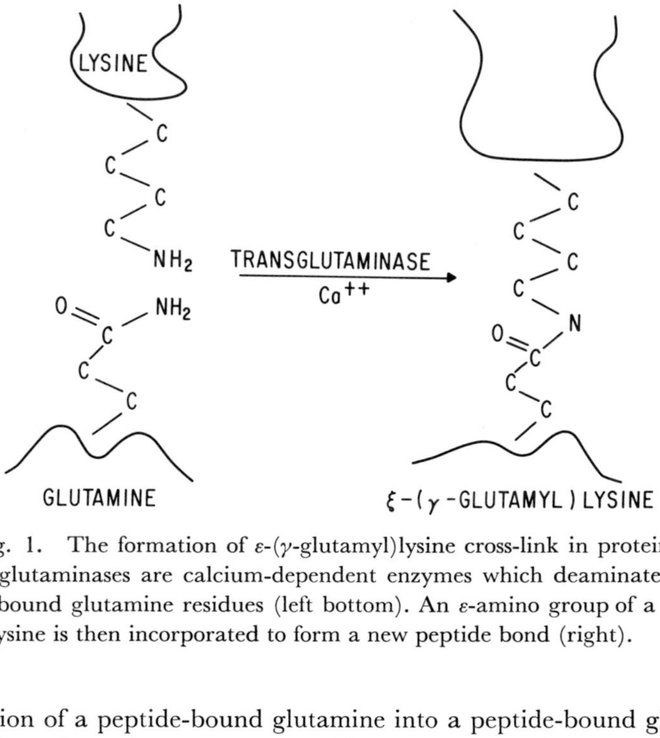

Fig. 1. The formation of ε-(γ-glutamyl)lysine cross-link in proteins. Transglutaminases are calcium-dependent enzymes which deaminate certain protein-bound glutamine residues (left bottom). An ε-amino group of a protein-bound lysine is then incorporated to form a new peptide bond (right).

conversion of a peptide-bound glutamine into a peptide-bound glutamic acid. Transglutaminases are assayed by measuring either the ammonia released,[2] the incorporation of amines into a protein substrate,[2] or the covalent polymerization of a protein substrate.[3] The amines may be measured directly by chemical means. Frequently radioactive amines or fluorescent amines (dansylcadaverine) are used and are detected by measuring radioactivity or fluorescence incorporated into a protein substrate. A common substrate for measuring transglutaminases is radioactive putrescine as shown in Fig. 2.

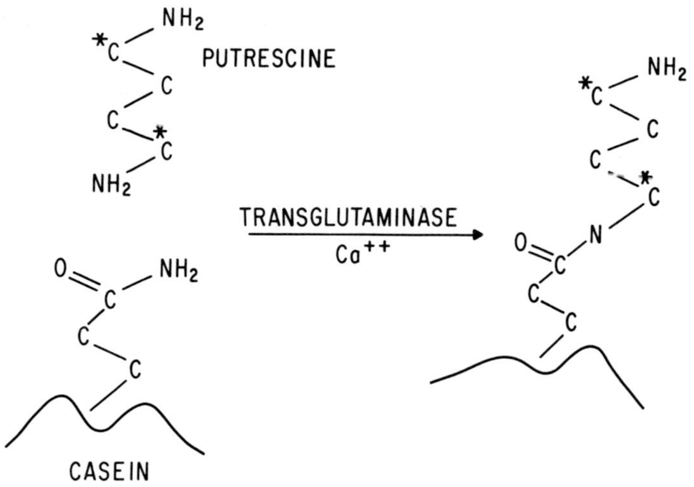

Fig. 2. ε-amino analogue used to assay transglutaminases.
One kind of assay for transglutaminases utilizes their ability to covalently incorporate primary amines into proteins. In this example the radioactive (*) amine putrescine is incorporated covalently into a protein-bound glutamyl residue formed by deamination of a protein-bound glutamine by a transglutaminase.

2. Liver transglutaminase

Liver transglutaminase, although it is the most studied transglutaminase, is still not assigned a definite biological role. It is a single polypeptide chain, with a molecular weight of 80,000–90,000 depending on the method of analysis. It is enzymatically active without pretreatment with proteolytic enzymes. It contains no disulfide bonds, but has 17 ± 1 sulfhydryl groups per molecule; one sulfhydryl in the active site of the enzyme is absolutely necessary for enzyme activity.[1] A variety of amines, esters, and proteins are substrates for this enzyme.[1] It, like all the transglutamines, has no activity toward protein-bound asparagines due to stringent structural characteristics of the glutamine binding site.[4] A chemically and immunologically identical enzyme is present in the placenta, liver, prostate and uterus, lung, heart, spleen, and kidney.[5]

3. Plasma transglutaminase (Plasma factor XIII)

Plasma factor XIII cross-links the γ chains of fibrin to form γ-dimers and cross-links the α chains of fibrin to form α polymers. A genetic deficiency of this enzyme, inherited as an autosomal recessive trait, is associated with umbilical cord bleeding, intracranial bleeding, prolonged bleeding from wounds, and slow or poor wound healing.[6] The enzyme is present as an inactive zymogen which is activated by thrombin. The enzyme exists as a tetramer of 2 catalytic subunits (A) and 2 noncatalytic (regulatory) subunits (B); there is calcium-dependent dissociation of the regulatory and catalytic subunits which is required for catalytically active enzyme.[7] Platelet factor XIII is identical to the catalytic dimer (a_2) of plasma factor XIII and does not contain regulatory subunits.[3] Placental transglutaminase is probably identical with plasma factor XIII.[5]

4. Hair follicle transglutaminases

Hair follicle transglutaminase has been previously studied in sheep,[8] guinea pigs,[8,9] rats,[8] and cows.[10] In the guinea pig[9] the molecular weight of the enzyme by gel filtration is reported as 54,000 $\pm$ 2,000 and SDS electrophoresis of an impure preparation suggested a subunit molecular weight of 27,000. Hair follicle transglutaminase cross-links the citrulline rich medullary proteins and possibly the internal root sheath proteins of hair. Hair transglutaminase is immunologically[10] distinct from other transglutaminases.

5. Epidermal transglutaminases

Both cow and human epidermal transglutaminase have been described. A component of the epidermis, possibly the cell envelope, is cross-linked by ε-(γ-glutamyl)lysine bonds.[11] The cow enzyme has a molecular weight of 50,000 to 55,000 by gel filtration[12,13] and sodium dodecyl sulfate (SDS) electrophoresis.[12] An earlier molecular weight of 100,000 by SDS electrophoresis has been revised after electrophoresis was performed under more stringent dissociating conditions (M. Buxman, personal communication, May 1976). Cow and human epidermal transglutaminases can cross-link fibrin forming γ-dimers and α-polymers.[12,14] Antibody to the cow epidermal enzyme did not react with cow hair follicle transglutaminase, suggesting differences between the two enzymes.[10]

Human epidermal transglutaminase has a reported molecular weight of 52,000 $\pm$ 2,500 by gel filtration and it cross-linked fibrin.[14] As detailed below, it has been purified to homogeneity and several chemical and immunological studies of its nature have begun.

RECENT STUDIES OF HUMAN EPIDERMAL TRANSGLUTAMINASE

1. Enzyme isolation

Human epidermal transglutaminase was prepared from human hyperkeratotic soles and assayed as previously described.[14] A summary of the purification is in Table 1.

Purification using affinity chromatography techniques using either

Table 1. Purification of epidermal transglutaminase.

Step	Total Enzyme Unit*	Yield	Specific Activity	Purification
1. Initial Homogenate	5350	100	0.60	1.00
2. Amicon (PM-30) Concentrate	2833	53	3.3	5.6
3. DEAE-cellulose Fraction	3188	60	56.1	93.5
4. Sephadex G-75 Fraction	1360	25	129.5	215.8
5. CMC-cellulose Fraction	778	15	185.2	308.7
6. Agarose 0.5 m Fraction	620	12	344.3	573.9

*Units: nmoles [^{14}C] putrescine incorporated into casein at pH 9.5 per 30 minutes at 37°C.

commercial Agarose-casein (Miles) or Agarose-casein prepared according to the techniques of Cuatrecasas and Anfinsen[15] were attempted. This system was tested under a variety of conditions and failed to show an affinity chromatography-like purification. The casein acted only as a polyionic substrate and had no special advantage over DEAE cellulose. Affinity techniques using neodynium as a calcium antagonist[16] and Agarose-casein were also not successful.

The greatest increase in specific activity occurred after DEAE cellulose ion exchange chromatography (Fig. 3). The majority of enzyme activity was eluted with a low concentration of NaCl, and a very small percentage of the total enzyme activity with higher concentrations of sodium chloride. Some enzyme did not stick to the column. Reequilibration of the active enzyme which was not absorbed to the column and rechromatography showed the enzyme activity still did not absorb to the column. It was immunologically identical to the bulk of the enzyme. Tubes were pooled from the main peak of enzyme activity and were purified further on Sephadex G-75 (Fig. 4). The material from the peak of activity was equilibrated for carboxymethyl cellulose column chromatography (Fig. 5). All of the enzyme absorbed to the column under the conditions used and was eluted at low NaCl concentrations. The eluted material was concentrated on an Amicon filter and was chromatographed on Agarose 0.5 m and the protein and enzyme activity peaks essentially coincided (Fig. 6). The peak tubes were used for immunization and further studies. The insert on Fig. 6 shows the result of a sodium-dodecyl sulfate electrophoresis analysis of the purified enzyme showing that it migrated as a single band.

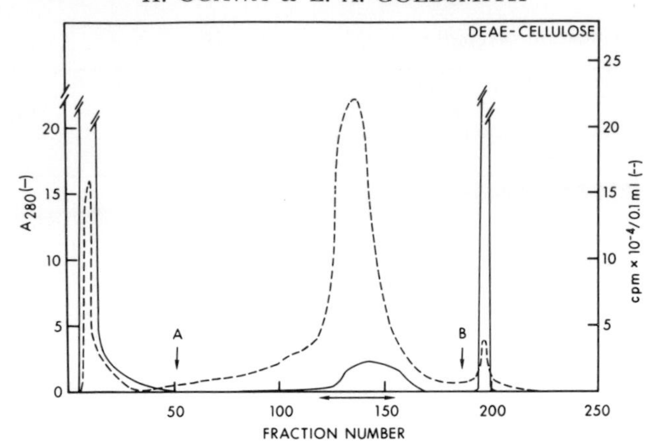

Fig. 3. Elution pattern of DEAE cellulose column.

The supernatant of the concentrated homogenate was applied to a column (1.2 × 15 cm) of DEAE cellulose that had been equilibrated with 5 mM Tris-acetate (pH 7.5) and 1 mM EDTA. The column was washed with 120 ml of equilibrating buffer and was eluted with a 200 ml linear gradient of NaCl (0–50 mM) in same buffer. After elution a second 200 ml linear gradient of NaCl (50 mM-1M) was applied. The enzyme activity appeared as a single peak which eluted between 20 and 30 mM NaCl.

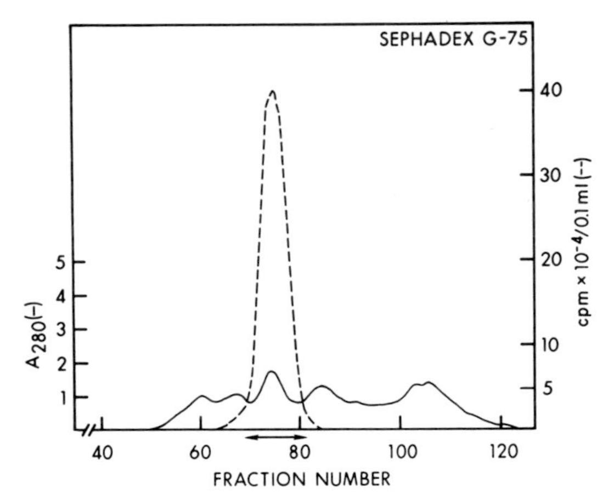

Fig. 4. Elution pattern of Sephadex G-75 column.

After elution for DEAE cellulose the enzyme was concentrated on an Amicon PM-30 filter and the enzyme solution was dialyzed overnight against 3 liters of 10 mM Tris-acetate (pH 7.5), 1 mM EDTA and 0.15 M NaCl, and was applied to a Sephadex G-75 column (2.5 × 96 cm) equilibrated with this buffer. The active fractions were combined, concentrated to 7.5 ml, adjusted to pH 6.0 with 0.1M acetic acid, and dialyzed for 20 minutes against 6 liters of 5 mM Tris-acetate (pH 6.0), with 1 mM EDTA.

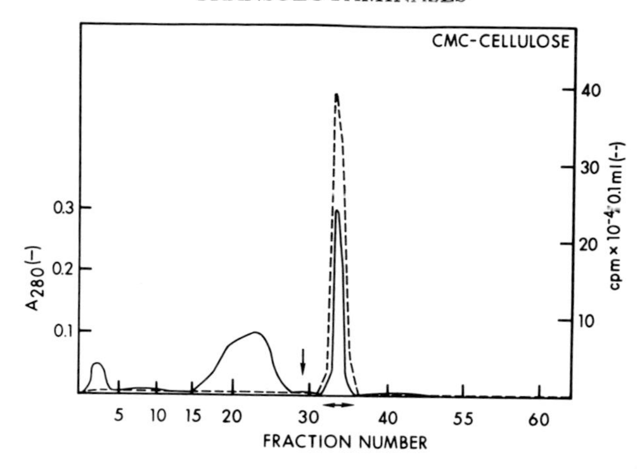

Fig. 5. Elution pattern of CMC column.

The material after Sephadex elution and concentration (legend of Fig. 4) was applied to a 1.2 × 13 cm CMC cellulose column which was equilibrated with 5 mM Tris-acetate (pH 6.0) with 1 mM EDTA. The column was washed with 100 ml of equilibrating buffer and then eluted with a 200 ml linear gradient of NaCl (0 to 0.5M) in the same buffer. Enzyme activity eluted as a single peak between 0.1 and 0.2 M NaCl. The peak tubes were concentrated to 3.5 ml dialyzed against 3 liters of 10 mM Tris-acetate pH 7.5, 1 mM EDTA and 0.15 M NaCl for 12 hours.

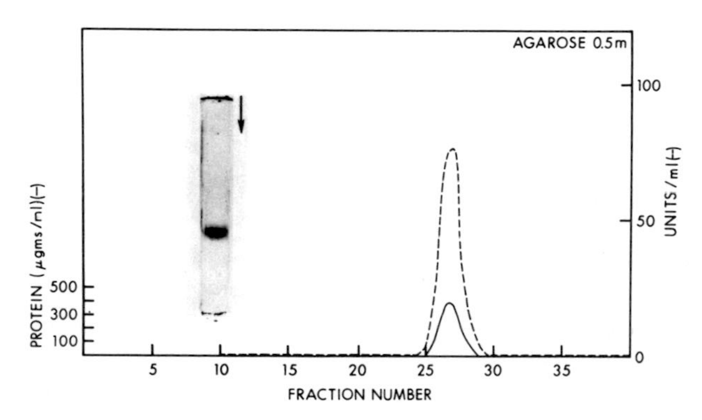

Fig. 6. Elution pattern of agarose 0.5 M column and SDS electrophoresis of purified enzyme.

The enzyme was prepared as in the legend of Fig. 5, and was chromatographed on a column (1.7 × 72 cm) of Bio-Gel A-0.5m in 10 mM Tris-acetate (pH 7.5) and 0.15 M NaCl. Fractions were assayed for protein and enzyme activity. The peak tubes had constant specific activity. The insert shows an SDS-electrophoresis gel of the purified enzyme. The arrow indicates the direction of migration.

2. Characterization of epidermal transglutaminase

Human epidermal transglutaminase required either calcium or strontium for activity; Mn, Mg, Hg, or Ba could not substitute for these.[17] The enzyme was inhibited by iodoacetamide (Table 2). The apparent molecular weight of the enzyme was 50,000 ± 20,000 by gel filtration chromatography in Agarose 0.5 m with various molecular weight markers, 51,000 ± 2,000 by sodium dodecyl sulfate electrophoresis, and 100,000 ± 5,000 by discontinuous electrophoresis.[17] Epidermal transglutaminase cross-linked fibrinogen forming dimers and multimers (Fig. 7). Thrombin activation of fibrinogen was not necessary for cross-linking.

Table 2. Effect of calcium, Me_2So and iodoacetamide on human epidermal transglutaminase.

Human epidermal transglutaminase with 10 mM $CaCl_2$ in 50 mM Tris-Cl (pH 7.5) was incubated with 33.3 mM iodoacetamide for various periods of time; 40 mM dithiothreitol was then added for 15 minutes at 37°; and then Me_2SO to a final concentration of 8.0 M was added and the mixture incubated at 37° for 10 minutes. The mixture was then assayed in the standard way. Controls with no iodoacetamide, no Me_2SO, and Me_2SO without iodoacetamide were performed as indicated.

Preincubation Time (mins) with Reagent	Enzyme Activity (cpm × 10^{-3}) Reagent: No additions	Iodoacetamide & Me_2SO	Me_2SO
5	39.1	12.0	101.0
10	—	4.5	104.1
21	—	3.1	113.3
33	—	1.4	108.4
60	64.3	4.7	114.8

3. Enzyme activation

During experiments to determine if thrombin could activate or increase epidermal transglutaminase activity it was noted that incubation with calcium or Me_2SO (dimethylsulfoxide) could increase the enzyme's activity.[17] Preincubation with calcium at 37° for 60 minutes at low calcium concentrations could double enzyme activity (Table 2). Iodoacetamide blocked the activation by Me_2SO, suggesting the active sites of the activated and nonactivated enzyme were accessible to iodoacetamide under the experimental conditions studied (Table 2). Both heat-calcium and Me_2SO activation were relatively independent of the pH of the pretreatment solution (Table 3). The pH-activation of the enzyme is a mixture of several functions: activation, and the possible degradation of native and activated enzyme at each of several pHs.

Me_2SO activation as a function of preincubation temperature was

Table 3. Effect of pH on the heat-calcium and dimethylsulfoxide activation of epidermal transglutaminase.

10 ul of epidermal transglutaminase in 50 mM Tris-Cl (pH 7.5) was mixed with 100 ul of 0.1 M glycine-NaOH at the indicated pHs and incubated at 37° for 10 minutes. At that time Me_2SO to a final concentration of 3.1 M was added and all tubes incubated for 10 minutes at 37°. 1.0 ml of standard assay mixture was added and the pH adjusted to pH 9.3 when necessary. The assay was then completed in the usual manner. The calcium-heated activation was studied in a similar manner. To the enzyme (10 ul) in 10 mM $CaCl_2$ with 25 mM Tris-Cl (pH 7.5), 100 ul of buffer was added, and the mixture heated to 56°C for 15 minutes. The pH was then adjusted and the mixture assayed. Control tubes were tubes containing the enzyme which were neither heated or treated with Me_2SO.

	Enzyme Activity (% Control)	
pH	Me_2SO Activation	Calcium-Heat Activation
5	484	120
6	412	124
7	473	113
8	520	183
9	416	161
10	448	89
11	420	94

studied (Table 4). The final column in the table shows that adding Me_2SO to the complete assay mixture does not significantly activate the enzyme or change the enzyme's rate enough to explain the activation phenomenon. At 37° degradation of the activated enzyme may be a factor in decreasing the percent of activation compared to lower temperatures. Further studies of activation rates as a function of temperature may lead to an analysis of the thermodynamics of the activation process.

To determine the nature of the activation process a number of studies were performed on the native and activated enzymes. Table 5 summarizes the results of these studies. Kinetic analysis showed there was a change in the Vmax of the enzyme without a Km change. The activation occurred in the presence of serine protease inhibitors; no demonstrable molecular weight change was demonstrated by various physical studies. Gel chromatography studies suggests a possible conformational change in the enzyme may have occurred. Other studies using techniques to determine N-terminal amino acids, and other direct conformational measures will be necessary to directly determine the cause of the change in enzyme activity. We consider the change in enzyme activity to be more than a laboratory curiosity since many of the agents which can alter this enzyme's activity such as Me_2SO can penetrate the epidermis and can modify enzyme activity

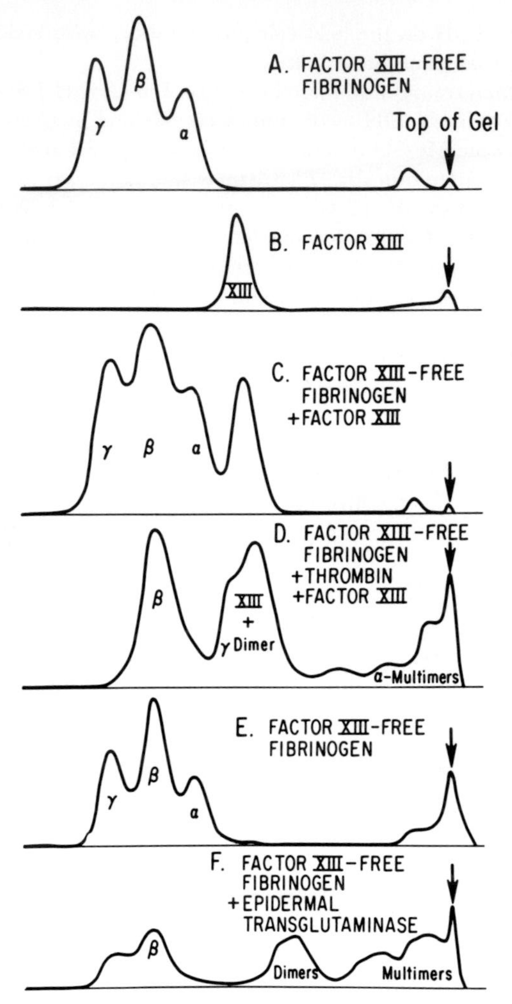

Fig. 7. Cross-linking of human fibrinogen by purified human epidermal transglutaminase and factor XIII.

After SDS electrophoresis the stained gels were scanned in a gel densitometer. The panels are the scans of different tubes.

A. Factor XIII free fibrinogen alone
B. Factor XIII alone
C. Mixture of A and B
D. C plus thrombin
E. Same as A, control for F
F. Factor XIII free fibrinogen plus activated epidermal transglutaminase.

Activated epidermal transglutaminase in the amount used had almost no protein staining.

Table 4. Effect of temperature on the activation of human epidermal transglutaminase by dimethylsulfoxide.

Epidermal transglutaminase in 10 mM Tris-Cl (pH 7.5) was mixed with an equal volume of 20 mM $CaCl_2$; Me_2SO to give various indicated final concentrations at this step was added and the mixtures incubated for 20 minutes in a water bath (37°), constant temperature bath (11°), cold room (4°), ice slush (0°) or a NaCl-ice slush (–10°). At that time, the complete reaction mixture, with additional water to balance the volumes when necessary, was added, and the mixture assayed at 37° for 30 minutes as in Methods. Controls in which Me_2SO was added to the enzyme already in the complete reaction mixture gave only the slight activation expected from the differences in the volume of the reaction mixture and the enzyme solution and hence the marked decrease in Me_2SO concentration. The control for each assay was the enzyme assayed in the standard fashion.

	%Control					
Me_2SO Concentration	Temperature (°C)					
(M)	–10	0	4	11	37	37*
3.3	206	168	166	250	454	116
5.6	370	391	510	649	468	118
7.1	—[a]	483	805	855	658	137
7.9	666	603	821	732	540	129
8.3	713	513	847	769	555	115

*Me_2SO added to the assay mixture instead of the enzyme. Final concentration in the mixture is the same as in the assay tubes.

[a]Laboratory accident.

In A, C, D, E, and F the concentration of factor XIII free fibrinogen was adjusted to 25 ug per gel. The concentration of factor XIII in B, C, and D was also adjusted to 10 ug per gel; thrombin concentration in D was 2.5 N.I.H. units.

Epidermal transglutaminase was preincubated in 50 mM Tris-HCl (pH 7.5) containing 100 mM NaCl, 0.66 mM EDTA, 6.7 mM $CaCl_2$ and 2 mM dithiothreitol for 15 minutes at 56°C to maximally activate the enzyme. The protein concentration of epidermal transglutaminase was too low to appear as a protein band in this experiment.

Factor XIII cross-linking in D showed a γ dimer and α polymers. The β chain of fibrinogen remained, and the factor XIII band is present. Epidermal transglutaminase cross-linking showed a γ dimer and also another band with similar mobility (? α-γ dimer), α-polymers, more γ band remaining than in Factor XIII induced cross-linking.

Note that epidermal transglutaminase cross-linked fibrinogen while factor XIII required fibrin as a substrate.

Table 5. Comparison of activated and nonactivated epidermal transglutaminases (17,18).

Type of Study	Result
Physical-Chemical	
SDS-electrophoresis for molecular weight	No difference
Disc electrophoresis in different percentage acrylamide gels	No difference
Agarose 0.5 m and Sephadex G-75 Gel Filtration	No difference in calcium free media; larger apparent Stokes' radius in calcium containing media.
Fluorescent Spectra (Intrinsic Tryptophan)	No difference
Immunological	
Ouchterlony Precipitation Reaction	Reaction of Identity
Neutralization Titer	Titer 4 times higher with activated than with native
Kinetic	
Km for casein	Identical
Km for putrescine	Identical

Table 6. Effect of Me_2SO and heating on human epidermal and hair follicle transglutaminases.

Enzyme samples in 10 mM Tris-acetate (pH 7.5) were assayed immediately after the addition of Me_2SO by adding the complete assay mixture to the enzyme plus Me_2SO solutions.

Enzyme samples were activated by heating at 56°C for 15 minutes in 50 mM Tris-Cl (pH 7.5), 2 mM dithiothreitol, 0.1 M NaCl, 0.6 mM EDTA and 6.7 mM $CaCl_2$. After the heating activities were measured in a standard reaction mixture at pH 9.5. Control enzymes were assayed in the standard assay at pH 9.5.

	Relative Activity	
Conditions	Hair follicle free epidermis	Hair follicle tissue
Control Assay	1	1
After heating in calcium containing buffer	26	1.9
After 8.3 M Me_2SO	10.3	2

in vivo potentially in a pharmacological fashion. Since this enzyme is active without previous proteolytic activation, the activity of the enzyme must be closely controlled *in vivo* in order to regulate the time and location of its action. We would hope that the understanding of factor controlling enzyme activity *in vivo* will lead to understanding of the *in vivo* control factors.

As previously noted, cow hair follicle transglutaminase and cow epidermal transglutaminase were immunologically different. In our studies human hair follicle and human epidermal transglutaminase were immunologically different, a precipitating antibody to human epidermal transglutaminase did not cross-react with human hair follicle transglutaminase.[18] Biochemical studies (Table 6) also suggested differences between the epidermal and hair follicle transglutaminase, as the degree of activation by calcium heating and Me_2SO was much lower in the hair follicle transglutaminase than in the epidermal transglutaminase.

Acknowledgements

The technical assistance of Ms. Judith Thorpe and the secretarial assistance of Ms. Susan Boos is gratefully acknowledged.

REFERENCES

1. Folk, J. E. and Chung, S. I. Molecular and catalytic properties of transglutaminases. *Adv. Enzymol.*, **38**: 109–91, 1973.
2. Clarke, D. D., Mycek, M. J., Neidle, A., and Waelsch, H. The incorporation of amines into protein. *Arch. Biochem. Biophys.*, **79**: 338–354, 1959.
3. Schwartz, M. L., Pizzo, S. V., Hill, R. L., and McKee, P. A. Human Factor XIII from plasma and platelets. Molecular weights, subunit structures, proteolytic activation and cross-linking of fibrinogen and fibrin. *J. Biol. Chem.*, **248**: 1395–1407, 1973.
4. Gross, M. and Folk, J. E. Mapping of the active sites of transglutaminases I. Activity of the guinea pig liver enzyme toward aliphatic amides. *J. Biol. Chem.*, **248**: 1301–1306, 1973.
5. Chung, S. I. Multiple molecular forms of transglutaminases in human and guinea pig. *Isozymes*, vol. 1, (C. L. Markert, Ed.), Academic Press, N. Y., 1975, pp. 259–274.
6. Duckert, F. Documentation of the plasma factor XIII deficiency in man. *Ann. N. Y., Acad. Sci.*, **202**: 190–199, 1972.
7. Cooke, R. D. Calcium-induced dissociation of human plasma factor XIII and the appearance of catalytic activity. *Biochem., J.*, **141**: 683–691., 1974.
8. Harding, H. W. J. and Rogers, G. E. Formation of ε-(γ-glutamyl)lysine cross-link in hair proteins. Investigation of transamidases in hair follicles. *Biochemistry*, **11**: 2858–2863, 1972.
9. Chung, S. I. and Folk, J. E. Transglutaminase from hair follicle of guinea pig. *Proc. Nat. Acad. Sci., USA*, **69**: 303–307, 1972.
10. Buxman, M. M., Buehner, G. E., and Wuepper, K. D. Isolation of substrates for epidermal transglutaminase from bovine epidermis (abstr). *Clin. Res.*, **24**: 262A, 1976.
11. Goldsmith, L. A. ε-(γ-glutamyl)lysine cross-links in proteins. *In* Biochemistry of Cutaneous Epidermal Differentiation, (M. Seiji, ed), Univ. of Tokyo Press, 1976.
12. Goldsmith, L. A., Baden, H. P., Roth, S. I., Colman, R., Lee, L., and Flem-

ing, B. Vertebral epidermal transamidase. *Biochim. Biophy. Acta*, **351**: 113–125, 1974.

13. Buxman, M. M. and Wuepper, K. D. Keratin cross-linking and epidermal transglutaminase. *J. Invest. Derm.*, **65**: 107–112, 1975.
14. Goldsmith, L. A. and Martin, C. M. Human epidermal transamidase. *J. Invest. Derm.*, **64**: 316–321, 1975.
15. Cuatrecasas, P. and Anfinsen, C. B. Affinity Chromatography. *Methods in Enzymol.*, **22**: 345–378, 1971.
16. Furie, B. C. and Furie, B. Purification of proteins involved in Ca (II) dependent protein-protein interactions (Coagulant protein of Russell's viper venom). *Methods in Enzymol.*, **34**: (part B): 592–4, 1974.
17. Ogawa, H. and Goldsmith, L. Human epidermal transglutaminase 1. Preparation and properties. Accepted for publication in *J. Biol. Chem.*, 1976.
18. Ogawa, H. and Goldsmith, L. Human epidermal transglutaminase 2. Immunological properties. Accepted for publication in *J. Invest. Derm.*, 1976.

Discussion

Dr. Baden: One question I would like to ask is whether or not you looked at whole epidermis, or only extracted the callus? I wonder if what you call "negative" is a change which is a result of drying out in the stratum corneum or whether it is the way the enzyme looks in the viable epidermis.

Dr. Ogawa: We extract this enzyme from human callus though it is pretty dried out. It has not been extracted from whole human epidermis. We did not use whole epidermis because we could not avoid contamination from the hair follicle material. It should be done, however, so we can make a comparison of the enzyme from the viable epidermis and from the callus.

Dr. Sugawara: Did you try to determine whether or not there is any natural inhibitor of this enzyme?

Dr. Ogawa: There are low molecular weight substances, which are separated in the procedure of concentration using an Amicon filter and which appeared to be natural inhibitors of this system. After the concentration by the use of the filter, the total activity of the enzyme increased.

Dr. Sugawara: We have isolated the glutaminase from the newborn rat epidermis and have found after gel filtration a large increase in total activity, so there must be a natural inhibitor in the tissue.

Dr. Goldsmith: We have been working with something that looks like a natural inhibitor in our human material. We think it is different from the activation system which essentially involves a change in V_{max}. We find that the natural inhibitor affects the K_m.

Epidermal Protein Synthesis*

Irwin M. Freedberg and Mary E. Gilmartin

Department of Dermatology, Harvard Medical School and the Thorndike Research Laboratories of the Harvard Medical School at the Beth Israel Hospital, Boston, Massachusetts, U.S.A.

ABSTRACT

Our recent studies of epidermal protein synthesis have focused upon the appropriate conditions for isolation of ribosomes from epidermal homogenates and upon a definition of the initiation factors of protein synthesis which are present in the tissue. Monomeric ribosomes are the predominant species which can be isolated although polymeric ribosomes and subunits can be identified. Detergent extraction is required for maximal recovery of epidermal ribosomes and yield is greatest in buffers with potassium concentrations between 100 and 250 mM. Yield is decreased in buffers with magnesium concentrations above 5 mM. *In vitro* amino acid incorporation is not affected by potassium and is highest at a magnesium concentration of 3 mM. The isolated ribosomes are unstable at 4°C and this instability may be due to the multiple ribonucleases which have been identified in the tissue.

Although the initiation reactions of protein synthesis are inhibited in epidermal homogenates, initiation factors can be identified in a salt wash of epidermal ribosomes. IF-MP-like activity has been identified and partially purified. IF-M3 and IF-M2A- and IF-M2B-like activities have been shown to be present and an inhibitor of initiation is present in the epidermal ribosomal salt wash. The pathways and controls of epidermal protein synthesis were reviewed in detail several years ago,[1,2] and the pathways of protein synthesis in hair root cells have been reviewed more recently.[3] These previous studies have indicated that the overall mechanism of protein synthesis in epidermal cells is similar to that in most other protein synthesizing tissues of prokaryotic or eukaryotic origin. Our current work in this area has focused upon two facets of the protein synthesis problem: the conditions necessary for isolation of active ribosomal particles from epidermal homogenates and a definition of the initiation

* Our laboratory is supported in part by grant AM 16262, National Institute of Arthritis, Metabolism and Digestive Diseases.

factors present in the tissue. In this communication we shall review the state of the art in thesc areas. More data on these topics may be found in two of our recent papers.[4,5]

EPIDERMAL RIBOSOMES

Several authors have proven that ribosomes can be successfully isolated from epithelial tissues and subsequently used for studies of epidermal or hair root protein synthesis. Steinert and Rogers[6] prepared an homogenate of guinea pig hair root cells and demonstrated that the mechanism of epithelial protein synthesis was similar to that in other tissues. Kumaroo and his coworkers[7] were able to isolate polyribosomes from a sodium desoxycholate homogenate of newborn rat epidermis, and Argyris and his associates[8,9] have defined the distribution of free and membrane-bound ribosomes in mouse epidermis. Studies from our laboratory have dealt with the subcellular pathways of protein synthesis in guinea pig epidermis and hair root cells. Since it was apparent that a systematic investigation of the conditions affecting the isolation and characterization of epidermal ribosomes was necessary if we were to be able to define the pathways and controls of protein synthesis in this tissue, we undertook the following studies.

Epidermis was obtained by the stretch technique from latex- or wax-epilated male, albino guinea pigs of 150–200 grams body weight. The tissue was placed in ice-cold buffer and all subsequent procedures were carried out at 4° C. The tissue was homogenized; unbroken cells, nuclei, mitochondria and debris were removed by centrifugation; and the supernatant fraction was used for amino acid incorporation studies or for density gradient centrifugation.

The density gradient patterns indicated that monomeric ribosomes were predominant in the homogenates although 40S and 60S ribosomal subunits, dimers, trimers and larger polyribosomes were also present. Both the yield of ribosomal particles and the protein synthetic capacity of each ribosome were greater in epidermis prepared from latex-epilated animals as compared to those which were epilated by the molten-wax method. The yield and activity of the ribosomal particles were greater when the tissue was extracted with a detergent such as desoxycholate, and these data are illustrated in Fig. 1. The gradient pattern of epidermal ribosomes obtained without desoxycholate is characterized by a visible 80S ribosomal peak and some heavier unresolved ribosomal material. When desoxycholate was present at a concentration of 0.25% to 0.5% in the extracting buffer, a clear pattern of subunits, monomeric and polymeric ribosomes became visible. As is apparent in Table 1, the protein-synthetic activity of the

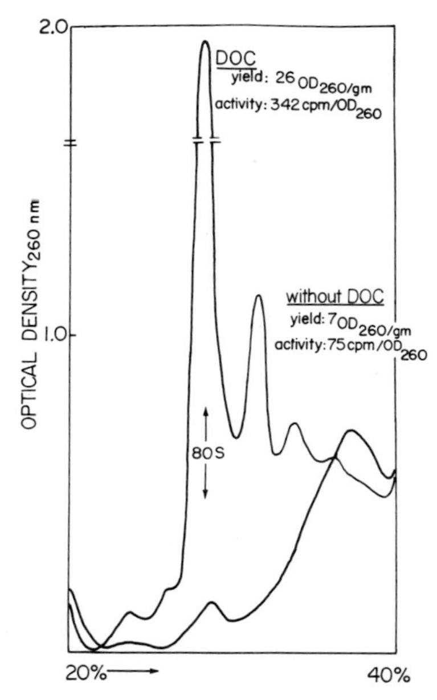

Fig. 1. Density-gradient centrifugation pattern of latex-epilated epidermis homogenized either in basic buffer or buffer made 0.5% with respect to sodium desoxycholate. (By permission, *The Journal of Investigative Dermatoloy*)

Table 1

Epilation method	DOC % (W/V)	Yield (mg RNA/gm)	Specific activity (CMP/mg RNA)
WAX	0.5	1.6	1720
	0.25	1.1	—
	0.13	0.2	—
	0.0	0.1	*
LATEX	0.5	2.7	3950
	0.0	0.7	1100

* Yield too low to permit incorporation studies.

particles extracted with desoxycholate was significantly greater than that of particles extracted without detergent.

The effects of the potassium and magnesium concentrations of the extracting media were also analyzed and we found that the yield of ribosomes was twice as great if the tissue was extracted in a buffer containing 250 mM potassium as compared to one in which the potassium concentration was 10 mM. The synthetic activity of ribosomes extracted in low and high potassium were the same, however. In contrast, magnesium had effects upon both the density gradient patterns and the activity of the isolated ribosomes. At 1 mM magnesium the concentration of 40 and 60S ribosomal subunits was increased while at 5 and 10 mM magnesium, ribosomal aggregates formed. Amino acid incorporation and relative proportion of nonaggregated, nondegraded ribosomes were highest at 3 mM magnesium.

Figure 2 summarizes data which we obtained concerning the stability of epidermal ribosomes. If isolated ribosomes were stored at −20° C, they retained *in vivo* amino acid incorporating activity for at least 22 hours. If the particles were stored at 4° C, in contrast, major losses of activity occurred by 5 hours, and at 18 hours the particles were almost completely inactive. We attempted to define the reason for this rapid inactivation by documenting the amount of runoff which took place in the epidermal cell-free protein synthesizing system. Using cycloheximide we showed that runoff was not the reason for either the inactivation or the unusual density gradient pattern. Ribonuclease could explain the gradient pattern and the rapid inactivation, and recent studies in our laboratory[13] have shown that there are at least 9 distinct ribonucleases in guinea pig epidermis.

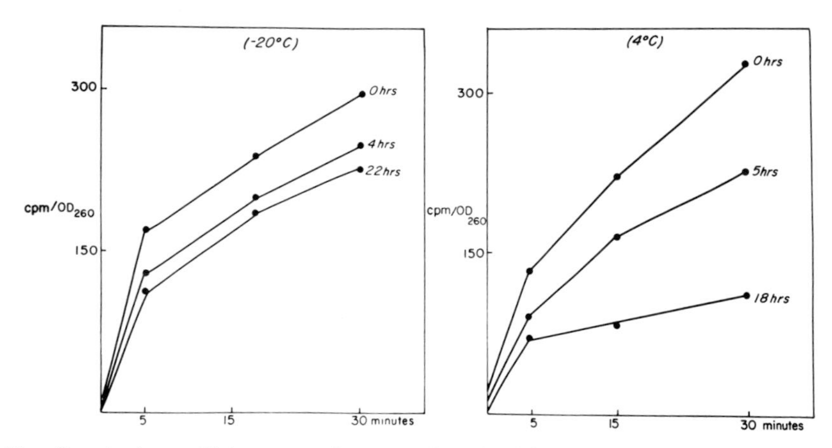

Fig. 2. Amino acid incorporating capacity of epidermal ribosomes prepared in basic buffer and incubated in the *in vitro* system either immediately after preparation or after storage for the indicated time periods (hr) at 4°C or −20°C. (By permission, *The Journal of Investigative Dermatoloy*)

INITIATION OF PROTEIN SYNTHESIS

An additional explanation for the unusual epidermal ribosomal sedimentation pattern and the rapidly inhibited *in vitro* amino acid incorporation could be a deficiency or inhibition of the components involved in the initiation of protein synthesis. These components are currently under study in many systems in an attempt to define the mechanisms by which tranlational control of protein synthesis is accomplished. The current state of our knowledge of initiation of protein synthesis is summarized in Fig. 3, which we have prepared by using data available from a variety of sources.[14–23]

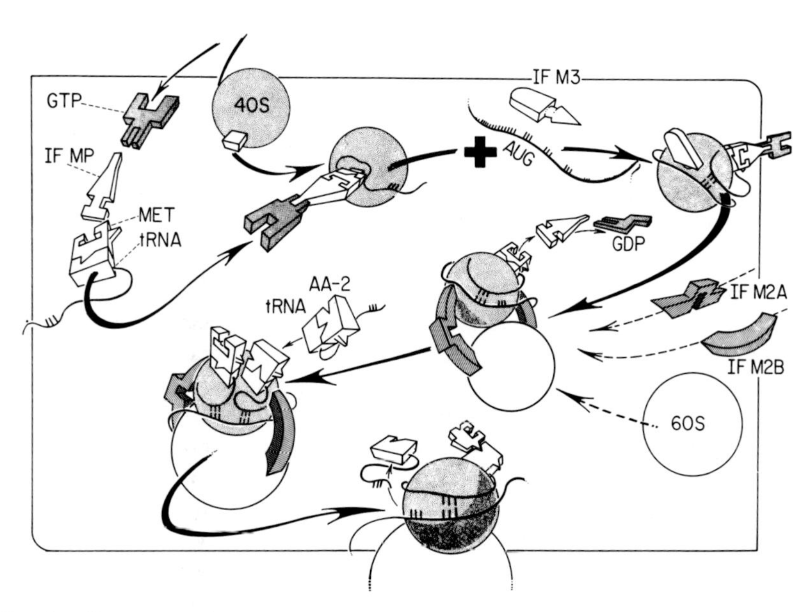

Fig. 3. Summary of the pathways of initiation of protein synthesis in mammalian epidermis.

The process of translation of messenger RNA begins with the formation of an active complex of initiator tRNA (Met-tRNA$_f$ in eukaryotic organisms), the smaller ribosomal subunit and the initiation codon of the messenger RNA. A number of protein factors which participate in the formation of this complex have been defined: IF-MP mediates the formation of a ternary complex with GTP and the binding of Met-tRNA$_f$ to the smaller ribosomal subunit; IF-M3 participates in binding of the small subunit to the initiation codon of the messenger RNA; and IF-M2A and IF-M2B appear to stabilize the 80S ribosome by functioning as joining factors. The factors are present either in the high-speed supernatant fraction of tissue homogenates or are bound to the native 40S ribosomal subunit from which they can be dissociated with a high salt buffer.

It has been proposed that these initiation reactions are a site for *in vivo* control of gene expression at the level of translation. In a reticulocyte system, for example, Clemens *et al.* have shown that globin synthesis is regulated through the activity of IF-MP.[19] The possibility that control at the initiation reactions is a general phenomenon is suggested by the fact that alteration of initiation factor activity has been demonstrated in such diverse areas as the development of brine shrimp and the carcinogenic transformation which leads to a hepatoma.[20–21]

We initially prepared partially fractionated homogenates from guinea pig liver and epidermis and proved that they actively incorporated ^{3}H-lableled amino acids into acid precipitable peptides over a 60-minute period. We found, as indicated in Table 2, that several inhibitors of the initiation and elongation reactions of protein synthesis had effects upon amino acid incorporation in these systems. Among the inhibitors of elongation which we studied, emetine inhibited the movement of ribosomes with respect to messenger RNA, NaF disaggregated polysomes, ribonuclease A degraded mRNA by cleaving at cytidine and uridine residues, and, at the high concentration noted in the table, aurintricarboxylic acid inhibited elongation. Each of these elongation inhibitors suppressed the epidermal and the liver protein synthesizing systems. In contrast, pactamycin, an inhibitor of initiation, dramatically reduced amino acid incorporation in the liver system but was without demonstrable effect upon the epidermal system. Edeine and low concentrations of aurintricarboxylic acid, which are also inhibitors of initiation, were also without effect in the epidermal system. Both antibiotics inhibit formation of the initiation complex, presumably by interfering with binding of initiator tRNA to the small ribosomal subunit.

Table 2. Effects of inhibitors of protein synthesis upon incorporation of ^{3}H-amino acids into protein by the epidermal and liver lysate systems

Incubation conditions	Incorporation of [^{14}C]-amino acids (cpm/mg lysate protein)	
	Epidermis	Liver
S-20 control	2166	7083
+ pactamycin	2666	333
+ edeine	2166	—
+ emetine	1083	—
+ sodium fluoride	1000	656
+ ribonuclease A	566	420
+ aurintricarboxylic acid	333	330

Duplicate aliquots of epidermal and liver S-20 supernatant fractions were incubated at 30 °C for 60 minutes in a [^{14}C]-amino acid incorporating system. Inhibitors of protein synthesis were added in the following concentrations: 0.002 mM pactomycin, 0.005 mM edeine, 0.001 mM emetine, 100 mM sodium floride, 0.5 mM aurintricarboxylic acid. Ribonuclease A was added to the epidermal assay system at 25 μg/tube and at 50 μg/tube to the liver system.

These results can be interpreted in several ways. They could indicate that the epidermal system lacks initiation factors or they could mean that the factors are present although in some way inactivated. As a first step in differentiating between these possibilities, we obtained a KCl wash from epidermal ribosomes and assayed its capacity to promote binding of [^{35}S] Met-tRNA$_f$ to nitrocellulose filters. We found that [^{35}S] Met-tRNA$_f$ was bound to nitrocellulose filters in a complex which required both epidermal KCl wash and GTP. In the absence of either of these components only background amounts of radioactivity were bound. Binding was increased by a nucleotide triphosphate generating system.

Since these results made it apparent that factors in the epidermal KCl wash participated in ternary complex formation ([^{35}S] Met-tRNA$_f$-initiation factor-GTP), we undertook further studies to standardize the binding reaction and its components. Binding was increased with increasing amounts of KCl wash and it was also stimulated by increasing amounts of GTP and tRNA. Ternary complex formation was strongly inhibited by increasing concentrations of magnesium, a phenomenon seen in other similar systems.

The initiation reaction mixture was analyzed on CsCl equilibrium density gradients, and the density gradient centrifugation patterns shown in Fig. 4 support our proposal that the presumptive epidermal initiation factors, present in the KCl wash, mediate the binding of [^{35}S] Met-tRNA$_f$

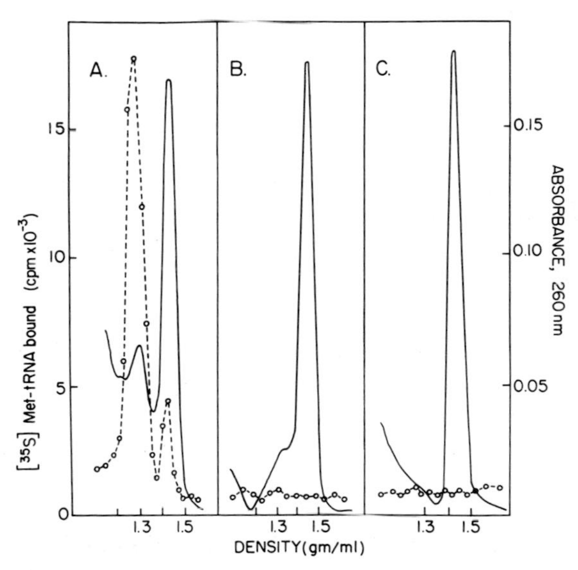

Fig. 4. Cesium chloride density gradient centrifugation patterns. (—) = absorbance at 260 nm. (○-○-○) = [^{35}S]Met-tRNA binding. A. Complete system. B. GTP omitted. C. KCl wash omitted.

to 40S ribosome subunits. When either KCl wash or GTP is omitted (4B & 4C) only a single species of 40S ribosomal subunit is apparent in the gradient and no [^{35}S] Met-tRNA$_f$ is associated with these subunits. If GTP and KCl wash are included, two species of 40S subunits are seen (4A) with buoyant densities of 1.30 and 1.44. [^{35}S] Met-tRNA$_f$ co-sediments with the lighter subunit.

Partial purification of the epidermal initiation factor activity has been accomplished by using DEAE ion-exchange chromatography, as shown in Fig. 5. The crude activity was adsorbed to the column and eluted with increasing concentrations of potassium chloride. The peak eluting at a KCl

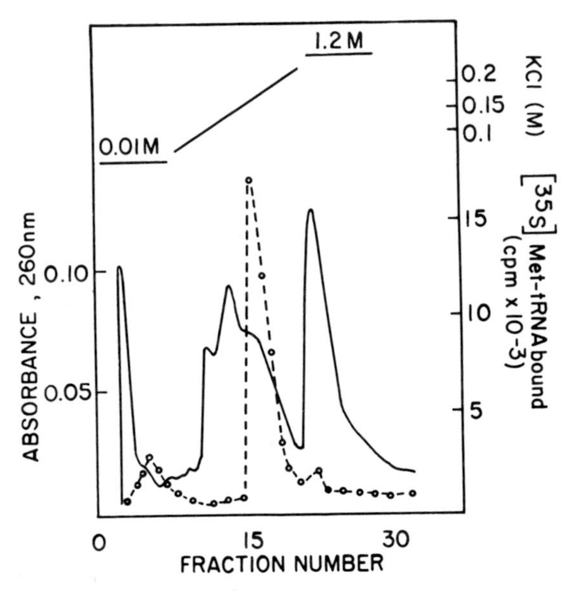

Fig. 5. Fractionation of epidermal ribosomal KCl wash on DEAE-cellulose. Absorbance at 260 nm (——); [^{35}S]Met-tRNA bound to nitrocellulose filters (O–O–O).

concentration of 0.17 M promoted GTP binding of [^{35}S] Met-tRNA$_f$ when assayed in an initiation factor-dependent system. The binding activity was stable for at least 10 days when the column eluates were stored at 0° C. An additional fraction with Met-tRNA$_f$ binding activity did not bind to the resin and eluted in the void volume. A fraction which strongly inhibited protein synthesis in an initiating system has been identified as eluting between 0.1 and 0.15 M KCl. The role of this inhibitor is not yet clear but it may be of importance in the overall control of epidermal protein synthesis.

REFERENCES

1. Bernstein, I. A., Chakrabarti, S. G., Kumaroo, K. K., and Sibrack, L. A. Synthesis of protein in mammalian epidermis. *J. Invest. Dermatol.*, **55**: 291–000, 1970.
2. Freedberg, I. M. Pathways and controls of epithelial protein synthesis. *J. Invest. Dermatol.*, **59**: 56–65, 1972.
3. Freedberg, I. M. Biochemistry of keratinization in hair: Protein synthesis, pathways and controls. *In Biology and Disease of Hair*. University of Tokyo Press, Tokyo, 1976, p. 45–52.
4. Gilmartin, M. E. and Freedberg, I. M. Isolation and characterization of epidermal ribosomes. *J. Invest. Derm.*, **64**: 90–95, 1975.
5. Gilmartin, M. E. and Freedberg, I. M. Mammalian epidermal protein synthesis: Initiation factors. *J. Invest. Derm.*, **67**: 240–245, 1976.
6. Steinert, P. M. and Rogers, G. E. Protein biosynthesis in cell-free systems prepared from hair follicle tissue of guinea pigs. *Biochim. Biophys. Acta*, **232**: 556–572, 1971.
7. Kumaroo, K. K., Gray, R. H., Kamen, I. L., and Bernstein, I. A. Isolation and characterization of polyribosomes from the epidermis of newborn rat. *J. Invest. Dermatol.*, **59**: 305–312, 1972.
8. Argyris, T. S., Nevar, C., Mueller, S., De Young, L., and Gordon, G. Ribosome fractions from normal and methylcholanthrene-treated mouse epidermis. *J. Invest. Dermatol.*, **63**: 262–267, 1974.
9. Mueller, S. N. and Argyris, T. S. Free and membrane-bound ribosomes in normal and methylcholanthrene-treated mouse epidermis. *Lab Invest.*, **32**: 209–216, 1975.
10. Freedberg, I. M., Fine, I. H., and Cordelle, F. H. Cell-free protein synthesis in mammalian skin. *J. Invest. Dermatol.*, **48**: 55–66, 1967.
11. Freedberg, I. M. Mammalian epidermal and hair root protein synthesis: Subcellular localization of the synthetic site. *Biochem. Biophys. Acta*, **224**: 219–231, 1970.
12. Freedberg, I. M. Hair root cell-free protein synthesis. *J. Invest. Dermatol.*, **54**: 108–120, 1970.
13. Melbye, S. W. and Freedberg, I. M. Epidermal nucleases: II. The multiplicity of ribonucleases in guinea pig epidermis. In press, 1976.
14. Jay, T. and Kaempfer, R. Initiation of protein synthesis: Binding of messenger RNA. *J. Biol. Chem.*, **250**: 5742–5748, 1975.
15. Benne, R., Ebes, F., and Voorma, H. O. Sequence of events in initiation of protein synthesis. *Eur. J. Biochem.*, **38**: 265–273, 1973.
16. Zasloff, M. and Ochoa, S. Polypeptide chain initiation in eukaryotes: Functional identity of supernatant factor from various sources. *Proc. Nat. Acad. Sci.*, **69**: 1796–1799, 1972.
17. Gupta, N. K., Woodley, C. L., Chem, Y. C., and Bose, K. K. Protein synthesis in rabbit retuculocytes: Assay, purification and properties of different ribosomal factors and their role in peptide chain initiation. *J. Biol. Chem.*, **248**: 4500–4511, 1973.
18. Seal, S. N., Bewley, J. D., and Marcus, A. Protein chain initiation in wheat

embryo: Resolution and function of the soluble factors. *J. Biol. Chem.*, **247**: 2592–2597, 1972.
19. Clemens, M. J., Henshaw, E. C., Rakaminoff, H., and London, I. M. Met-tRNA$_{fmet}$ binding to 40S ribosomal subunits: A site for the regulation of initiation of protein synthesis by hemin. *Proc. Nat. Acad. Sci.*, **71**: 2946–2950, 1974.
20. Sierra, J. M., Meier, D., and Ochoa, S. Effect of development on the translation of messenger RNA in Artemia salina embryos. *Proc. Nat. Acad. Sci.*, **71**: 2693–2697, 1974.
21. Mruty, C. N., Verney, E., and Sidransky, H. Studies of initiation factors in protein synthesis of host liver and transplantable hepatoma. *Cancer Research*, **34**: 410–418, 1974.
22. Weissbach, H. and Brot, N. The role of protein factors in the biosynthesis of proteins. *Cell*, **2**: 137–144, 1974.
23. Anderson, W. F. Cell-free synthesis of globin chains: An overview. *Ann. N.Y. Acad. Sci*, **241**: 142–155, 1974.

Discussion

Dr. Seiji: Are there polyribosomes larger than dimers, for example, pentamers, present in the epidermal preparations which you use? Which of the ribosomal particles are most active in protein synthesis?

Dr. Freedberg: Yes, there are polyribosomes in epidermis. In our hands using newborn guinea pig epidermis and working as rapidly as we can, this is as well as we can do. Dr. Bernstein has published data which he interprets as showing more polyribosomes than this. From electron microscopic studies we know that there are many large polyribosomes in the epidermis. This then means that the polyribosomal profiles which we get, are artifacts. Probably because of the high content of ribonuclease in the tissue, they are broken down during the preparative procedures which are used. In answer to your second question, the polysomes are more active in an *in vivo* protein synthesizing system than are the monomeric ribosomes. Furthermore, Dr. Argyris has recently reported the presence of both bound and free ribsomes and the separation of the two. The bound ones are the ones with the "action."

Dr. Karasek: Is the "100 mM" inhibitory factor specific for the skin ribosomes or can one find a similar kind of thing with the ribosomes from liver?

Dr. Freedberg: I do not know. Dr. Henshaw has unpublished evidence for such an inhibitor in ribosomes from Ehrlich's ascites tumor cells but I know of no other effort to identify such a substance. There is a hemin-sensitive inhibitor which is involved in control of initiation in the reticulocyte system. That is an inhibitor which is not bound to ribosomes. In the case where you have a 0.25 M KCl wash which was activated by a slight amount, it is almost certainly EF-1 or EF-2 (elongation factors).

DR. BERNSTEIN: I am sure that you are well aware that the data you have presented are consistent with the hypothesis but do not necessarily prove the hypothesis that these initiation factors are indeed present. I wonder if you have given that any thought. The reason that I raise the issue is that philosophically we tend to "follow the leader," or as currently designated in the United States, "get on the bandwagon" but in fact, the demonstration is not so ubiquitous but what it may not be general.

DR. FREEDBERG: Let me make an important point if I may. I showed you pictures of things moving in and out as though we were absolutely certain. This is a result of a combination of a large number of "molecular biological" minds and a good medical artist. The scheme is something like that shown and is a framework within which I think one can work. I believe that your point is well taken.

The Mechanism of Assembly of Bovine Epidermal Keratin Filaments *in vitro*

Peter M. Steinert

Dermatology Branch, National Cancer Institute, Bethesda, Maryland, U.S.A.

ABSTRACT

The seven bovine epidermal α-keratin polypeptides which comprise the subunits of the *in situ* keratin filaments polymerize *in vitro* into filaments that have the same structure as the *in situ* filaments. Recombinations of two of these polypeptides also form filaments of the same structure. As such filaments contain the two polypeptides in the precise molar ratios of 1:2, it is concluded that the epidermal keratin filament is composed of a basic three-chained structural unit. Further information on the structure of the epidermal keratin filament was obtained by a study of the mechanism of assembly of the filaments *in vitro*. Turbidometric measurements of filament assembly revealed a biphasic mechanism, involving the initial rate-limiting formation of a hexamer nuclear particle, followed by a more rapid rate of polymerization to filaments. The nuclear particle may consist of a pair of triple-chained units and thus is structurally very similar to a multichain particle previously defined as prekeratin. As the formation of the nuclear particle is the most difficult aspect of filament assembly, it is probably a structural building block of the keratin filament. Preliminary electron microscopic studies suggest that the second fast stage of filament polymerization occurs by end-to-end and side-by-side aggregation of the nuclear particles to form filaments that are uniformly 70–80 Å wide. Very long filaments (> 15 μm) apparently assemble by end-to-end joining of shorter filaments.

INTRODUCTION

The major structural protein of mammalian epidermis is an α-fibrous protein, α-keratin, which appears intracellularly as keratin filaments. These are produced by the developing epidermal cells and form the bulk of the terminally differentiated ("keratinized") stratum corneum cells.[1] When seen in thin sections in the electron microscope, the keratin filaments appear as rods of uniform width of 7–8 nm but their length is unknown since they are not well aligned in the plane of the section.[1,2]

Over the years considerable effort has been directed toward the elucidation of the structure of the epidermal keratin filament. Biochemical studies have concerned the analysis of the chemistry and structure of proteins extracted from the epidermis. The work of Matoltsy and Skerrow has shown that prekeratin, extracted from the living epidermal cells with a citrate buffer, consists of a pair of triple-chained units, each of which contains regions of coiled-coil α-helix interspersed with regions of non-α-helix.[3–6] They have concluded that prekeratin is a "structural building block" of the keratin filament. In other studies, denaturing solvents such as NaOH,[7] formic acid,[8] and alkaline urea buffers[9–15] extract the keratin filaments of both the living and stratum corneum cells in the form of their constituent polypeptide chain subunits. In all cases, the polypeptides appear to have molecular weights within the range of 47–70,000[11–15] and consist of a number of different components.

We[14,15] have studied the chemistry and structure of the α-keratin polypeptides extracted from bovine epidermis of the hoof. They consist of seven distinct species of monomer molecular weights of 45–58,000 and all possess very similar chemical and biophysical properties. Perhaps their most interesting property is their marked propensity for aggregation in the absence of denaturing solvents. Indeed, in aqueous solution, they spontaneously polymerize *in vitro* into filaments that appear to have the same structure as the *in situ* epidermal keratin filaments.[14,16] It would appear that a detailed study of these filaments polymerized *in vitro* provides the most useful approach to the further delineation of the structure of the keratin filament. By analogy, studies of the structure and mechanism of assembly of other structural proteins which polymerize *in vitro* into particles similar to their native forms, such as collagen,[17] actin,[18,19] tubulin,[20,21] and flagellin,[22] have in each case led to a thorough understanding of the structure of the fibrous protein. As with these four proteins, preliminary experiments showed that the polymerization of the α-keratin polypeptides also follows orderly kinetics. Therefore, a study of the structure of these filaments, together with a combined kinetic and electron microscopic study of their formation, was initiated to further the understanding of the structure of the epidermal keratin filament.

MATERIALS AND METHODS

Preparation of α-keratin polypeptides

The epidermis of the posterior region of the bovine hoof is about 1 cm thick, of which the living cell layers comprise a distinct layer 1–2 mm wide and the stratum corneum may vary up to 10mm thick.[23] Tissue containing only the living cells was used in this work. Strips, about 1mm wide, were removed with a razor blade, chopped into small pieces and extracted for 3 hr at 37° with stirring in a buffer of 8 M urea, 0.05 M Tris-HCl (pH 9.0) and 0.025 M 2 mercaptoethanol. Cellular debris was removed by centrifugation at 30,000 × *g* for 30 min and then at 250,000 × *g* for 2 hr to remove smaller particles.[14] The final protein concentration was about 10 mg/ml. More than 90% of the protein in solution consists of the α-keratin polypeptides.

Separation and purification of the polypeptides was effected by the two-step procedure involving chromatography of DEAE-cellulose followed by preparative polyacrylamide gel electrophoresis of selected fractions as described elsewhere.[15]

Filament polymerization *in vitro*

Previously,[16] it was shown that the α-keratin polypeptides polymerize into filaments with maximum yields within 18 hr by dialysis of solutions (1–2 mg/ml in urea buffer) against 1000 vol of 5 mM Tris-HCl (pH 8.0) containing 0.025 M 2–mercaptoethanol, and these conditions were used here. About 80–85% of the protein in solution present as "initial" filaments assembled under these conditions could be pelleted by ultracentrifugation at 250,000 × *g* for 1 hr. The transparent pellet of filaments was redissolved (depolymerized) in the same urea buffer used for extraction of the polypeptides and then centrifuged at 250,000 × *g* for 2 hr to remove any large particles. More than 98% of the protein remained in the supernatant. Upon dialysis into the same salt solution as above, virtually 100% of the polypeptides repolymerized into filaments as determined by pelleting after ultracentrifugation.[16] Repolymerized filaments were used in preference to initial filaments as this procedure removed the traces of nonkeratin proteins also present in the extracts.

For studies on the time dependence of polymerization, the urea was removed within 3 min at 23° by chromatography on a 25 × 1.0 cm column of Sephadex G–25 (coarse) equilibrated in the salt solution.

Electron microscopy

Filaments polymerized *in vitro* were examined only by negative-staining with 0.7% uranyl acetate on either holey ("lacey film") grids for short particles or on carbon-coated ("stress-free") grids (Ladd Research Industries Inc., Burlington, Vt.) for long particles. A Siemens Elmiskop lA electron microscope was used. The magnification of the machine was repeatedly checked by using germanium-shadowed carbon replicas of diffraction gratings and varied only by about ± 5%. When the lengths of filaments were to be determined, numerous adjacent photographs of minimum overlap were assembled to form a square of side corresponding to about 20 μm.

Analytical procedures

Turbidity (light scattering) measurements were made at 300 nm in a Beckman DK 2–A recording spectrophotometer and at a temperature of 23 ± 0.5°.

Polyacrylamide gel electrophoresis (on 9% T, 3% C gels) was performed using the discontinuous system.[14,15] All samples in urea were equilibrated by dialysis for 18 hr in the cathode buffer containing 0.5% sodium decyl sulfate, and heated at 98° for 2 min to effect complete binding. Gels were loaded with about 50 μg of sample, electrophoresed for 3 hr at 2 mA/gel and stained with Fast Green. When necessary, the gels were scanned with an Isco Model 659 gel scanner.

X-ray diffraction was done on fibers drawn from pellets of filaments as described previously.[16]

Protein was estimated by the method of Bramhall *et al.*[24] using the unfractionated mixture of polypeptides as a standard, which have $E^{1\%}_{277nm}$ = 5.6.[14]

The sedimentation assay for filament formation was performed by centrifugation at 250,000 $\times g$ (125,000 $\times g_{av}$) for 1 hr. The amount of protein in the pellet, determined from the difference in protein concentration of the supernatant solutions before and after centrifugation, was considered to be due to the filaments.

RESULTS

Structure of epidermal keratin filaments polymerized *in vitro* ultrastructure

"Initial" and repolymerized filaments assembled *in vitro* were of uniform width of 70–80 Å and their length varied within the range of 1–40 μm (Fig. 1). Measurement of numerous (N = 281) repolymerized filaments revealed an average ± s.d. length of 15.7 ± 3.5 μm. The filaments appeared as solid flexible rods and there was no visible evidence by negative staining of ultrastructure or of repeating structural units (Fig. lb). In addition, the filaments showed no tendency to form wider fibrils. The ends of the filaments terminated abruptly ("square") with no evidence of fraying into finer units, but in several cases, terminated against other filaments at oblique or perpendicular angles suggestive of branching, or, more likely, of random juxtaposition of "sticky" ends.

Polypeptide composition On polyacrylamide gels containing sodium decyl sulfate, the extracted α-keratin polypeptides separated as six bands

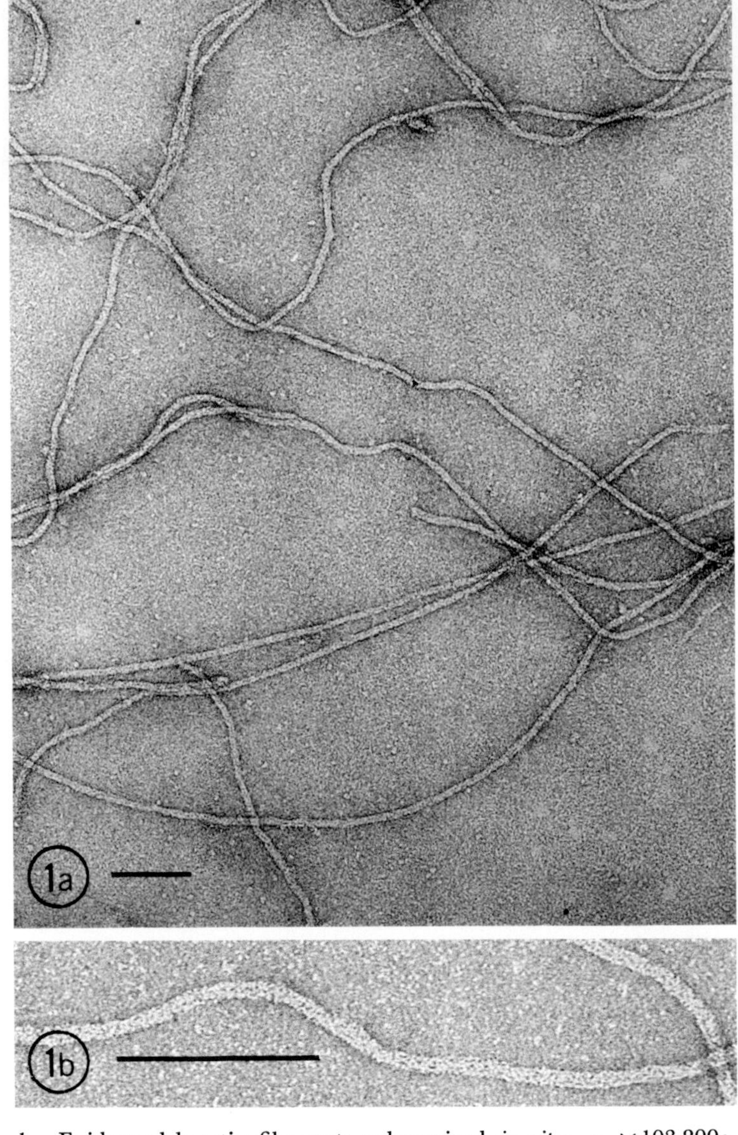

Fig. 1 Epidermal keratin filaments polymerized *in vitro*. a, ×103,800; b, × 260,500. The bar is 0.1 μm.

(Fig. 2, gel a). Band 1 has been shown to consist of two components, designated la and lb, of slightly different charge.[15] The assembled filaments also contained the same bands of polypeptides, and, moreover, the relative

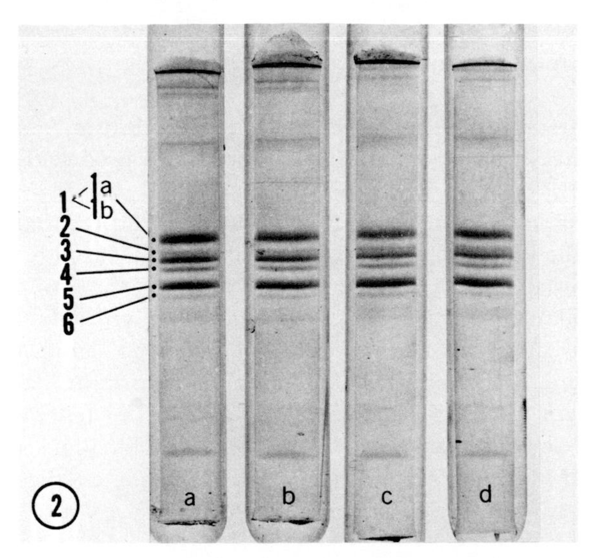

Fig. 2. Polypeptide composition of epidermal keratin filaments. All samples were pelleted by the centrifugation assay and depolymerized in the urea buffer before preparation for electrophoresis. a, extracted α-keratin polypeptides: the numbers refer to the polypeptide numbers used in the text; b, "initial" filaments; c, repolymerized filaments; d, filaments repolymerized a second time.

amount of each band was the same (Fig. 2). These patterns were unchanged through two cycles of depolymerization and repolymerization.

X-ray diffraction The X-ray diffraction patterns displayed by these filaments revealed a sharp meridional reflection at 5.15 Å and a prominent equatorial reflection at 9.8 Å (Fig. 3). This pattern is typical of epidermal

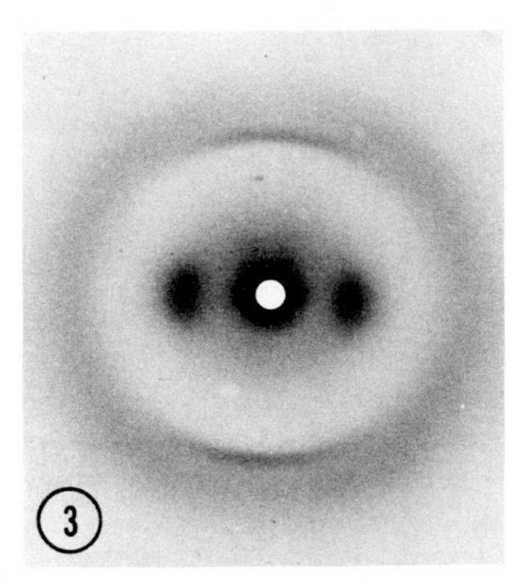

Fig. 3. X-ray diffraction pattern of repolymerized filaments.

α-keratin and indicates the presence within the filaments of regions of segmented triple-chained coiled-coil α-helix.[1,25–28]

Recombination of the purified polypeptide chains

None of the purified polypeptides by themselves produced filaments *in vitro*.

Many combinations of two purified chains were prepared, and most resulted in the formation of filaments (Table 1). The yields of filaments produced varied widely; most mixtures resulted in the formation of relatively few filaments, but in four combinations, a virtually 100% yield was obtained. As the low yields were not improved significantly on repolymerization, it is likely that some combinations form filaments more efficiently than others. Since some combinations did not produce even traces of filaments, it would appear that such combinations are "incompatible" with filament polymerization. All filaments tested gave the classic α-type X-ray diffraction pattern (Table 1).

Table 1. Mixtures of two polypeptides tested for filament polymerization *in vitro*.

Mixture	Filaments	α-type X-ray pattern	Yield	Polypeptide composition (molar ratio)
1 + 2	+	n.d.	15	1:2
1 + 3	+	+	25	1:2
1 + 4	+	+	30	1:2
1 + 5	+	+	95	1:2
1 + 6	+	n.d.	95	1:2
2 + 3	—			
2 + 4	—			
2 + 5	+	n.d.	15	1:2 to 2:1
3 + 4	—			
3 + 5	+	+	95	1:2 to 2:1
3 + 6	+	n.d.	95	1:2 to 2:1
4 + 5	+	+	15	1:2 to 2:1
4 + 6	+	+	35	1:2 to 2:1
5 + 6	—			

Mixtures containing up to 0.5 mg (in 0.2–0.5 ml of 8 M urea-0.1 M ammonium acetate buffer) of the two polypeptides were used. The presence of filaments was tested by electron microscopy. Filaments were pelleted from suspension to determine yields and for X-ray diffraction analysis. To determine the polypeptide compositions, different mixtures of the two polypeptides with relative amounts varying between 0.1 to 0.9 were prepared and the filaments were pelleted for polyacrylamide gel electrophoresis. The relative molar amounts of the two polypeptides in the filaments was then calculated after scanning the stained gels.

A notable feature of the filaments was the stoichiometry of their polypeptide composition (Table 1). For example, all mixtures of polypeptides la and 5 resulted in the formation of filaments containing one mole of polypeptide la and two moles of polypeptide 5, irrespective of the relative amounts of the polypeptides in the mixtures (Fig. 4). Similar results were obtained in all other combinations, including polypeptide 1 (or la or lb).

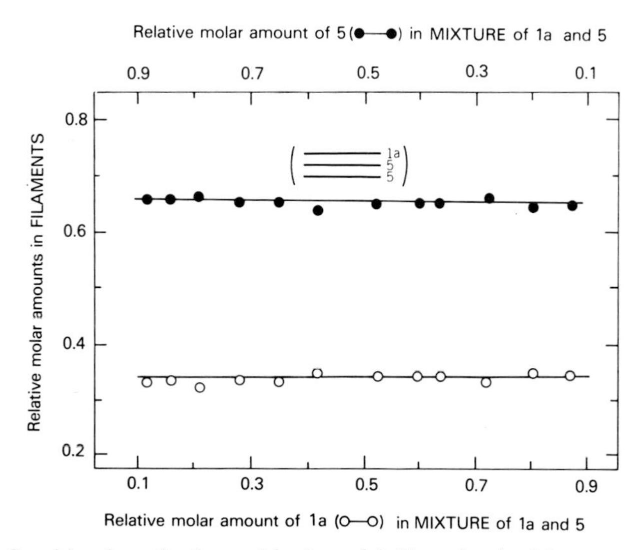

Fig. 4. Combination of polypeptides la and 5. Note that in this case, polypeptide 1a was used and not polypeptide 1, which contains 1a and 1b. Other details are as in Table 1. In parenthesis is shown the probable molar ratio of the two polypeptides in the filaments.

In mixtures of polypeptides 3 and 5, their molar ratios in the filaments varied between one extreme of 1:2 when one component in the mixture was in excess to 2:1 when the other was in excess (Fig. 5). The shapes of the curves imply that intermediate mixtures of these two polypeptides resulted in the formation of mixtures of filaments of these two extremes of composition. Several other combinations of two purified polypeptides also resulted in the formation of filaments of variable composition (Table 1).

The properties of filaments produced from a combination of polypeptides 3 and 5 are shown in Fig. 6. The filaments appeared as solid flexible rods, 7–8 nm wide, were very long (average > 10 μm) and displayed no visible ultrastructure (Fig. 6a). Their polypeptide composition (Fig. 6b) and α-type X-ray diffraction pattern (Fig. 6c) are also shown.

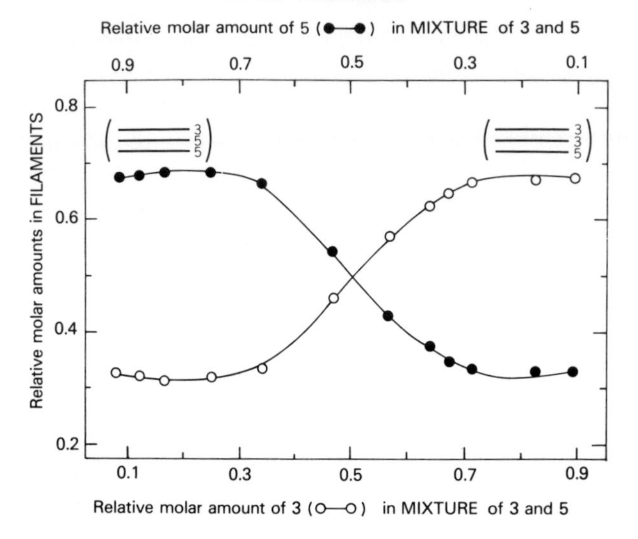

Fig. 5. Combination of polypeptides 3 and 5. The details are as in Table 1. The two probable extremes of molar ratios of the polypeptides in the filaments are shown in parenthesis.

Kinetic analysis of filament polymerization

In order to follow the kinetics of filament polymerization, changes in the turbidity ("light scattering") of solutions were monitored. The chosen wavelength was 300 nm to maximize sensitivity.

The kinetics of polymerization at 23° at different concentrations of polypeptides are shown in Fig. 7. In all cases, a sigmoidal curve was seen with time, and the time course of attainment of final polymerization was quite sensitive to polypeptide concentration. The shape of the curves is indicative of a multiphasic reaction: there is an initial rate-limiting step followed by a more rapid rate of polymerization.

When the final turbidity values at equilibrium polymerization (after 15 hr) were plotted against total polypeptide concentration (C_c), a straight line was observed which on extrapolation did not pass through the origin (Fig. 8). Below a concentration of about 50 μg/ml, the turbidity of the solution was the same as the buffer; that is, no filaments had formed. This concentration is called the critical concentration (C_c). At all concentrations above C_c, the turbidity increased and was linear with concentration (Fig. 8).

Such a phenomenon resembles a gas-liquid condensation. Dispersed monomers correspond to gas molecules and polymers correspond to liquid. When the number of gas molecules (density) increases gradually, at a critical density, a liquid phase begins to appear and with further increases in

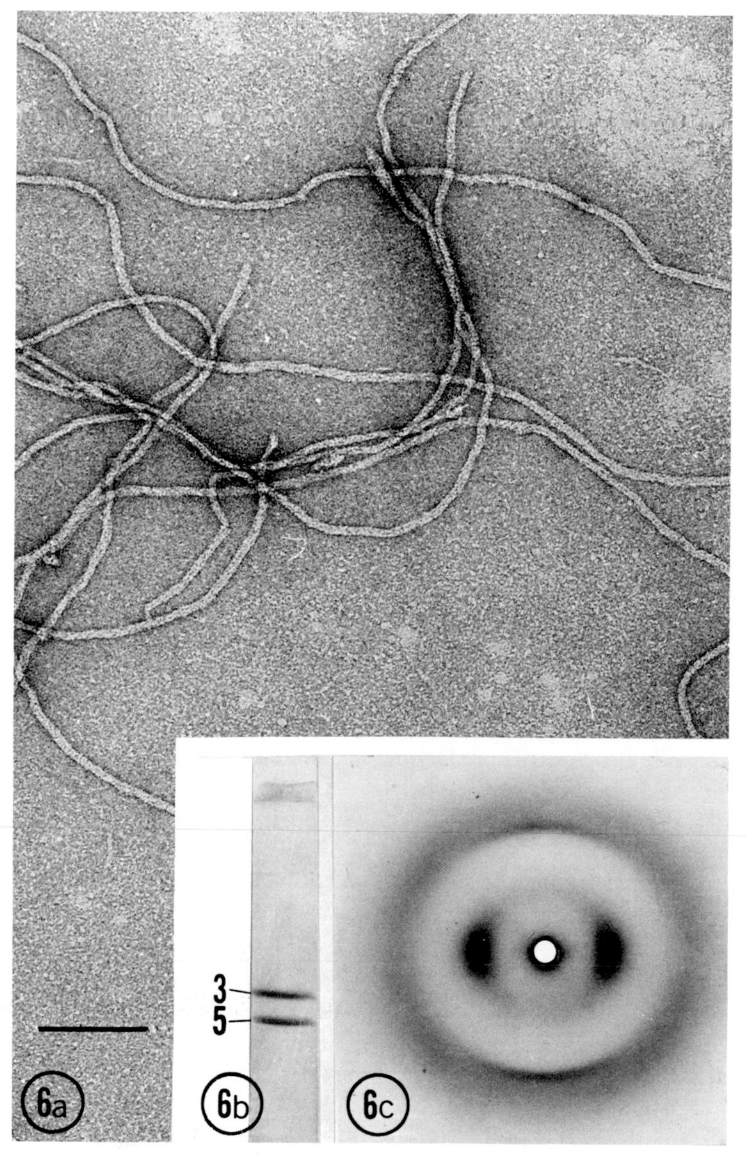

Fig. 6. Structure of filaments containing polypeptides 3 and 5, repolymerized from an approximately equimolar mixture of the two. a, filaments, × 137,300; the bar is 0.1 μm; b, a polyacrylamide gel of a sample of the pelleted filaments; c, an X-ray diffraction profile of a sample of these filaments.

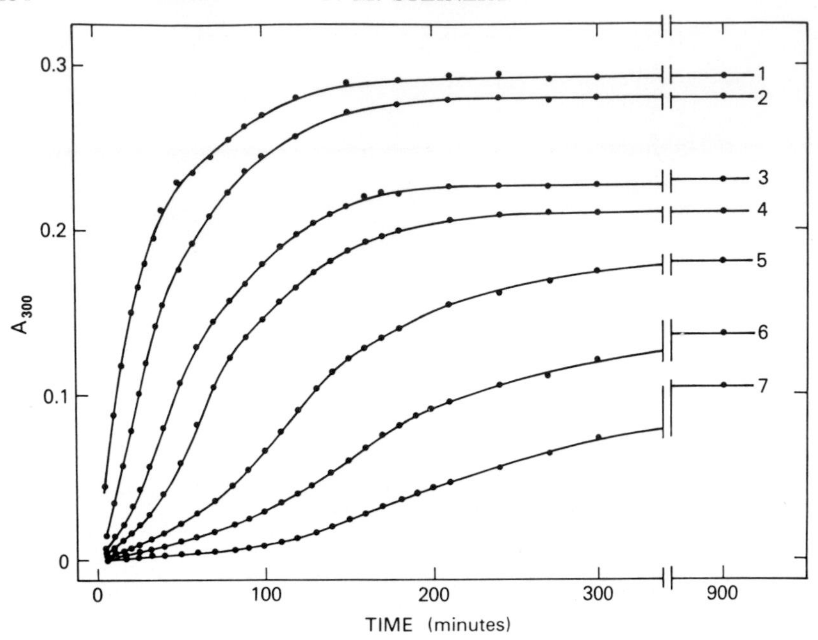

Fig. 7. The dependence of polypeptide concentration on the rate of polymerization into filaments, using repolymerized filaments. Before use, samples in urea buffer were heated at 50° for 10 min to effect complete dissociation to monomer polypeptides. The concentrations used were: 1, 1.31 mg/ml; 2, 1.20 mg/ml; 3, 1.11 mg/ml; 4, 1.05 mg/ml; 5, 0.97 mg/ml; 6, 0.90 mg/ml; and 7, 0.81 mg/ml.

the total number of molecules, the amount of liquid phase increases under a constant density of gas.

Accordingly, the observation of a critical concentration implies that the α-keratin polypeptides are in equilibrium with long polymers (filaments) at total concentrations above C_c. If this is so, it should be possible to pellet the filaments by centrifugation. This was done by the centrifugation assay ("Methods"). A plot of the weight of protein in the pellet, due to filaments, (C_p), versus C_0 gave essentially the same slope and intercept as the turbidity curves, and C_p/C_0 was constant (Fig. 8).

The occurrence of a critical concentration is the result of a slow or thermodynamically unfavorable initiation process for polymerization. Specifically, the behaviour demonstrated in Fig. 8 is expected for a self-nucleating polymerization mechanism.[29,30] That is, the aggregation or addition of the first few polypeptides to each other is much more difficult than subsequent additions or overall filament polymerization. An analysis

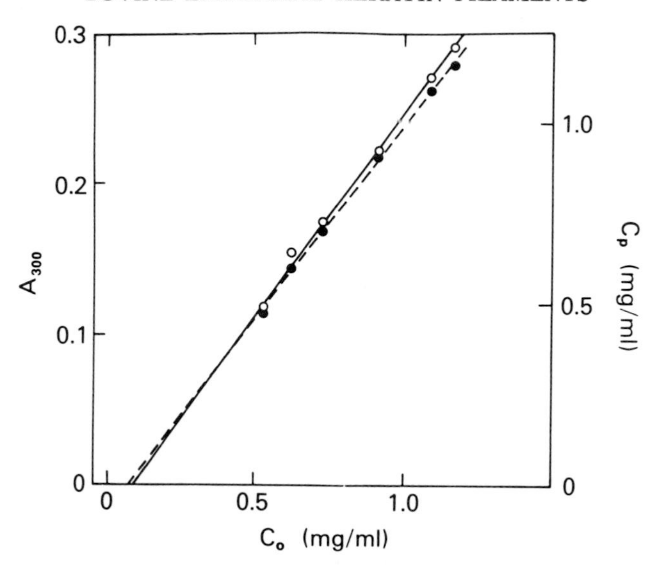

Fig. 8. The critical concentration of filament polymerization. This was determined from the final turbidity (equilibrium polymerization) values of Fig. 5 (–○–○–) and by pelleting the filaments in the same samples by the centrifugation assay (–●–●–). C_0 is the total polypeptide concentration; C_p is the concentration of the protein in the filaments (pellet). Extrapolation of the line to zero C_p reveals the critical concentration (C_c) value, about 50 μg/ml. The range over several different experiments was 40–70 μg/ml.

of the data of Fig. 8 provides information on the number of monomeric polypeptide chains involved in the nucleation process.

Theories for the polymerization of macromolecules have been elegantly developed by Oosawa[29,30] and will not be described here. For the self-nucleating "cooperative initiation" mechanism found:

$$a_1 + a_1 \underset{k_{-1}}{\overset{k_1}{\rightleftharpoons}} a_2 + a_1 \rightleftharpoons \ . \ . \ . \ n_{-1} + a_1 \overset{k_n}{\rightleftharpoons} n \rightarrow \text{filament}$$

where a_1 = monomer and n is the nuclear particle consisting of n monomers.

The initial rate of polymerization $(dC_p/dt)_{t=0}$ was shown[29] to be proportional to C_0^{n+1}. Determination of the initial polymerization rate is difficult because accurate measurements of the amounts of polymer formed at very early times are not possible. However, the half-polymerization time of filament assembly ($t_{1/2}$) is a function of both the initiation step and the subsequent propagation rate (filament growth) for which Oosawa[29] has also shown that $C_p \propto C_0^{-n/2}$. The $C_p(t_{1/2})$ values, being the times required to reach half the final equilibrium polymerization, are readily obtained from Fig. 7. By solving the simultaneous equations of $C_p(t_{1/2})$ at the different values of C_0, a value of $n = 5.7$—6.1 was obtained.

Therefore, the "nuclear particle," the formation of which is the rate-limiting step in the self-assembly of the epidermal keratin filaments, is probably a hexamer.

Electron microscopic analysis of filament polymerization

Figure 7 showed that equilibrium polymerization of filaments from polypeptide concentrations of about 1.0—1.3 mg/ml was obtained within 2–5 hr, and Figs. 1 and 6a showed the final structures of the assembled filaments. It was of interest to examine what types of particles are visible in the electron microscope at much earlier times, before attainment of equilibrium polymerization. Experiments similar to those of Fig. 7 were performed, and samples were withdrawn and stopped at various early times by application to and staining on grids. At a concentration of about 1.0 mg/ml, few particles were seen until about 20 min, whereupon large numbers of fine elongated particles became apparent (Fig. 9). The average $\pm$ s.d. dimensions of these particles ($N = 161$) is 41 $\pm$ 7 Å by 960 $\pm$ 150 Å. Very rapidly after the appearance of these particles, larger, that is, longer and wider, structures appeared. Attempts to analyze the order or manner by which larger structures formed were difficult because of the rapidity at which events occurred; long filaments of uniform width of 70–80 Å appeared within a few minutes. In addition, further aggregation at this stage was exquisitely sensitive to protein concentration; dilution of the aggregating particles to a concentration more suitable for electron microscopic examination resulted in a reduced rate or termination of further polymerization. Figure 10, however, reveals some information on the assembly phenomenon. At about 35 min, a wide variety of particles were present. Numerous structures 40 Å wide and about 2000 Å long (Fig. 10b) or 3000 Å long (Fig. 10c) were present, suggestive of simple end-to-end aggregation of the small particles. Elsewhere, the small particles appeared to associate in an approximately aligned manner to form an 80 Å-wide structure (Fig. 10d). In most cases, a mixture of structures encompassing these apparent modes of aggregation were present (Fig. 10a). By about 1 hr, the filamentous particles were of uniform width of 70–80 Å, showed no visible ultrastructure and were of variable but relatively short length (Fig. 11). By 2 hr, very long filaments were visible, as shown in Figs. 1 and 6a.

These electron microscopic observations support the kinetic data outlined above. The substantial delay in appearance of significant amounts of the 40 $\times$ 1000 Å particle indicate that its formation is slow or rate-limit-

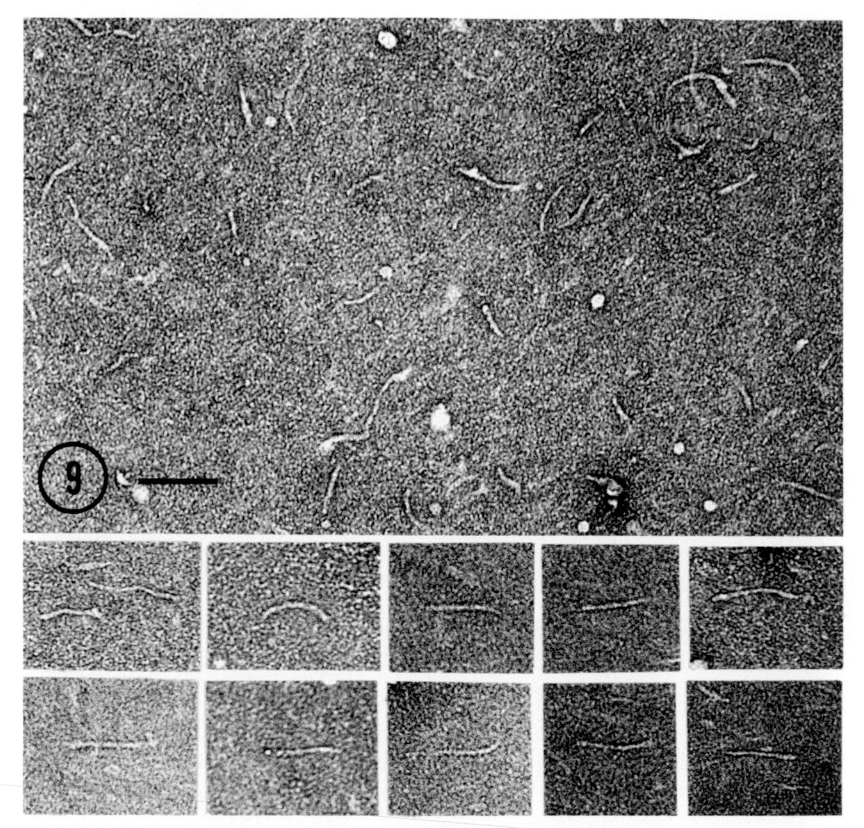

Fig. 9. Particles seen in the electron microscope at the earliest times of filament polymerization. In this experiment, C_0 was 0.96 mg/ml and $t = 25$ min after removal of the urea. ×105,000; the bar is 1000 Å (0.1 μm).

ing. Therefore, this may be the nuclear particle defined above from the kinetic experiments. Interestingly, it has dimensions similar to those reported for prekeratin in solution (37 ×1050 Å; ref. 3). Examination of prekeratin isolated by standard procedures from bovine hoof epidermis by electron microscopy showed particles that are about 30–50 Å wide and 700–1100 Å long (Fig. 12).

DISCUSSION

A basic assumption in this work is that the filaments polymerized *in vitro* from the α-keratin polypeptides have the same structure as the *in situ* epidermal keratin filaments. Three criteria support this view. Firstly, the filaments contained the polypeptides in the same relative molar amounts

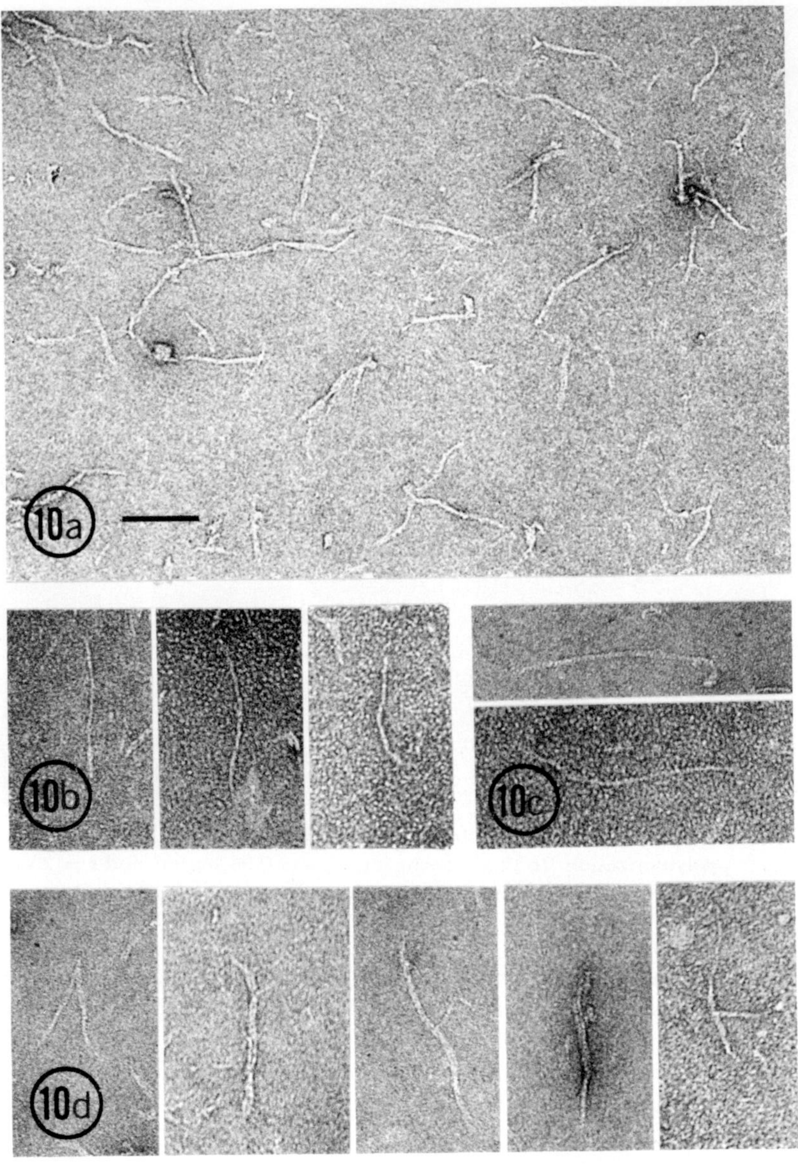

Fig. 10. Particles seen at later stages of filament polymerization. In a, $C_0 =$ 1.05 mg/ml; $t =$ 35 min. b–d, possible different simultaneous forms of aggregation. Particles which displayed structure clearly were collated from different experiments of about the same time during polymerization. b, elongated particles about 40 ×2000 Å; c, elongated particles about 40 ×3000 Å; d, branching, showing various dispositions of the small particles leading to the formation of the 80 Å-wide filament. In all cases, the magnification is about 100,000 ×; the bar is 1000 Å.

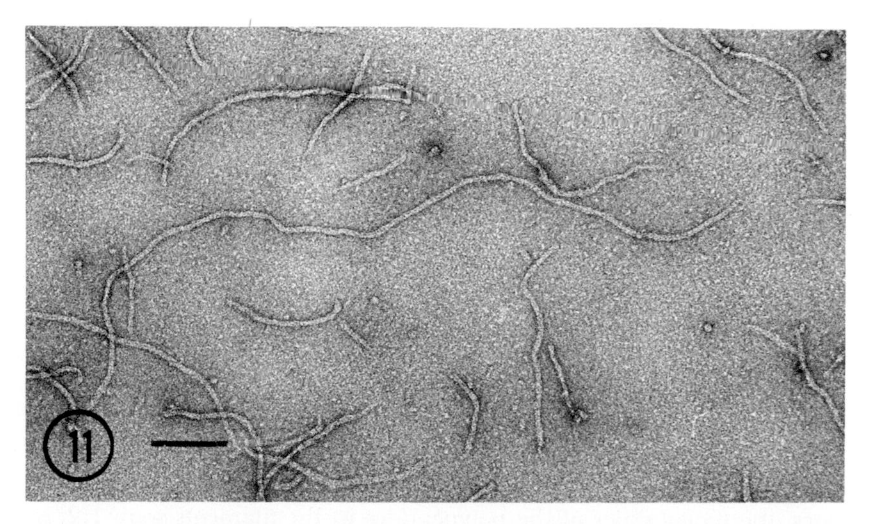

Fig. 11. Particles seen at an advanced stage of polymerization. $C_0 = 0.96$ mg/ml; $t = 55$ min. $\times 110{,}000$; the bar is 1000 Å.

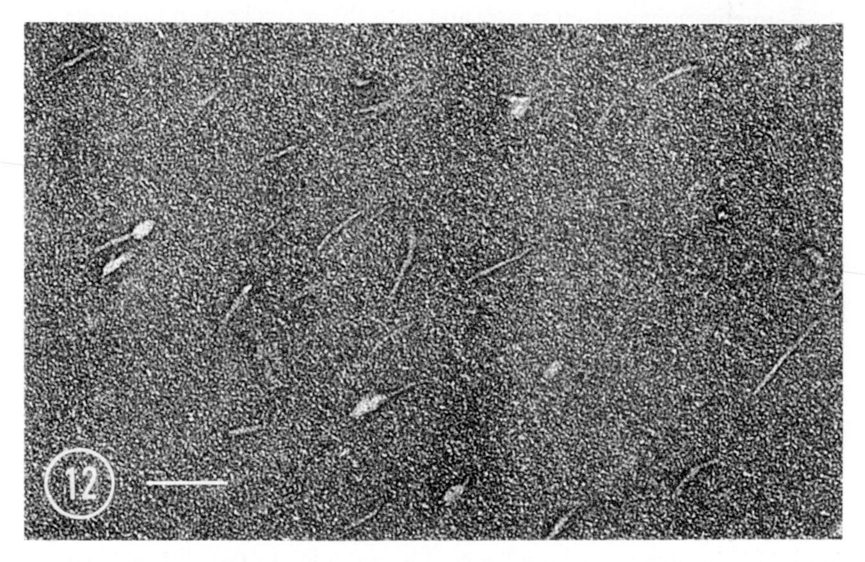

Fig. 12. Bovine epidermal prekeratin, prepared by the published method (3). $\times 96{,}000$; the bar is 1000 Å.

as the extracted α-keratin polypeptides, and moreover this composition was unchanged through two cycles of depolymerization and repolymerization *in vitro*. Thus the polypeptide composition is conserved by basic structural requirements. Secondly, the filaments gave the classic α-type X-ray diffraction pattern which is indicative of the presence within the filaments of segmented regions of triple-chained coiled-coil α-helix.[1,23–26] This pattern is diagnostic of α-keratin from mammalian epidermis and epidermal appendages. Thirdly, the filaments had similar physical dimensions as the *in situ* keratin filaments. The latter appeared in thin sections as solid rods 70–80 Å wide but their length is unknown, although they may be as long as the cell (up to 30 μm) or longer.[2] The polymerized filaments were also solid flexible rods, devoid of visible ultrastructure as seen after negative staining in the electron microscope, and were of uniform width of 70–80 Å and very long (range of 1–40 μm).

Most combinations of two of the purified α-keratin polypeptides also resulted in the formation of filaments *in vitro* and these had the same physical dimensions and gave the same X-ray diffraction patterns as the filaments polymerized from the unfractionated mixture of seven polypeptides. The most striking finding in these recombination experiments was the observed stoichiometry of polypeptide composition of the filaments. In all cases, the molar ratios of the polypeptides in the filaments were 1:2, irrespective of the amounts of the two polypeptides in the original mixtures.

The simplest model to account for these precise stoichiometries is that the filaments polymerized *in vitro* and thus the *in situ* keratin filaments are composed of a basic three-chained unit structure. The X-ray diffraction patterns which provide information on the physical ultrastructure of the filaments afford independent support for this hypothesis.

The concept of a basic three-chained unit structure for the keratin filament was initially proposed by Crick[25,26] and Pauling and Corey[27] from X-ray diffraction data. More recently, Skerrow[6] has suggested that prekeratin consists of a pair of triple-chained units. The presently observed polypeptide stoichiometry of the filaments polymerized from two electrophoretically homogeneous α-keratin polypeptides provides direct evidence for the existence of a triple-chained unit structure for the epidermal keratin filament.

The ultimate goal of these studies is the elucidation of the structure of the epidermal keratin filament. The approach outlined in this paper concerns a study of the mechanism of self-assembly of the keratin filaments *in vitro*. Such an experimental approach has been very useful in the determination of the structure of other fibrous proteins, such as actin[18,19] and tubulin.[20,21] In contrast, unlike collagen, for example, a more direct structural study at the electron microscope level is of limited value as the epidermal keratin filaments display no visible ultrastructure which might provide clues on the alignment of the polypeptides or of structural units within the filament. The kinetic experiments described demonstrated that the assembly of the α-keratin polypeptides is an orderly concentration- and time-dependent process involving an initial rate-limiting step followed by a more rapid rate of polymerization. The observation of a critical concentration below which filaments do not form, even at extended times, enabled classification of the type of mechanism, based on the previous theoretical considerations of Oosawa.[29,30] Specifically, the type of mechanism observed is a self-nucleating "cooperative initiation" process. The rate-limiting step involved the thermodynamically difficult formation of a "nuclear" particle, which is probably a structural intermediate. Subsequent filament assembly occurred more rapidly by either aggregation of the nuclear particles or by the further addition of polypeptide chains to these nuclei. An analysis of the half-polymerization time of filament polymerization as a function of the total polypeptide concentration revealed that the nuclear particle is a hexamer. In view of the evidence for a three-chained structural unit demonstrated above, it is possible that the nuclear particle consists of a pair of the triple-chained units. A schematic representation of the stages of filament polymerization consistent with these findings and possibilities is shown in Fig. 13.

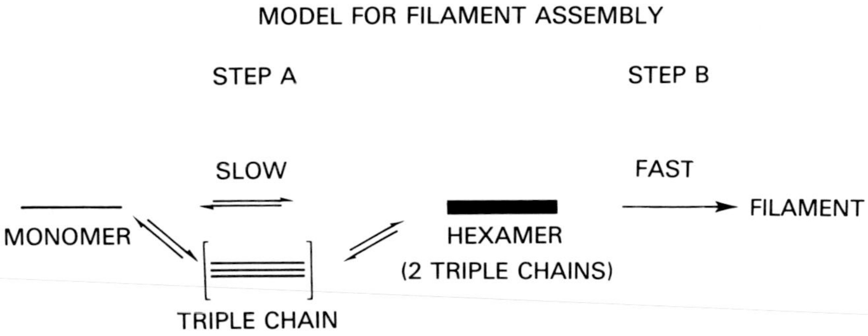

MODEL FOR FILAMENT ASSEMBLY

STEP A
STEP B
SLOW
FAST
MONOMER
FILAMENT
HEXAMER
(2 TRIPLE CHAINS)
TRIPLE CHAIN

Fig. 13. Schematic model of bovine epidermal keratin filament polymerization. This takes into account the results from the recombination and kinetic experiments. Two steps are envisaged; the slow rate-limiting stage, step A, leads to the formation of the hexamer nuclear particle; the formation of a triple-chained unit may be an intermediate step. The fast stage, step B, leads to the formation of the 80 Å-wide filament.

It is noteworthy that Skerrow[6] adduced evidence that prekeratin is a hexamer and it is possible that the nuclear particle defined here and prekeratin are equivalent. In support of this idea, the 40 × 1000 Å particle observed in the electron microscope at the earliest times of filament assembly is most likely the hexamer nuclear particle, and there is a marked similarity between its structure and that of prekeratin (Figs. 9 and 12). Therefore, the present findings are consistent with and support the view of Skerrow that a six-chained molecule may serve as a structural building block of the epidermal keratin filament.

The preliminary electron microscopic observations of the second or fast stage of filament polymerization provide some further information on the structural organization of the filament. Precise information is lacking at this time, primarily because of the rapidity at which the second stage occurs. A simple model to account for the particles seen in the electron microscope (Fig. 10) is that the 70–80 Å wide filament forms by aggregation of the nuclear particles rather than by addition of single polypeptide chains to the nuclei. The complete filament may be formed by end-to-end and side-by-side arrangement of the nuclei. It is not clear whether two or more nuclear particles are involved in forming the 70–80 Å-wide filament and whether they are arranged in a linear or twisted manner. From Fig. 11, it is possible that the eventual formation of very long filaments involves joining of shorter 70–80 Å-wide filaments.

The salient feature of this work is the presentation of a new approach to the study of the structure of the epidermal keratin filament. Clearly, only a broad outline of the important aspects of filament structure and ultrastructure has been possible, but the present findings provide useful bases for further experimentation.

Acknowledgements

This work would not have been possible without the continued support, encouragement, and assistance of Dr. Marvin Lutzner, Chief, Dermatology Branch, Dr. Gary Peck, Dr. Vincent Hearing, Dr. Marisa Gullino, Dr. Steven Zimmerman, and Mr. William Idler.

REFERENCES

1. Fraser, R. D. B., MacRae, T. M., and Rogers, G. E. The Keratins, Their Chemistry, Structure and Biosynthesis. C. C. Thomas, Springfield, Illinois, 1972.
2. Brody, I. The Ultrastructure of the tonofibrils in the keratinization process of normal human epidermis. *J. Ultrastruct. Res.*, **4**: 264–297, 1960.
3. Matoltsy, A. G. Soluble prekeratin. *In* Biology of the Skin and Hair Growth. (Lyne, A. G. and Short, B. F., eds.), Angus and Robertson, Sydney, Australia, 1965, pp. 291–305.
4. Matoltsy, A. G. Desmosomes, Filaments and keratohyalin granules: Their role in the stabilization and keratinization of the epidermis. *J. Invest. Dermatol.*, **65**: 127–142, 1975.
5. Skerrow, D., Matoltsy, A. G., and Matoltsy, M. N. Isolation and characterization of the α helical regions of epidermal prekeratin. *J. Biol. Chem.*, **248**: 4820–4826, 1973.
6. Skerrow, D. The structure of prekeratin. *Biochem. Biophys. Res. Commun.*, **59**: 1311–1316, 1974.
7. Crounse, R. G. Epidermal keratin and epidermal prekeratin. *Nature*, **211**: 1301–1302, 1966.

8. O'Donnell, I. J. The search for a simple keratin—The precursor keratin of cow's lip epidermis. *Aust. J. Biol. Sci.*, **24**: 1219–1234, 1971.
9. Baden, H. P. and Bonar, L. The α-fibrous protein of epidermis. *J. Invest. Dermatol.*, **51**: 178 483, 1968.
10. Tezuka, T. and Freedberg, I. M. Epidermal structural proteins. II. Isolation and purification of the tonofilaments of the new-born rat. *Biochim. Biophys. Acta*, **263**: 382–396, 1972.
11. Baden, H. P., Goldsmith, L. A., and Fleming, B. C. A comparative study of the physicochemical properties of human keratinized tissues. *Biochim. Biophys. Acta*, **322**: 269–278, 1973.
12. Shimizu, Y., Fukuyama, K., and Epstein, W. L. Partial purification of proteins isolated from mammalian cornified epidermis. *Biochim. Biophys. Acta*, **359**: 389–400, 1974.
13. Huang, L. Y., Stern, I. B., Claggett, J. A., and Chi, E. Y. Two polypeptide chain constituents of the major protein of the cornified layer of new-born rat epidermis. *Biochemistry*, **14**: 3573–3580, 1975.
14. Steinert, P. M. Extraction and characterization of bovine epidermal α-keratin. *Biochem. J.*, **149**: 39–48, 1975.
15. Steinert, P. M. and Idler, W. W. The Polypeptide composition of bovine epidermal α-keratin. *Biochem. J.*, **151**: 603–614, 1975.
16. Steinert, P. M. and Gullino, M. I. Bovine epidermal keratin filament assembly *in vitro*. *Biochem. Biophys. Res. Commun.*, **70**: 221–227, 1976.
17. Hodge, A. J. and Petruska, J. A. Recent studies with the electron microscope on ordered aggregates of the tropocollagen macro-molecule. *In* Aspects of Protein Structure (Ramachandran, G. N., ed.), Academic Press, New York, 1963, pp. 280–300.
18. Hanson, J., and Lowy, S. *J. Mol. Biol.*, **6**: 46–60, 1963.
19. Wegner, A. and Engel, J. Kinetics of the cooperative association of actin to actin filaments. *Biophys. Chem.*, **3**: 215–225, 1975.
20. Shelanski, M. L., Gaskin, F., and Cantor, C. R. Microtubule assembly in the absence of added nucleotides. *Proc. Natl. Acad. Sci., U.S.A.*, **70**: 765–768, 1973.
21. Gaskin, F., Cantor, C. R., and Shelanski, M. L. Turbidometric studies of the *in vitro* assembly and disassembly of porcine neurotubules. *J. Mol. Biol.*, **89**: 737–758, 1974.
22. Asakuri, S., Eguchi, G., and Iino, T. Unidirectional growth of *Salmonella* flagella *in vitro*. *J. Mol. Biol.*, **35**: 227–236; and: A kinetic study of *in vitro* polymerization of flagellin. *ibid*, 237–239, 1968.
23. Ugel, A. R. and Idler, W. W. Stratum granulosum: Dissection from cattle hoof epidermis. *J. Invest. Dermatol.*, **55**: 350–353, 1970.
24. Bramhall, S., Noack, N., Wu, M., and Lowenberg, J. R. A simple colorimetric method for the determination of protein. *Anal. Biochem.*, **31**: 146–148, 1969.
25. Crick, F. H. C. Is α-keratin a coiled-coil? *Nature*, **170**: 882–883, 1952.
26. Crick, F. H. C. The Fourier transform of a coiled-coil. *Acta Crystallogr.*, **6**: 685–688, 1953.
27. Pauling, L. and Corey, R. B. Compound Helical configurations of polypep-

tide chains: Structure of proteins of the α-helical type. *Nature*, **171**: 59–61, 1953.

28. Fraser, R. D. B., MacRae, T. M., Millward, G. R., Parry, D. A. D., Suzuki, E., and Tulloch, P. A. The Molecular Structure of Keratins. Applied Polymer Symp. No. 18, Wiley, New York, 1971, pp. 65–83.
29. Oosawa, F. and Kasai, M. A theory of linear and helical aggregation of macromolecules. *J. Mol. Biol.*, **4**: 10–21, 1962.
30. Oosawa, F. and Higashi, S. Statistical thermodynamics of polymerization and polymorphism of proteins. *Prog. Theoret. Biol.*, **1**: 79–164, 1967.

Discussion

Dr. Odland: I'd like to ask two small questions. Did all of these micrographs which you showed, use some negative staining methodology? Would you expect this to be the optimal technique for showing internal structure at this order of resolution?

Dr. Steinert: The answer to your first question is, that's correct. I had a problem doing this work. The filaments don't stain very well and I found that the only negative staining technique I could use was with uranyl acetate on carbon-coated grids. In other words, I do not think there is a better way.

Dr. Odland: Shadowing would be very crude, I suppose.

Dr. Steinert: I think it would be much less informative than the technique I used.

Dr. Odland: I just want this information for me. I looked at your X-ray diagrams and somewhere in the remote recesses of my educational history, I had the notion that you had to have things oriented and I didn't know that you could get anything more than a powder pattern out of filaments. Could you comment on this point?

Dr. Steinert: I did the X-ray diffraction work on a pellet of filaments. The fibers were drawn from the sedimented filaments. These were about 20–30 nanometers in diameter and were highly birefringent, indicating a high degree of orientation along the long axis of the fiber.

Dr. Matoltsy: One question has bothered us for several years. Does this chain which you are isolating have a helical and a non-helical segment? Did you get any information on the arrangement of the α-helical and non-helical regions in the filaments?

DR. STEINERT: This has not been studied to this time. The techniques described in my paper do indeed indicate that the α-helical segments are present within the filaments but do not provide any information on the orientation within the filaments.

DR. MATOLTSY: Do you have a speculation on the orientation of the peptides in the filament? Would they be parallel within the filaments?

DR. STEINERT: These proteins have an average molecular weight of around 50,000 to 55,000. From our optical rotatory dispersion measurements of the separated chains, there is a 45 to 50% content of α-helix. If you postulate that the peptide chains are fully extended then the length of the chain is approximately 900 Å. I would like to think, therefore, that in the hexamer of the hexamer particle which I have seen, the six chains are fully elongated and aligned exactly with each other. However, I am not sure of this and so am continuing ultrastructural studies to determine this fact.

DR. McGUIRE: I have never noticed it before in looking at the gels as you purify the material and then reapply it to the gels. There is always some heavy molecular weight bands that appear to run very slowly. They are not very prominent, and I wonder whether it is at all possible that you are co-purifying a nucleation factor or a higher molecular weight material that could serve as a nucleation factor in a manner analogous to microtubular assembly where heavy molecular weight molecules are important? A second question concerns whether you have looked at the hexamer. Have you sedimented the material after 30 minutes of polymerization before the long fibers appear when you only have the hexamer and have you applied that material to the analytic gel?

DR. STEINERT: I cannot exclude the presence of a nucleating factor such as occurs in microtubular assembly. What I can say is that the experimental condition required to polymerize these filaments is totally different from anything else so far defined, so I think it is unlikely. I have tried your proposed experiment and obtained fibers at a very low ionic strength, as low as about 0.005 ionic strength. At that concentration, actin, microtubulin, flagellan, collagen, and a whole range of other things just will not polymerize. Secondly, the pH is a very narrow range of 7 to 8. I think that the crucial point here is that in the case of tubulin, for example, the protein is never totally dissociated by SDS or urea. I would expect that a complicated nucleation factor such as occurs with tubulin would be irreversibly denatured and could not withstand such denaturation and renaturation. I do not know. I cannot exclude your suggested possibility. As for your second question, the early form is extremely unstable. Its formation is dependent on time and concentration. It exists in effect in solution and it exists at a concentration below 50 μg of protein per ml. If I were to sediment it and increase the concentration, I would imagine that it would "zipper-up" to a filament within a very short time, so I do not believe that there is an easy way for me to study this early particle.

DR. BADEN: The question I have is relative to the combination experiments involving the various chains in the X-ray diffraction patterns. I wonder whether you looked or measured these to see if any of these combinations gave you a cross-beta which superficially looks very much like the α-helix in that it has meridional reflections but the distances are slightly different?

DR. STEINERT: The answer to that question is yes, and the presence of the cross-beta is known because in all the experiments, I take a picture of the fiber first and then dust it with NaCl crystals to enable me to make a precise measurement. These figures are accurate within 0.03 Å.

DR. BERNSTEIN: I am not a mathematician but I intuitively feel that the concentration curve should not have a sharp initiation point if it took time for a few of the molecules to get together. I wonder if some kind of autocatalytic phenomenon is not involved here, and once you get that, then the thing "kicks off."

DR. STEINERT: It is in a sense an autocatalytic reaction, that is why I call it a self-nucleating reaction which is concentration dependent. At low concentrations it takes awhile for the monomers to get together.

DR. BERNSTEIN: But should not that initial "take-off" be a bend and not a sharp curve? If that were a time-dependent phenomenon?

DR. STEINERT: You only see the pronounced motile effect at the lower concentrations. Another point is the experimental design. The reason why the initially observed rate is slow is that formation of filaments which scatter light cannot occur until at least some "nuclear particle" has formed. The nuclear peptides themselves are too small to scatter light significantly.

The Histidine-rich, Matrix Proteins of Human Plantar Stratum Corneum

Tadashi TEZUKA

Department of Dermatology, School of Medicine, Kinki University, Osakafu, Japan

ABSTRACT

A high-speed supernatant fraction, which was extracted from the human plantar stratum corneum with Tris-DOC solution, was directly subjected to molecular sieve chromatography on a Sepharose 6 B. There were three peaks obtained and the second peak fraction was positively stained with Pauly reagent and also reacted with the antibody forming a single precipitin line, which had been produced in the purified material of keratohyalin granule origin in my previous studies. This second peak fraction was further subjected to Sephadex G-200 column chromatography and only a single peak was obtained. In contrast, a disc gel electrophoretic profile of this initial peak fraction of the Sephadex G-200 column in the presence of sodium dodecyl-sulfate showed that it was heterogeneous, and consisted of one slowly migrating protein band and two fast-migrating protein bands. The slowly migrating band was identical to the protein bands of the purified material of keratohyalin granule origin. When the initial peak fraction of the Sephadex G-200 column was fractionated by the preparative isoelectric focusing technique, three precipitated zones were obtained. Isoelectric points were 4.6 to 5.1, 7.1 and 12. Results of amino acid analyses of these fractions showed that the contents of histidine, aspartic acid, and glycine were very high, but the content of cystine was very low in all three fractions.

INTRODUCTION

There are matrix proteins in the stratum corneum which I believe play an important functional role. In many skin diseases, including certain types of ichthyosis and psoriasis, there is decreased formation of keratohyalin granules and the superficial layers of the epidermis are abnormal. A number of studies[1–7] suggest that keratohyalin granules are the major precursors of the matrix proteins of the stratum corneum. Approximately 40% of the protein of the stratum corneum can be solubilized with 50 mM Tris-HCl buffer, pH 8.6, containing 3% sodium deoxycholate [Tris-DOC], and I have previously presented data on the purification and characterization of the Tris-DOC soluble material.[6,7] When it was dialysed against redistilled water, a precipitate in the Tris-DOC extract was strongly positive with Pauly's reagent. The precipitated material was solubilized and subjected to chromatography on a Sephadex G-200. The Pauly positive initial peak was then chromatographed on a Sepharose 6 B column. During this step the Pauly positivity was lost, although the purified material was found by immunological criteria to be of keratohyalin granular origin. The purified material consisted of two major and one minor protein band by disc gel electrophoresis. The lack of Pauly positivity indicated that the histidine-rich proteins of the keratohyalin granules were not in the material which was purified. The purpose of the studies reported in this paper therefore was to identify and purify the Pauly-positive histidine-rich proteins from the high-speed supernatant fraction of the Tris-DOC soluble proteins of human plantar stratum corneum.

EXPERIMENTAL PROCEDURES

Materials

Human plantar skin was obtained following amputation from six patients with osteosarcoma. Sodium deoxycholate was purchased from Difco Laboratories, Detroit, Mich.; Sephadex G-200, Sepharose 6 B and dextran blue 2000 from Pharmacia Fine Chemicals Inc.,Fair Lawn, N.J., Ampholine pH 3.5–10 from LKB Produkter AB, Bromma, Sweden. Glass homogenizers were purchased from Kontes Glass Company, Vineland, N.J. and a Polytron was obtained from Kinematica, Steinhofhalde, Switzerland. Bovine serum albumin and ribonuclease-a from bovine pancreas were purchased from Sigma Chemical Company. All other chemicals were of reagent grade or analytic grade obtained from various commercial sources.

Preparation and extraction of tissue

Human plantar skin was excised immediately after surgery and cut into thin strips (approx. 7 × 1 cm) after removal of subcutaneous tissue. The strips were placed on a stretching apparatus under moderate tension and were sliced carefully with a razor blade to obtain sheets consisting only of stratum corneum.[6] The tissue was minced and homogenized in 50 mM Tris-HCl buffer, pH 8.6, containing 3% deoxycholate and stirred for 48 hours at 4°C. The extraction procedure was repeated three times and the extracting solution was changed every 48 hours. The homogenate was centrifuged at 15,000 $\times g$ for 10 minutes (Tominaga refrigerated automatic centrifuge), and the resulting pooled deoxycholate-supernatant fractions were recentrifuged at 270,000 $\times g$ for 60 minutes (Beckman Spinco Ultracentrifuge Model L–5–65). The lipid layer was discarded and the high-speed supernatant fraction was dialysed against 10 mM Tris-HCl buffer, pH 8.8, for 24 hours at 4° C, concentrated by lyophilization and stored at —80° C.

Histochemical studies

These studies were carried out on a variety of materials including normal plantar skin and materials which were precipitated in redistilled water or ethanol. The precipitates and tissue were sectioned in a cryostat at 6 μ and fixed in ethanol for 5 minutes. The sections were stained with diazotized sulfanilic acid[8] and Harris hematoxylin-eosin. Counter stains were not employed.

Gel filtration chromatography

Two milliliters of the concentrated high-speed supernatant fraction were subjected to gel filtration on a column (2.5 $\times$ 32 cm) of Sepharose 6 B. The profile was monitered at 280 nm during elution with 50 mM Tris-HCl buffer, pH 8.8, and, of the several peaks described below, the second peak was further fractionated on a Sephadex G-200 column (2.5 $\times$ 63 cm) using the same eluant. These studies were carried out at 4° C. For the histochemical studies described above, the three peak fractions from the Sepharose 6 B column chromatography were dialysed against 2 liters of redistilled water at 4° C for 72 hours. The dialysis fluid was changed every 24 hours.

Five ml of 95% ethanol were added to one ml of the Sepharose 6 B fractions and the samples were left at 4°C for 4 to 5 days. Aggregates which formed were harvested by centrifugation at 700 $\times g$ for 10 minutes.

Disc gel electrophoretic studies

Various fractions were subjected to disc gel electrophoresis in the presence of sodium dodecylsulfate using the technique of Weber and Osborn.[9]

Preparative isoelectro-focusing electrophoresis

Several milliliters of the concentrated, initial peak fraction of the Sephadex G-200 column chromatography were dialysed against 1% glycine redistilled water at 4° C overnight and mixed with sucrose and ampholine.[10] Electrophoresis was carried out overnight, at which time the current was less than 4 mA. The fractions were collected using an LKB Ultra Rac 7000 fraction collector.

Amino acid analysis

The three fractions from isoelectric focusing were hydrolysed in 6 N redistilled HCl at 100°C for 18 hours. Aliquots of the fractions were oxidized with performic acid prior to hydrolysis to permit detection of cystic acid residues. Amino acid analyses were performed on a Hitachi KLA-5 amino acid analyser.

Immunological studies

The antibody used in this study had been previously produced by injecting rabbits with purified protein fraction of keratohyalin granule origin.[6] For the Ouchterlony double diffusion technique[11] 1% ion-agar in 50 mM Tris-HCl buffer, pH 7.4, was used. The antigen-antibody reaction was developed in a humid chamber at 4°C for 48 hours. To remove nonreacted, diffusable protein, the slide was washed for 24 hours in physiological saline with several changes of the solution, and the washed slide was stained in amido black 10 B (100 mg of dye in 100 ml of 9:1 methanol-acetic acid solution). After immersion in the solution and rapid washing in water, the slide was air dried.

RESULTS AND DISCUSSION

Pauly-positive material was present in the concentrated high-speed supernatant fraction which could be extracted from human plantar stratum corneum, since the pellet which forms following dialysis of this high-speed supernatant fraction stained positively (Fig. 2-B). When this high-speed supernatant fraction was subjected to column chromatography on a Sepharose 6 B, three major peaks (A, B, and C) were obtained (Fig. 1). These peaks were either dialysed against redistilled water or precipitated in cold ethanol and the resulting precipitates were stained with Pauly's reagent. Only the second peak was Pauly positive (Fig. 2-C) when the water precipitates were studied, although the second and third peak fractions were positive when ethanol precipitates were used. The dialysis precipitates demonstrated the same tinctorial properties as those of both the stratum corneum and keratohyalin granules *in vivo*. These histochemical examinations showed that these precipitates were materials of keratohyalin granule origin and were transferred to stratum corneum since the granular cell was not contained in the original material.[6]

In order to determine in which fraction of these the protein of keratohyalin granule origin was present, immunological techniques were used. Precipitin reactions were run against the antibody which had previously been produced to the purified material of keratohyalin granule origin. A precipitin line was observed only between the antibody and the second

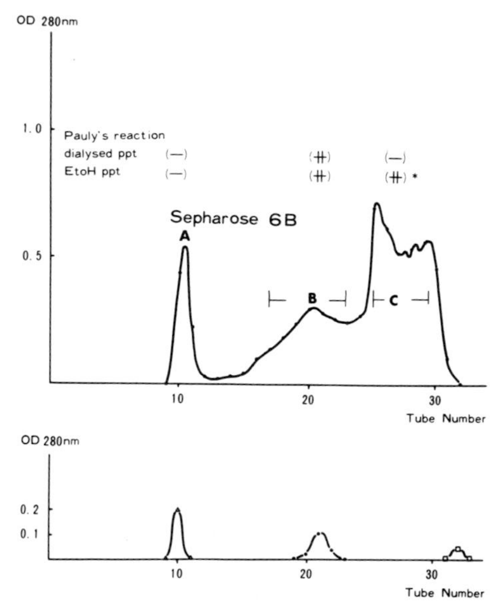

Fig. 1. Molecular sieve chromatography (Sepharose 6 B) of the high-speed supernatant fraction (upper figure). Elution from the column was monitored at 280 nm and the eluate was 50 mM Tris-HCl buffer pH 8.6. Standard proteins were subjected to the same column. △–△: blue dextran 2000; ●–●: bovine serum albumin; □–□: ribonuclease-a from bovine pancreas. Pauly's reaction was performed on the dialysed precipitates from three peak fractions (A, B, and C).[8]

peak (Fig. 3). This indicates that the material which was of keratohyalin granule origin and Pauly positive was in the second fraction. This second peak was further purified on a Sephadex G-200 column in order to eliminate the proteins of small molecular size. Only an initial large peak was obtained at the void volumn (Fig. 4). Disc gel electrophoresis in the presence of sodium dodecylsulfate showed that the initial peak fraction of the Sephadex G-200 column was heterogeneous (Fig. 5-B) and that it consisted of one slowly migrating protein band and two fast-migrating protein bands. The slowly migrating band was identical to the protein band of the purified material of keratohyalin granule origin (Fig. 5-A).[6] The molecular weights of these latter protein bands were estimated to be less than 30,000. In contrast, the second peak fraction of the Sepharose 6 B column was eluted at the same position as bovine serum albumin, which has a molecular weight of 68,000, and only a single peak was eluted when this material was further chromatographed on a Sephadex G-200. These observations suggest that the

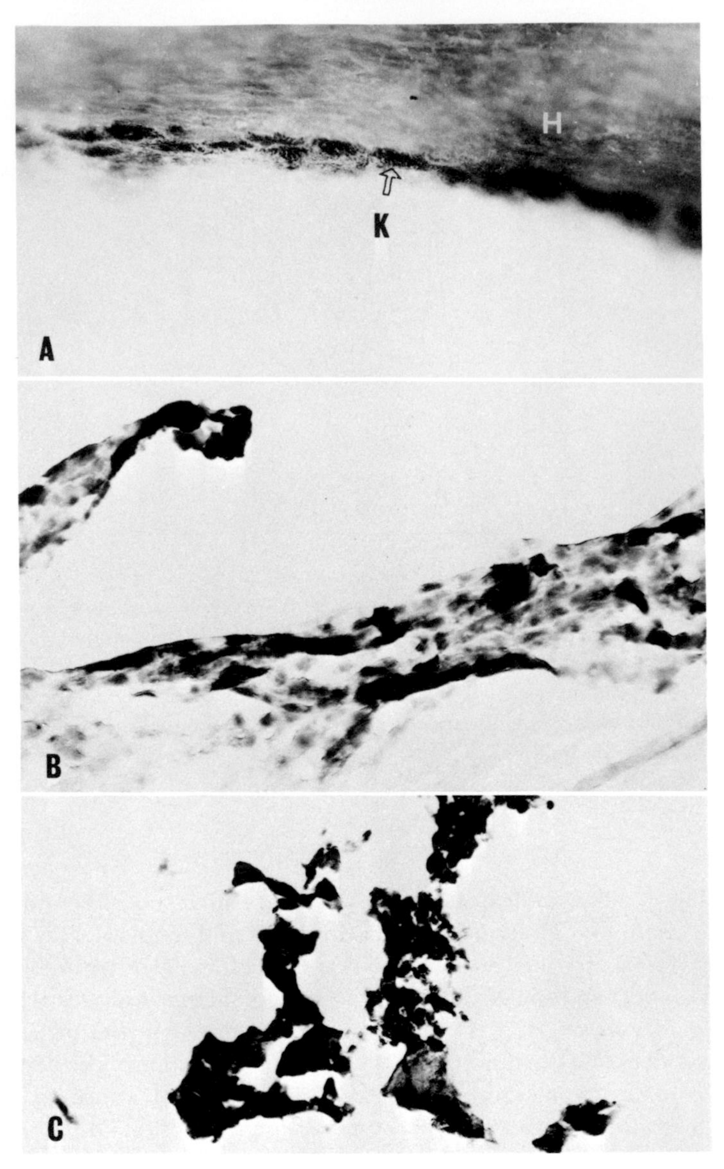

Fig. 2. Normal human plantar skin and the precipitates obtained by dialysis were stained with Pauly's reagent. Keratohyalin granules were stained deep orange yellow and the stratum corneum was stained light orange yellow (A). Both the precipitates from the high-speed supernatant fraction (B) and the second peak fraction of the Sepharose 6 B column were stained deep orange yellow (C), showing the same tinctorial characteristics as keratohyalin granules *in vivo* (A).

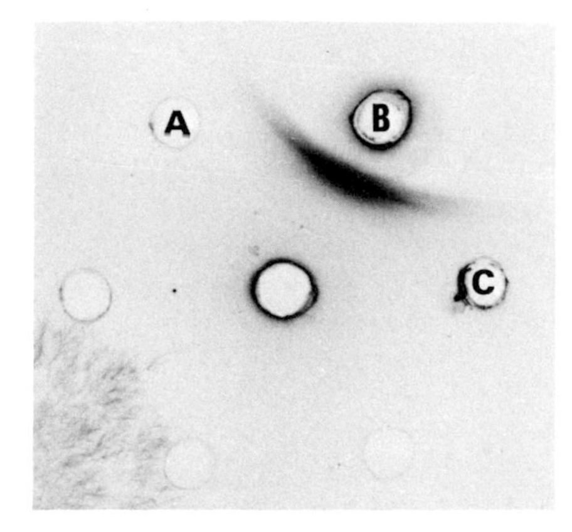

Fig. 3. Analysis by double diffusion in agar of the protein of the three peak fractions (A, B, and C) obtained from the column chromatography on the Sepharose 6 B (Fig. 1). The center well contains rabbit antiserum raised to the protein of the purified material of keratohyalin granule origin.[6] The peripheral well contains the materials of the initial (A), second (B), and third (C) peaks from the Sepharose 6 B column. A single precipitin line occurred between B well and the center well.

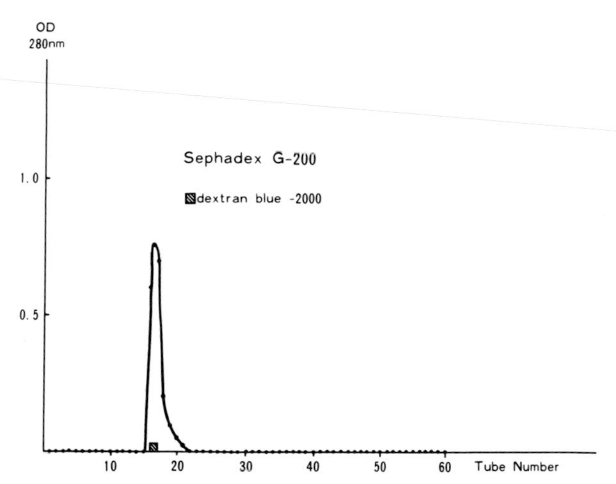

Fig. 4. Molecular sieve chromatography (Sephadex G-200) of the second peak fraction of the Sepharose 6 B column chromatography. The dialysed, lyophilized, resolubilized second peak fraction from the Sepharose 6 B column was rechromatographed on a Sephadex G-200 column. Optical density at 280 nm. Eluate was 50 mM Tris-HCl buffer, pH 8.6.

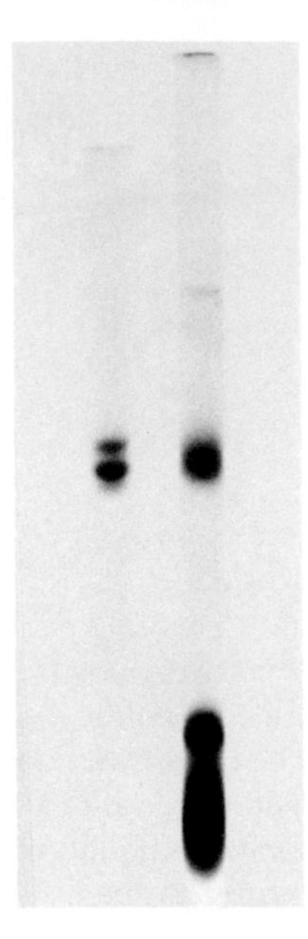

Fig. 5. Acrylamide gel electrophoretic patterns of the second peak fraction from the Sepharose 6 B column chromatography. The purified material of keratohyalin granule origin [6](50 μg) (A) and 100 μg of the second peak fraction (B) were analysed in the presence of sodium dodecylsulfate using acrylamide gels of 10%.

three protein bands were part of a single protein molecule which was dissociated in the presence of sodium dodecylsulfate.

In my previous studies[6,7] the protein of keratohyalin granule origin was separated from the histidine-rich proteins during the purification procedures with sodium deoxycholate. Detergents may cause this kind of dissociation.

When the initial peak fraction of the Sephadex G-200 column was further fractionated by the preparative isoelectric focusing technique, three precipitated zones (A, B, and C) were obtained (Fig. 6). The isoelectric points of these precipitates were 4.6 to 5.1, 7.1 and 12. Disc gel electrophoretic studies of these three fractions are under investigation. The results of amino acid analyses of these fractions are shown in Table 1. The fraction with an acidic isoelectric point contains large amounts of aspartic acid, glycine, serine, glutamic acid, histidine, and lysine. The neutral protein fraction contains a high content of histidine, aspartic acid, and glycine. The basic fraction has a very high content of lysine, aspartic acid, and glycine. The cystine content was very low in all fractions.

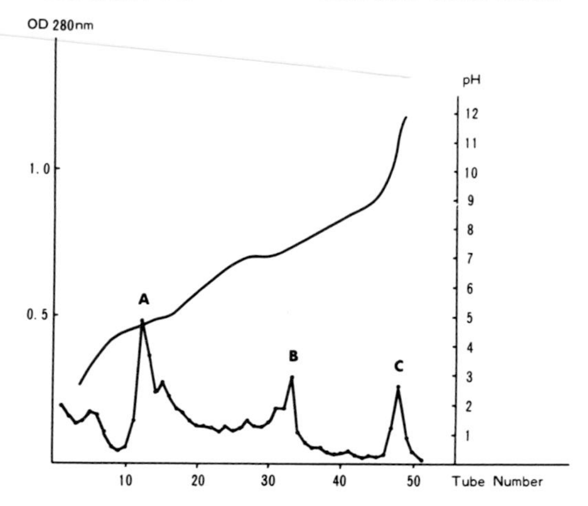

Fig. 6. Isoelectric focusing electrophoresis. The concentrated, initial peak fraction from the Sephadex G-200 column, which had previously been dialysed against 1% glycine redistilled water overnight, was subjected to the column. Electrophoresis was performed for 12 hours until the current became less than 4 mA. Optical density at 280 nm and pH values were measured.

Table 1. Amino acid analyses. The three fractions (A, B, and C in Fig. 6) were dialysed for 3 days at 4°C against 50 mM Tris-HCl buffer, pH 8.6, following dialysis against redistilled water, and were subjected to amino acid analyses using the techniques described in the text. Results are expressed as residues/100 residues. Hydrolysis was performed for 18 hours.

	I	II	III	*
Lys	9.9	4.0	18.9	5.5
His	10.5	11.2	5.6	1.2
Arg	2.8	—	0.8	5.9
Asp	15.6	10.7	17.7	10.1
Thr	5.6	—	2.4	4.4
Ser	12.8	4.5	8.8	9.2
Glu	12.1	2.9	4.4	15.2
Pro	—	—	—	—
Gly	13.3	61.4	26.1	19.1
Ala	—	—	7.2	6.7
Val	4.3	—	1.6	3.2
Met	—	—	—	1.8
Ile	4.2	5.4	3.6	3.3
Leu	5.6	—	2.8	7.6
Tyr	—	—	—	4.1
Phe	1.9	—	—	2.7
Cys	1.2	—	1.6	

*, human plantar matrix protein of keratohyalin granule origin

REFERENCES

1. Fukuyama, K., Nakamura, T., and Bernstein, I. A. Differential incorporation of amino acids in relation to epidermal keratinization in the newborn rat. *Anat. Rec.*, **152**: 525–536, 1965.
2. Fukuyama, K. and Epstein, W. L. Epidermal keratinization, localization of isotopically labelled amino acids. *J. Invest. Dermatol.*, **47**: 551–560, 1966.
3. Fukuyama, K. and Epstein, W. L. Ultrastructural autoradiographic studies of keratohyalin granule formation. *J. Invest. Dermatol.*, **49**:595–604, 1967.
4. Fukuyama, K., Buxman, M. M., and Epstein, W. L. The preferential extraction of keratohyalin granules and interfilamentous substances of the horny cell. *J. Invest. Dermatol.*, **51**: 355–364, 1968.
5. Tezuka, T. and Freedberg, I. M. Epidermal structural proteins. I. Isolation and purification of keratohyalin granules of the newborn rat. *Biochim. Biophys. Acta.*, **261**: 402–417, 1972.
6. Tezuka, T. The extraction and partial purification of the deoxycholate-soluble matrix protein from human plantar horny layers. *Acta Dermatovener.*, **55**: 401–412, 1975.
7. Tezuka, T. Electron microscopic, histochemical and disc gel electrophoretic studies on the deoxycholate soluble proteins from human plantar horny layers. *Tohoku J. Exp. Med.*, **11**: 209–221, 1976.
8. Reaven, E. P. and Cox, A. J. The histochemical localization of histidine in the human epidermis and its relationship to zinc binding. *J. Histochem. Cytochem.*, **11**: 782–790, 1963.
9. Weber, K. and Osborn, M. The reliability of molecular weight determinations by dodecyl sulfate polyacrylamide gel electrophoresis. *J. Biol. Chem.*, **244**: 4406–4412, 1969.
10. Instruction Manual [Ampholine 8100, electrofocusing Equipment] LKB-Produktor AB, Sweden, page 29–39.
11. Ouchterlony, O., Antigen-Antibody reaction in gels, IV. Types of reaction co-ordinated systems of diffusion. *Acta Path Microbiol. Scand.*, **32**: 231–240, 1953.

Discussion

Dr. Freedberg: What do you believe is the function of these materials which you have proven to be heterogeneous and to originate from the keratohyalin granule?

Dr. Tezuka: I am not certain of the function but I believe that the material which originates from the keratohyalin granule may play an important role in the barrier function of the epidermis or in the elasticity of epidermis.

Dr. Baden: Have you been able to find this material in the stratum corneum from other cutaneous surfaces such as the forearm or the back? Is it unique to the palm or is it seen in other areas of the body?

DR. TEZUKA: I am not sure, but I am doing the same experiment on newborn rat skin and I have found the same materials when I subjected the extracted material to SDS-gel electrophoresis. So I think these materials are not specific to the palm and sole

DR. GOLDSMITH: Does your antibody only react to the highest molecular weight component of those three main proteins?

DR. TEZUKA: In my previous experiment, I purified only the high molecular weight proteins. This was the material I injected to obtain antibodies.

DR. BERNSTEIN: This material which you are finding may be similar to the material that I referred to as "degradation product." Dr. Ball, in my laboratory, has shown that the material I call "degradation protein," having a molecular weight on the order of 10^4 daltons, is labeled in newborn rats about 24 hours after the initial labeling of the histidine-protein and is localized in the epidermal material remaining after all the viable cells are scraped off.

DR. TEZUKA: Well, I have another histidine-rich fraction which was not precipitated by dialysis against distilled water. This may be the material you mention, but I don't know.

The Structure of Epidermal Keratin

Howard P. Baden and Loretta D. Lee

Harvard Medical School, Department of Dermatology, Massachusetts General Hospital, Boston, Massachusetts, U.S.A.

ABSTRACT

The fibrous proteins of the malpighian layer of cow snout epidermis can be isolated with citrate buffer, pH 2.65, and consist of three major and one minor polypeptide chains. As these become transformed into stratum corneum proteins they require urea and urea and mercaptoethanol containing buffers for solubilization. Comparison of the polypeptides of prekeratin isolated from cow and pig epidermis indicates that there are significant interspecies differences. X-ray diffraction analysis and SDS polyacrylamide electrophoresis of formalin cross-linked cow prekeratin suggest that the molecule consists of three different polypeptides. Both monomer and dimer units have been detected.

Although keratins were among the first structural proteins to be studied our knowledge of them is still rather incomplete. As a consequence of their limited solubility they were more amenable to study by physical means, such as X-ray diffraction, than by chemical techniques. However, our understanding of most macromolecules has come about through detailed biochemical studies in which the molecules have been purified, their polypeptide components isolated and the number and sequence of their amino acid constituents identified. By combining such data with physical parameters and other chemical studies it has been possible to develop a more complete model of molecular structure.

The demonstration by Rudall[1] that fibrous proteins could be isolated from cow snout epidermis using buffers containing urea provided an opportunity to use the techniques of modern protein chemistry to study this keratin. Matoltsy[2] subsequently showed that the fibrous proteins could be isolated from the malpighian layer with citrate buffer, pH 2.65, and purified by isoelectric precipitation. More recent investigations[3,4,5,6] have been directed to identifying the polypeptide chains of the fibrous proteins and to elucidating their structure and composition. Polyacrylamide SDS electrophoresis has proved to be the most useful method for identifying the polypeptide chains. The initial studies of purified prekeratin (Table 1)

Table 1. Preparation of prekeratin.

1. Sliced epidermis is homogenized in a Virtis homogenizer in 0.25 M sucrose at 4°C and spun at 30,000 $\times g$ for 1/2 hour.

2. The pellet is homogenized and extracted in 0.1 M citrate buffer, pH 2.65, at 4°C and the suspension clarified by centrifugation at 30,000 $\times g$ for 1/2 hour.

3. Prekeratin protein is then purified by isoelectric precipitation at pH 7.0, 6.0, 5.0 and 4.5.

suggested the presence of 2 or 3 major chains,[4,7] but with improved electrophoretic techniques three principle components and at least one minor one have been identified (Fig. 1). These results clearly indicate that prekeratin is composed of different polypeptide chains rather than multiple single subunits as suggested earlier.[2]

Since the fibrous proteins of the malpighian layer are soluble in acid buffer while the stratum corneum proteins are not, it seems likely that pre-

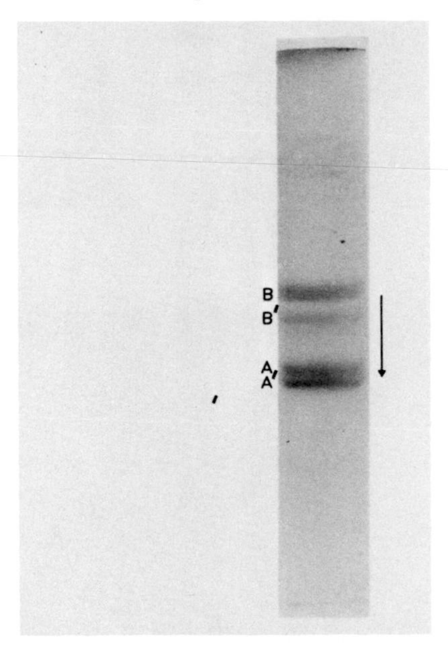

Fig. 1. SDS polyacrylamide electrophoresis of cow snout prekeratin. The molecular weight of A′ is 45,000, A 47,000, B′ 59,000 and B 67,000.

keratin is a precursor of the fully differentiated proteins of the cornified layer. 6 M urea containing 0.1 M Tris, pH 9.0 (Tris-urea) was used (Table 2) to isolate stratum corneum proteins[8] and these were found to be similar to prekeratin by SDS electrophoresis with one exception. The Tris-urea soluble proteins required treatment with a reducing agent prior to equilibration with SDS in order to resolve the polypeptide chains, but this was not true for prekeratin (Fig. 2). After exhaustive extraction of the stratum

Table 2. Preparation of stratum corneum proteins.

1. The pellet remaining after extraction with citrate buffer is rehomogenized and extracted with citrate buffer two additional times.

2. The pellet is homogenized and then extracted with 6 M urea in 0.1 M Tris, pH 9.0, under nitrogen for 18 hours. After centrifugation of the suspension at 30,000 $\times g$ for 1/2 hour, the pellet is rehomogenized and extracted 2 additional times.

3. The remaining pellet is extracted with 6 M urea and 0.1 M mercaptoethanol in 0.1 M Tris, pH 9.0, under nitrogen.

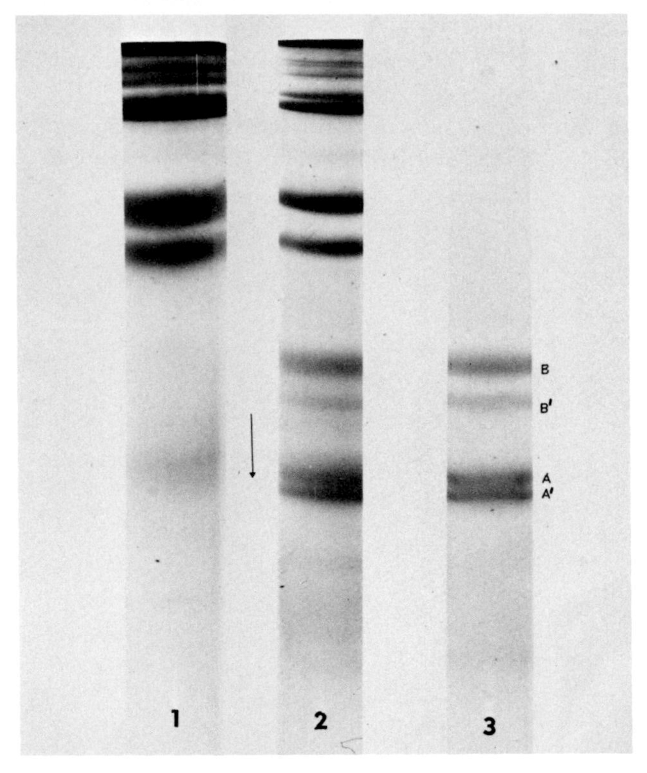

Fig. 2. SDS polyacrylamide electrophoresis of the crude urea-Tris extract[1,2] and prekeratin.[3] In 2 the specimen was reduced with mercaptoethanol before equilibration with SDS while in the others it was not.

corneum an additional component was solubilized with Tris-urea containing 0.1 M mercaptoethanol (Tris-urea-mercaptoethanol). This material had an SDS electrophoretic pattern similar to prekeratin (Fig. 3) but consistent differences in the relative intensities and mobilities of the bands were observed. Confirmation that the Tris-urea and the Tris-urea-mercaptoethanol soluble proteins were related to prekeratin was obtained using an antibody prepared to prekeratin. When the antibody was tested against the two urea extracts by the Ouchterlony technique, it gave precipitin lines identical to those observed for prekeratin (Fig. 4).

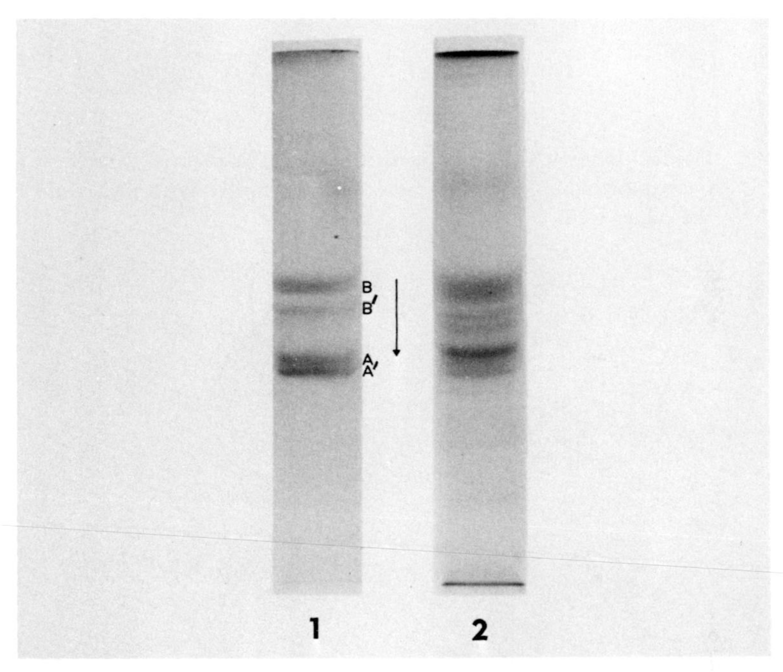

Fig. 3. SDS polyacrylamide electrophoresis of prekeratin[1] and the urea-Tris-mercaptoethanol extract (stratum corneum).[2]

These studies suggested that a series of modifications occurred in the fibrous proteins which altered their solubility and mobility in SDS electrophoresis. However, it was felt that an alternative method of studying the polypeptides was desirable to support this idea. This was accomplished by preparing S-carboxymethylated (SCM) and citraconylated (CC) prekeratin (Fig. 5) and doing disc polyacrylamide electrophoresis in the presence of 6 M urea. These modifications were necessary to open up the protein structure so that urea electrophoresis would give resolution of the polypeptide chains.

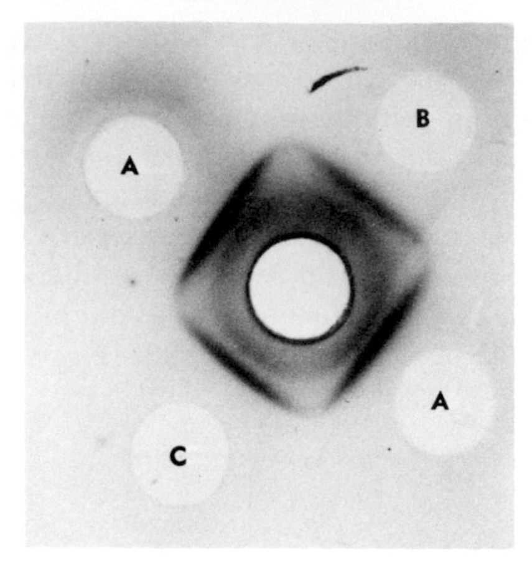

Fig. 4. Double diffusion with prekeratin (A) and Tris-urea (B) and Tris-urea-mercaptoethanol (C) soluble proteins. An antibody to prekeratin is in the center well.

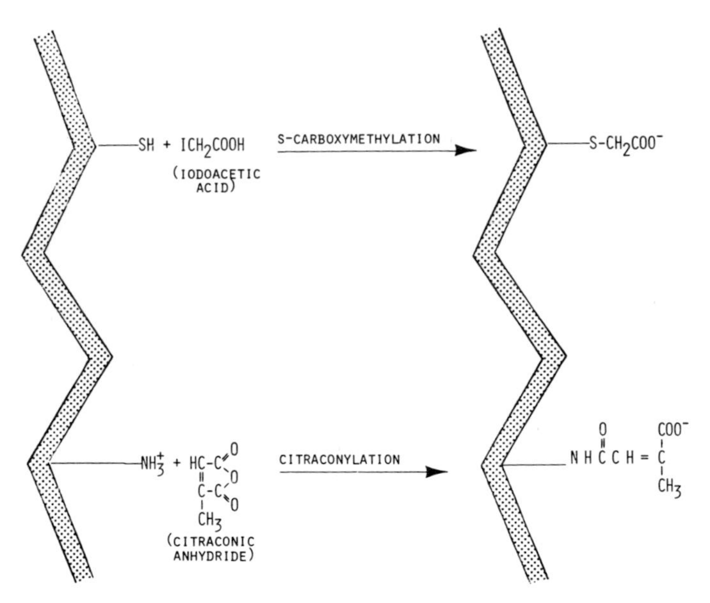

Fig. 5. S-carboxymethylation and citraconylation of a protein.

The electrophoretic pattern of SCM and CC prekeratin is shown in Fig. 6 and is compared with the SDS electrophoretic pattern of the native protein. The identity of the bands in the two patterns was established by doing these studies with purified components. The minor B′ chain in this system has the same charge to mass ratio as B while A and A′ migrate differently.

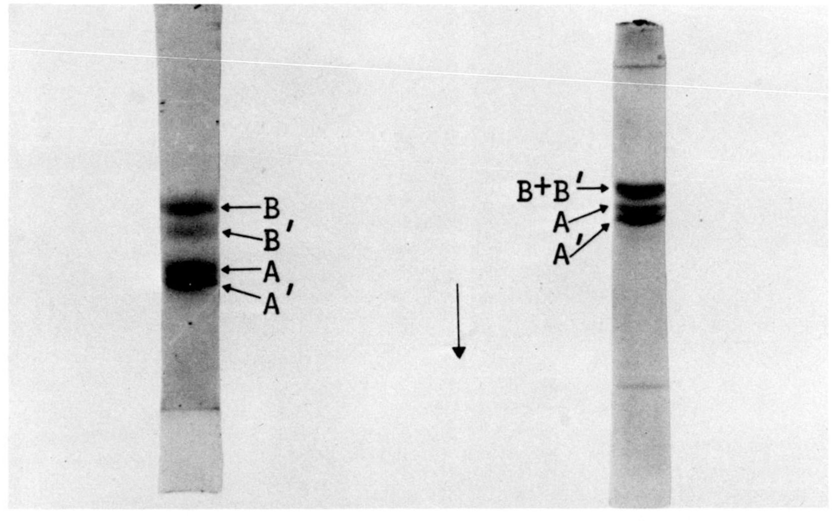

SDS electrophoresis of unmodified prekeratin

Urea electrophoresis of SCM-CC prekeratin

Fig. 6. Polyacrylamide disc electrophoresis of prekeratin. The native protein was run in SDS while the SCM and CC protein was run in 6 M urea.

When SCM and CC prekeratin and Tris-urea-mercaptoethanol soluble proteins are compared, the number of bands and their mobilities are found to be different (Fig. 7). Although immunological studies have shown that the A and A′ chains share a common antigenic determinant, it is apparent that they differ both in molecular weight and in their charge to mass ratio. The B and B′ chains, on the other hand, are similar both antigenically and in their charge to mass ratio.

In order to determine if the polypeptide composition of prekeratin was universal citrate buffer extracts were prepared from the epidermis of a number of animals. In most instances these reacted with the antibody to prekeratin. Differences in the SDS electrophoretic patterns were observed, but the small yield of solubilized proteins did not permit satisfactory purification. This problem was overcome by using pig snout epidermis where large amounts of tissue were available. The polyacrylamide SDS electrophoretic pattern of pig prekeratin (Fig. 8) is different from that of cow prekeratin but the proteins are immunologically identical. As in the case of cow snout epidermis, the SDS electrophoretic pattern of pig prekeratin and the Tris-urea-mercaptoethanol soluble protein show some differences (Fig. 9) and this is further substantiated by comparing urea polyacrylamide disc electrophoretic patterns of the SCM and CC proteins (Fig. 10). These

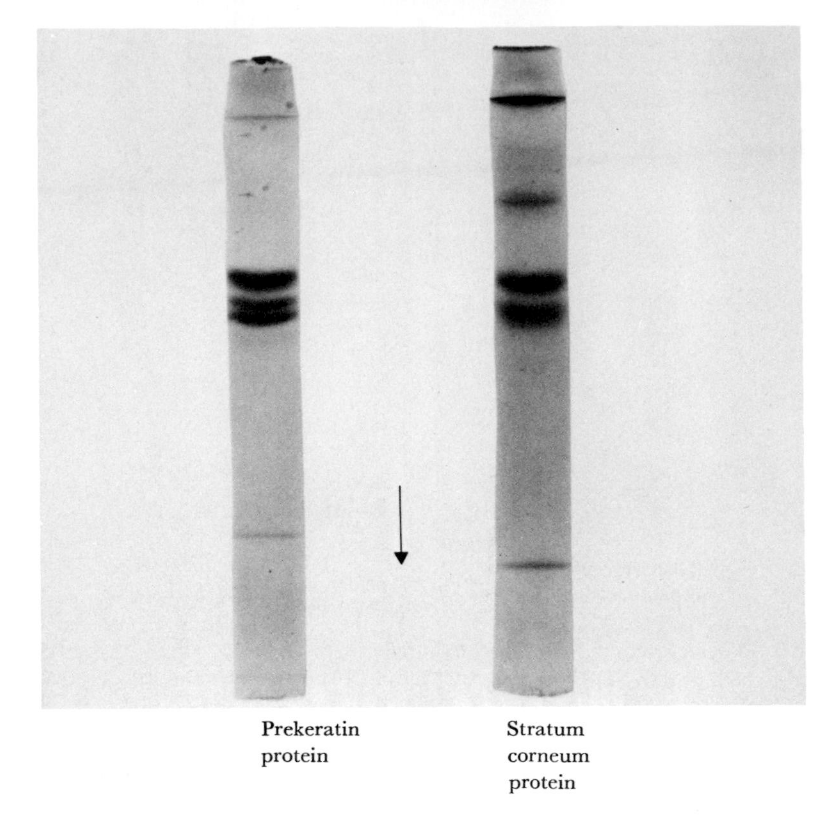

Prekeratin protein

Stratum corneum protein

Fig. 7. Polyacrylamide disc electrophoresis in 6 M urea of SCM and CC prekeratin and Tris-urea-mercaptoethanol soluble protein (stratum corneum).

results clearly indicate that there are significant interspecies variations in the polypeptide composition of epidermal fibrous proteins.

From the data we have seen on the polypeptide composition of cow and pig prekeratin it is clear that the number and ratio of bands would not fit a single molecule. In the case of cow prekeratin the ratio of bands suggests there may be a principle prekeratin with A, A′ and B chains and a minor prekeratin with A, A′ and B′ chains. This is supported by the finding that the separate A and B type chains would not give an α-fibrous protein by X-ray diffraction analysis but only a combination of them would (Table 3). It is of interest in this regard that minor collagens[1] and fibrinogens[10] also have been described. A triple-stranded helix model for prekeratin is also suggested by the results of X-ray diffraction studies of various keratins and the similarity of prekeratin and fibrinogen. This theory has been tested by experiments involving intramolecular cross-linking of prekeratin with formaldehyde (Fig. 11).[11] Although we have hypothesized that prekeratin

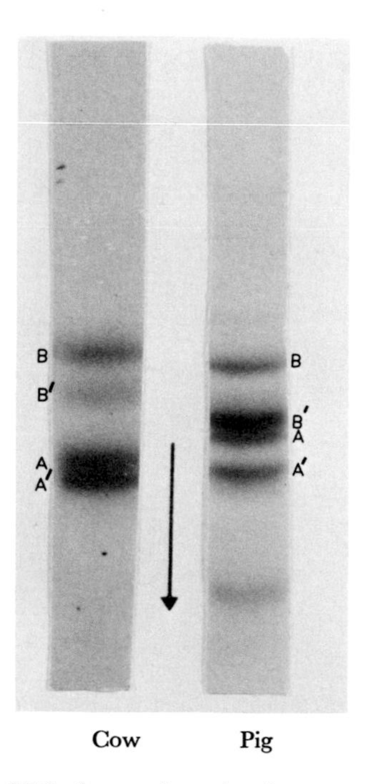

Fig. 8. Polyacrylamide SDS electrophoresis of cow snout and pig snout epidermal prekeratins. The molecular weight of pig A′ is 44,000, A 50,000, B′ 52,000 and B 63,000.

consists of A and B chains, an alternative possibility is that the individual chains form helical structures (Fig. 12). The SDS electrophoretic patterns of formalin cross-linked prekeratin that would result from either of these possibilities is depicted in Fig. 13. The observed pattern is shown in Fig. 14 and it is clear that the single monomer and dimer peaks fit best with the multichain hypothesis. The minor multichain component could not be discerned as a distinct peak because of its similarity in molecular weight to the major one. The molecular weights of the cross-linked components are 144,000 daltons for the monomer and 288,000 for the dimer, which are somewhat smaller than the 159,000 and 318,000 expected from simple addition. Because of the restriction imposed by cross-linking, unfolding and binding of SDS would be decreased and one would anticipate lower apparent molecular weights by SDS electrophoresis. The model for dimers shows the extra pieces of the B chains reacting with each other, but the molecules could also be aligned side by side. In the case of fibrinogen the dimerization results from interaction at the amino ends of the molecule and the extra pieces are free at the carboxyl end. However, proteolytic excision of peptides from the carboxyl ends occurs and this has not been found in the

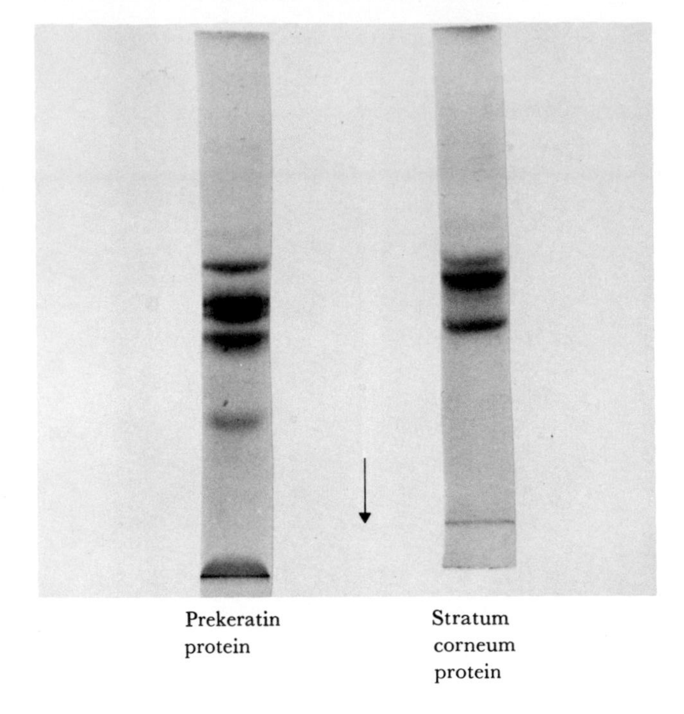

Fig. 9. Polyacrylamide SDS electrophoresis of pig prekeratin and Tris-urea-mercaptoethanol soluble protein.

Table 3. X-Ray diffraction data

Sample	Presence of α Pattern
Prekeratin	Yes
A	No
B	No
A + A′	No
B + B′	No
A + B	No
2A + B	Yes
A + A′ + B	Yes
2A′ + B	Not done

case of the keratins. How the polypeptides are arranged in supermolecular aggregations is of great interest and will follow from an increased understanding of the prekeratin molecule.

Our model for cow prekeratin is the simplest that could be derived since it assumes that an A and A′ chain are present in each molecule. Since we have found that 2 A chains plus a B chain can also give an α-fibrous X-ray diffraction pattern it is also possible that 2 A′ chains plus a B chain could

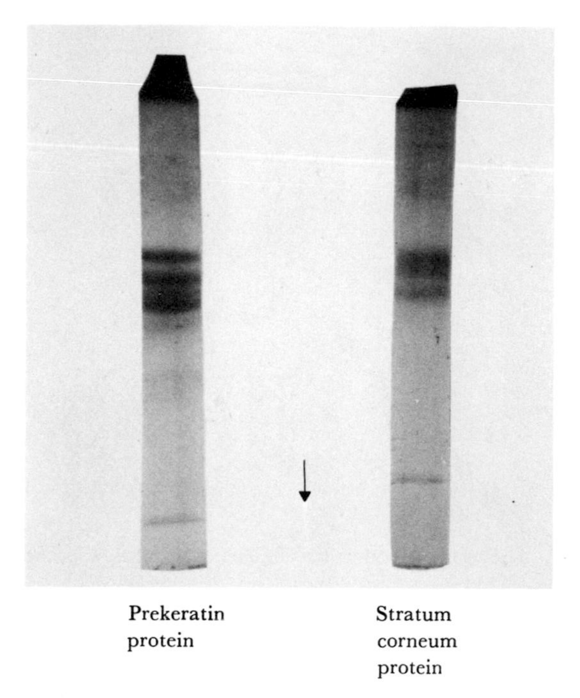

Fig. 10. Polyacrylamide disc electrophoresis in 6 M urea of SCM and CC pig prekeratin and Tris-urea-mercaptoethanol soluble protein (stratum corneum).

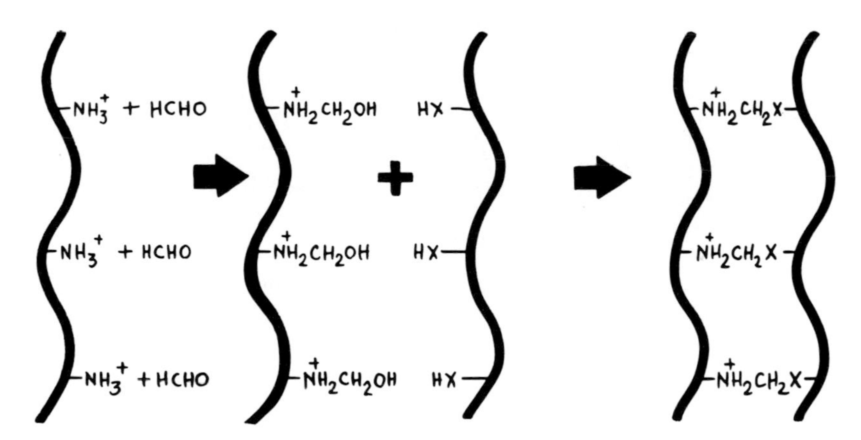

Fig. 11. Formaldehyde cross-linking of protein. Prekeratin was dialyzed against 0.1 M acetic acid and formaldehyde added to a final concentration of 6%. After incubation at 4°C for varying times the solution was dialyzed to remove free formaldehyde.

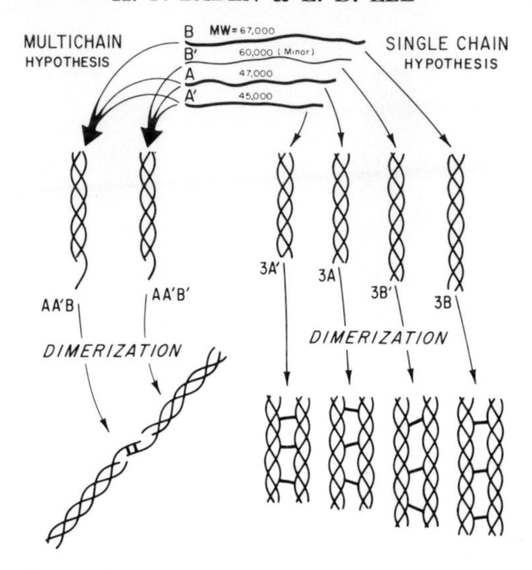

Fig. 12. Theoretical models for the structure of cow prekeratin.

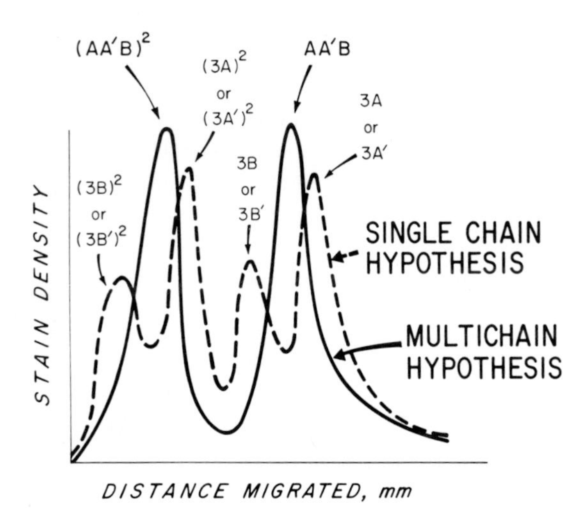

Fig. 13. Predicted densitometer tracings of the SDS polyacrylamide electrophoretic patterns of formalin cross-linked prekeratin. Only the monomer and dimer units are shown. The superscript 2 indicates the dimer.

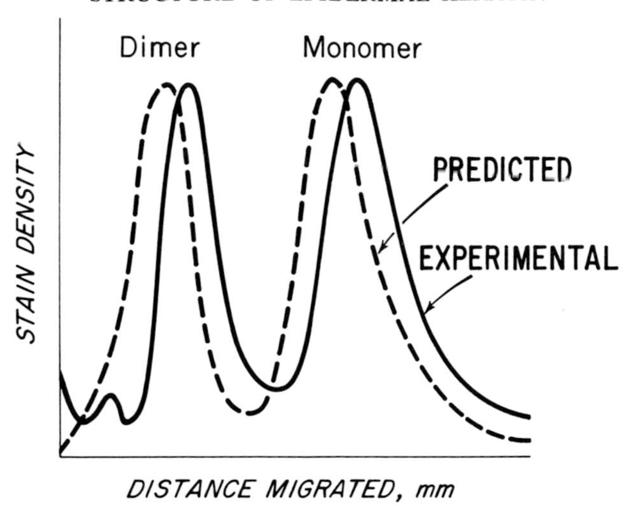

Fig. 14. Densitometer tracings of the SDS polyacrylamide electrophoretic pattern of formalin cross-linked prekeratin. The predicted curve for a multichain structure, assuming a molecular weight of 159,000 for the monomer, is also shown. Some highly aggregated material was seen at the top of the gel.

do this. Thus we could have AAB, AA′B, A′A′B, AAB′, AA′B′ and A′A′B′ molecules. This problem would be further complicated if microheterogeneity existed which was not detected by our electrophoretic techniques. For the present our model should be considered as an average structure until further data is available. The differences in the polypeptide chains within a species and between species may not be as great as they appear from the electrophoretic analyses. There may be very similar helical cores, as described by Skerrow,[12] in the polypeptide chains which would explain the similarity in immunological responsiveness despite differences in size and charge. This concept can be tested by comparative studies of the helical segments of the polypeptide chains of various prekeratins.

REFERENCES

1. Rudall, K. M. The proteins of the mammalian epidermis. In Advances in Protein Chemistry, vol. 7 (M. L. Anson, K. Bailey, and J. T. Edsall, eds.) Academic Press, New York, 1952, pp. 253–290.
2. Matoltsy, A. G., Soluble prekeratin. In Biology of Skin and Hair Growth (A. G. Lyne and B. F. Short, eds.), Angus and Robertson, Sydney, 1965, pp. 291–305.
3. Lee, L. D., Fleming, B. C., Waitkus, R. F., and Baden, H. P. Isolation of the

polypeptide chains of prekeratin. *Biochim. Biophys. Acta*, **412**: 82–90, 1975.
4. Skerrow, D. The structure of prekeratin. *Biochem. Biophys. Res. Commun.*, **59**: 1311–1316, 1974.
5. Steinert, P. M. The extraction and characterization of bovine epidermal α-keratin. *Biochem. J.*, **149**: 39–48, 1975.
6. O'Donnell, I. J. A search for a simple keratin—the precursor keratin proteins from cow's lip epidermis. *Aust. J. Biol. Sci.*, **24**: 1219–1234, 1971.
7. Baden, H. P., Goldsmith, L. A., and Fleming B. The polypeptide composition of epidermal prekeratin. *Biochim. Biophys. Acta*, **137**: 303–311, 1973.
8. Baden, H. P., Bonar, L. C., and Katz, E. The fibrous proteins of epidermis, *J. Invest. Derm.*, **50**: 391, 1968.
9. Epstein, E. H., Jr. and Munderloh, N. H. Isolation and characterization of CNBr peptides of human $[\alpha 1(III)]_3$ collagen and tissue distribution of $[\alpha 1(I)]_2 \alpha 2$ and $[\alpha 1(III)]_3$ collagens. *J. Biol. Chem.*, **250**: 9304–9312, 1975.
10. Doolittle, R. F. Structural aspects of the fibrinogen to fibrin conversion. In Advances in Protein Chemistry. vol. 27 (C. B. Anfinsen, J. T. Edsall, and F. M. Richards, eds.) Academic Press, New York, 1973, pp. 1–109.
11. Lee, L. D. and Baden, H. P. Organization of the polypeptide chains in keratin, *Nature*, 1976, in press.
12. Skerrow, D. A repeating subunit of soluble prekeratin. *Biochim. Biophys. Acta*, **257**: 398–403, 1972.

Discussion

Dr. Steinert: I am impressed by the similarity in the results of our independent experiments and I am quite pleased about it. It gives me great confidence in my own work. One thing about which I am not sure, is the basis for your hypothesis that you form this triple-chain structure.

Dr. Baden: The reason we proposed this structure is the similarity to the data on the structure of fibrinogen. In our cross-linking experiments, we found both monomers (3 chains) and dimers (6 chains).

Dr. Steinert: From your cross-linking experiments with formaldehyde, can you exclude the possibility that single chains in solution randomly cross-linked rather than that only those chains which were structurally related, associated with each other?

Dr. Baden: We found that single (unassociated) chains did not cross-link in solution.

Dr. Steinert: So, if I understand you correctly, you think from these cross-linking experiments a single purified chain can form a triple or "6-chain" molecule, is that correct?

DR. BADEN: No, we were not able to demonstrate that this chain alone could form a helix. Your data showed a certain percent of helix and I think it is possible that under the right conditions you may get a percent of helix which may be an association between pieces of a chain but that may not be the most desirable configuration involved here. I cannot answer precisely until I have a chance to digest the data about how much of the helix you obtained when you mixed one or the other. In our hands one does not obtain ''observable'' helices unless one has both chains.

DR. STEINERT: What I didn't mention in my talk is that my purified chains will not form filaments by themselves either.

DR. BADEN: Ours will not either. It takes a combination to do that.

DR. MATOLTSY: I am very impressed, if I understand you correctly, that you have here one chain which didn't give you an α-X-ray diffraction pattern. Is it possible that this chain is something else, for example, a contaminant or something or do we have here in this prekeratin, or whatever we call it, an extra chain which is not α-helical?

DR. BADEN: No, I think that we have immunologic data which account for all the chains. A single chain, A alone, or A′alone does not do it. B does not give it to you. AA′B is the main molecule and AA′B′ is the minor molecule.

DR. MATOLTSY: Dr. Steinert, did not you get an actual pattern for each of your chains?

DR. STEINERT: No, neither of our proteins gave a fiber or X-ray pattern by itself.

DR. BADEN: We found that A and B chains alone did not give the α-pattern, you needed both.

DR. ODLAND: Can either you or Dr. Steinert speculate on the meaning of that?

DR. BADEN: I think that molecules are programmed to suppress stability and I think that the complementation that occurs when these bands come together lend them stability which allows you to get this pattern. I am sure that there are thermodynamic explanations for how this occurs, but I do not know what they are.

DR. GOLDSMITH: On one of your slides of the urea-Tris-soluble fraction of stratum corneum, even after reduction, there were still very large amounts of high molecular weight material. Would you like to comment on what this is?

DR. BADEN: We have no data on the nature of the high molecular weight material.

Solubilization of Hard Keratins

Morizo Tsuda, Tadao Ogawa, Toshio Ono and Yasuhiro Kawanishi

Department of Biochemistry and Biophysics, Faculty of Hygienic Sciences, Kitasato University, Tokyo, Japan

SUMMARY

For a chemical and physical investigation of keratin structure, an approach to a new method of the solubilization of keratins should be developed. We found that treatment with trimethylalkylammonium hydroxide, especially tetramethyl-ammonium hydroxide (TMAH) solution, was very suitable for the purpose.

Human scalp hair and merino wool were homogenized in 10 percent aqueous TMAH solution with a Potter homogenizer for 20 minutes under ice-water cooling. After centrifugation at 20,000 $\times g$ at 4°C, a brownish yellow solution was obtained. Its pH was immediately lowered to about 10. Samples were solubilized up to 80% during the two successive treatments.

To elucidate how keratins were solubilized by TMAH, disc electrophoresis and N-terminal analysis by the DNP method were carried out on the solubilizates. The fact that newly-produced N-terminal amino acids were scarcely detected indicates little destructive change at the peptide bond.

Solubilizates from human hair and merino wool were eluted through a Sephadex G-200 column using 50 mM borate buffer (pH 8.8). They were fractionated into three fractions, I, II, and III. Fraction I contains about 50% of the solubilizate. Sedimentation velocity measurements of fraction I, from any sample, revealed a single and somewhat broad peak centering at about 7S. Comparative studies were also made of human nail, bovine hoof and horn, and pigeon feather.

Keratineous substances are readily solubilized in 10% TMAH solution with negligible degradation, giving a possibility for use in comparative studies.

INTRODUCTION

The word "hard keratins," which has been used in a rather unexact manner, means in the present paper proteins that exist in tissues of ectodermic origin, such as hair, wool, horn, and nail. They are characterized by a high content of disulfide cross-links and by considerably strong resistance against usual solubilizers such as solutions of neutral salts, detergents, dilute alkali hydroxide, and some acids and proteinases. Consequently, the solubilization of hard keratins cannot be achieved unless it is accompanied by cleavage of disulfide bonds due to oxidation[1,2] or reduction followed by carboxymethylation of sulfhydryls thereby formed.[3,4] Progress in appropriate use of S-carboxymethylated (SCM) keratins for protection of sulfhydryl groups from their recombination, has made it possible to proceed in comparative studies of hair keratins from various species[5] and of keratins from various tissues of human beings.[6] Much information is now available for clarifying the structure of SCM-fragments of keratins. No information, however, has been practically available for the techniques to determine the interfragmental sequence. This could be achieved by the use of another solubilizer that disintegrates keratin molecules to produce disulfide bonds containing fragments. An attempt has thus been made to find a solubilizer suited for this purpose.

It has been reported that anhydrous hydrazine can readily solubilize many "insoluble" proteins such as bone insoluble collagen.[7] Some amines and trimethylalkylammonium ions penetrate readily into animal tissues and often disintegrate the tissues. Tetramethylammonium hydroxide (TMAH) is used to homogenize biological substances for liquid scintillation counting[8] and atomic absorption analysis.[9]

In the present investigation, some amines and trimethylalkylammonium ions were examined for this purpose.

MATERIALS

Hair

A human scalp hair sample (3 mm long) was rinsed repeatedly with water and acetone, and cut with a "Polytron" high speed homogenizer under ice-water cooling. After centrifugation at 10,000 $\times g$ for 20 minutes at 4°C, the sample was lyophilized and stored in a desiccator over $CaCl_2$. Water content of the sample was found to be 11.8%, by the micro Abderhalden method (at 110°C over P_2O_5, *in vacuo*).

Wool

The sample of merino wool, which had been washed with a detergent solution for about 15 minutes at 45 to 50°C, rinsed twice in water at 45° and finally dried at 80°C, was kindly given by Daido Keori Inc. It was cut manually into small pieces as it was too soft to be homogenized with "Polytron."

Other keratineous tissues

Human nail, fresh bovine hoof and horn, and pigeon feather were treated in a manner similar to the case of the hair sample.

METHODS

Solubilization of keratins with 10% TMAH

About 500 mg of sample was homogenized with 8 ml of 10% TMAH in a 50 ml Potter homogenizer in an ice-water bath. A period of 10 minutes was required for complete homogenization. The homogenate was centrifuged at 20,000 ×*g* for 10 minutes at 4°C. The residue was homogenized in the same manner as before with an additional 8 ml of 10% TMAH and centrifuged. Supernatant at each step was immediately cooled in an ice-water bath, and its pH was lowered to about 10 by addition of HCl solution. (See Scheme 1.)

Scheme 1. TMAH-Treatment of keratin

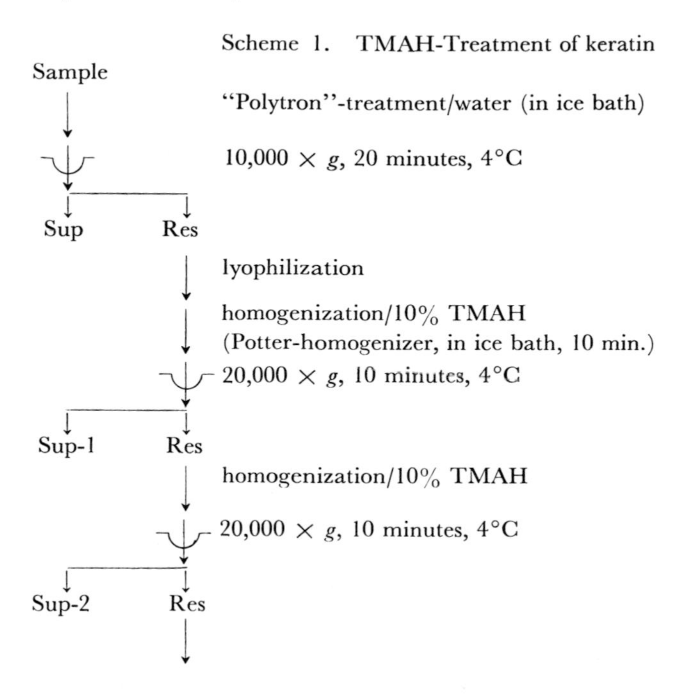

Preparation of SCM kerateines

Reduction and carboxymethylation of human scalp hair were performed by the method of O'Donnell *et al.*[4] (See Scheme 2.) About 1 g of sample was suspended in 20 ml of 0.8 M 2-mercaptoethanol solution (in Tris-NaOH buffer, pH 10.5, containing 8 M urea) at room temperature overnight. Care was taken to keep pH 10.5 by occasional addition of

NaOH solution. Iodoacetate solution (pH 8) freshly prepared from 3 g of monoiodoacetic acid and 6N NaOH solution was added to the reaction mixture, the pH of which had been adjusted to 8. After stirring the mixture for 30 minutes, it was centrifuged at 20,000 $\times g$ for 20 minutes. The supernatant was dialysed against water and 0.8 M acetate buffer (pH 4.4). The resulting precipitate, after redissolution in Tris-HCl-8M urea buffer (pH 7.6) and reprecipitation, was taken for SCMK-A as a part of the keratin fractions, whereas the supernatant was considered as SCMK-B.

Scheme 2. Preparation of SCM-kerateine

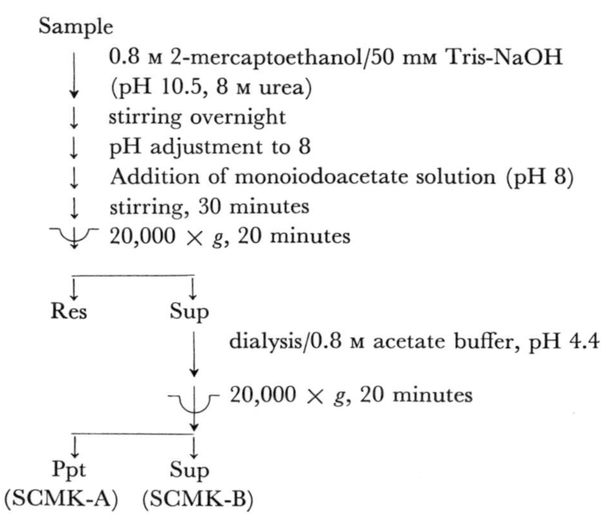

Gel filtration through Sephadex G-200

The solubilizate in TMAH was fractionated through a column of Sephadex G-200 (2.5 $\times$ 45 cm) using 50 mM borate buffer (pH 8.8, containing 50 mM KCl) for the eluate to be monitored at 280 nm. The SCM kerateines were fractionated through another column of Sephadex G-200 (2.5 $\times$ 67 cm) using 50 mM Tris-HCl buffer (pH 7.6, containing 0.2 M KCl, and 1 mM EDTA and 8 M urea) for their eluate.

Disc electrophoresis

Disc electrophresis was carried out with 7.5% acrylamide gel (pH 9.4, containing 6 M urea) and 50 mM Tris-glycine buffer (pH 8.3), under the condition of 5mA/tube.

N-terminal analysis

Sanger's method[10] was employed for N-terminal analysis by dinitrophenylating hair and wool keratins, treated or intact.

Detection of TMAH

TMAH, both in solution and on paper, was detected by the use of Dragendorff's reagent.[11] Presence of more than 0.01% of TMAH can be readily detected by an orange coloration, immediately after the reagent is sprayed over a filter paper on which sample solution has been spotted.

Observation under polarized light microscope

Hair samples soaked in aqueous TMAH solutions were observed under an Olympus polarized light microscope Type POM with crossed Nichols. Time required for extinction of birefringence was taken as a criterion of a structural change of hair. Swelling of samples was also observed without crossed Nichols.

Estimation of solubilizability

50 mg of samples were homogenized in 4 ml of 10% TMAH for 15 minutes in an ice-water bath. After centrifugation at 5,000 $\times g$ for 10 minutes, absorbance of the supernatant was measured at 280 nm on a Hitachi 356 spectrophotometer.

Circular dichroism (c.d.) measurement

Circular dichroism spectra were recorded at 25°C on a Jasco Model J-20 c.d. spectrophotometer.

Sedimentation velocity measurement

Ultracentrifugation was performed, with the sample solution in 50 mM borate buffer (pH 8.8, containing 50 mM KCl), using a Beckman Spinco Model E equipped with a schlieren optical system.

RESULTS AND DISCUSSIONS

The samples of human scalp hair and merino wool were treated with 10% solutions of ethylenediamine, trimethylamine, trimethylcetylammonium bromide (TMCAB), TMAH (pH being adjusted to 11 with sulfuric acid), TMAH, tetrabutylammonium hydroxide (TBAH) and ammonia, or 1N solution of NaOH. It is obvious from Table 1 that 10% TMAH without inorganic electrolyte is readily solubilizable to the keratin samples among the reagents used in the present experiments. Figure 1 shows concentration dependence of 280 nm absorbance of the supernatant of the solubilizate. As seen in the figure, the solubilizability of TMAH is almost saturated at a concentration of about 8%. Hair swells to almost double size in diameter 5 to 7 minutes after soaking in 10% TMAH. In 1N NaOH, on the other hand, it takes as long as 15 minutes to merely begin swelling. Figure 2 shows a microphotograph taken 30 minutes after soaking in 10%

Table 1. Solubilization of hair and wool in various reagents

Reagents	Concentration	Absorbance at 280 nm	
		Human scalp hair	Merino wool
Control (H_2O)		0.08	0.11
$NH_3.H_2O$	10%	0.38	0.28
$C_2H_4(NH_2)_2$	10%	0.20	0.38
$N(CH_3)_3$	10%	0.56	0.42
TMAH	10%	6.39	8.20
TBAH	10%	1.59	—
TMCAB	10%	0.79	—
TMAH + H_2SO_4	10%	0.29	0.20
NaOH	1 N	2.28	4.54

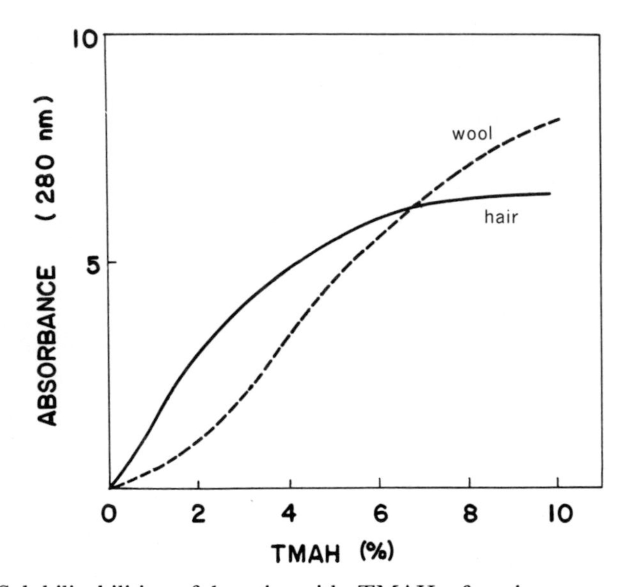

Fig. 1. Solubilizabilities of keratin with TMAH of various concentrations.

TMAH. Hair swells so extensively that cuticle begins to open unfolding the hair cortex. When extinction of birefringence of hair was observed under a polarized light microscope, it took 15 to 20 minutes after soaking in 10% TMAH for complete extinction, whereas in 1N NaOH, time as long as 10 hours was required. Figure 3-a shows a photograph taken after 6 minutes, polarized light being still observed clearly, and Figure 3-b shows one after 10 minutes, taken under the same condition as before. It can be seen clearly that the extinction has progressed considerably. It should be mentioned that the swelling was followed always by the extinction of the bire-

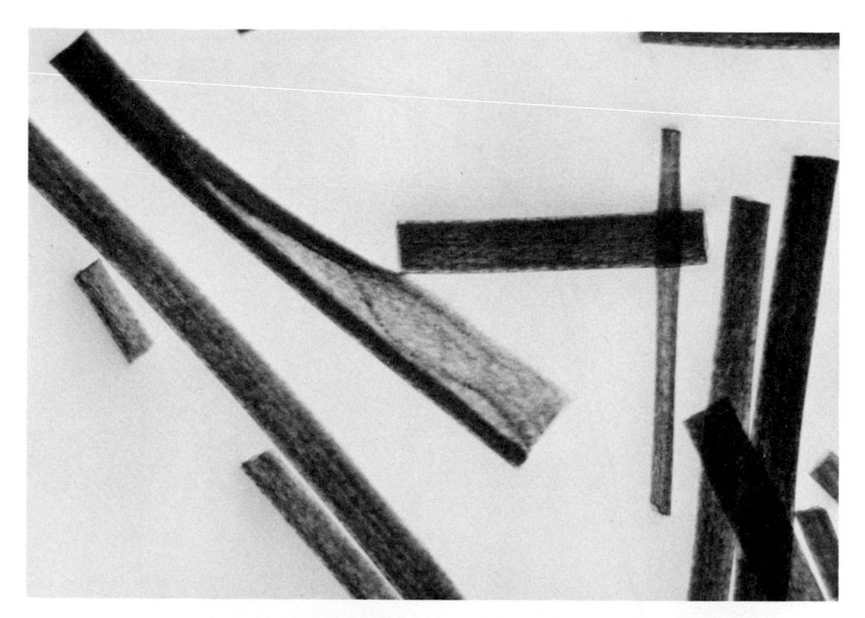

Fig. 2. Microphotograph of swollen hair.

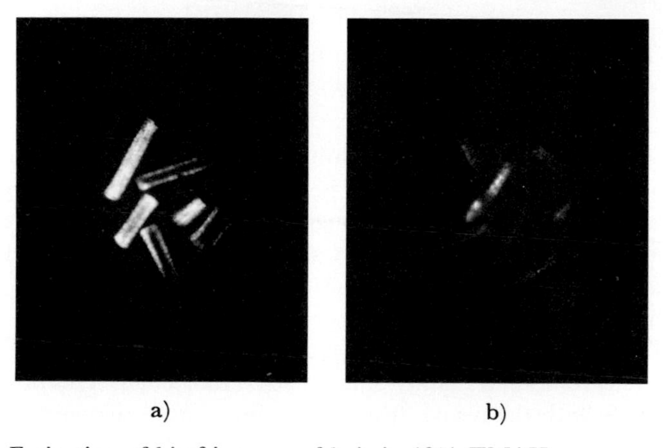

a) b)

Fig. 3. Extinction of birefringence of hair in 10% TMAH.
a) after 6 minutes
b) after 10 minutes

fringence and that difference between actions of 10% TMAH and of 1N NaOH observed in this case is in rough conformity with the result shown in Table 1.

To elucidate the solubilizing process of keratin upon treatment with 10% TMAH, stepwise homogenization was carried out to give Fig. 4 for the percent solubilization as a function of the homogenization step number. It

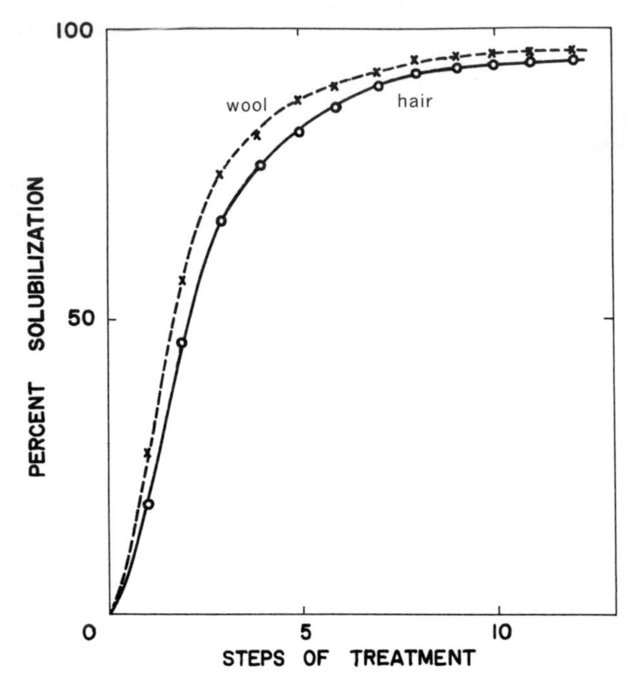

Fig. 4. Percent solubilization as a function of treatment steps for 3 mm-long samples of hair and wool keratins.

should be noted that the percent solubilization at the first two steps is increased to about 80% by the use of powdered samples in place of 3 mm-long ones. Upon gel filtration through a Sephadex G-200 column using borate buffer (pH 8.8), the solubilizates at the 1st, 2nd, and 3rd step always gave three fractions I, II, and III, as is exemplified in Fig. 5. The patterns were similar to each other, giving a gradual increase of peak III and decrease of peak I with the increasing step number. It is remarkable that the solubilizate is composed mainly of large-sized molecules. This observation accords fairly with the results from the sedimentation experiment which gave 7S for the size of the molecules in Fraction I. Results from disc electrophoresis also indicate that the molecules in Fraction I are composed of largesized ones, being distributed among several rather broad bands. In fact, the newly born N-terminal in the solubilizate was always less than 0.1% to the original (TMAH-untreated) protein. Amino acid analysis of Fraction I revealed a very similar amino acid composition to that of original hair except for cysteinyl residue, the content of which being somewhat decreased in Fraction I. The amount of S^{2-} ion was estimated for the solubilizate by the Methylene Blue method,[12] but it could not account for the lost amount of the cysteinyl residue. These observations suggest some unknown

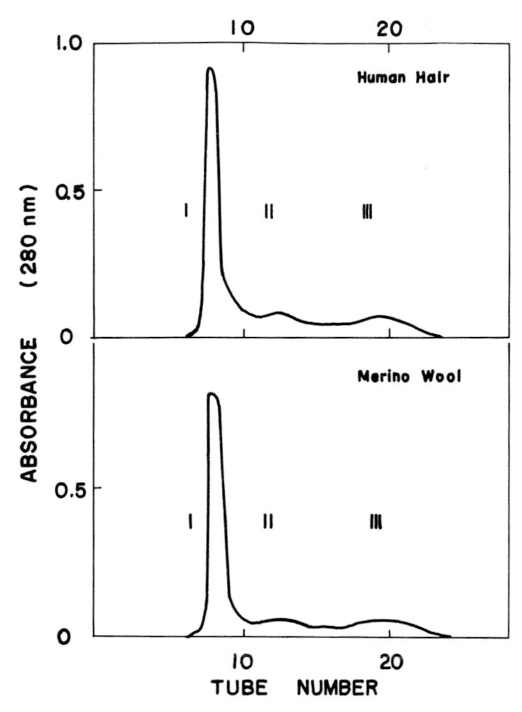

Fig. 5. Gel chromatography of TMAH-solubilized keratins through a Sephadex G-200 column (2.5 × 45 cm).
50 mM Borate Buffer (pH 8.8, containing 50 mM KCl) 20ml/hr 10ml/tube

but not so destructive change occurred at the disulfide bond. Incidentally, TMA ion was detected in practically none of these fractions. In Fig. 6 are shown three elution patterns, which were obtained under the same conditions through the Sephadex G-200 column using 50 mM Tris-HCl buffer (pH 7.6, containing 6 M urea). The two of them are for SCM-kerateines from hair and wool, and the rest are for TMAH-solubilizate. Since the SCM-kerateines were not dialysed in this case, there appeared excess peaks due to monoiodoacetate. It is obvious from the figure that the elution pattern for TMAH-solubilizate is practically the same as the one using borate buffer as an eluant.

In Table 2 are shown the α-helix contents based on the mean molar ellipticity at 222 nm for the TMAH-solubilizate, as well as for the hair SCMK-A and B for comparison. Because of the absence of β-structure in any sample (see Fig. 7.), one could assume $-40,000$ degree·cm^2/decimole for 100% α-helix and 0 degree·cm^2/decimole for 100% random coil.[13] The

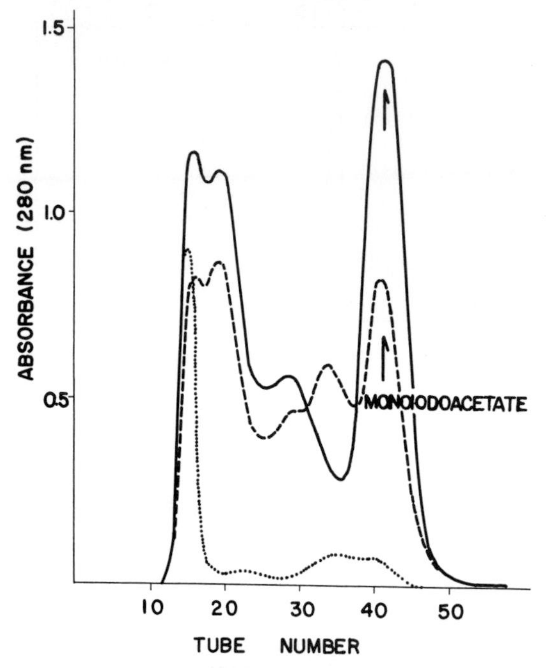

Fig. 6. Gel chromatography of SCM-kerateines and TMAH-solubilized hair through a Sephadex G-200 (2.5 × 67 cm).
50 mM Tris-HCl (pH 7.6, containing 50 mM KCl and 8M urea)
10ml/hr 9ml/tube

—— Hair
---- Wool
········ TMAH-solubilized Hair

Table 2. α-Helix content of TMAH-treated hair and SCM-hair (in percentages)

SCMK	A	45
	B	0
TMAH-treated hair	Fraction I	18
	Fraction II	18
	Fraction III	9

α-helix content of TMAH-solubilizate is always less than 20%, compared to 45% for SCMK-A. This implies that TMAH-treated hair keratin still holds fibre-matrix linkages, since SCMK-A and B correspond to microfibril and matrix, respectively.

Since the method of solubilizing keratin with 10% TMAH was found to be satisfactory, it was applied to a comparative study of keratins from human nail, bovine hoof and horn, and pigeon feather. Solubilizabilities

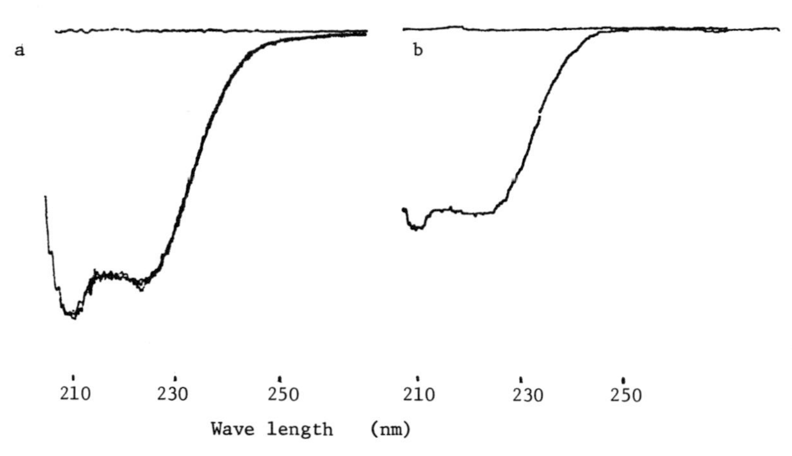

Fig. 7. C. D. Spectra of TMAH-solubilized hair (a) and hair SCMK-A(b) in 50 mM borate buffer (pH 8.8, containing 50 mM KCl).
0.01°/cm, optical length = 1 mm

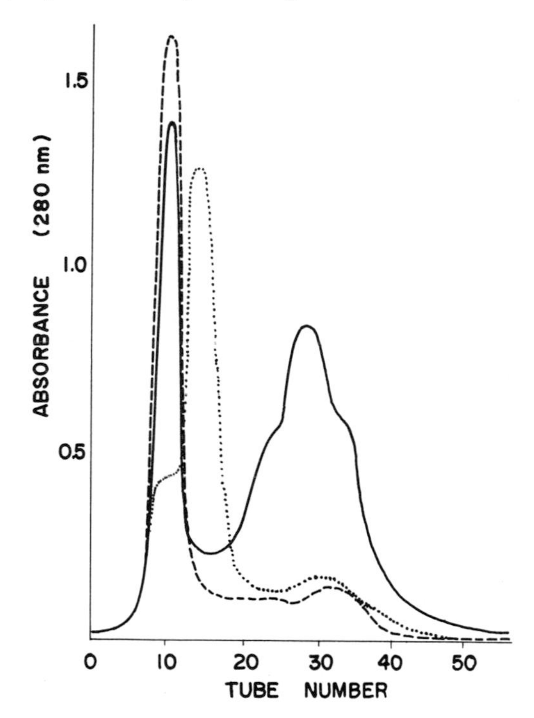

Fig. 8. Gel chromatography of TMAH-solubilized keratins through a Sephadex G-200 column (2.5 × 36 cm) 50 mM borate buffer (pH 8.8, containing 50 mM KCl) 20 ml/hr 5ml/tube

---- Hoof (Bovine) and Nail (Human)

········ Horn (Bovine)

—— Feather (Pigeon)

were similar to those for human hair and merino wool. As is illustrated in Fig. 8, for gelfiltration patterns of the four specimens, the feather keratin seemed to be markedly different from the others, elucidating well their structural characteristics.

Acknowledgements

The authors wish to thank Professor Mitsuo Muramatsu of Tokyo Metropolitan University and Professor Kazuo Hata of Kitasato University, for their kind advice on revising the manuscript.

REFERENCES

1. Alexander, P. and Earland, C. *Nature*, **166**: 396, 1950.
2. Blackburn, S. and Lowther, A. G. *Biochem. J.*, **49**: 554, 1951.
3. Gillespie, J. M. and Lennox, F. G. *Biochim. Biophys. Acta*, **12**: 481, 1953.
4. O'Donnell, I. J. and Thompson, E. O. P. *Aus. J. Biol. Sci.*, **17**: 973, 1964.
5. Schechter, Y., Landau, J. W., and Newcomer, V. D. *J. Invest. Derm.*, **52**: 57, 1969.
6. Baden, H. P., Goldsmith, L. A., and Fleming, B. *Biochim. Biophys. Acta*, **322**: 269, 1973.
7. Tsuda, M., Ogawa, T., Ono, T., and Kawanishi, Y. Proceedings of 30th Association of Japan Chemical Society, Part II, p. 1073, 1974 In Japanese.
8. Hansen, D. L. and Bush, E. T. *Anal. Biochem.*, **18**: 320, 1967.
9. Murthy, L., Menden, E. E., Eller, P. M., and Petering, H. G., *Anal. Biochem.*, **53**: 365, 1973.
10. Porter, R. R. and Sanger, F. *Biochem. J.*, **42**: 287, 1948.
11. Bregoff, H. M., Roberts, E., and Delwiche, C. C. *J. Biol. Chem.*, **205**: 565, 1953.
12. Kuratomi, K., Ohno, K., and Akabori, S., *J. Biochem.* (Tokyo), **44**: 183, 1957.
13. Hamaguchi, K. and Takesada, H. "Optical Activity of Proteins," Tokyo University Press, Tokyo, 1971, p. 63. In Japanese.

DISCUSSION

DR. MATOLTSY: The hair consists of cuticle and cortex. The cortex consists of α-helical protein while cuticle has non-helical protein. Did you find a small non-helical fraction in your preparation?

DR. TSUDA: No, I have not found such a fraction yet. It seems that the fractions obtained from Sephadex G-200 were very heterogeneous. Further fractionation of these fragments would be required to answer your question. We are now in the process of doing such experiments.

DR. STEINERT: I am not quite sure what kind of experiments you are trying to do. Are you trying to do comparative studies of different types of keratin? What I don't understand, therefore, is why you are using a method which is so chemically degradative to the tissue. For so many years now, we have had the "urea-reducing" (urea-thiol) method which, beyond any reasonable doubt, does not result in covalent bond breakage in the tissue and in proteins. It seems to me that this type of methodology, using urea and thiol reagents, could be far superior to something such as TMAH which can break covalent bonds. I just want to know why you are using this reagent for these particular experiments?

DR. TSUDA: I have been studying the keratin from hair and felt that a new approach for the disintegration of keratins was required. We found that TMAH was effective for this purpose.

INDEX

D

E

F

G

N

O

P

R

S

T

U

V